Enterprise Internetworking Technology (Bilingual)

Zhang Chunrong

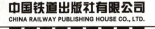

Introduction

The book has designed three network interconnection projects with different situation, which originated from real projects. The first project is an enterprise branch internetworking project, which includes four tasks: basic configuration and management of routers, static routing and default routing, configuration and management of DHCP and access control list. The second project is a small -sized enterprise internetworking project, which includes seven tasks: basic configuration and management of switches, VLAN, communications between the VLANs, spanning tree, PPP, RIP and NAT. The third project is a medium-sized enterprise internetworking project, which includes three tasks: planning and configuration of IP, VLAN and NAT, high reliability links, OSPF.

The book not only focuses on cultivating students' professional knowledge and skills, but also focuses on cultivating the ability for students to apply professional knowledge and skills to actual network interconnection projects.

This book is suitable for teaching courses such as network interconnection technology in higher vocational colleges, as well as reference book for computer practitioners.

Preface

This textbook is the result from such teaching reform projects as of the three-level clustered project curriculum system construction and the research and practice of the network engineering curriculum group construction integrating vocational certification and professional competition in the "Software Technology Professional Group" of the "Double High Plan" professional group. The textbook selects three real network projects with different situation. After extraction, improvement, and combination of the knowledge and skills of professional certification and network professional competition such as Huawei 1+X Network System Management and Maintenance, three projects and a total of 14 tasks were formed in this textbook..

The features of this textbook are as follows:

(1) Accurate positioning of talent cultivation goals. This textbook provides training programs to network application professionals such as network engineers, network security operation and maintenance engineers and network administrators. We hope that we can form an accurate positioning of the above-mentioned network application professionals and training specifications, through deeper demand analysis and technical research, clarifies the requirements of such talents in terms of technical knowledge, skills and engineering quality of network interconnection, and establishes the teaching framework of this textbook on this basis.

(2) Innovated development of teaching materials. We work with school teachers and senior

engineers in the industry to develop textbooks. Senior engineers in the industry provide real networking projects, the school teachers and the enterprise engineers analyze and refine the knowledge points and skills of teaching materials together, and the school teachers are responsible for compiling teaching materials. This mode effectively ensures that the teaching materials not only reflects the characteristics and structure of vocational college curriculum teaching, but also fully integrate with mainstream network technology and engineering practice.

(3) Innovation in writing. The textbook has designed three projects which are different in range and orientated from industries. Each project is divided into corresponding tasks based on the knowledge points and skills involved. Each task is composed of task description, task objectives, related knowledge, related skills, task completion and exercises. After all the tasks in the project are completed, you can complete the planning and deployment of the whole project. This writing can not only help students to master the knowledge and skills related to internetworking, but also teach them to master the application of internetworking expertise and technology in practical projects.

This textbook is edited by Zhang Chunrong, with An Ning, Peng Tianwei and Song Mu as deputy editors. Task 1 and Task 2 of Project 1 were written by An Ning. Tasks 1 to Task 3 of Project 2 were written by Peng Tianwei. Tasks 1 and Task 2 of Project 3 were written by Song Mu and Zhang Chunrong. The rest of the tasks of the project were written by Zhang Chunrong, who is also responsible for revision and finalization. Han Jiang, an engineer from Huawei Technologies Co., Ltd., gave great help in the compilation of this textbook, and I would like to express my heartfelt thanks.

Based on the initiative of cultivating applied network technical talents, this textbook adopts some innovative methods in the selection of teaching content, arrangement and design of teaching methods. We would appreciate it if readers would put forward amendments and suggestions to this textbook. E-mail address of editor: 289302109@qq.com.

Zhang Chunrong

May, 2023

Contents

Project 1   **Enterprise Branch Internetworking Project** 1

     Task 1   Basic Configurations and Management of Routers 3

     Task 2   Static Routing and Default Routing 33

     Task 3   Configuration and Management of DHCP 60

     Task 4   Access Control List 74

Project 2   **Small-sized Enterprise Internetworking Project** 93

     Task 1   Basic Configuration and Management of Switches 95

     Task 2   Planning and Configuration of VLAN 111

     Task 3   Planning and Configuration of Communications Between
                 the VLANs 143

     Task 4   Planning and Configuration of Spanning Tree 168

     Task 5   Planning and Configuration of PPP 191

     Task 6   Planning and Configuration of RIP 201

     Task 7   Planning and Configuration of NAT 215

Project 3   **Medium-sized Enterprise Internetworking Project** 223

     Task 1   Planning and Configuration of IP, VLAN and NAT 225

     Task 2   Planning and Configuration of High Reliability Links 253

     Task 3   Planning and Configuration of OSPF 275

Project 1

Enterprise Branch Internetworking Project

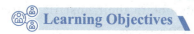

Learning Objectives

Knowledge Objectives

◎ Understand the internal composition and physical interfaces of routers, and master the basic configuration and management of routers.

◎ Understand the composition and insertion rules of the routing table, and master the planning and configuration of static routing, default routing, floating routing and aggregated static routing on routers.

◎ Understand the working process of DHCP and master the configuration and management of DHCP on routers.

◎ Understand the working process of ACL and master the planning and configuration of ACL.

Competence Objectives

◎ Competent to deploy and implement the branch internetworking project based on routers, specifically: basic configuration and management of routers, the configuration of DHCP services, the configurations of static routing, default routing, floating routing and aggregated static routing and the configurations of ACLs.

◎ Somewhat competent for troubleshooting.

Quality Objectives

◎ Competent for teamwork.

◎ Competent for problem analysis and solutions.

Project Description

A certain enterprise branch plans to build up a network topology as shown in Fig. 1.1 to connect its budget department, management department and server to three different ports of the router through the access layer switch to achieve the branch internal interconnection. The branch router is connected to the router of the headquarters via two links, namely, a Gigabit Ethernet link and a serial link. The IP addresses assigned to the branch include two IP network segments: "172.16.128.0/24" and "172.16.129.0/24". The IP addresses of the two access network segments between the branch export router and the headquarters router are "172.16.254.0/30" and "172.16.254.4/30". The networking requirements are as follows:

◎ The branch budget department has 140 hosts, the management department has 100 hosts, and the branch company has 8 servers for businesses. The data from the budget and management departments to servers needs to be isolated and cannot communicate directly but through routers.

◎ The branch intranet and the branch are interconnected with the headquarters, and the branch is connected to the Internet through the headquarters network.

◎ The branch budget and management departments obtain their host IP addresses automatically via DHCP, and the branch router prompts for DHCP services.

◎ Introduce ACL to the outlet and inlet of the branch router connected to the headquarters, and disable the 135-139, 445, and 3389 high-risk ports of TCP and UDP protocols.

◎ Only PC3 from the branch management department and PC1 from the headquarters can use SSH to log in to the "fenbu" router for remote management.

◎ Finally, save the configurations of the branch and headquarters routers and backup them to the FTP server at the headquarters.

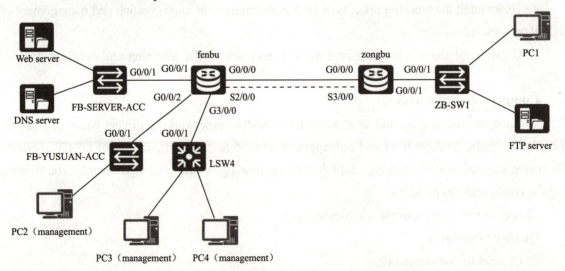

Fig. 1.1 An enterprise branch network topology

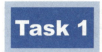 Basic Configurations and Management of Routers

Task Description

To know the physical interfaces of routers. Set up a network topology as shown in Fig.1.1 in the Huawei ENSP simulation environment or using Huawei's real routers and switches. Complete the planning of the network IP address and test the IP configuration of the host PC and server. Use the Console port to accomplish the basic configurations and testing on routers in the network, including host name, interface, Console login authentication, and SSH configuration. Manage the backup and restoration of router configuration files, password recovery, and other related tasks.

Task Objectives

- Master the physical interface types and functions of routers.
- Know how to connect routers for initialization configuration.
- Know how to use the command line interface (CLI) for routers.
- Master the online help for router commands, know how to use the router shortcut keys, and understand the view commands of routers.
- Know how to configure the router interfaces, control port authentication, Telnet, and SSH.
- Know how to manage router configuration files and clear the Console passwords.

Related Knowledge

1. Introduction to routers

A router is an internetworking device working on the third layer of the OSI model, mainly for choosing the best path, commonly known as routing. Routers can help select the best path for data packet transmission in the third layer through the protocol address of the network layer.

Routers are essentially computers with special functions, so they have the same components as ordinary computers. A router mainly consists of such components as follows: motherboard, central processing unit (CPU), flash memory, random access memory (RAM), read only memory (ROM), operating system (OS), power supply, baseboard, metal casing, and network interface.

2. Physical interfaces of routers

Routers can interconnect heterogeneous networks. Routers provide different types of physical ports to support various LAN, MAN, and WAN technologies. Common LAN ports include fast Ethernet port, GE (Gigabit Ethernet) port, 10GE port and 40GE port, etc. The WAN interfaces include high-speed A/S (asynchronous/synchronous) serial interfaces, etc. The interfaces of routers can be fixed (for example, the G0/0/1 interface in Fig. 1.2), or modular (such as the S0/1/1 interface

in Fig. 1.3) according to user needs. Usually, low-end routers use fixed interfaces, while high-end routers provide modular ones. Fig. 1.2 and Fig. 1.3 show the physical interfaces of Huawei AR1220 and NetEngine AR6000 Series routers, respectively.

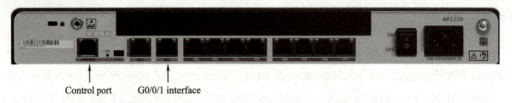

Fig. 1.2 Physical interface of Huawei AR1220 router

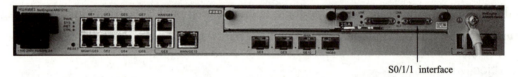

Fig. 1.3 Physical interface of Huawei NetEngine AR6000 series routers

In order to distinguish the physical interfaces on routers, a naming convention for router interfaces is introduced to assign each physical interface on the router a unique identifier for easy identification of router interfaces.

The router interfaces are named by "interface type slot number/module number/interface number". Specifically, the slot number, module number and interface number generally start with "0", namely, the first number is "0", and the second number is "1" and so on.

For routers with fixed interfaces, the number in the interface name only includes the interface number, for example, "Ethernet0 (can be abbreviated as E0)" represents the first Ethernet interface, and "Serial1 (can be abbreviated as S1)" represents the second serial port.

For a router that supports "online plugging and deletion" or can have physical interface configurations dynamically changed, its interface name should include at least two digits: slot number and interface number. For example, in the Huawei AR2220 router, "Gigabit Ethernet 0/1 (abbreviated as G0/1)" represents the "1" Gigabit Ethernet port located in slot "0".

For a router with a module in its graphics card, its interface name should include a slot number, a module number and a interface number. For example, in the Huawei AR2220 router, "Serial 2/0/0 (abbreviated as S2/0/0)" refers to the "0" serial port of the "0" module on slot "2".

3. Methods for router access

Routers can be configured as follows: firstly, configured via its management port (such as the Console port); secondly, configured by means of remote login to the router online, for example, telnet, secure shell (SSH) protocol, or simple network management protocol (SNMP) are used for router access only if the router is connected via IP with the Telnet client, SSH client, or SNMP

network management workstation. See Fig. 1.4 for schematic router access.

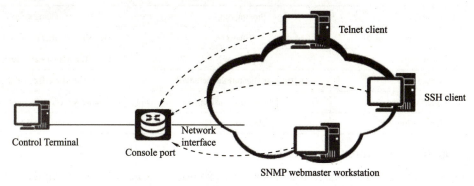

Fig. 1.4 Schematic router access

4. Command view

The versatile routing platform (VRP) is a universal operating system platform for Huawei's data communication products. VRP adopts a hierarchical command structure, providing users with different command views. When users are in a certain view, they can only execute the specific commands and operations allowed by that view. Common command views include:

① User view: After entering the configuration interface of the VRP system, the first view that appears on the VRP is the user view. In this view, the user can view the basic operating status and statistical information of the router.

② System view: It is such a view that configures global common parameters of the system. To modify the system parameters, you must enter the system view. Moreover, you can also access other functional configuration views through the system view, such as interface view, user interface view, and protocol view.

③ Interface view: It is such a view that configures the interface parameters.

④ User interface view: It manages asynchronous interfaces that work in the streaming mode. The user interface view can be divided into Console user interface view, AUX user interface view, TTY user interface view, and VTY interface view.

⑤ Routing protocol view: Most of the parameters for routing protocols are configured in the routing protocol view. Routing protocols will be introduced in subsequent projects, such as OSPF protocol view and RIP protocol view.

Each command of the router can only be executed in a specific view. You can decide the current view through the router prompt, for example, "<>" represents the user view while "[]" represents views other than the user view. Table 1.1 provides common command views and prompts for Huawei routers.

Table 1.1 Common command views and prompts for Huawei routers

Command view	Prompt sample	View switching sample
User view	<Huawei>	Default view after router startup
System view	[Huawei]	<Huawei>system-view
Interface view	[Huawei-GigabitEthernet0/0/0]	[Huawei]interface G0/0/0
User interface view	[Huawei-ui-console0]	[Huawei]user-interface console0
Routing protocol view	[Huawei-ospf-1]	[Huawei]ospf 1

5. Online help for router commands

To achieve various functions, a router not only provides a large set of commands, but a single command may also provide different operation parameters, so it is almost impossible to rely solely on the user's memory for command use. Therefore, the router provides a function of online help for using router commands to help users complete relevant configuration commands. The online help basically can be used as follows:

① ? // It can display all the command sets and simple descriptions available in the current
// view, for example, enter "?" in the user view and you can see all available command sets and
// simple descriptions in the user view

② String +? // It can display all command sets and simple descriptions starting with the
// string at the current prompt, for example, enter "dis?" and it can display all
// commands and simple descriptions starting with the string "dis" in the current view

③ Command name + space +? // It can display all keywords or parameters and simple
// descriptions of this command at the current prompt, for
// example, enter "display ?" and it can display all keywords and
// simple descriptions of the "display" command in the current view

6. Use of command history function

When configuring a router or checking its status, it is common to encounter situations where users need to enter the same command or enter similar commands repeatedly. For reducing the workload of entering commands when using a router, it generally provides a command history function, which can store several commands that you have just used. The maximum number of command records that can be stored in command history depends on the maximum storage space set by the system for this purpose. Within the maximum number of records limited by the system, users can configure the number of command records in command history as needed. For command history, you can use the view commands provided by the system, usually including UP (Forward) and DOWN (Backward) commands. To view the history commands, use the 【Ctrl+P】 or 【Ctrl+N】 combination key, or use the 【↑】 or 【↓】 key on the keypad.

7. Router shortcut keys

Except for help commands, the router operating system also provides relevant shortcut keys in

order to facilitate router configuration, monitoring, and troubleshooting. See Table 1.2 for the list of the router shortcut keys.

Table 1.2   Router shortcut keys

Shortcut keys	Functions
Tab	Used to complete some input command items
↑ (or Ctrl+P)	Scroll backward in the list of previously used commands
↓ (or Ctrl+N)	Scroll forward in the list of previously used commands
Ctrl+C	Abandon the current command and exit
Ctrl+Z	Return to the user view directly

When using the shortcut key 【Tab】, as required, only if the input abbreviation command or abbreviation parameter contains enough letters to distinguish it from any other commands or parameters currently available can it be used to supplement the remaining part of the abbreviation command or abbreviation parameter automatically.

8. Router view commands

Before router configuration or configuration changes, the function for router status check shall be provided to verify the configuration effect and complete router troubleshooting. For this reason, all routers provide a series of commands for router status view. Table 1.3 provides the commonly used view commands for Huawei routers.

Table 1.3   Commonly used view commands for Huawei routers

View commands	Functions
display version	Display the system version
display current-configuration	Display the contents of the current configuration file
display saved-configuration	Display the contents of the configuration file saved in Flash
display interface	Display the interface information
display ip interface brief	Display the brief interface configuration information, including IP address and interface status
display ip routing-table	Display the routing table information
display this	Display the running configuration of the current view
dir flash:	Display the contents in Flash

9. Methods for router file management

All configuration information about the router exists in the form of a router configuration file. The current configuration file (current-configuration) of the router is saved in its RAM. The configuration information in the RAM will be lost when the router is powered down or restarted. Therefore, the configuration file needs to be saved in the router Flash, and the router will call the configuration file from Flash for running in the RAM every time it is turned on or restarted. In addition to the above methods for configuration file saving, you can also use TFTP, FTP, SFTP, SCP

and FTPS to save the configuration file in a reliable disk on other hosts.

In the process of file management, the router can work as a server or a client.

Related Skills

1. Router service ABC

Complete the initialization, configuration and connection of the router, know the command view of the router well, and use the history and help commands to configure the host name of the router.

(1) Router initialization, configuration and connection

① Use a USB to RJ-45 control cable to connect the host as the control terminal to the Console port of the router. Fig. 1.5 shows how to connect the laptop as a router console. The USB terminal of the USB to RJ-45 control cable is plugged into the USB port of the laptop, and the RJ-45 terminal of the USB to RJ-45 control cable is plugged into the Console port of the router.

② After the USB to RJ-45 control line driver is installed on the host, right click the "This Computer" icon on the desktop, select

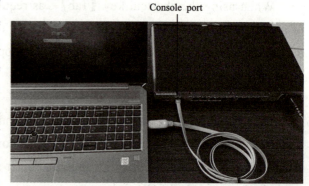

Fig. 1.5   Connection of laptop as router console

the "Manage" command from the pop-up shortcut menu, and then click "Device Manager" in the pop-up "Computer Management" window. Next, click to display the "ports (COM and LPT)" in the right list, and observe the COM port values of USB-to-Serial Comm Port (COM3) in parentheses, as shown in Fig. 1.6. The COM port of the USB to serial port shown in the figure is the port COM3.

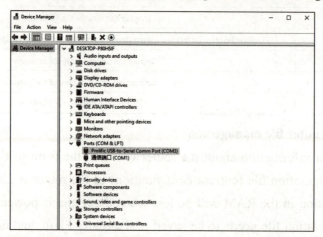

Fig. 1.6   COM port number of "USB-to-Serial Comm Port" on control terminal

③ Run the SecureCRT software on the router console. In the pop-up "SecureCRT" window, click the "File" menu and select the "Quick Connect" command. In the pop-up "Quick Connect" window, select "Serial" for the protocol and "COM3" for the port (**Note**: The COM port number should be identical to the COM port of the USB serial port queried in the "Device Manager"). The baud rate is 9600, there are 8 data bits, and no parity check exists; there is one stop bit, but no flow control exists, namely, all three check boxes below the Flow Control are unchecked, as shown in Fig. 1.7.

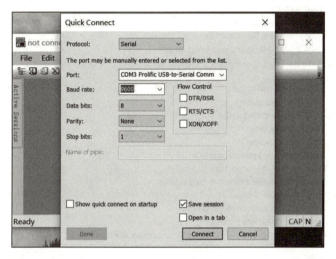

Fig. 1.7 Setting of control terminal port properties

④ Click the "Connect" button, and then press the 【Enter】 key to enter the default username and password for access to the command view interface of the router, as shown in Fig. 1.8.

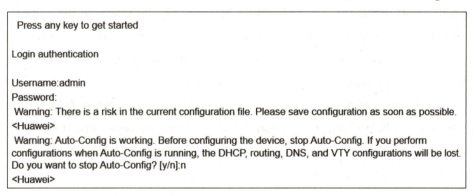

Fig. 1.8 Router initialization configuration interface

(2) Router command view ABC

Usually, after started up, the router will provide the user view for by default. The prompt for the user view of Huawei router is "<Huawei>", where "Huawei" between "<" and ">" is the system host name of the router, and the default host name is "Huawei". The host name of the router can be

changed. Enter "?" in the user view and you can view the allowed commands and their functions in that view, as shown in Fig. 1.9.

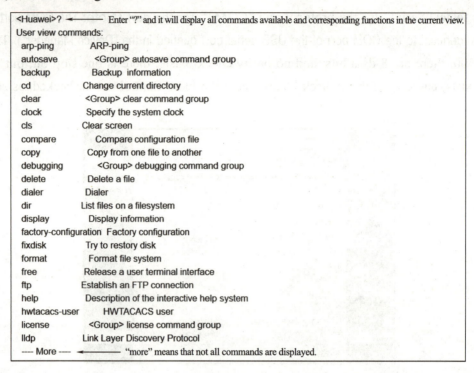

Fig. 1.9 User view of router

In Fig. 1.9, the "--- More ---" at the bottom of the screen indicates that not all screen commands are displayed. At this time, you can press the 【Enter】 key or space bar to display the remaining commands. Press the 【Enter】 key to display one line down on the screen: press the space bar to display one screen down on the screen; press the letter 【Q】 key on the keyboard to exit directly and no longer display commands that are not displayed. Press the space bar to display the command on the next screen, then press the 【Enter】 key to find "system-view" for system view access, view the brief description of the command, press the letter 【Q】 key to exit the help, and then use the help command "system-view?" to view its keywords. If there is "<cr>" in the keywords, it means "Enter", and you can execute the command to enter the system view. The prompt for the system view is "[Huawei]", as shown in Fig. 1.10.

Fig. 1.10 Use "Help" to view the keywords of the command "system-view" and enter the system view

Project 1 Enterprise Branch Internetworking Project   11

In the system view, use the help command "?" to view the commands for interface view access, and then use the help command "interface?" to view the keywords of the "interface" command. Follow the steps as follows to enter the interface view of the G0/0/0 interface. Take care to observe and record the interface view prompt of the G0/0/0 interface.

```
[Huawei]?                              //List all commands available in the current view
[Huawei]interface ?                    //View the "interface" keywords or parameters
[Huawei]interface GigabitEthernet 0/0/0 //Enter the interface view of the G0/0/0 interface
[Huawei-GigabitEthernet0/0/0]          //interface view prompt for the G0/0/0 interface
[Huawei-GigabitEthernet0/0/0]q?        //List all commands starting with the letter "q"
[Huawei-GigabitEthernet0/0/0]quit
        //Quit the interface view of the G0/0/0 interface and return to the system view
[Huawei]quit                           //Quit the system view and return to the user view.
<Huawei>                               //User view prompt
```

(3) Using history and help commands to configure router host names

In any view, you can use the 【↑】 key on the keypad to recall the command you just used. Please refer to the following commands to practice using history and help commands to configure router host names:

```
<Huawei> ↑
```

Press the 【↑】 key on the keypad repeatedly to recall the "system-view" command that has been used, and then press the 【Enter】 key to execute.

```
[Huawei]? //List the commands available in the system view
```

After the commands are displayed, you can press the space bar repeatedly to find the "sysname" command.

```
[Huawei]sysname ?                //View keywords or parameters for sysname
[Huawei]sys ?                    //View all commands starting with the string "sys"
[Huawei] sys, press Tab again    //Complete the command starting with "sys"
[Huawei]sysname AR1              //Configure the host name of the router as "AR1"
[AR1]quit                        //Quit the system view and return to the user view
<AR1>                            //User View Prompt
```

(4) Router view command training

In the user view or at the system view prompt, enter the "display version" command to display the system version of the router, as shown in Fig. 1.11.

```
[AR1]display version
Huawei Versatile Routing Platform Software
VRP (R) software, Version 5.130 (AR2200 V200R003C00)
Copyright (C) 2011-2012 HUAWEI TECH CO., LTD
Huawei AR2220 Router uptime is 0 week, 0 day, 0 hour, 51 minutes
BKP 0 version information:
1. PCB      Version : AR01BAK2A VER.NC
2. If Supporting PoE : No
3. Board    Type    : AR2220
4. MPU Slot Quantity : 1
5. LPU Slot Quantity : 6

MPU 0(Master) : uptime is 0 week, 0 day, 0 hour, 51 minutes
MPU version information :
1. PCB      Version : AR01SRU2A VER.A
2. MAB      Version : 0
3. Board    Type    : AR2220
4. BootROM  Version : 0
```

Fig. 1.11 Display the system version of the router via the "display version" command

2. Router basic configurations

Build a network topology as shown in Fig. 1.12, configure router interfaces, router Console ports, Telnet services, and SSH services, and then carry out verification and testing.

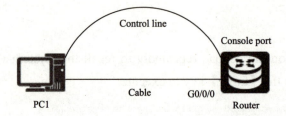

Fig. 1.12 "Router basic configuration" network topology

See Table 1.4 for the IP planning for router interface IP and PC.

Table 1.4 Router interface IP and PC planning

Device	Interface	IP address	Gateway
AR1	G0/0/0	192.168.0.254/24	
PC1	NIC	192.168.0.10/24	192.168.0.254

According to the network topology shown in Fig. 1.12, use a USB to RJ-45 control cable to connect the USB port of the host PC1 to the Console port of the router AR1, and use a network cable to connect the network card of the host PC1 to the G0/0/0 interface of the router. The host PC1 serves as both the control terminal of the router and the Telnet and SSH clients.

(1) Router interface configurations and view

The router interface configurations typically include such tasks as follows:

① Enter the interface view where the router needs to configure its interfaces;

② In the interface view, configure the interface IP address, subnet mask, interface description, and enabled interface, etc.

Note: Huawei router ports are enabled by default.

The detailed steps are as follows for configuring the IP address, subnet mask, interface description, and enabled interface of the router G0/0/0 interface:

```
[AR1]interface g0/0/0      //Enter the interface view of the interface G0/0/0
[AR1-GigabitEthernet0/0/0]ip address 192.168.0.254 24
//Configure the IP address of the G0/0/0 interface as "192.168.0.254", and the length
//of the subnet mask as "24"
```

Note: When configuring the subnet mask, you can also use dotted decimal.

```
[AR1-GigabitEthernet0/0/0]description linkto801room  //Configure the G0/0/0 interface as "linkto
                                                     //801room"
[AR1-GigabitEthernet0/0/0]undo shutdown              //Enable interface
```

After completing the configuration, use the following view commands to view the configuration and detailed information of the interface:

① In the interface view of the G0/0/0 interface, use the "display this" command to display the running configuration of the current interface view, as shown in Fig. 1.13.

```
[AR1-GigabitEthernet0/0/0]display this
[V200R003C00]
#
interface GigabitEthernet0/0/0
 description linkto801room
 ip address 192.168.0.254 255.255.255.0
#
return
[AR1-GigabitEthernet0/0/0]
```

Fig. 1.13 Use the "display this" command to display the running configuration of the current interface view

② In any view, use the "display ip interface brief" command to display the brief configuration information of the router interface as shown in Fig. 1.14. See whether the IP address of the G0/0/0 interface in the figure is correctly configured, and whether both "Physical" and "Protocol" are "up". If the "Physical" and "Protocol" of the G0/0/0 interface are both "down", it usually indicates that the network cable of the interface is damaged or the device connected to the network cable is not powered on.

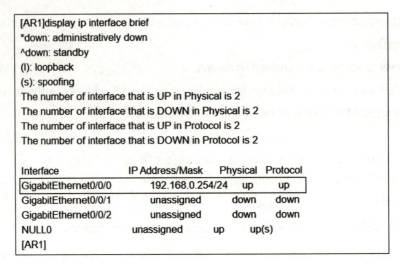

Fig. 1.14 Use the "display ip interface brief" command to display the brief configuration information of the router interface

③ In any view, use the "display interface g0/0/0" command to display the detailed information about the router G0/0/0 interface, as shown in Fig. 1.15. Pay attention to the boxed contents in Fig. 1.15 and see whether the current status and protocol state of the G0/0/0 interface are UP, and whether the IP address and subnet mask of the interface are configured correctly.

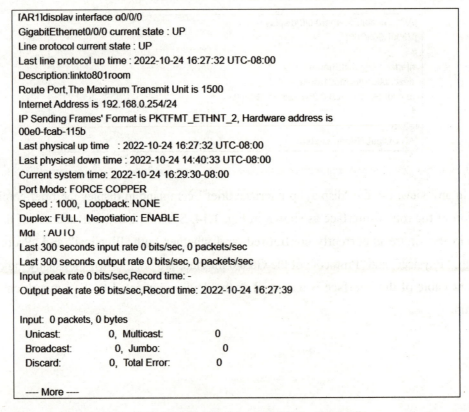

Fig. 1.15 "display interface g0/0/0" display results

(2) Configuration and testing of control ports using password authentication

When access to the router through the control port, users need to configure authentication via their passwords or AAA (authentication authorization accounting).

The control port configuration shall have such tasks as follows when subject to password authentication:

① Enter the user interface view of Console0.
② Configure the authentication mode to password authentication.
③ Configure the password.

Please refer to the following steps to configure the control port subject to password authentication:

```
[AR1]user?                      //View all commands starting with the string "user"
[AR1]user-interface ?           //View keywords or parameters for "user-interface"
[AR1]user-interface console ?   //View keywords or parameters for "user-interface console"
[AR1]user-interface console 0   //Enter the console0 user interface view
[AR1-ui-console0]?              //List the commands available in the console0 user interface view
[AR1-ui-console0]authentication-mode?   //View the keywords or parameters for
                                        //"authentication-mode"
[AR1-ui-console0]authentication-mode password   //Configure the authentication mode of
                                                //console0 as password authentication
```

In the following interactive information, enter the password to be configured as "Chengdu123", as shown below:

```
Please configure the login password (maximum length 16):Chengdu123
[AR1-ui-console0]quit   //Quit the console0 user interface view and return to the user view
[AR1]quit               //Quit the system view and return to the user view
<AR1>quit               //Quit the user view
```

After configuring the password authentication of the Console, it will prompt for Console login authentication when using the Console to enter the user view of the router. Only after entering the authentication password "Chengdu123" configured by the Console, can you enter the user view. See Fig. 1.16 for Console password login authentication.

```
<AR1>quit
 Configuration console exit, please press any key to log on

Login authentication

Password:  ←—— Enter the Console password configured "Chengdu123" here.
<AR1>           Note that the password entered will not be displayed.
```

Fig. 1.16 Console password login authentication

(3) Configuration and testing of control ports using AAA authentication

Password authentication for Console login can only verify the password but not identify the user. Local users can be configured for Console login authentication.

For control port configuration via AAA authentication, the following configuration tasks need to be completed:

① Configure a valid AAA user for Console login, specifically including a valid username and password, a service type of the user authorized (as "terminal"), and the privilege level of the user;

② Enter the Console user interface view and configure the authentication mode to AAA.

Please refer to the following steps to configure the control port subject to AAA authentication:

```
[AR1]aaa                             //Enter the AAA view
[AR1-aaa]local-user zhangcr password cipher Chengdu123
                                     //Create a local user with the username "zhangcr",
                                     //The password is "Chengdu123"
[AR1-aaa]local-user zhangcr service-type terminal
                                     //The service type of the authorized user "zhangcr" is
                                     //"terminal"
[AR1-aaa]local-user zhangcr privilege level 15  //Configure the privilege level of
                                     //the user "zhangcr" to 15
[AR1-aaa]quit                        //Return to the system view
[AR1]user-interface console 0        //Enter the console0 user interface view
[AR1-ui-console0]authentication-mode aaa
                                     //Configure the authentication mode of console0 to use
                                     //AAA authentication
[AR1-ui-console0]quit                //Quit the console0 UI view and
                                     //return to the user view
[AR1]quit                            //Quit the system view and return to the user view
<AR1>quit                            //Quit the user view
```

After completing the control port configuration for AAA authentication, it will prompt for Console login authentication when using Console for router access. Enter a valid Console username and password to enter the user view, as shown in Fig. 1.17.

(4) Telnet configuration and testing

A remote host can use Telnet to log in to the router remotely. The remote host is a Telnet client, and the router is configured as a Telnet server. Before configuring the Telnet service on the router, it is necessary to configure the router interface in advance. If there are other routing devices between the router and the Telnet client, the routing also needs to be configured to ensure IP connectivity between the Telnet client and the router.

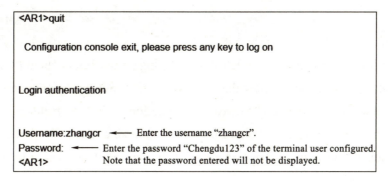

Fig. 1.17 Console local user authentication

The Telnet service for routers subject to AAA authentication includes such configuration tasks as follows:

① Enable the Telnet service.

② Configure the valid user to log in to Telnet, including entering AAA view, creating username and password, configuring the user's service type as "Telnet", and configuring the user's privilege level.

③ Enter the VTY user interface view, configure the protocol as Telnet, and set the authentication mode as AAA.

Please refer to the following steps to configure Telnet on the router. The planning is as follows: the number of Telnet sessions is VTY 0 to VTY 4, with a total of five Telnet sessions. The AAA authentication is adopted, with the Telnet username as "zhangcr", password as "Chengdu123", and privilege level as "15".

```
[AR1]Telnet server enable   //Enable the Telnet server
[AR1]aaa                    //Enter the AAA view
[AR1-aaa]local-user zhangcr password cipher Chengdu123
        //Create a local user with the username "zhangcr" and password "Chengdu123"
[AR1-aaa]local-user zhangcr service-type telnet
                    //Authorize the user "zhangcr" service type as Telnet service
[AR1-aaa]local-user zhangcr privilege level 15
                    //Configure the privilege level of the user "zhangcr" to 15
[AR1-aaa]quit       //Return to the system view
[AR1]user-interface vty 0 4   //Enter the user interface view from VTY0 to VTY4
[AR1-ui-vty0-4]protocol inbound telnet   //Configure the inbound protocol to Telnet protocol
[AR1-ui-vty0-4]authentication-mode aaa   //Configure the authentication mode of the five
                            //virtual terminal sessions from VTY0 to VTY4
                            //as AAA authentication
[AR1-ui-vty0-4]quit      //Quit the user interface view of VTY0 to VTY4, and return to
                         //the user view
```

```
[AR1]quit                   //Quit the system view and return to the user view
<AR1>quit                   //Quit the user view
```

Refer to the following steps to complete Telnet testing on the Telnet client PC1:

① Open the "Control Panel", select the "Change Adapter Settings" command in the "Network and Sharing Center", in the pop-up "Network Connection" window, right-click the "Ethernet" network card, select the "Network and Sharing Center" command under "Properties", and configure the IP address, subnet mask and default gateway of the host PC1 according to the parameters in Table 1.4, as shown in Fig. 1.18.

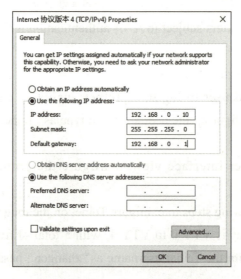

Fig. 1.18   Configure the IP address of host PC1

② After the configuration is completed, use the "ipconfig /all" command in the DOS command line to check if the configured IP address has taken effect, as shown in Fig. 1.19.

Fig. 1.19   View the IP information of host PC1 via the "ipconfig /all" command

③ Use the "ping 192.168.0.254" command in the command line of PC1 to test the IP connectivity between the host PC1 and the router AR1 interface G0/0/0. If you receive a response from the target host "192.168.0.254", it indicates that the G0/0/0 interface of the router has IP connectivity. Fig. 1.20 indicates successful ping via "192.168.0.254".

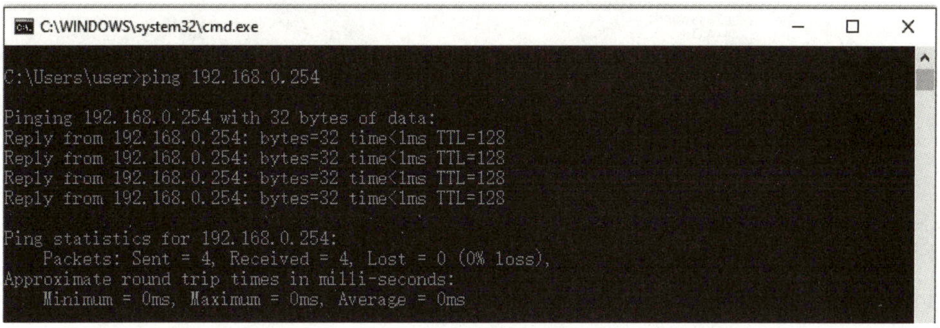

Fig. 1.20   Test IP connectivity between PC and router AR1

④ On PC1, run the SecureCRT virtual terminal software. In the SecureCRT software window, select the "Quick Connect" command from the "File" menu. In the "Quick Connect" window that pops up, select "Telnet" as the protocol, enter the IP address "192.168.0.254" for the router AR1 interface G0/0/0 for the host name, and the default port is "23", as shown in Fig. 1.21.

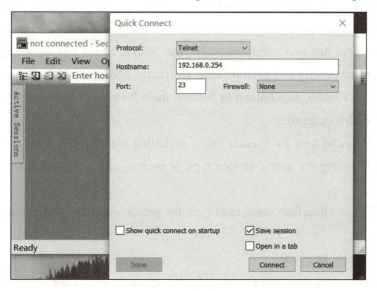

Fig. 1.21   PC1 telnet to router AR1

⑤ Click on the "Connect" button in Fig. 1.21, and then enter the username "zhangcr" and password "Chengdu123" according to the prompts (note that the password will not be displayed). Press the 【Enter】 key to enter the user view prompt of the router, as shown in Fig. 1.22.

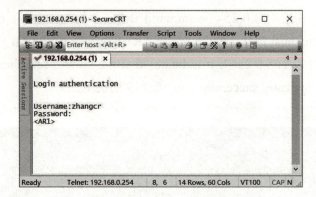

Fig. 1.22　Successfully Telnet to router AR1

(5) SSH configuration and testing

Various security issues may occur in case of router access via Telnet. The most prominent issue is as follows: the username and password required for Telnet authentication are transmitted in plaintext when transmitted online and then hackers can use Wireshark and other package capture software to capture the Telnet username and password transmitted on the line.

SSH is currently a relatively reliable protocol designed to provide security for remote login and other network services. In time of router access via SSH, digital certificates are used to authenticate the connection between the SSH client and the router, and encrypt the transmission of identity authentication passwords.

The SSH protocol has two versions: SSH1 and SSH2. SSH2 has advantages over SSH1 in terms of security, functionality and performance. Currently, SSH2 is widely used.

SSH on Huawei routers, also known as Stelnet, includes such configuration tasks as follows:

① Enable the Stelnet service.

② Configure a valid user for Stelnet login, including entering AAA view, creating username and password, configuring the user's service type as Stelnet, and configuring the privilege level of the user.

③ In the VTY user interface view, configure the protocol to SSH and the authentication mode to AAA.

④ Generate a local key pair.

Taking Huawei router AR2200 (Version 5.130) as an example, please refer to the following steps to configure Stelnet on the router AR1. The planning is as follows: use AAA for Stelnet authentication, and set the Stelnet username as "zhangcr", password as "Chengdu123", and privilege level as "15".

```
[AR1]stelnet server enable                    //Enable the Stelnet server
[AR1]aaa                                      //Enter the AAA view
[AR1-aaa] local-user zhangcr password cipher Chengdu123
```

```
                              //Create a local user with the username "zhangcr" and password "Chengdu123"
[AR1-aaa]local-user zhangcr service-type ssh
                              //Authorize the user "zhangcr" SSH service
[AR1-aaa]local-user zhangcr privilege level 15
                              //Configure the privilege level of the user "zhangcr" to 15
[AR1-aaa]quit                 //Return to the system view
[AR1]user-interface vty 0 4           //Enter the user interface view from VTY0 to VTY4
[AR1-ui-vty0-4]protocol inbound SSH   //Configure the protocol to SSH protocol
[AR1-ui-vty0-4]authentication-mode aaa //Configure the authentication mode of the five
                                      //virtual terminal sessions from VTY0 to VTY4
                                      //as AAA authentication
[AR1-ui-vty0-4]quit                   //Quit the user interface view of VTY0 to VTY4,
                                      //and return to the user view
[AR1]rsa local-key-pair create        //Create a local key pair
```

After completing IP configuration and successfully ping the router AR1 on the host PC1 serving as an SSH client, refer to the following steps to complete the SSH login test:

① Run the SecureCRT virtual terminal software on PC1. In the SecureCRT software window, select the "Quick Connect" command from the "File" menu. In the pop-up "Quick Connect" window, select "SSH2" for the protocol, enter the IP address "192.168.0.254" for the router AR1 interface G0/0/0 for the host name, set the default port "22" for the port, and then enter the configured SSH username "zhangcr" for the username, as shown in Fig. 1.23.

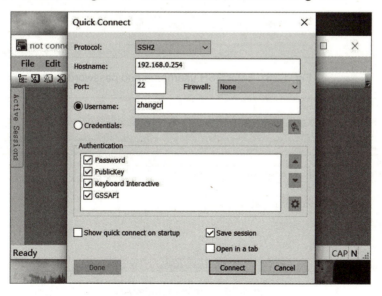

Fig. 1.23 Connect to a router via SSH

② Click the "Connect" button in Fig. 1.23, and in the pop-up "New Host Key" window, click the "Accept & Save" or "View Host Key" button, as shown in Fig. 1.24.

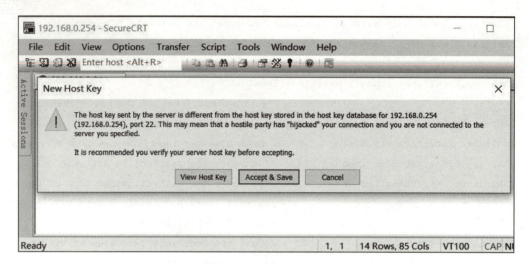

Fig. 1.24   SSH "New Host Key" window

③ According to the prompt, enter the SSH username "zhangcr" and password "Chengdu123", as shown in Fig. 1.25. Click the "OK" button to log in to the router AR1, as shown in Fig. 1.26.

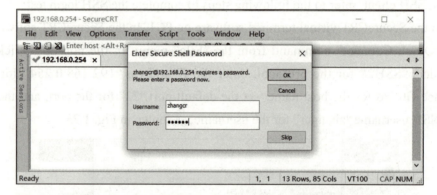

Fig. 1.25   Enter SSH username and password

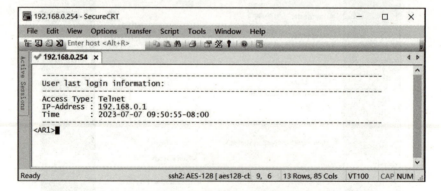

Fig. 1.26   Log in to router AR1

3. Router management

After completing the basic configuration of the router according to Fig. 1.27 and Table 1.5, perform such management tasks as follows: save the router configuration file, back up the configuration file to the FTP server, clear the router configuration, and clear the router Console password.

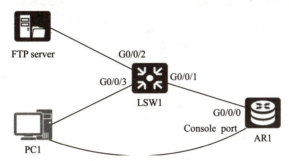

Fig. 1.27　Router basic management topology structure

Table 1.5　IP planning

Device	Interface	IP address	Gateway
AR1	G0/0/0	192.168.0.254/24	
PC1	NIC	192.168.0.10/24	192.168.0.254
FTP	NIC	192.168.0.1/24	192.168.0.254

(1) Router configuration file management

① Complete the configuration of the G0/0/0 interface on the router AR1 according to the IP planning in Table 1.5. The configuration steps are as follows:

```
[AR1] interface g0/0/0
[AR1-GigabitEthernet0/0/0] ip address 192.168.0.254 24
[AR1-GigabitEthernet0/0/0] quit
```

② Use the "dir" command in the user view of the router AR1 to view the list of files in the current directory.

Fig. 1.28 shows an example of viewing the list of files in the current directory on AR1.

```
<AR1>dir
Directory of flash:/

  Idx  Attr   Size(Byte)   Date         Time(LMT)   FileName
   0   drw-            -   Mar 08 2023  09:41:22    dhcp
   1   -rw-      121,802   May 26 2014  09:20:58    portalpage.zip
   2   -rw-        2,263   Mar 08 2023  09:41:18    statemach.efs
   3   -rw-      828,482   May 26 2014  09:20:58    sslvpn.zip

1,090,732 KB total (784,464 KB free)
```

Fig. 1.28　Example of viewing the list of files in the current directory on AR1

③ In the command view of AR1, use the "save" command to save the configuration, and then use the "dir flash:" command to view the configuration files saved in Flash. See Fig. 1.29 for the steps and display results.

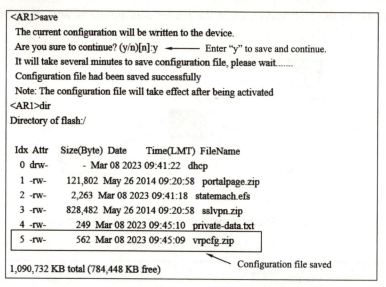

Fig. 1.29 Configuration save and view

④ Use the "save" command on AR1 to save the current configuration and name it "AR1.cfg", then use the "dir flash:" command to view the files in Flash, as shown in Fig. 1.30.

Note: The name of the configuration file must have an extension of ".cfg" or ".zip".

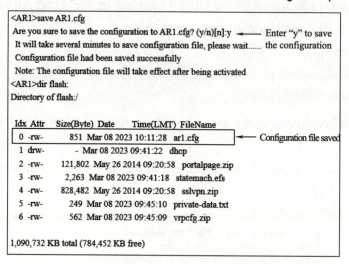

Fig. 1.30 Save the configuration and name it "ar1.cfg"

⑤ Use the "startup saved-configuration" command to set the configuration file to be used for the next startup of the router, and then verify the settings. The specific steps are as follows:

```
<AR1>startup saved-configuration?  //View the "startup saved-configuration" command
                                   //keywords or parameters
<AR1>startup saved-configuration ar1.cfg
        //Configure the configuration file to be used for the next startup as "ar1.cfg"
<AR1>display startup    //Display the operating system, configuration files and other
                        //information for the current and next startup of the device
```

⑥ According to Table 1.5, complete the IP configuration of the FTP server and the establishment and configuration of the FTP service. Configure the input "ftp 192.168.0.1" on the router AR1 to connect to the FTP server as an FTP client, and upload the "ar1.cfg" configuration file from Flash to the FTP server. See Fig. 1.31 for the steps.

```
<AR1>ftp 192.168.0.1
Trying 192.168.0.1 ...

Press CTRL+K to abort
Connected to 192.168.0.1.
220 FtpServerTry FtpD for free
User(192.168.0.1:(none)):anonymous   ←—— Enter an anonymous account "anonymous"
331 Password required for anonymous .
Enter password:   ←—— Enter any e-mail address
230 User anonymous logged in , proceed
[AR1-ftp]put flash:/ar1.cfg
200 Port command okay.
150 Opening BINARY data connection for ar1.cfg
226 Transfer finished successfully. Data connection closed.
FTP: 0 byte(s) sent in 0.170 second(s) 0.00byte(s)/sec.
```

Fig. 1.31 Upload the configuration file from Flash to the FTP server

⑦ Enter the "reset saved-configuration" command, and then enter the "reboot" command to restart the router and clear its configuration. See Fig. 1.32 for the steps.

```
<AR1>reset saved-configuration
This will delete the configuration in the flash memory.

The device configuratio
ns will be erased to reconfigure.

Are you sure? (y/n)[n]:y   ←—— Enter "y" to delete the configuration
 Clear the configuration in the device successfully.
<AR1>reboot
Info: The system is comparing the configuration, please wait.
Warning: All the configuration will be saved to the next startup configuration.
Continue ? [y/n]:n   ←—— Enter "n" to clear the configuration
System will reboot! Continue ? [y/n]:y   ←—— Enter "y" to reboot and continue
```

Fig. 1.32 Clear router configuration

(2) Clear router console password

Use a control cable to connect the console with the Console port of the router reliably. Afterwards, first disconnect the power supply of the router and then turn on the power supply.

Observe the startup process of the router and clear the Console password as follows:

① When "Press Ctrl+B to break auto startup" appears as shown in the box in Fig. 1.33, press the 【Ctrl】 and letter 【B】 keys on the keyboard immediately.

```
cold init done
Press Ctrl+B to break auto startup ... 2
Enter Password:************
Default Password,Please Set New Password.
          BootLoader Menu
    1. Default Startup
    2. Serial Menu
    3. Network Menu
    4. Startup Select
    5. File Manager
    6. Reboot
    7. Password Manager
Enter your choice(1-7):7
          Password Manager
    1. Modify the menu password
    2. Clear the console login password
    0. Return
Enter your choice(0-2):2
Clear the console login password Succeed!

          Password Manager
    1. Modify the menu password
    2. Clear the console login password
    0. Return
Enter your choice(0-2):0
          BootLoader Menu
    1. Default Startup
    2. Serial Menu
    3. Network Menu
    4. Startup Select
    5. File Manager
    6. Reboot
    7. Password Manager
Enter your choice(1-7):6
```

Fig. 1.33   Steps for clearing console password of Huawei AR1220

② Enter the BootLoad password at "Enter password" (by default, the default password is "Admin@huawei" for Huawei AR1220C).

③ Under the "BootLoader" menu, enter "7" to select "Password Manager".

④ Under "Password Manager", enter "2" and select "Clear the console login password".

⑤ Enter "0" under "Password Manager" and select "Return" to return to the "BootLoader" menu.

⑥ Under the "BootLoader" menu, enter "6" and select "Reboot" to restart the device. See Fig. 1.33 for the specific steps.

Note: The steps for password recovery vary slightly among different routing devices.

Task Completion

1. Task planning

According to the network topology and construction requirements of Project 1 shown in Fig.1.1, the branch budget department has 140 hosts, the management department has 100 hosts, and the branch company has 8 servers for businesses. The data from the budget and management

departments to servers needs to be isolated and cannot communicate directly but through routers. This task uses the IP planning as shown in Table 1.6.

Table 1.6 IP planning

Device	Interface	IP address	Gateway
fenbu	G0/0/0	172.16.254.1/30	
	G0/0/1	172.16.128.129/25	
	G0/0/2	172.16.129.1/24	
	G3/0/0	172.16.128.1/25	
	S2/0/0	172.16.254.5/30	
zongbu	G0/0/0	172.16.254.2/30	
	G0/0/1	172.16.0.1/24	
	S3/0/0	172.16.254.6/30	
PC1	NIC	172.16.0.10/24	172.16.0.1
PC2	NIC	172.16.129.10/24	172.16.129.1
PC3	NIC	172.16.128.13/25	172.16.128.1
PC4	NIC	172.16.128.14/25	172.16.128.1
Web server	NIC	172.16.128.131/28	172.16.128.129
DNS server	NIC	172.16.128.130/28	172.16.128.129
FTP server	NIC	172.16.0.254/24	172.16.0.1

① The Console authentication of the branch router is planned as follows: Use local user authentication with the username of cdp, password of Chenghua123 and user command line level of 15.

② The SSH authentication of the branch router is planned as follows: Use local user authentication with the username of cdp, password of Chenghua123 and user command line level of 15.

③ The Console authentication of the headquarters router is planned as follows: Use local user authentication with the username of zhang, password of Chengdu123 and user command line level of 15.

④ The SSH authentication of the headquarters router is planned as follows: Use local user authentication with the username of zhang, password of Chengdu123 and user command line level of 15.

2. Task implementation

(1) Configuration of router "fenbu"

```
[Huawei]sysname fenbu
//Configure the host name of the router as "fenbu"
```

```
[fenbu]interface g0/0/0
[fenbu-GigabitEthernet0/0/0]ip address 172.16.254.1 30
[fenbu-GigabitEthernet0/0/0]description linktozongbu
[fenbu-GigabitEthernet0/0/0]quit
//Configure the G0/0/0 interface of the router "fenbu"
[fenbu]interface s2/0/0
[fenbu-Serial2/0/0]ip address 172.16.254.5 30
[fenbu-Serial2/0/0]description linktozongbu
[fenbu-Serial2/0/0]quit
//Configure the S2/0/0 interface of the router "fenbu"
[fenbu]interface g0/0/1
[fenbu-GigabitEthernet0/0/1]ip address 172.16.128.129 28
[fenbu-GigabitEthernet0/0/1]description server
[fenbu-GigabitEthernet0/0/1]undo shutdown
[fenbu-GigabitEthernet0/0/1]quit
//Configure the G0/0/1 interface of the router "fenbu"
[fenbu]interface g0/0/2
[fenbu-GigabitEthernet0/0/2]ip address 172.16.129.1 24
[fenbu-GigabitEthernet0/0/2]description yusuanbumen
[fenbu-GigabitEthernet0/0/2]quit
[fenbu]
//Configure the G0/0/2 interface of the router "fenbu"
[fenbu]interface g3/0/0
[fenbu-GigabitEthernet3/0/0]description manager
[fenbu-GigabitEthernet3/0/0]ip address 172.16.128.1 25
[fenbu-GigabitEthernet3/0/0]quit
[fenbu]
// Configure the G0/0/3 interface of the router "fenbu"
[fenbu]aaa
[fenbu-aaa]local-user cdp password cipher Chenghua123
[fenbu-aaa]local-user cdp service-type terminal ssh
//Configure the local user "cdp" of the router "fenbu" with the password "Chenghua123"
//and the authorized service types as end users and SSH users
[fenbu]user-interface console 0
[fenbu-ui-console0]authentication-mode aaa
[fenbu-ui-console0]user privilege level 15
[fenbu-ui-console0]quit
//Configure the Console authentication of the router "fenbu" to AAA local user authen-
//tication, with the user command line level of 15
[fenbu]stelnet server enable
```

```
[fenbu]user-interface vty 0 4
[fenbu-ui-vty0-4]protocol inbound ssh
[fenbu-ui-vty0-4]authentication-mode aaa
[fenbu-ui-vty0-4]user privilege level 15
[fenbu-ui-vty0-4]quit
[fenbu]rsa local-key-pair create
//Configure SSH of the router "fenbu", enable the Stelnet service, use SSH as the virtual
//terminal protocol, use AAA local user authentication, set the user command line
//level of 15 and generate a local key pair
[fenbu]quit
<fenbu>save                    //Save the configuration
```

(2) Configuration of router "zongbu"

```
[Huawei]sysname zongbu
//Configure the host name of the router as "zongbu"
[zongbu]interface g0/0/0
[zongbu-GigabitEthernet0/0/0]ip address 172.16.254.2 30
[zongbu-GigabitEthernet0/0/0]quit
//Configure the G0/0/0 interface of the router "zongbu"
[zongbu]interface s3/0/0
[zongbu-Serial3/0/0]ip address 172.16.254.6 30
[zongbu-Serial3/0/0]quit
//Configure the S3/0/0 interface of the router "zongbu"
[zongbu]interface g0/0/1
[zongbu-GigabitEthernet0/0/1]ip address 172.16.0.1 24
[zongbu-GigabitEthernet0/0/1]quit
//Configure the G0/0/1 interface of the router "zongbu"
[zongbu]aaa
[zongbu-aaa]local-user zhang password cipher Chengdu123
[zongbu-aaa]local-user zhang service-type terminal ssh
[zongbu-aaa]quit
//Configure the local user "zhang" of the router "zongbu" with the password "Chenghua123"
//and the authorized service types as end users and SSH users
[zongbu]user-interface console 0
[zongbu-ui-console0]authentication-mode aaa
[zongbu-ui-console0]user privilege level 15
[zongbu-ui-console0]quit
//Configure the Console authentication of the router "zongbu" to AAA local user
//authentication, with the user command line level of 15
[zongbu]stelnet server enable
```

```
[zongbu]user-interface vty 0 4
[zongbu-ui-vty0-4]protocol inbound ssh
[zongbu-ui-vty0-4]authentication-mode aaa
[zongbu-ui-vty0-4]user privilege level 15
[zongbu-ui-vty0-4]quit
[zongbu]rsa local-key-pair create
//Configure SSH of the router "zongbu", enable the Stelnet service, use SSH as the
//virtual terminal protocol, use AAA local user authentication, set the user command
//line level of 15 and generate a local key pair
[zongbu]quit
<zongbu>save
//Save the configuration
```

(3) Router view and testing

① After configuring the host names and IP addresses for the branch router "fenbu" and the headquarters router "zongbu", use "display IP interface brief" to check if the interface IP of the router has been correctly configured and whether Physical and Protocol have been "UP" with all network links functioning properly. See Fig.1.34 and Fig. 1.35 respectively for the interface brief information about the router "fenbu" and router "zongbu".

```
[fenbu]display ip interface brief
*down: administratively down
^down: standby
(l): loopback
(s): spoofing
The number of interface that is UP in Physical is 6
The number of interface that is DOWN in Physical is 2
The number of interface that is UP in Protocol is 6
The number of interface that is DOWN in Protocol is 2

Interface              IP Address/Mask      Physical  Protocol
GigabitEthernet0/0/0   172.16.254.1/30      up        up
GigabitEthernet0/0/1   172.16.128.129/28    up        up
GigabitEthernet0/0/2   172.16.129.1/24      up        up
GigabitEthernet3/0/0   172.16.128.1/25      up        up
GigabitEthernet4/0/0   unassigned           down      down
NULL0                  unassigned           up        up(s)
Serial2/0/0            172.16.254.5/30      up        up
Serial2/0/1            unassigned           down      down
```

Fig. 1.34 Brief interface information of router "fenbu" under conditions of normal links

② Test the Console login on the branch router "fenbu" and the headquarters router "zongbu" separately, with the test results as shown in Fig. 1.36 and Fig. 1.37 respectively.

③ Complete the IP configuration on the host PC1 as planned in Table 1.6, and use ping on the command line to test the IP connectivity with the router "zongbu" G0/0/1 interface IP. After ensuring connectivity, run the SecureCRT virtual terminal software on PC1, select the "Quick Connect" command from the "File" menu, and in the "Quick Connect" window that pops up, select "SSH2" for the protocol, and enter the IP address "172.16.0.1" of the router zongbu interface

G0/0/1 as the host name, with the default port "22". Enter the configured SSH username "zhang" for the username, click the "Connect" button, and in the "New Host Key" window that pops up, click the "Accept & Save" button. In the pop-up "Enter Secure Shell Password" window, enter the password "Chengdu123" for the user "zhang", as shown in Fig. 1.38. Use SSH to log in to the router "zongbu" remotely.

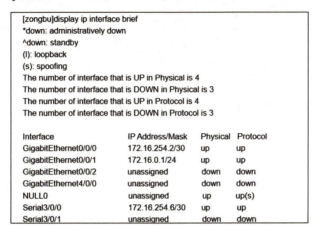

Fig. 1.35 Brief interface information of router "zongbu" under conditions of normal links

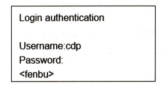   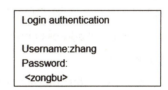

Fig. 1.36 Router "fenbu" Console login authentication   Fig. 1.37 Router "zongbu" Console login authentication

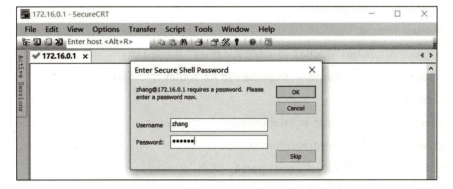

Fig. 1.38 PC1 SSH to router "zongbu"

④ Complete the IP configurations on the hosts PC2, PC3 and PC4 as planned in Table 1.6, and use ping on the command line to test the IP connectivity from PC2 to router "fenbu" G0/0/2 interface IP (172.16.129.1), and from PC3 and PC4 to router "fenbu" G3/0/0 interface IP (172.16.128.1) to ensure reliable connectivity. Then test whether SSH can be used to log in to the

router "fenbu" on PC2, PC3, and PC4. Fig. 1.39 shows that the host PC2 uses the IP (172.16.128.1) of the router "fenbu" G3/0/0 as the target address of the remote host to log in to the router "fenbu" via SSH.

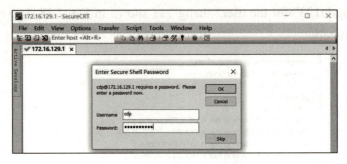

Fig. 1.39　PC2 SSH to router "fenbu"

Exercises

Multiple choice

(1) The contents of the router storage device (　　) below will be lost after power outage or restart.

　　A. RAM　　　　　B. ROM　　　　　C. NVRAM　　　　　D. Flash

(2) Configure the IP address and mask of the router G0/0/0 interface at the prompt (　　).

　　A. [Huawei]　　　　　　　　　　B. <Huawei>

　　C. [Huawei-GigabitEthernet0/0/0]　　D. [Huawei-aaa]

(3) The following command (　　) is used to clear router configurations.

　　A. reset saved-configuration　　　B. save

　　C. display version　　　　　　　D. dir

(4) The router command "display ip interface brief" is used to (　　).

　　A. display brief configuration information for router interfaces

　　B. configure interfaces

　　C. check if the connection is established

　　D. enter the interface configuration mode

(5) A router shall be initialized via its port (　　).

　　A. G0/0/0　　　B. S0/0/0　　　C. G0/0　　　　D. Console

(6) The following command (　　) can modify the device name of the Huawei router.

　　A. hostname　　B. rename　　　C. sysname　　　D. domain

(7) After successfully Telnet to the router, the network administrator found that the router interface IP address could not be configured. The possible reason should be (　　).

　　A. Telnet user level configuration error

B. Configuration error of Telnet user's authentication mode

C. The network administrator host cannot ping the router

D. The Telnet client software used by the network administrator prohibits corresponding operations

 Static Routing and Default Routing

Task Description

After completing Task 1 of this project, plan and configure the static routing, floating static routing, aggregated static routing and default routing on the "fenbu" router and "zongbu" router in the network shown in Fig 1.1. Achieve the networking requirements of "network interconnection between the internal branch and the enterprise headquarters, and branch access to the Internet through the headquarters". On the basis of interconnection, reduce the routing table entries on the routers "fenbu" and "zongbu" as possible, with no routing loop allowed. Data packets should normally go through the gigabit Ethernet link between the routers "fenbu" and "zongbu" when interconnected. Only when the gigabit Ethernet link fails can they go through the serial link. Perform network testing and troubleshooting after completing the configuration.

Task Objectives

- Understand the composition of routing tables and insertion rules.
- Master the planning and configuration of static routing.
- Master the planning and configuration of aggregated static routing.
- Master the configuration of default routing.
- Master the planning and configuration of floating routing.

Related Knowledge

1. Routing table

A router can be manually configured by administrators or exchange routing information with other routers through dynamic routing protocols to get the optimal path to the target network or node, and the information of the optimal path obtained can be saved in the RAM of the router in the form of a table, which is called a routing table. Each router in the network can independently make a decision for data packet forwarding based on the routing information in its own routing table.

Fig. 1.40 shows an example of a routing table.

① Destination/Mask: It indicates the destination network address and subnet mask length of

the route entry.

② Proto (Protocol): It indicates the protocol type of the routing entry. It indicates the routing protocol through which the router learns about the route.

③ Pre (Preference): It indicates the routing protocol preference of the routing entry.

④ Cost: Routing cost.

⑤ Flags: Routing flags. "D" indicates that the route is downloaded to FIB.

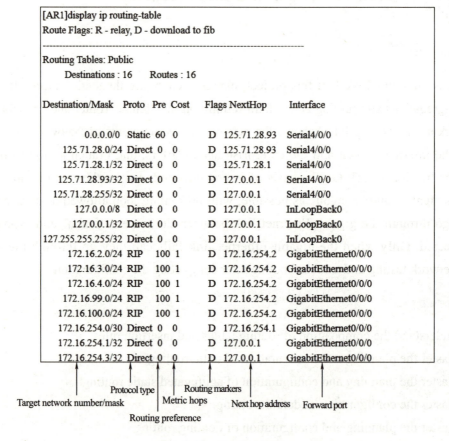

Fig. 1.40 Routing table

⑥ NextHop: It indicates the next hop, whose value is the IP address, and the value is the IP address of the next router interface directly connected to this router on the way to the target network.

⑦ Interface: It refers to the forwarding port of the routing entry.

2. Routing preference

In the process of building a routing table, when a router can obtain multiple routing information about the same target network from different sources (such as various dynamic routing protocols, static routes, or direct routes), the router uses the routing preference to identify the credibility of different routing sources and select the routing entry with the highest credibility to

insert into the routing table.

The routing preference is defined as an integer value between 1 and 255. The smaller the routing preference value, the higher the credibility of the routing source, and the higher the preference inserted into the routing table. By default, the routing preference of direct routes is the smallest and is usually set to "0". The routing preference of direct routes cannot be manually changed. The routing preferences of static routes and dynamic routing protocols are higher than those of direct interconnection networks and are assigned different routing preference values.

For the routing devices from different manufacturers, the default route preference specified for each routing protocol may be different. Table 1.7 shows the default administrative distance (AD) values of some routing protocols for Huawei routing devices. According to the needs of network management, the routing preference values of static routing and various dynamic routing protocols can be manually modified.

Table 1.7 Default administrative distance values of some routing protocols for Huawei routing devices

Route source	Administrative distance value
Direct routing	0
Static routing	60
RIP	100
OSPF	10
ISIS	15

3. Principle for routing table insertion

A router can obtain the routing information to the destination network through various approaches, but not all such routing information may be inserted into the routing table as a basis for packet forwarding. The principles for inserting routing information into the routing table are as follows:

① Determine the source of the routing information, and insert the route obtained from the route source with the lowest preference value into the routing table first.

② If the routing preference values are the same, determine the relevant cost value and insert the path with a lower cost value into the routing table.

4. Static, default and floating static routing

Static routing refers to such routing information that is manually configured and managed by network administrators. Static routing requires network administrators to select the best path based on the network topology and manually configure the selected route. When the topology or link status of the network changes and routing information needs to be updated, the network

administrator must manually change the relevant static routing entries.

The default route is a special static route. The default route gives the router forwarding interface or next hop information corresponding to the data packets whose destination addresses are not explicitly listed in the routing table. Usually, the default route is introduced to effectively reduce the size and maintenance cost of the routing table. The target network and subnet mask of the default route is 0.0.0.0/0 in the routing table.

The floating static routing is a type of static routing that exists as a backup route. After configuring one or more ordinary routes on the router, a floating static route is configured as a static route with a higher preference value than one or more main routes. Since the preference value of a floating static route is higher than that of the main route, the floating static route can be inserted into the routing table when and only when the main route fails.

5. Implementation of routing and data forwarding

Packet forwarding involves two processes, namely "routing" and "switching", where "routing" refers to selecting the best path to the destination for packets passing through the router; "switching" refers to the process by which a router receives data packets from one interface and forwards them from another interface. When a router receives a packet from an interface that needs to reach another network, the router routing and data forwarding process is as follows:

① The router determines whether to receive a data frame based on its destination address. If received, the router will hand it over to the IP processing module for frame unpacking and separate the corresponding IP packets and hand them over to the routing module.

② The routing module extracts the target network number from the IP packet through the "AND" operation of the destination address and the subnet mask, and matches the target network number with the routing entries in the routing table. In the matching process, the longest matching principle is adopted, that is, if multiple routing entries in the routing table match the target network number, then it will select the route with the most matching digits from the leftmost side of the destination IP address of the packet in the routing table as the preferred route. If all routing entries in the routing table do not match the target network number, the router will abandon the corresponding IP packet; if some match, the router will encapsulate the data packet into the data frame format as the exit requires from the port determined based on the preferred route and then forward such packet accordingly.

Related Skills

1. Planning and configuration of static routing

Build a topology as shown in Fig. 1.41, and complete the IP configuration of the router interface and host according to the IP addresses planned in Table 1.8; complete the planning

and configuration of the static routes on the routers AR1 and AR2, achieve interconnection and interoperability across the entire network, and conduct network connectivity testing and troubleshooting.

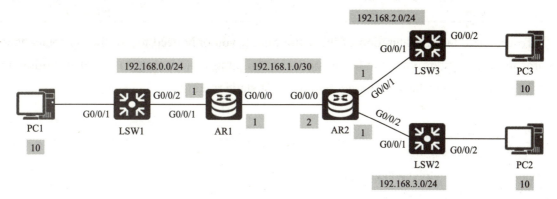

Fig. 1.41 Topology for "planning and configuration of static routing"

Table 1.8 IP planning

Device	Interface	IP address	Gateway
AR1	G0/0/0	192.168.1.1/30	
	G0/0/1	192.168.0.1/24	
AR2	G0/0/0	192.168.1.2/30	
	G0/0/1	192.168.2.1/24	
	G0/0/2	192.168.3.1/24	
PC1	NIC	192.168.0.10/24	192.168.0.1
PC2	NIC	192.168.3.10/24	192.168.3.1
PC3	NIC	192.168.2.10/24	192.168.2.1

(1) Planning and configuration of static routing for router AR1

As can be seen from Fig. 1.41 and Table 1.8, there are four network segments in the entire network. For AR1, there are two directly connected network segments (192.168.0.0/24, 192.168.1.0/30) and two remote network segments (192.168.2.0/24, 192.168.3.0/24). According to the principle for inserting routing tables, after the IP address and subnet mask are configured for the interface of the router connecting to the direct network segment, and when the physical interface and link protocol state are both "UP", the network where the interface is located will be directly inserted into the routing table, with no additional configuration required. For remote network segments "192.168.2.0/24" and "192.168.3.0/24", the router should use static or dynamic routing to obtain its routing information.

Therefore, the router AR1 needs to use static routing for such two remote network segments "192.168.2.0/24" and "192.168.3.0/24", as shown in Table 1.9 for planning.

Table 1.9　Static route planning for router AR1

Target network	Target network mask/Mask length	Next hop
192.168.2.0	24	192.168.1.2
192.168.3.0	24	192.168.1.2

To complete the configuration of the static routes, you only need to use the "ip route-static" command in the system view mode for manual configuration according to the plan. Follow the steps below to complete the configuration of the AR1 static routing:

```
<Huawei>sys
[Huawei]sysname AR1
[AR1]interface g0/0/1
[AR1-GigabitEthernet0/0/1]ip address 192.168.0.1 24
[AR1-GigabitEthernet0/0/1]quit
[AR1]interface g0/0/0
[AR1-GigabitEthernet0/0/0]ip address 192.168.1.1 30
[AR1-GigabitEthernet0/0/0]quit
//Configure the host name and interface IP address of the router AR1
[AR1]ip route-static 192.168.2.0 24 192.168.1.2
//Configure the static route to the target network "192.168.2.0/24", with the selected
//route represented by the next hop at "192.168.1.2"
[AR1]ip route-static 192.168.3.0 24 192.168.1.2
//Configure the static route to the target network "192.168.3.0/24", with the selected
//route represented by the next hop at "192.168.1.2"
[AR1]quit
<AR1>save
//Save configurations
```

(2) Planning and configuration of static routing for router AR2

For AR2, there are three directly connected network segments (192.168.1.0/30, 192.168.2.0/24, 192.168.3.0/24) and one remote network segment (192.168.0.0/24). According to the principle for inserting routing tables, after the IP address and subnet mask are configured for the interface of the router connecting to the direct network segment, and when the physical interface and link protocol state are both "UP", the network where the interface is located will be directly inserted into the routing table, with no additional configuration required. For the remote network segment, the router should use static or dynamic routing to obtain its routing information.

Therefore, the router AR2 needs to use static routing for the remote network segment "192.168.0.0/24", as shown in Table 1.10 for planning.

Table 1.10 Static route planning for router AR2

Target network	Target network mask/Mask length	Next hop
192.168.0.0	24	192.168.1.1

To complete the configuration of the static routes, you only need to use the "ip route-static" command in the system view mode for manual configuration according to the plan. Follow the steps below to complete the configuration of the AR2 static routing:

```
<Huawei>sys
[Huawei]sysname AR2
[AR2]interface g0/0/0
[AR2-GigabitEthernet0/0/0]ip address 192.168.1.2 30
[AR2-GigabitEthernet0/0/0]quit
[AR2]interface g0/0/1
[AR2-GigabitEthernet0/0/1]ip address 192.168.2.1 24
[AR2-GigabitEthernet0/0/1]quit
[AR2]interface g0/0/2
[AR2-GigabitEthernet0/0/2]ip address 192.168.3.1 24
[AR2-GigabitEthernet0/0/2]quit
//Configure the host name and interface IP address of the router AR2
[AR2]ip route-static 192.168.0.0 24 192.168.1.1
//Configure the static route to the target network "192.168.0.0/24", with the selected
//route represented by the next hop at "192.168.1.1".
[AR2]quit
<AR2>save
//Save configurations
```

(3) Network connectivity testing and troubleshooting

Refer to the IP planning in Table 1.8 to configure the IP address, subnet mask and default gateway of all PCs in the network. After the configuration is completed, use "ipconfig" on the DOS command line to check whether the IP configuration information is correct. Fig. 1.42 shows the IP configuration information displayed by the "ipconfig" command on the host PC1.

```
PC>ipconfig

Link local IPv6 address..........: fe80::5689:98ff:fe25:3e0f
IPv6 address....................: :: / 128
IPv6 gateway....................: ::
IPv4 address....................: 192.168.0.10
Subnet mask.....................: 255.255.255.0
Gateway.........................: 192.168.0.1
Physical address................: 54-89-98-25-3E-0F
DNS server......................:
```

Fig. 1.42 IP configuration information displayed by the "ipconfig" command on host PC1

On any host in the network, enter the DOS command line and use the ping utility to test the IP connectivity to other target hosts. Fig. 1.43 shows that the host PC1 can ping the IP addresses of the hosts PC2 and PC3.

```
PC>ping 192.168.2.10

Ping 192.168.2.10: 32 data bytes, Press Ctrl_C to break
From 192.168.2.10: bytes=32 seq=1 ttl=126 time=78 ms
From 192.168.2.10: bytes=32 seq=2 ttl=126 time=78 ms
From 192.168.2.10: bytes=32 seq=3 ttl=126 time=78 ms
From 192.168.2.10: bytes=32 seq=4 ttl=126 time=94 ms
From 192.168.2.10: bytes=32 seq=5 ttl=126 time=78 ms

--- 192.168.2.10 ping statistics ---
  5 packet(s) transmitted
  5 packet(s) received
  0.00% packet loss
  round-trip min/avg/max = 78/81/94 ms

PC>ping 192.168.3.10

Ping 192.168.3.10: 32 data bytes, Press Ctrl_C to break
From 192.168.3.10: bytes=32 seq=1 ttl=126 time=47 ms
From 192.168.3.10: bytes=32 seq=2 ttl=126 time=78 ms
From 192.168.3.10: bytes=32 seq=3 ttl=126 time=94 ms
From 192.168.3.10: bytes=32 seq=4 ttl=126 time=94 ms
From 192.168.3.10: bytes=32 seq=5 ttl=126 time=78 ms
```

Fig. 1.43   Host PC1 can ping the IP addresses of hosts PC2 and PC3

Moreover, the extended ping command on the router can also be used to test the network connectivity, as shown in Fig. 1.44. Use "ping - a 192.168.0.1 192.168.3.1" in AR1 to display the results. Therein, the parameter "- a 192.168.0.1" indicates that the source address used for the ICMP packet sent by ping is the G0/0/0 interface address "192.168.0.1" of the router AR1 and that the destination address is the G0/0/2 interface address "192.168.3.1" of the router AR2.

```
<AR1>ping -a 192.168.0.1 192.168.3.1
  PING 192.168.3.1: 56  data bytes, press CTRL_C to break
    Reply from 192.168.3.1: bytes=56 Sequence=1 ttl=255 time=50 ms
    Reply from 192.168.3.1: bytes=56 Sequence=2 ttl=255 time=20 ms
    Reply from 192.168.3.1: bytes=56 Sequence=3 ttl=255 time=30 ms
    Reply from 192.168.3.1: bytes=56 Sequence=4 ttl=255 time=30 ms
    Reply from 192.168.3.1: bytes=56 Sequence=5 ttl=255 time=30 ms

  --- 192.168.3.1 ping statistics ---
  5 packet(s) transmitted
  5 packet(s) received
  0.00% packet loss
  round-trip min/avg/max = 20/32/50 ms
```

Fig. 1.44   The router source address "192.168.0.1" specified by extended ping can ping "192.168.3.1"

If the network cannot be connected during testing, please follow the steps below for troubleshooting:

① Use the "ipconfig" command on the DOS command line of all hosts to check whether the IP address, subnet mask, and default gateway of each host are configured correctly.

② Use the "display ip interface brief" command on the routers AR1 and AR2 to check if the interface IP of the router is configured correctly, and whether the physical interface and link protocol state have been "UP".

③ On the routers AR1 and AR2, use the "display current-configuration | include ip route-static" command to check if the static routing configuration is correct in the currently running configuration file, as shown in Fig. 1.45 and Fig. 1.46.

```
[AR1]display current-configuration | include ip route-static
ip route-static 192.168.2.0 255.255.255.0 192.168.1.2
ip route-static 192.168.3.0 255.255.255.0 192.168.1.2
```

Fig. 1.45 Check the static routing configuration in the currently running configuration file on AR1

```
[AR2]display current-configuration | include ip route-static
ip route-static 192.168.0.0 255.255.255.0 192.168.1.1
```

Fig. 1.46 Check the static routing configuration in the currently running configuration file on AR2

Pay attention to the static routing configuration in the running configuration file and avoid any wrong or redundant static routing configuration. If there is any wrong or redundant static routing configuration, then add the "undo" command before the original static routing configuration command to delete the redundant or incorrect static routing configuration, and then use the "ip route-static" command to configure a correct static routing again. Here follows an example for deleting an incorrect static routing configuration:

```
[AR1]undo ip route-static 192.168.3.0 24 192.168.0.2
```

④ Use the "display ip routing-table" command on the router to display the routing table contents, check whether all network segments in the network have been selected, and confirm whether the routed paths are correct, as shown in Fig. 1.47 and Fig. 1.48.

```
[AR1]display ip routing-table
Route Flags: R - relay, D - download to fib
------------------------------------------------------------
Routing Tables: Public
     Destinations : 12      Routes : 12

Destination/Mask    Proto   Pre  Cost    Flags NextHop     Interface
       127.0.0.0/8         Direct  0    0       D   127.0.0.1    InLoopBack0
       127.0.0.1/32        Direct  0    0       D   127.0.0.1    InLoopBack0
 127.255.255.255/32  Direct  0    0       D   127.0.0.1    InLoopBack0
     192.168.0.0/24       Direct  0    0       D   192.168.0.1  GigabitEthernet0/0/1
     192.168.0.1/32       Direct  0    0       D   127.0.0.1    GigabitEthernet0/0/1
   192.168.0.255/32    Direct  0    0       D   127.0.0.1    GigabitEthernet0/0/1
     192.168.1.0/30       Direct  0    0       D   192.168.1.1  GigabitEthernet0/0/0
     192.168.1.1/32       Direct  0    0       D   127.0.0.1    GigabitEthernet0/0/0
     192.168.1.3/32       Direct  0    0       D   127.0.0.1    GigabitEthernet0/0/0
     192.168.2.0/24       Static  60   0       RD  192.168.1.2  GigabitEthernet0/0/0
     192.168.3.0/24       Static  60   0       RD  192.168.1.2  GigabitEthernet0/0/0
 255.255.255.255/32  Direct  0    0       D   127.0.0.1    InLoopBack0
```

Fig. 1.47   Contents of AR1 routing table

```
[AR2]display ip routing-table
Route Flags: R - relay, D - download to fib
------------------------------------------------------------
Routing Tables: Public
     Destinations : 14      Routes : 14

Destination/Mask    Proto   Pre  Cost    Flags NextHop     Interface
       127.0.0.0/8         Direct  0    0       D   127.0.0.1    InLoopBack0
       127.0.0.1/32        Direct  0    0       D   127.0.0.1    InLoopBack0
 127.255.255.255/32  Direct  0    0       D   127.0.0.1    InLoopBack0
     192.168.0.0/24       Static  60   0       RD  192.168.1.1  GigabitEthernet0/0/0
     192.168.1.0/30       Direct  0    0       D   192.168.1.2  GigabitEthernet0/0/0
     192.168.1.2/32       Direct  0    0       D   127.0.0.1    GigabitEthernet0/0/0
     192.168.1.3/32       Direct  0    0       D   127.0.0.1    GigabitEthernet0/0/0
     192.168.2.0/24       Direct  0    0       D   192.168.2.1  GigabitEthernet0/0/1
     192.168.2.1/32       Direct  0    0       D   127.0.0.1    GigabitEthernet0/0/1
   192.168.2.255/32    Direct  0    0       D   127.0.0.1    GigabitEthernet0/0/1
     192.168.3.0/24       Direct  0    0       D   192.168.3.1  GigabitEthernet0/0/2
     192.168.3.1/32       Direct  0    0       D   127.0.0.1    GigabitEthernet0/0/2
   192.168.3.255/32    Direct  0    0       D   127.0.0.1    GigabitEthernet0/0/2
 255.255.255.255/32  Direct  0    0       D   127.0.0.1    InLoopBack0
```

Fig. 1.48   Contents of AR2 routing table

2. Planning and configuration of floating static routing

Build a topology as shown in Fig. 1.49, and complete the IP configuration of the router and host according to the IP addresses planned in Table 1.11; plan and configure a floating static route to the remote target network on the routers AR1 and AR2. The data packet will normally run on the gigabit Ethernet link between AR1 and AR2. When the gigabit Ethernet link between AR1 and AR2 fails, the data packet will run on the serial link between AR1 and AR2 to achieve interconnection and interoperability across the entire network, and complete network connectivity testing and troubleshooting.

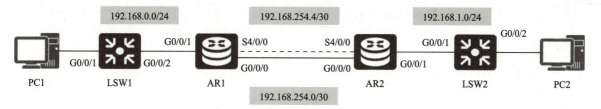

Fig. 1.49 Topology for "planning and configuration of floating static routing"

Table 1.11 IP planning

Device	Interface	IP address	Gateway
AR1	G0/0/0	192.168.254.1/30	
	G0/0/1	192.168.0.1/24	
	S4/0/0	192.168.254.5/30	
AR2	G0/0/0	192.168.254.2/30	
	G0/0/1	192.168.1.1/24	
	S4/0/0	192.168.254.6/30	
PC1	NIC	192.168.0.10/24	192.168.0.1
PC2	NIC	192.168.1.10/24	192.168.1.1

According to the task requirements, the gigabit Ethernet link between AR1 and AR2 is the primary route for AR1 and AR2 floating static routing, and the serial link between AR1 and AR2 is the backup route for AR1 and AR2 floating static routing. According to the principle for inserting routing tables, it can be achieved only by setting the static routing preference value of the main route lower than that of the backup route.

See Table 1.12 for the floating static routing planning of AR1 and AR2.

Table 1.12 Floating static routing planning of AR1 and AR2

Device	Target network/Mask length	Next hop or forward port	Route preference value	Remarks
AR1	192.168.1.0/24	192.168.254.2	Default value 60	Main route
	192.168.1.0/24	S4/0/0	70	Backup route
AR2	192.168.0.0/24	192.168.254.1	Default value 60	Main route
	192.168.0.0/24	S4/0/0	70	Backup route

(1) Configuration of router AR1

Refer to the following steps to configure AR1 and achieve network interconnection.

```
<Huawei>sys
[Huawei]sysname AR1
[AR1]interface g0/0/0
[AR1-GigabitEthernet0/0/0]ip address 192.168.254.1 30
```

```
[AR1-GigabitEthernet0/0/0]quit
[AR1]interface g0/0/1
[AR1-GigabitEthernet0/0/1]ip address 192.168.0.1 24
[AR1-GigabitEthernet0/0/1]quit
[AR1]interface s4/0/0
[AR1-Serial4/0/0]ip address 192.168.254.5 30
[AR1-Serial4/0/0]quit
//Configure the host name and interface IP address of the router AR1
[AR1]ip route-static 192.168.1.0 24 192.168.254.2
[AR1]ip route-static 192.168.1.0 24 s4/0/0 preference 70
//Configure the floating static route to the target network "192.168.1.0/24", with the
//next hop address of the main route path being "192.168 254.2"; the forwarding port
//of the backup route is the S4/0/0 interface
[AR1]quit
<AR1>save
//Save configurations
```

(2) Configuration of router AR2

```
<Huawei>sys
[Huawei]sysname AR2
[AR2]
[AR2]interface g0/0/0
[AR2-GigabitEthernet0/0/0]ip address 192.168.254.2 30
[AR2-GigabitEthernet0/0/0]quit
[AR2]interface g0/0/1
[AR2-GigabitEthernet0/0/1]ip address 192.168.1.1 24
[AR2-GigabitEthernet0/0/1]quit
[AR2]interface S4/0/0
[AR2-Serial4/0/0]ip address 192.168.254.6 30
[AR2-Serial4/0/0]quit
//Configure the host name and interface IP address of the router AR2
[AR2]ip route-static 192.168.0.0 24 192.168.254.1
[AR2]ip route-static 192.168.0.0 24 s4/0/0 preference 70
//Configure the floating static route to the target network "192.168.0.0/24", with the
//next hop address of the main route path being "192.168 254.1"; the forwarding port
//of the backup route is the S4/0/0 interface
[AR2]quit
<AR2>sa
<AR2>save
//Save configurations
```

(3) Routing testing and troubleshooting

Use the "display ip routing-table" command on the routers AR1 and AR2 to display the routing table contents and check if the static routes displayed in the routing table are the primary routing path configured. Fig. 1.50 shows that the path to the target network "192.168.1.0/24" in the AR1 routing table is the next hop address "192.168.254.2", and the port is the "GigabitEthernet 0/0/0" interface. Fig. 1.51 shows that the path to the target network "192.168.0.0/24" in the AR2 routing table is the next hop address "192.168.254.1", and the port is the "Gigabit Ethernet 0/0/0" interface.

```
[AR1]display ip routing-table
Route Flags: R - relay, D - download to fib
------------------------------------------------------------
Routing Tables: Public
         Destinations : 15    Routes : 15

Destination/Mask    Proto  Pre  Cost     Flags NextHop       Interface

       127.0.0.0/8  Direct 0    0          D   127.0.0.1     InLoopBack0
      127.0.0.1/32  Direct 0    0          D   127.0.0.1     InLoopBack0
  127.255.255.255/32 Direct 0   0          D   127.0.0.1     InLoopBack0
     192.168.0.0/24 Direct 0    0          D   192.168.0.1   GigabitEthernet0/0/1
     192.168.0.1/32 Direct 0    0          D   127.0.0.1     GigabitEthernet0/0/1
   192.168.0.255/32 Direct 0    0          D   127.0.0.1     GigabitEthernet0/0/1
     192.168.1.0/24 Static 60   0          RD  192.168.254.2 GigabitEthernet0/0/0
   192.168.254.0/30 Direct 0    0          D   192.168.254.1 GigabitEthernet0/0/0
   192.168.254.1/32 Direct 0    0          D   127.0.0.1     GigabitEthernet0/0/0
   192.168.254.3/32 Direct 0    0          D   127.0.0.1     GigabitEthernet0/0/0
   192.168.254.4/30 Direct 0    0          D   192.168.254.5 Serial4/0/0
   192.168.254.5/32 Direct 0    0          D   127.0.0.1     Serial4/0/0
   192.168.254.6/32 Direct 0    0          D   192.168.254.6 Serial4/0/0
   192.168.254.7/32 Direct 0    0          D   127.0.0.1     Serial4/0/0
  255.255.255.255/32 Direct 0   0          D   127.0.0.1     InLoopBack0
```

Fig. 1.50 Static routing path entries in AR1 routing table

```
[AR2]display ip routing-table
Route Flags: R - relay, D - download to fib
------------------------------------------------------------
Routing Tables: Public
         Destinations : 15    Routes : 15

Destination/Mask    Proto  Pre  Cost     Flags NextHop       Interface

       127.0.0.0/8  Direct 0    0          D   127.0.0.1     InLoopBack0
      127.0.0.1/32  Direct 0    0          D   127.0.0.1     InLoopBack0
  127.255.255.255/32 Direct 0   0          D   127.0.0.1     InLoopBack0
     192.168.0.0/24 Static 60   0          RD  192.168.254.1 GigabitEthernet0/0/0
     192.168.1.0/24 Direct 0    0          D   192.168.1.1   GigabitEthernet0/0/1
     192.168.1.1/32 Direct 0    0          D   127.0.0.1     GigabitEthernet0/0/1
   192.168.1.255/32 Direct 0    0          D   127.0.0.1     GigabitEthernet0/0/1
   192.168.254.0/30 Direct 0    0          D   192.168.254.2 GigabitEthernet0/0/0
   192.168.254.2/32 Direct 0    0          D   127.0.0.1     GigabitEthernet0/0/0
   192.168.254.3/32 Direct 0    0          D   127.0.0.1     GigabitEthernet0/0/0
   192.168.254.4/30 Direct 0    0          D   192.168.254.6 Serial4/0/0
   192.168.254.5/32 Direct 0    0          D   192.168.254.5 Serial4/0/0
   192.168.254.6/32 Direct 0    0          D   127.0.0.1     Serial4/0/0
   192.168.254.7/32 Direct 0    0          D   127.0.0.1     Serial4/0/0
  255.255.255.255/32 Direct 0   0          D   127.0.0.1     InLoopBack0
```

Fig. 1.51 Static routing path entries in AR2 routing table

Refer to the parameters planned in Table 1.11 to configure the IP address, subnet mask, and default gateway of the test hosts PC1 and PC2. After the configuration is completed, use the "ipconfig" command on the DOS command line to check whether the IP configuration information is correct.

Use "tracert PC2's IP address" on the DOS command line of the host PC1 to track the path through which PC2 accesses the IP packet of PC1, as shown in Fig. 1.52. As can be seen, the data packets from PC1 to PC2 follow the path of "192.168.0.1" → "192.168.254.2" → "192.168.1.10", which is the same as the main route.

```
PC>tracert 192.168.1.10

traceroute to 192.168.1.10, 8 hops max
(ICMP), press Ctrl+C to stop
 1  192.168.0.1     62 ms  47 ms  47 ms
 2  192.168.254.2   62 ms  47 ms  32 ms
 3  192.168.1.10    78 ms  78 ms  62 ms
```

Fig. 1.52 Test results from Host PC2 "tracert" PC1

Delete the gigabit Ethernet link between routers AR1 and AR2, and then use the "display ip routing-table" command to display the routing table contents. Check whether the static routing displayed in the routing table has changed to the configured backup routing path, as shown in Fig. 1.53. At this time, the path forwarding port to the target network "192.168.1.0/24" in the AR1 routing table is the "Serial4/0/0" interface. As shown in Fig. 1.54, the path forwarding port in the AR2 routing table to the target network "192.168.0.0/24" is the "Serial4/0/0" interface.

```
[AR1]display ip routing-table
Route Flags: R - relay, D - download to fib
-----------------------------------------------------------
Routing Tables: Public
        Destinations : 15    Routes : 15

Destination/Mask    Proto  Pre  Cost    Flags NextHop       Interface

      127.0.0.0/8   Direct  0    0       D   127.0.0.1      InLoopBack0
      127.0.0.1/32  Direct  0    0       D   127.0.0.1      InLoopBack0
127.255.255.255/32  Direct  0    0       D   127.0.0.1      InLoopBack0
    192.168.0.0/24  Direct  0    0       D   192.168.0.1    GigabitEthernet0/0/1
    192.168.0.1/32  Direct  0    0       D   127.0.0.1      GigabitEthernet0/0/1
  192.168.0.255/32  Direct  0    0       D   127.0.0.1      GigabitEthernet0/0/1
    192.168.1.0/24  Static  70   0       D   192.168.254.5  Serial4/0/0
   192.168.254.0/30 Direct  0    0       D   192.168.254.1  GigabitEthernet0/0/0
   192.168.254.1/32 Direct  0    0       D   127.0.0.1      GigabitEthernet0/0/0
   192.168.254.3/32 Direct  0    0       D   127.0.0.1      GigabitEthernet0/0/0
   192.168.254.4/30 Direct  0    0       D   192.168.254.5  Serial4/0/0
   192.168.254.5/32 Direct  0    0       D   127.0.0.1      Serial4/0/0
   192.168.254.6/32 Direct  0    0       D   192.168.254.6  Serial4/0/0
   192.168.254.7/32 Direct  0    0       D   127.0.0.1      Serial4/0/0
255.255.255.255/32  Direct  0    0       D   127.0.0.1      InLoopBack0
```

Fig. 1.53 AR1 routing table

```
[AR2]display ip routing-table
Route Flags: R - relay, D - download to fib
------------------------------------------------------------
Routing Tables: Public
        Destinations : 15    Routes : 15

Destination/Mask      Proto  Pre  Cost    Flags  NextHop       Interface
        127.0.0.0/8   Direct  0    0       D     127.0.0.1     InLoopBack0
        127.0.0.1/32  Direct  0    0       D     127.0.0.1     InLoopBack0
127.255.255.255/32    Direct  0    0       D     127.0.0.1     InLoopBack0
     192.168.0.0/24   Static  70   0       D     192.168.254.6 Serial4/0/0
     192.168.1.0/24   Direct  0    0       D     192.168.1.1   GigabitEthernet0/0/1
     192.168.1.1/32   Direct  0    0       D     127.0.0.1     GigabitEthernet0/0/1
   192.168.1.255/32   Direct  0    0       D     127.0.0.1     GigabitEthernet0/0/1
   192.168.254.0/30   Direct  0    0       D     192.168.254.2 GigabitEthernet0/0/0
   192.168.254.2/32   Direct  0    0       D     127.0.0.1     GigabitEthernet0/0/0
   192.168.254.3/32   Direct  0    0       D     127.0.0.1     GigabitEthernet0/0/0
   192.168.254.4/30   Direct  0    0       D     192.168.254.6 Serial4/0/0
   192.168.254.5/32   Direct  0    0       D     192.168.254.5 Serial4/0/0
   192.168.254.6/32   Direct  0    0       D     127.0.0.1     Serial4/0/0
   192.168.254.7/32   Direct  0    0       D     127.0.0.1     Serial4/0/0
255.255.255.255/32    Direct  0    0       D     127.0.0.1     InLoopBack0
```

Fig. 1.54   AR2 routing table

Again, use "tracert PC2's IP address" on the DOS command line of the host PC1 to track the path through which PC2 accesses the IP packet of PC1, as shown in Fig. 1.55. As can be seen, the data packets from PC1 to PC2 follow the path of "192.168.0.1" → "192.168.254.6" → "192.168.1.10", which is the same as the backup route.

```
PC>tracert 192.168.1.10

traceroute to 192.168.1.10, 8 hops max
(ICMP), press Ctrl+C to stop
 1  192.168.0.1      31 ms   47 ms   47 ms
 2  192.168.254.6    31 ms   47 ms   47 ms
 3  192.168.1.10     63 ms   62 ms   63 ms
```

Fig. 1.55   Test Results from Host PC2 "tracert" PC1

Use a network cable to connect the G0/0/0 interface of the router AR1 to the G0/0/0 interface of AR2, and then use the "display ip routing-table" command to display the routing table contents. Check whether the static routes displayed in the routing table have been restored to the primary routing path already configured.

If the network cannot be connected during testing, or if there is something wrong found in routing during path tracking, please follow the steps below for troubleshooting:

① Use the "ipconfig" command on the DOS command line of all hosts to check whether the IP address, subnet mask, and default gateway of each host are configured correctly.

② Use the "display ip interface brief" command on the routers AR1 and AR2 to check if the interface IP of the router is configured correctly, and whether the physical interface and link protocol state have been "UP".

③ If the routing table has no correct route, please use the "display current-configuration | include ip route-static" command on the routers AR1 and AR2 to check if the floating static routing configuration in AR1 and AR2 is correct, as shown in Fig. 1.56 and Fig. 1.57.

```
[AR1]display current-configuration | include ip route-static
ip route-static 192.168.1.0 255.255.255.0 192.168.254.2
ip route-static 192.168.1.0 255.255.255.0 Serial4/0/0 preference 70
```

Fig. 1.56 Floating static routing configuration in the currently running configuration file on AR1

```
[AR2]display current-configuration | include ip route-static
ip route-static 192.168.0.0 255.255.255.0 192.168.254.1
ip route-static 192.168.0.0 255.255.255.0 Serial4/0/0 preference 70
```

Fig. 1.57 Floating static routing configuration in the currently running configuration file on AR2

Pay attention to the floating static routing configuration in the running configuration file and avoid any wrong or redundant static routing configuration. If there is any wrong or redundant static routing configuration, use the "undo" command to delete the redundant or incorrect static routing configuration, and then use the "ip route-static" command to configure a correct floating static routing again.

3. Planning and configuration of aggregated static routing and default routing

Build a topology as shown in Fig. 1.58, and complete the IP configuration of the router and host according to the IP addresses planned in Table 1.13; plan and configure the default route to the Internet on the router AR1; then plan and configure the aggregated static routing to the enterprise intranet on the ISP router on the Internet to realize interconnection across the network. Minimize the number of routing table entries on the router AR1 and ISP as much as possible, and avoid routing loops; moreover, complete network connectivity testing and troubleshooting.

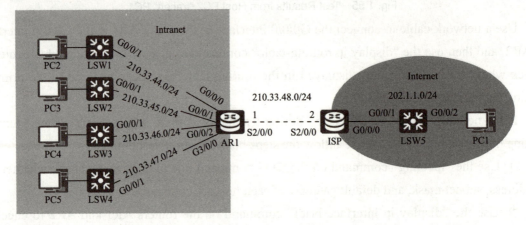

Fig. 1.58 Topology for "planning and configuration of aggregated static routes and default routes"

Table 1.13 IP planning

Device	Interface	IP address	Gateway
AR1	G0/0/0	210.33.44.1/24	
	G0/0/1	210.33.45.1/24	
	G0/0/2	210.33.46.1/24	
	G3/0/0	210.33.47.1/24	
	S2/0/0	210.33.48.1/30	
ISP	S2/0/0	210.33.48.2/30	
	G0/0/0	202.1.1.1/24	
PC1	NIC	202.1.1.10/24	202.1.1.1
PC2	NIC	210.33.44.10/24	210.33.44.1
PC3	NIC	210.33.45.10/24	210.33.45.1
PC4	NIC	210.33.46.10/24	210.33.46.1
PC5	NIC	210.33.47.10/24	210.33.47.1

For AR1, "210.33.44.0/24", "210.33.45.0/24", "210.33.46.0/24", "210.33.47.0/24", and "210.33.48.0/24" are all direct interconnection networks. It only needs to be planned and configured to the default route on the Internet (**Note: only one network segment on the Internet is listed in the subtask topology, but the actual number of the target networks on the Internet is very large, and it is impossible to route them one by one in the router**). The routed path can be represented by the local forwarding port "S2/0/0" interface or the next hop address "210.33.48.2". Here the local forwarding port "S2/0/0" is planned.

For ISP routers on the Internet, there are four remote networks: "210.33.44.0/24", "210.33.45.0/24", "210.33.46.0/24", and "210.33.47.0/24", and static routing to such four remote networks needs to be configured. Moreover, the four remote networks are routed the same, the forwarding ports are all serial "S2/0/0" interfaces, and the four target networks are numbered continuously, so the network numbers of the four remote networks can be aggregated into "210.33.44.0/22", thereby reducing the routing table entries of the router ISP and lowering the maintenance cost of the routing table.

See Table 1.14 for the default route and aggregated static route planning of AR1 and ISP.

Table 1.14 Default route and aggregated static route planning of AR1 and ISP

Device	Target network	Target network mask/Mask length	Next hop or forward port
AR1	0.0.0.0	0.0.0.0	S2/0/0
ISP	210.33.44.0	22	S2/0/0

(1) Configuration of router AR1

Refer to the following steps to configure AR1 and achieve network interconnection.

```
<Huawei>sys
[Huawei]sysname
[AR1]interface G0/0/0
[AR1-GigabitEthernet0/0/0]IP address 210.33.44.1 24
[AR1-GigabitEthernet0/0/0]quit
[AR1]interface g0/0/1
[AR1-GigabitEthernet0/0/1]IP address 210.33.45.1 24
[AR1-GigabitEthernet0/0/1]quit
[AR1]interface g0/0/2
[AR1-GigabitEthernet0/0/2]IP address 210.33.46.1 24
[AR1-GigabitEthernet0/0/2]quit
[AR1]interface g3/0/0
[AR1-GigabitEthernet3/0/0]IP address 210.33.47.1 24
[AR1-GigabitEthernet3/0/0]quit
[AR1]interface s2/0/0
[AR1-Serial2/0/0]IP address 210.33.48.1 30
[AR1-Serial2/0/0]description linktointernet
[AR1-Serial2/0/0]quit
//Configure the host name and interface IP address of the router AR1
[AR1]ip route-static 0.0.0.0 0.0.0.0 s2/0/0
//Configure the default route to the Internet. The routed path is represented by the
//local forwarding port, namely, the S2/0/0 interface
```

(2) Configuration of router ISP

```
<Huawei>sys
[Huawei]sysname ISP
[ISP]interface s2/0/0
[ISP-Serial2/0/0]ip address 210.33.48.2 30
[ISP-Serial2/0/0]quit
[ISP]interface g0/0/0
[ISP-GigabitEthernet0/0/0]ip address 202.1.1.1 24
[ISP-GigabitEthernet0/0/0]quit
//Configure the host name and interface IP address of the router AR2
[ISP]ip route-static 210.33.44.0 22 s2/0/0
//Configure an aggregated static route to the enterprise intranet "210.33.44.0/22",
//with the forwarding port being the S2/0/0 interface
```

(3) Network testing and troubleshooting

Configure the IP address, subnet mask, and default gateway configuration of a test host PC1 on the Internet and an enterprise intranet test host PC2 according to the parameters in Table 1.13. After

the configuration is completed, use the "ipconfig" command on the DOS command line to check if its IP configuration information is correct.

On the DOS command line of the host PC1, use the ping utility to test the IP connectivity of the enterprise intranet (four gigabit Ethernet ports G0/0/0, G0/0/1, G0/0/2 and G3/0/0 of the router AR1). Fig. 1.59 shows the test results from G0/0/0 and G0/0/1 of the host PC1 ping the router AR1.

```
PC>ping 210.33.44.1

Ping 210.33.44.1: 32 data bytes, Press Ctrl_C to break
From 210.33.44.1: bytes=32 seq=1 ttl=254 time=62 ms
From 210.33.44.1: bytes=32 seq=2 ttl=254 time=31 ms
From 210.33.44.1: bytes=32 seq=3 ttl=254 time=16 ms
From 210.33.44.1: bytes=32 seq=4 ttl=254 time=15 ms
From 210.33.44.1: bytes=32 seq=5 ttl=254 time=31 ms

--- 210.33.44.1 ping statistics ---
  5 packet(s) transmitted
  5 packet(s) received
  0.00% packet loss
  round-trip min/avg/max = 15/31/62 ms

PC>ping 210.33.45.1

Ping 210.33.45.1: 32 data bytes, Press Ctrl_C to break
From 210.33.45.1: bytes=32 seq=1 ttl=254 time=32 ms
From 210.33.45.1: bytes=32 seq=2 ttl=254 time=31 ms
From 210.33.45.1: bytes=32 seq=3 ttl=254 time=62 ms
From 210.33.45.1: bytes=32 seq=4 ttl=254 time=16 ms
From 210.33.45.1: bytes=32 seq=5 ttl=254 time=31 ms
```

Fig. 1.59 Test results from G0/0/0 and G0/0/1 of host PC1 ping router AR1

On the DOS command line of the host PC2, use the ping utility to test the IP connectivity to the Internet. Fig. 1.60 shows that the host PC2 can ping the test host PC1 on the Internet.

```
PC>ping 202.1.1.10

Ping 202.1.1.10: 32 data bytes, Press Ctrl_C to break
From 202.1.1.10: bytes=32 seq=1 ttl=126 time=93 ms
From 202.1.1.10: bytes=32 seq=2 ttl=126 time=78 ms
From 202.1.1.10: bytes=32 seq=3 ttl=126 time=79 ms
From 202.1.1.10: bytes=32 seq=4 ttl=126 time=62 ms
From 202.1.1.10: bytes=32 seq=5 ttl=126 time=78 ms

--- 202.1.1.10 ping statistics ---
  5 packet(s) transmitted
  5 packet(s) received
  0.00% packet loss
  round-trip min/avg/max = 62/78/93 ms
```

Fig. 1.60 Host PC2 ping test host PC1 on the Internet

On the DOS command line of the host PC2, use the "tracert" routing tracking utility to determine the path through which PC2 accesses the IP packet of PC1, as shown in Fig. 1.61.

```
PC>tracert 202.1.1.10

traceroute to 202.1.1.10, 8 hops max
(ICMP), press Ctrl+C to stop
 1  210.33.44.1   47 ms   47 ms   47 ms
 2  210.33.48.2   31 ms   47 ms   47 ms
 3  202.1.1.10    62 ms   63 ms   62 ms
```

Fig. 1.61 Test results from host PC2 "tracert" PC1

If the network cannot be connected during testing, please follow the steps below to complete troubleshooting and ensure that the network IP connectivity is normal.

① Use the "ipconfig" command on the DOS command line of all hosts to check whether the IP address, subnet mask, and default gateway of each host are configured correctly.

② Use the "display ip interface brief" command on the router AR1 and ISP to check if the interface IP of the router is configured correctly, and whether the physical interface and link protocol state have been "UP".

③ Use the "display ip routing-table" command on the router to display the contents of the routing table, check whether all network segments in the network have been routed, and whether the routed path is correct, as shown in Fig. 1.62 and Fig. 1.63.

```
[AR1]display ip routing-table
Route Flags: R - relay, D - download to fib
------------------------------------------------------------
Routing Tables: Public
        Destinations : 21    Routes : 21
Destination/Mask    Proto   Pre  Cost    Flags NextHop        Interface
        0.0.0.0/0   Static  60   0        D    210.33.48.1    Serial2/0/0
      127.0.0.0/8   Direct  0    0        D    127.0.0.1      InLoopBack0
     127.0.0.1/32   Direct  0    0        D    127.0.0.1      InLoopBack0
 127.255.255.255/32 Direct  0    0        D    127.0.0.1      InLoopBack0
    210.33.44.0/24  Direct  0    0        D    210.33.44.1    GigabitEthernet0/0/0
    210.33.44.1/32  Direct  0    0        D    127.0.0.1      GigabitEthernet0/0/0
  210.33.44.255/32  Direct  0    0        D    127.0.0.1      GigabitEthernet0/0/0
    210.33.45.0/24  Direct  0    0        D    210.33.45.1    GigabitEthernet0/0/1
    210.33.45.1/32  Direct  0    0        D    127.0.0.1      GigabitEthernet0/0/1
  210.33.45.255/32  Direct  0    0        D    127.0.0.1      GigabitEthernet0/0/1
    210.33.46.0/24  Direct  0    0        D    210.33.46.1    GigabitEthernet0/0/2
    210.33.46.1/32  Direct  0    0        D    127.0.0.1      GigabitEthernet0/0/2
  210.33.46.255/32  Direct  0    0        D    127.0.0.1      GigabitEthernet0/0/2
    210.33.47.0/24  Direct  0    0        D    210.33.47.1    GigabitEthernet3/0/0
    210.33.47.1/32  Direct  0    0        D    127.0.0.1      GigabitEthernet3/0/0
  210.33.47.255/32  Direct  0    0        D    127.0.0.1      GigabitEthernet3/0/0
    210.33.48.0/30  Direct  0    0        D    210.33.48.1    Serial2/0/0
    210.33.48.1/32  Direct  0    0        D    127.0.0.1      Serial2/0/0
    210.33.48.2/32  Direct  0    0        D    210.33.48.2    Serial2/0/0
    210.33.48.3/32  Direct  0    0        D    127.0.0.1      Serial2/0/0
 255.255.255.255/32 Direct  0    0        D    127.0.0.1      InLoopBack0
```

Fig. 1.62 Contents of AR1 routing table

```
[ISP]display ip routing-table
Route Flags: R - relay, D - download to fib
------------------------------------------------------------
Routing Tables: Public
         Destinations : 12    Routes : 12

Destination/Mask      Proto  Pre Cost    Flags NextHop      Interface

         127.0.0.0/8  Direct  0   0       D    127.0.0.1    InLoopBack0
        127.0.0.1/32  Direct  0   0       D    127.0.0.1    InLoopBack0
  127.255.255.255/32  Direct  0   0       D    127.0.0.1    InLoopBack0
        202.1.1.0/24  Direct  0   0       D    202.1.1.1    GigabitEthernet0/0/0
        202.1.1.1/32  Direct  0   0       D    127.0.0.1    GigabitEthernet0/0/0
      202.1.1.255/32  Direct  0   0       D    127.0.0.1    GigabitEthernet0/0/0
       210.33.44.0/22 Static 60   0       D    210.33.48.2  Serial2/0/0
      210.33.48.0/30  Direct  0   0       D    210.33.48.2  Serial2/0/0
      210.33.48.1/32  Direct  0   0       D    210.33.48.1  Serial2/0/0
      210.33.48.2/32  Direct  0   0       D    127.0.0.1    Serial2/0/0
      210.33.48.3/32  Direct  0   0       D    127.0.0.1    Serial2/0/0
  255.255.255.255/32  Direct  0   0       D    127.0.0.1    InLoopBack0
```

Fig. 1.63   Contents of ISP routing table

If the routing table has no correct route, please use the "display current-configuration | include ip route-static" command on the routers AR1 and ISP to check the currently running configuration file and see whether the default route in AR1 and the aggregated static route of ISP are configured correctly, as shown in Fig. 1.64 and Fig. 1.65.

```
[AR1]display current-configuration | include ip route-static
ip route-static 0.0.0.0 0.0.0.0 Serial2/0/0
```

Fig. 1.64   Configuration information of default route in configuration file running on AR1

```
[ISP]display current-configuration | include ip route-static
ip route-static 210.33.44.0 255.255.252.0 Serial2/0/0
```

Fig. 1.65   Configuration information of aggregated static routing in configuration file running on ISP

Pay attention to the configurations of the default route and aggregated static route in the running configuration file and avoid any wrong or redundant routing configuration. If there is any wrong or redundant routing configuration, use the "undo" command to delete the redundant or incorrect static routing configuration, and then use the "ip route-static" command to configure a correct default route and an aggregated static route.

Task Completion

1. Task planning

According to the network topology shown in Fig.1.1 and the networking requirements for interconnection in Project 1, for the branch router "fenbu", the remote network is only the network of the headquarters (but the branch router does not know the number of network segments of the headquarters, and only one network of the headquarters is exampled in the topology), and there

are two paths to the headquarters. The routing requires that the serial link be taken only when the gigabit Ethernet link fails, so it is necessary to plan a floating default path to the headquarters.

For the router "zongbu" in the headquarters, there are three remote networks "172.16.128.0/25", "172.16.128.129/28" and "172.16.129.0/24". The routes to such three networks are the same, namely, the gigabit Ethernet link acts as the main link, the serial link acts as the backup link, the three target networks have their serial numbers arranged continuously in bulk, and the routes can be aggregated via CIDR (classless inter-domain routing).

See Table 1.15 for the static default and aggregated static route planning of the routers "fenbu" and "zongbu".

Table 1.15 Static default and aggregated static route planning of routers "fenbu" and "zongbu"

Device	Target network	Target network mask/Mask length	Next hop or forward port	Route preference value
fenbu	0.0.0.0	0.0.0.0	172.16.254.2	Default value 60
	0.0.0.0	0.0.0.0	S2/0/0	70
zongbu	172.16.128.0	23	172.16.254.1	Default value 60
	172.16.128.0	23	S3/0/0	70

2. Task implementation

Note: This task is implemented and completed on the basis of Task 1 "Task Completion" of this project.

(1) Configuration of router "fenbu"

Please refer to the following steps for routing configuration to achieve network interconnection and interoperability.

```
<fenbu>sys
[fenbu]ip route-static 0.0.0.0 0.0.0.0 172.16.254.2
[fengbu]ip route-static 0.0.0.0 0.0.0.0 s2/0/0 preference 70
//Configure the floating default route to the headquarters network and set the next hop
//address of the main route path as "172.16.254.2", namely, the path is a Gigabit Ethernet
//link; the forwarding port for backup routing is the interface S2/0/0
```

(2) Configuration of router "zongbu"

```
<zongbu>sys
[zongbu]ip route-static 172.16.128.0 23 172.16.254.1
[zongbu]ip route-static 172.16.128.0 23 s3/0/0 preference 70
//Configure the floating aggregated static route to the branch network and set the next
//hop address of the main route path as "172.16.254.1", namely, the path is a Gigabit
//Ethernet link; the forwarding port for backup routing is the interface S3/0/0
```

(3) Network testing and troubleshooting

① On the routers "fenbu" and "zongbu", use the "display ip routing-table | include Static" command to display the static routing entries in the routing table, and check if the static routing information displayed in the routing table is consistent with the plan. As can be seen from Fig. 1.66, the default route in the routing table of the router "fenbu" is the next hop address of "172.16.254.2", and the port is GigabitEthernet 0/0/0. As can be seen from Fig. 1.67, in the routing table, the router "zongbu" is routed to the target network "172.16.128.0/23" via the next hop address "172.16.254.1", and the port is GigabitEthernet 0/0/0. Based on the routing situation in the routing table, analyze whether any routing loop occurs in the data forwarding logic.

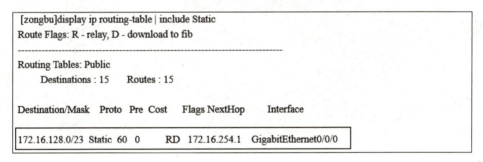

Fig. 1.66 Static routing entries in routing table of router "fenbu"

Fig. 1.67 Static routing entries in routing table of router "zongbu"

② Refer to the parameters planned in Table 1.6 to configure the IP address, subnet mask, and default gateway of the test hosts PC1, PC2, Web server, DNS server, and FTP server. After the configuration is completed, use the "ipconfig" command on the DOS command line to check whether the IP configuration information is correct.

③ On the DOS command line of the host PC1, use the ping utility to test the IP connectivity with other hosts and servers in the network. Fig. 1.68 shows that the branch host PC2 ping the host PC1 of the headquarters.

```
PC>ping 172.16.0.10

Ping 172.16.0.10: 32 data bytes, Press Ctrl_C to break
From 172.16.0.10: bytes=32 seq=1 ttl=126 time=78 ms
From 172.16.0.10: bytes=32 seq=2 ttl=126 time=62 ms
From 172.16.0.10: bytes=32 seq=3 ttl=126 time=47 ms
From 172.16.0.10: bytes=32 seq=4 ttl=126 time=47 ms
From 172.16.0.10: bytes=32 seq=5 ttl=126 time=47 ms

--- 172.16.0.10 ping statistics ---
  5 packet(s) transmitted
  5 packet(s) received
  0.00% packet loss
  round-trip min/avg/max = 47/56/78 ms
```

Fig. 1.68 The branch host PC2 ping the host PC1 of the headquarters

④ On the DOS command line of the branch host PC2, use "tracert PC's IP address" to track the path through which the IP data packet goes when the branch host PC2 accesses the head office host PC1, as shown in Fig. 1.69. As can be seen, the data packet from PC2 to PC1 follows the path of "172.16.129.1" → "172.16.254.2" → "172.16.0.10", which is the same as the path of the main route in the planned floating route.

```
PC>tracert 172.16.0.10

traceroute to 172.16.0.10, 8 hops max
(ICMP), press Ctrl+C to stop
 1  172.16.129.1    31 ms   47 ms   47 ms
 2  172.16.254.2    46 ms   32 ms   47 ms
 3  172.16.0.10     62 ms   94 ms   78 ms
```

Fig. 1.69 Test results from host PC2 "tracert" PC1

⑤ On the DOS command line of the branch host PC2, use the "tracert" command to track the path through which the data packet goes from the branch host PC2 to an IP address that does not exist in the network planning, and check if there is any routing loop (that is, the packet is constantly forwarded back and forth between two or more routers, forming a loop). The tracking result shown in Fig. 1.70 shows that there is no routing loop, but the tracking result shown in Fig. 1.71 shows that there is a routing loop (with the data packet forwarded back and forth between the two routing devices "172.16.254.1" and "172.16.254.2").

```
PC>tracert 172.16.100.1

traceroute to 172.16.100.1, 8 hops max
(ICMP), press Ctrl+C to stop
 1   172.16.129.1     47 ms   46 ms   47 ms
 2     *    *    *
 3     *    *    *
 4     *    *    *
 5     *    *    *
 6     *    *    *
 7     *    *    *
 8     *    *    *
```

Fig. 1.70   No routing loop occurs when the host PC2 "tracert" an IP address, which does not exist

```
PC>tracert 172.16.100.1

traceroute to 172.16.100.1, 8 hops max
(ICMP), press Ctrl+C to stop
 1   172.16.129.1     47 ms   31 ms   47 ms
 2   172.16.254.2     47 ms   46 ms   47 ms
 3   172.16.254.1     32 ms   46 ms   47 ms
 4   172.16.254.2     47 ms   47 ms   31 ms
 5   172.16.254.1     78 ms   63 ms   47 ms
 6   172.16.254.2     31 ms   63 ms   46 ms
 7   172.16.254.1     63 ms   78 ms   63 ms
 8   172.16.254.2     62 ms   63 ms   62 ms
```

Fig. 1.71   Routing loops occur when the host PC2 "tracert" an IP address, which does not exist

⑥ Delete the gigabit Ethernet link between the router "fenbu" and the router "zongbu", and then use the "display ip routing-table | include Static" command to display the static routing entries in the routing table. Check if the static routing displayed in the routing table has changed to the configured backup routing path. As shown in Fig. 1.72, "Serial2/0/0" is the path forwarding port of the default route in the routing table of the router "fenbu". As shown in Fig. 1.73, "Serial3/0/0" is the path forwarding port from the routing table of the router "zongbu" to the branch "172.16.128.0/23" at this moment.

```
[fenbu]display ip routing-table | include Static
Route Flags: R - relay, D - download to fib
------------------------------------------------------------
Routing Tables: Public
         Destinations : 18    Routes : 18

Destination/Mask   Proto   Pre  Cost      Flags NextHop       Interface

0.0.0.0/0          Static  70   0         D     172.16.254.5  Serial2/0/0
```

Fig. 1.72   Static routing entries in the routing table of the router "fenbu"

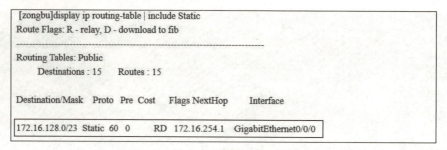

Fig. 1.73 Static routing entries in the routing table of the router "zongbu"

⑦ On the DOS command line of the branch host PC2, use "tracert PC1's IP address" to track the path through which the IP data packet passes when the branch host PC2 accesses the head office host PC1, as shown in Fig. 1.74. As can be seen, the data packet from PC2 to PC1 follows the path of "172.16.229.1" → "172.16.254.6" → "172.16.0.10", which is the same as the backup route in the planned floating route.

```
PC>tracert 172.16.0.10

traceroute to 172.16.0.10, 8 hops max
(ICMP), press Ctrl+C to stop
 1  172.16.129.1    46 ms   47 ms   47 ms
 2  172.16.254.6    31 ms   47 ms   47 ms
 3  172.16.0.10     78 ms   94 ms   78 ms
```

Fig. 1.74 Test Results from Host PC2 "tracert" PC1

If the network cannot be connected during testing or if any loop occurs in the routing, please refer to the following steps for troubleshooting, so that the network IP connectivity is normal and no routing loop occurs in the network:

① Use the "ipconfig" command on the DOS command line of all hosts to check whether the IP address, subnet mask, and default gateway of each host are configured correctly.

② Use the "display ip interface brief" command on the routers "fenbu" and "zongbu" to check if the interface IP of the router is configured correctly, and whether the physical interface and link protocol state have been "UP".

③ On the router, use the "display ip routing-table" command to display the routing table contents, and check if all network segments in the network have been routed, if the routed path is correct, and if there is a logical loop.

④ If the routing table has no correct route, please use the "display current-configuration | include ip route-static" command on the routers "fengbu" and "zongbu" to check if check the configuration file currently running and see whether the floating default route in the router "fenbu" and the floating aggregated static route in the router "zongbu" are correctly configured, as shown in

Fig. 1.75 and Fig. 1.76.

Pay attention to the configuration of the floating default route and floating aggregated static route in the running configuration file and avoid any wrong or redundant routing configuration. If there is any wrong or redundant routing configuration, use the "undo" command to delete the redundant or incorrect static routing configuration, and then use the "ip route-static" command to configure a correct floating static routing and a floating aggregated static route again.

```
[fenbu]display current-configuration | include ip route-static
ip route-static 0.0.0.0 0.0.0.0 172.16.254.2
ip route-static 0.0.0.0 0.0.0.0 Serial2/0/0 preference 70
```

Fig. 1.75 Floating routing configuration in the currently running configuration file on the router "fengbu"

```
[zongbu]display current-configuration | include ip route-static
ip route-static 172.16.128.0 255.255.254.0 172.16.254.1
ip route-static 172.16.128.0 255.255.254.0 Serial3/0/0 preference 70
```

Fig. 1.76 Floating aggregated static routing configuration in the currently running configuration file on the router "zongbu"

Exercises

Multiple choice

(1) The default routing preference value is (　　) for static routing on Huawei routers.

  A. 60   B. 70   C. 110   D. 10

(2) Huawei routers use the following (　　) command to configure static routing.

  A. ip routing B. ip route-static C. AAA   D. route

(3) The following (　　) command is used to display the routing table contents of Huawei routers.

  A. show ip route   B. display ip routing-table

  C. display interface   D. display ip interface brief

(4) When the default route is configured, the target network number/mask is (　　).

  A. 0. 0. 0. 0/32   B. 0. 0. 0. 0/0

  C. 255. 255. 255. 255/0   D. 255. 255. 255. 255/32

(5) The incorrect statement regarding the command "ip route-static 192.168.0.0 22 192.168.1.1 preference 70" is (　　).

  A. The target network number of this route is 192.168. 0. 0/24

  B. The target network number of this route is 192.168. 0. 0/22

  C. The next hop address of this route is 192 168. 1.1

D. The preference value of this route is 70

(6) On the VRP platform, the default protocol preference value is (   ) for static routing.

    A. 1   B. 10   C. 60   D. 70

Task 3 Configuration and Management of DHCP

Task Description

After completing Task 1 and Task 2 of this project, configure the DHCP services on the "fenbu" router in the network shown in Fig. 1.1, automatically assign IP addresses, default gateways, and DNS server addresses to the branch budget and management department hosts, bind the static IP address of "172.16.128.13" to the management department host PC3, achieve the networking requirements for "DHCP services" in Project 1, and complete DHCP testing and troubleshooting.

Task Objectives

- Understand the working process of DHCP.
- Master the planning and configuration of DHCP services on routers.
- Master the planning and configuration of DHCP relay.
- Competent for DHCP troubleshooting.

Related Knowledge

1. Working process of DHCP

DHCP (dynamic host configuration protocol) is a technology for centralized dynamic management and configuration of user IP addresses. DHCP adopts a client/server mode to configure the host that automatically obtains IP addresses as the DHCP client and the host that provides DHCP services as the server. Usually, routers, three-layer switches, hardware firewalls, wireless controllers, wireless routers, and other devices all can be configured as DHCP servers. See Fig. 1.77 for the working process of DHCP.

2. DHCP relay

Due to the fact that the DHCP client broadcasts and sends a DHCP DISCOVER message during the dynamic acquisition of the IP address, the router will not forward any broadcast message when the DHCP client and DHCP server are in different physical network segments, so it needs to use the DHCP relay function. The DHCP client can communicate with DHCP servers in other network segments through DHCP relays, ultimately acquiring IP addresses.

See Fig. 1.78 for the working process of DHCP relay.

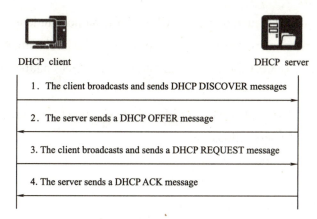

Fig. 1.77   DHCP working process

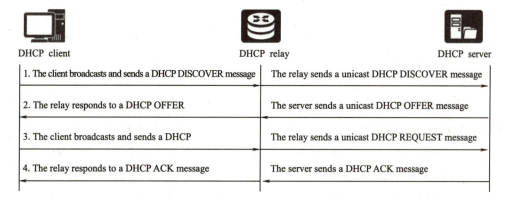

Fig. 1.78   Working process of DHCP relay

Related Skills

1. Planning and configuration of DHCP services

Build a topology as shown in Fig. 1.79. The IP address of the G0/0/0 interface on the router AR1 is 192.168.0.1/24. Please deploy DHCP services on AR1 and automatically assign IP addresses to the hosts in the network segments where PC1 and PC2 are located, and complete testing and troubleshooting.

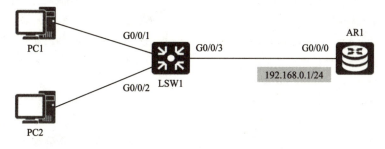

Fig. 1.79   "DHCP services planning and configuration" topology

The DHCP services are planned as follows:

① The IP address pool provided by the DHCP server is named student.

② The IP address network number assigned to the address pool "student" is 192.168.0.0/24; the gateway address is 192.168.0.1/24; the primary and backup DNS addresses are 172.16.5.25 and 172.16.5.26.

③ The addresses 192.168.0.240 to 192.168.0.254 are excluded from the address pool "student" and do not participate in automatic allocation.

④ The lease time is 4 hours.

⑤ The static IP bound for PC2 is 192.168.0.2.

(1) Configuration of DHCP services on routers

The DHCP services on routers are configured as follows:

① Enable the DHCP service.

② Create a planned IP address pool and configure its relevant parameters (address range, default gateway address, DNS address, address lease time, excluded addresses, static bound address, etc.).

③ DHCP Server functionality using global address pooling under interfaces.

Before completing DHCP services configuration on the router, first enter the DOS command line on the host PC2 that needs to bind a static IP address. Use the "ipconfig /all" command on the DOS command line to view the physical address (also known as MAC address) value of its network card, record the value, and then follow the steps below to complete DHCP service configuration on AR1.

```
[AR1]interface g0/0/0
[AR1-GigabitEthernet0/0/0]ip address 192.168.0.1 24
[AR1-GigabitEthernet0/0/0]quit
//Configure the IP address and subnet mask of the G0/0/0 interface of the router AR1
[AR1]dhcp enable  //Enable DHCP service
[AR1]ip pool student     //Create an IP address pool named "student"
[AR1-ip-pool-student]network 192.168.0.0 mask 24 //Configure the address pool range
[AR1-ip-pool-student]gateway-list 192.168.0.1     //Configure the default gateway address
[AR1-ip-pool-student]dns-list 172.16.5.25 172.16.5.26  //Configure the DNS server address
[AR1-ip-pool-student]excluded-ip-address 192.168.0.240 192.168.0.254
//Configure the exclusion address range from 192.168.0.240 to 192.168.0.254
[AR1-ip-pool-student] lease day 0 hour 4        //Configure the address lease time as 4 hours
[AR1-ip-pool-student]static-bind ip-address 192.168.0.2 mac-address
  5489-98a5-2eab   //Configure the IP address bound to host PC2 is 192.168.0.2 (Note:
              //"5489-98a5-2eab" is the MAC address of PC2)
[AR1]interface g0/0/0                    //Enable the G0/0/0 interface and select the
[AR1-GigabitEthernet0/0/0]dhcp select global   //DHCP Server function of the global address
[AR1-GigabitEthernet0/0/0]quit                //pool
```

```
[AR1]quit
<AR1>save
```

(2) DHCP testing and troubleshooting

On the hosts PC1 and PC2, configure their IP and DNS addresses to be automatically obtained. The specific steps are as follows: select "Control Panel" → "Network and Sharing Center" → "Change Adapter Settings", right-click "Ethernet Networking", select "Properties" → "Internet Protocol Version 4 (TCP/IPv4)", select "Get IP Address Automatically" and "Get DNS Address Automatically" in "Properties", and click the "OK" button to shut down the "Networking Properties" window.

Use the "ipconfig /all" command on the host PC1 to view the IP information, and check whether the correct IP address, subnet mask, default gateway, DNS server address, and lease time are obtained.

Note: In the Huawei simulation software ENSP, the "ipconfig /all" command can not be available for host testing, and you can only use the "ipconfig" command to view the IP information. Fig. 1.80 shows the IP information automatically obtained during host testing in ENSP.

```
PC>ipconfig

Link local IPv6 address..........: fe80::5689:98ff:fe2f:64e1
IPv6 address....................: :: / 128
IPv6 gateway....................: ::
IPv4 address....................: 192.168.0.239
Subnet mask.....................: 255.255.255.0
Gateway.........................: 192.168.0.1
Physical address................: 54-89-98-2F-64-E1
DNS server......................: 172.16.5.25
```

Fig. 1.80　IP information automatically obtained by host PC1

Use the "ipconfig /all" command on host PC2 to view the IP information and check if the obtained IP address is the bound IP address "192.168.0.2".

In the IP information shown in Fig. 1.81, the IP address obtained by PC2 is statically bound "192.168.0.2".

```
PC>ipconfig

Link local IPv6 address..........: fe80::5689:98ff:fea5:2eab
IPv6 address....................: :: / 128
IPv6 gateway....................: ::
IPv4 address....................: 192.168.0.2
Subnet mask.....................: 255.255.255.0
Gateway.........................: 192.168.0.1
Physical address................: 54-89-98-A5-2E-AB
DNS server......................: 172.16.5.25
                                  172.16.5.26
```

Fig. 1.81　IP information automatically obtained by host PC2

Moreover, you can also use "display ip pool name student used" on the router to view the

address used in the address pool "student", as shown in Fig. 1.82. As can be seen from the figure, "192.168.0.2" is statically bound to the host with the MAC address "5489-98a5-2eab", and "192.168.0.239" has been dynamically assigned to the host.

If the host cannot obtain an IP address during testing, refer to the following steps to complete troubleshooting:

① Check whether the IP and DNS addresses are configured as "Get Automatically" on all hosts.

② Use the "display ip interface brief" command on the router AR1 to check if the interface IP of the router has been correctly configured, and whether the physical interface and link protocol state have been "UP".

③ On the router AR1, use the "<AR1>display current-configuration | begin ip pool" command to check if the IP address pool "student" in the currently running configuration file is correctly configured, as shown in Fig. 1.83.

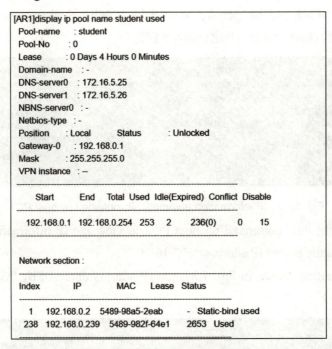

Fig. 1.82 Check the addresses used in the address pool "student"

Fig. 1.83 Check the configuration of the IP address pool "student" in the current configuration file on the router

④ On the router AR1, enter the interface view of the G0/0/0 interface and use the "display this" command to display the configurations in the current interface view. Check if the interface is configured with DHCP Server functionality using global address pooling, as shown in Fig. 1.84.

```
[AR1-GigabitEthernet0/0/0]display this
[V200R003C00]
#
interface GigabitEthernet0/0/0
 ip address 192.168.0.1 255.255.255.0
 dhcp select global
#
return
```

Fig. 1.84 Use the "display this" command to display the configuration in the interface view

2. Planning and configuration of DHCP relay

Build a topology as shown in Fig. 1.85 and provide DHCP services through the router AR2 to the hosts PC1 and PC2, with IP addresses dynamically assigned. Please configure DHCP services and relays, and complete testing and troubleshooting.

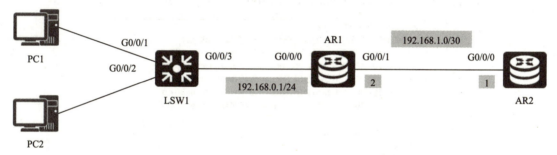

Fig. 1.85 Topology for "DHCP relay planning and configuration"

DHCP services are planned as follows:

① The IP address pool provided by the DHCP server is named student.

② The IP address network number assigned to the address pool "student" is 192.168.0.0/24; the gateway address is 192.168.0.1/24; the primary and backup DNS addresses are 172.16.5.25 and 172.16.5.26.

③ The addresses "192.168.0.240" to "192.168.0.254" are excluded from the address pool "student" and do not participate in automatic allocation.

④ The lease time is 4 hours.

⑤ The static IP bound for PC2 is "192.168.0.2".

DHCP relay is planned as follows:

① Configure the DHCP relay function on the G0/0/0 interface of the router AR1.

② The DHCP server address is 192.168.1.1.

(1) Configuration of router interface and routing

Refer to the following steps to configure the IP addresses of the interfaces on the routers AR1

and AR2, and configure static routing to the target network "192.168.0.0/24" on the router AR2, to achieve interconnection and interoperability of all network segments in the network.

```
[AR1]interface g0/0/0
[AR1-GigabitEthernet0/0/0]ip address 192.168.0.1 24
[AR1-GigabitEthernet0/0/0]quit
[AR1]interface g0/0/1
[AR1-GigabitEthernet0/0/1]ip address 192.168.1.2 30
[AR1-GigabitEthernet0/0/1]quit
[AR2]interface g0/0/0
[AR2-GigabitEthernet0/0/0]ip address 192.168.1.1 30
[AR2-GigabitEthernet0/0/0]quit
[AR2]ip route-static 192.168.0.0 24 192.168.1.2
```

After completing the configuration of the router interface IP and routing, ping the IP address of the G0/0/0 interface (192.168.0.1) on the router AR1 from the router AR2 to test the connectivity between the router AR2 and the remote network, as shown in Fig. 1.86.

```
[AR2]PING 192.168.0.1
  PING 192.168.0.1: 56  data bytes, press CTRL_C to break
    Reply from 192.168.0.1: bytes=56 Sequence=1 ttl=255 time=120 ms
    Reply from 192.168.0.1: bytes=56 Sequence=2 ttl=255 time=20 ms
    Reply from 192.168.0.1: bytes=56 Sequence=3 ttl=255 time=30 ms
    Reply from 192.168.0.1: bytes=56 Sequence=4 ttl=255 time=30 ms
    Reply from 192.168.0.1: bytes=56 Sequence=5 ttl=255 time=30 ms

  --- 192.168.0.1 ping statistics ---
    5 packet(s) transmitted
    5 packet(s) received
    0.00% packet loss
    round-trip min/avg/max = 20/46/120 ms
```

Fig. 1.86 AR2 ping G0/0/0 interface IP of AR1

(2) Configure the DHCP services on router AR2

Please refer to the following steps to configure the DHCP service on AR2.

```
[AR2]dhcp enable                    //Enable DHCP service
[AR2]ip pool student                //Create an IP address pool named "student"
[AR2-ip-pool-student]network 192.168.0.0 mask 24      //Configure the address pool range
[AR2-ip-pool-student]gateway-list 192.168.0.1
//Configure the default gateway address assigned to DHCP clients
[AR2-ip-pool-student]dns-list 172.16.5.25 172.16.5.26
//Configure a DNS server address for the DHCP client
[AR2-ip-pool-student]excluded-ip-address 192.168.0.240 192.168.0.254
//Configure the exclusion address range from 192.168.0.240 to 192.168.0.254
[AR2-ip-pool-student]lease day 0 hour 4 //Configure the lease time of the IP address
                                 //assigned to DHCP clients as 4 hours
```

```
[AR2-ip-pool-student]static-bind ip-address 192.168.0.2 mac-address 5489-985e-6cc6
// bind the static IP address "192.168.0.2" to the host PC2 with the MAC address "5489-985e-6cc6".
[AR2]interface g0/0/0
[AR2-GigabitEthernet0/0/0]dhcp select global   //Enable DHCP using global address
//pool for the G0/0/0 interface Server functions
[AR2-GigabitEthernet0/0/0]quit
[AR2]quit
<AR2>save
```

(3) Configure the router AR1 as a DHCP relay

To configure the router AR1 as a DHCP relay, the following configuration tasks shall be completed:

① Enable the DHCP service.

② Enter the three-layer port (configured with an IP address) that can receive DHCP DISCOVERY broadcast packets sent by DHCP clients, enable the DHCP relay function of the interface, and configure the IP address of the DHCP server.

Refer to the following steps to configure the router AR1 as a DHCP relay, so that the hosts in the network segments where PC1 and PC2 are located to automatically obtain IP addresses from the router AR2.

```
[AR1]dhcp enable                         //Enable the DHCP service
[AR1]interface g0/0/0
[AR1-GigabitEthernet0/0/0]dhcp select relay //Enable the DHCP relay function of the port
[AR1-GigabitEthernet0/0/0]dhcp relay server-ip 192.168.1.1
//Configure the address of the DHCP server as "192.168.1.1", which is the IP address
//of the G0/0/0 interface on the router AR2
[AR1-GigabitEthernet0/0/0]quit
[AR1]quit
<AR1>
```

(4) DHCP testing and troubleshooting

Refer to the aforementioned DHCP testing and troubleshooting methods for DHCP relay testing and troubleshooting.

If the host cannot obtain an IP address during testing, it is necessary to follow the steps below to complete troubleshooting on the router AR1 as a DHCP relay:

① On the routers AR1 and AR2, use the "display ip interface brief" command to check if the interface IP of the router has been correctly configured, and whether the physical interface and link protocol state have been "UP".

② On the router AR2, use the "display ip routing-table" command to display the routing table

contents, and check whether the remote network segment is routed, and whether the routed path is correct, as shown in Fig. 1.87.

```
[AR2]display ip routing-table
Route Flags: R - relay, D - download to fib
-------------------------------------------------------------------------
Routing Tables: Public
        Destinations : 8      Routes : 8

Destination/Mask      Proto  Pre  Cost    Flags  NextHop       Interface

       127.0.0.0/8    Direct  0    0       D     127.0.0.1     InLoopBack0
       127.0.0.1/32   Direct  0    0       D     127.0.0.1     InLoopBack0
 127.255.255.255/32   Direct  0    0       D     127.0.0.1     InLoopBack0
     192.168.0.0/24   Static  60   0       RD    192.168.1.2   GigabitEthernet0/0/0
     192.168.1.0/30   Direct  0    0       D     192.168.1.1   GigabitEthernet0/0/0
     192.168.1.1/32   Direct  0    0       D     127.0.0.1     GigabitEthernet0/0/0
     192.168.1.3/32   Direct  0    0       D     127.0.0.1     GigabitEthernet0/0/0
 255.255.255.255/32   Direct  0    0       D     127.0.0.1     InLoopBack0
```

Fig. 1.87　Routing table contents on router AR2

③ On the router AR1, use the "display dhcp relay interface g0/0/0" command to check if the DHCP relay of the G0/0/0 interface is correctly configured, as shown in Fig. 1.88.

```
[AR1]display dhcp relay interface g0/0/0
DHCP relay agent running information of interface GigabitEthernet0/0/0 :
Server IP address [01] : 192.168.1.1
Gateway address in use : 192.168.0.1
```

Fig. 1.88　DHCP relay configuration information of G0/0/0 interface on router AR1

④ On the router AR1, use the "display dhcp statistics" command to check DHCP packet statistics, as shown in Fig. 1.89.

```
[AR1]display dhcp statistics
Input: total 12 packets, discarded 0 packets
  Bootp request    :      0,  Bootp reply    :   0
  Discover         :      6,  Offer          :   2
  Request          :      2,  Ack            :   2
  Release          :      0,  Nak            :   0
  Decline          :      0,  Inform         :   0

Output: total 8 packets, discarded 0 packets
```

Fig. 1.89　Use the "display dhcp statistics" command to check DHCP sending and receiving message statistics

Task Completion

1. Task planning

According to the network topology shown in Fig. 1.1 and the DHCP service requirements in Project 1, provide two IP address pools to configure the DHCP services on the "fenbu" router. See

Table 1.16 for DHCP service planning.

Table 1.16 DHCP service planning

Address pool name	IP address network number	Gateway address	DNS address	Address lease time	Bound static IP
yusuan	172.16.129.0/24	172.16.129.1	172.16.128.253	8 h	
manage	172.16.128.0/25	172.16.128.1	172.16.128.253	8 h	Static IP "172.16.128.13" bound to host PC2

2. Task implementation

(1) Check and record the MAC address of host PC3

On the host PC3, enter the DOS command line and use the "ipconfig /all" command to check and record the MAC address of the host PC3. As shown in Fig. 1.90, the MAC address of the host PC3 is 54-89-98-5E-6B-B6.

```
PC>ipconfig

Link local IPv6 address...........: fe80::5689:98ff:fe5e:6bb6
IPv6 address......................:  :: / 128
IPv6 gateway......................:  ::
IPv4 address......................: 0.0.0.0
Subnet mask.......................: 0.0.0.0
Gateway...........................: 0.0.0.0
Physical address..................: 54-89-98-5E-6B-B6
DNS server........................:
```

Fig. 1.90 Check and record MAC address of host PC3

(2) Configuration of DHCP service on "fenbu" router

Refer to the following steps to configure the DHCP service on the "fenbu" router.

```
[fenbu]dhcp enable
//Enable the DHCP service
[fenbu]ip pool yusuan
[fenbu-ip-pool-yusuan]network 172.16.129.0 mask 24
[fenbu-ip-pool-yusuan]gateway-list 172.16.129.1
[fenbu-ip-pool-yusuan]dns-list 172.16.128.253
[fenbu-ip-pool-yusuan]lease day 0 hour 8
[fenbu-ip-pool-yusuan]quit
//Configure the DHCP address pool "yusuan" with the network number and gateway of
//172.16.129.0/24 and 172.16.129.1 respectively;DNS is 172.16.128.253, and the address
//lease time is 8 hours
[fenbu]ip pool manage
[fenbu-ip-pool-manage]network 172.16.128.0 mask 25
[fenbu-ip-pool-manage]gateway-list 172.16.128.1
[fenbu-ip-pool-manage]dns-list 172.16.128.253
```

```
[fenbu-ip-pool-manage]lease day 0 hour 8
[fenbu-ip-pool-manage]static-bind ip-address 172.16.128.13 mac-address 5489-985e-6bb6
[fenbu-ip-pool-manage]quit
//Configure the DHCP address pool "manage" with the network number, gateway, DNS and
//address lease time of 172.16.128.0/25, 172.16.128.1, 172.16.128.253 and 8 hours respectively,
//and bind the static IP address "172.16.128.13" to the host PC3 with the MAC address
//"5489-985e-6bb6".
[fenbu]interface g0/0/2
[fenbu-GigabitEthernet0/0/2]dhcp select global
[fenbu-GigabitEthernet0/0/2]quit
[fenbu]interface g3/0/0
[fenbu-GigabitEthernet3/0/0]dhcp select global
[fenbu-GigabitEthernet3/0/0]quit
//Enable the DHCP Server function using global address pooling for the interfaces
//G0/0/2 and G3/0/0
[fenbu]quit
<fenbu>save
//Save configurations
```

(3) DHCP testing and troubleshooting

① On the budget department host PC2 and the management department hosts PC3 and PC4, configure their IP and DNS addresses to be automatically obtained.

② On the hosts PC2, PC3 and PC4, use the "ipconfig /all" command to view the IP information, and check whether the correct IP address, subnet mask, default gateway, DNS server address, and lease time are obtained. Check whether the IP address obtained by the management department host PC3 is the statically bound address "172.16.128.13", as shown in Fig. 1.91.

```
PC>ipconfig

Link local IPv6 address.............: fe80::5689:98ff:fe59:55a6
IPv6 address........................: :: / 128
IPv6 gateway........................: ::
IPv4 address........................: 172.16.129.254
Subnet mask.........................: 255.255.255.0
Gateway.............................: 172.16.129.1
Physical address....................: 54-89-98-59-55-A6
DNS server..........................: 172.16.128.253
```

Fig. 1.91　IP address information automatically obtained on hosts PC2, PC3, and PC4

```
PC>ipconfig

Link local IPv6 address............: fe80::5689:98ff:fe5e:6bb6
IPv6 address......................: :: / 128
IPv6 gateway......................: ::
IPv4 address......................: 172.16.128.13
Subnet mask.......................: 255.255.255.128
Gateway...........................: 172.16.128.1
Physical address..................: 54-89-98-5E-6B-B6
DNS server........................: 172.16.128.253
```

```
PC>ipconfig

Link local IPv6 address............: fe80::5689:98ff:fe6d:6664
IPv6 address......................: :: / 128
IPv6 gateway......................: ::
IPv4 address......................: 172.16.128.126
Subnet mask.......................: 255.255.255.128
Gateway...........................: 172.16.128.1
Physical address..................: 54-89-98-6D-66-64
DNS server........................: 172.16.128.253
```

Fig. 1.91 IP address information automatically obtained on hosts PC2, PC3, and PC4(continued)

③ On the "fenbu" router, use the "display ip pool" command to view the address pool and address statistics, as shown in Fig. 1.92.

④ On the "fenbu" router, use the "display ip pool name yusuan used" command to view the address pool "yusuan" and used address statistics, as shown in Fig. 1.93.

⑤ On the "fenbu" router, use the "display ip pool name manage used" command to view the address pool "manage" and used address statistics, as shown in Fig. 1.94.

⑥ Finally, on the DOS command line of hosts PC2, PC3 and PC4, use "tracert PC1's IP address" to track the path through which the IP packet passes when accessing PC1. Fig. 1.95 shows the results of tracert PC1 on the host PC3, indicating that the data packet from PC3 to PC1 goes through "172.16.128.1" → "172.16.254.2" → "172.160.10".

If the hosts PC2, PC3 and PC4 cannot obtain their IP addresses during testing, refer to the following steps for troubleshooting:

① Check whether the IP and DNS addresses are configured for automatic acquisition on all hosts.

② Use the "display ip interface brief " command on the "fenbu" router to check if the interface (pay attention to the interfaces G0/0/0 and G3/0/0) IP of the router is configured correctly, and whether the physical interface and link protocol state have been "UP".

③ On the "fenbu" router, use the "display current-configuration | begin ip pool" command to check if the IP address pool in the currently running configuration file is correctly configured, as shown in Fig. 1.96.

```
[fenbu]display ip pool
─────────────────────────────────────────
  Pool-name     : yusuan
  Pool-No       : 0
  Position      : Local      Status      : Unlocked
  Gateway-0     : 172.16.129.1
  Mask          : 255.255.255.0
  VPN instance  : --

─────────────────────────────────────────
  Pool-name     : manage
  Pool-No       : 1
  Position      : Local      Status      : Unlocked
  Gateway-0     : 172.16.128.1
  Mask          : 255.255.255.128
  VPN instance  : --

IP address Statistic
  Total     :378
  Used      :3      Idle     :375
  Expired   :0      Conflict :0      Disable :0
[fenbu]
```

Fig. 1.92　Check address pool and address statistics

```
[fenbu]display ip pool name yusuan used
  Pool-name     : yusuan
  Pool-No       : 0
  Lease         : 0 Days 8 Hours 0 Minutes
  Domain-name   : -
  DNS-server0   : 172.16.128.253
  NBNS-server0  : -
  Netbios-type  : -
  Position      : Local      Status      : Unlocked
  Gateway-0     : 172.16.129.1
  Mask          : 255.255.255.0
  VPN instance  : --
─────────────────────────────────────────────────────
  Start         End           Total  Used  Idle(Expired)  Conflict  Disable
─────────────────────────────────────────────────────
  172.16.129.1  172.16.129.254  253    1       252(0)         0         0

Network section :
─────────────────────────────────────────────────────
  Index      IP              MAC             Lease    Status
─────────────────────────────────────────────────────
  253    172.16.129.254   5489-9859-55a6    3423     Used
─────────────────────────────────────────────────────
```

Fig. 1.93　Check used addresses in address pool "yusuan"

```
[fenbu]display ip pool name manage used
  Pool-name      : manage
  Pool-No        : 1
  Lease          : 0 Days 8 Hours 0 Minutes
  Domain-name    : -
  DNS-server0    : 172.16.128.253
  NBNS-server0   : -
  Netbios-type   : -
  Position       : Local        Status       : Unlocked
  Gateway-0      : 172.16.128.1
  Mask           : 255.255.255.128
  VPN instance   : --
  -----------------------------------------------------------
       Start         End      Total  Used  Idle(Expired)  Conflict  Disable
  -----------------------------------------------------------
   172.16.128.1  172.16.128.126  125    2      123(0)         0         0
  -----------------------------------------------------------

  Network section :
  -----------------------------------------------------------
   Index       IP          MAC         Lease   Status
  -----------------------------------------------------------
    12   172.16.128.13   5489-985e-6bb6    -     Static-bind used
   125   172.16.128.126  5489-986d-6664   945    Used
  -----------------------------------------------------------
```

Fig. 1.94 Check used addresses in address pool "manage"

```
PC>tracert 172.16.0.10

traceroute to 172.16.0.10, 8 hops max
(ICMP), press Ctrl+C to stop
 1  172.16.128.1   32 ms   47 ms   31 ms
 2  172.16.254.2   47 ms   47 ms   46 ms
 3  172.16.0.10    94 ms   78 ms   94 ms
```

Fig. 1.95 Test results from host PC3 "tracert" PC1

```
[fenbu]display current-configuration | begin ip pool
ip pool yusuan
 gateway-list 172.16.129.1
 network 172.16.129.0 mask 255.255.255.0
 lease day 0 hour 8 minute 0
 dns-list 172.16.128.253
#
ip pool manage
 gateway-list 172.16.128.1
 network 172.16.128.0 mask 255.255.255.128
 static-bind ip-address 172.16.128.13 mac-address 5489-985e-6bb6
 lease day 0 hour 8 minute 0
 dns-list 172.16.128.253
```

Fig. 1.96 Check the configuration of the IP address pool in the current configuration file

④ On the "fenbu" router, enter the interface view of the interfaces G0/0/2 and G3/0/0, and use the "display ip this" command to display the configuration in the current interface view. Check if the interface is configured with DHCP Server functionality using global address pooling.

Exercises

Multiple choice

(1) The destination IP address of the DHCP DISCOVER message is (   ).

    A.0.0.0.0   B.224.0.0.9   C.127.0.0.1   D.255.255.255.255

(2) In Huawei routers, the following (   ) command can display the assigned IP addresses and corresponding MAC addresses in the address pool.

    A. display ip pool name yusuan used   B. display ip pool

    C. display ip pool name yusuan   D. display dhcp

(3) The following (   ) command is used to display the IP address pool of Huawei routers.

    A. display ip pool   B. display dhcp pool

    C. display interface   D. display ip interface brief

(4) A certain router has its DHCP address pool configured as follows:

```
#
ip pool jwc
    network 192.168.10.0 mask 255.255.255.0
    gateway-list 192.168.10.1
dns-list 172.16.5.25
#
```

The following statement (   ) is incorrect.

    A. The gateway address of this address pool is 192.168.10.1

    B. The DNS address of this address pool is 192.168.10.1

    C. The DNS address of this address pool is 172.16.5.25

    D. The network address of this address pool is 192.168.10.0/24

Task 4　Access Control List

Task Description

After completing the Task 1 to Task 3 of this project, complete the ACL (access control list) planning and configuration tasks on the branch "fenbu" router in the network topology shown in Fig. 1.1, and back up the configurations of the branch router to the FTP server of the headquarters, so that the networking of Project 1 can satisfy the requirements of "introducing ACL at the out and in ports of the branch router connected to the headquarters", "disabling the high-risk ports 135-139, 445, and 3389 of TCP and UDP protocols", "only allowing PC3 of the branch management department and PC1 of the headquarters to remotely log in to the 'fenbu' router for remote

management via SSH", and "saving the configurations of the branch and headquarters routers and backing up to the FTP server of the headquarters after the configurations are completed".

Task Objectives

- Understand the working principle of ACL.
- Master the planning and configuration of basic ACL.
- Master the planning and configuration of advanced ACL.
- Competent for testing and troubleshooting.

Related Knowledge

1. Working principle of ACL

As a list of instructions that act on such device interfaces as routers, ACL (access control list) determines whether to allow or reject packets based on the conditions in the packet header, also known as packet filtering. ACL is not only a means of controlling network communication traffic, but also a component of network security policies and one of the important mechanisms for achieving network perimeter security. ACL can be implemented on both routers and devices such as firewalls.

ACL is usually applied to such device interfaces as routers. According to the location of the interface where the ACL is located, it can filter both packets entering the router interface and packets going out from the router interface. The former is called "inbound" packet filtering, while the latter is called "outbound" packet filtering. Fig. 1.97 shows the working principle of ACL.

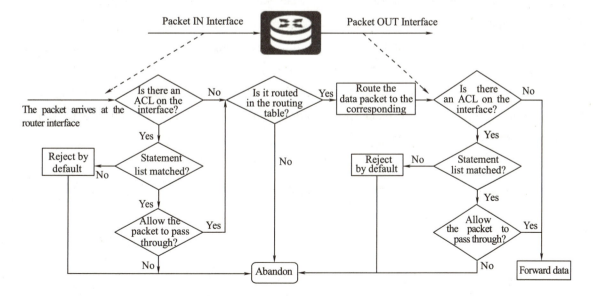

Fig. 1.97 Working principle of ACL

2. ACL category and identification

According to the different access control functions, ACLs are divided into basic ACLs, advanced ACLs and Layer-2 ACLs. The basic ACLs define rules based on source IP address, sharding information, and effective time period to filter IPv4 messages. If you only need to filter messages based on the source IP address, you can configure a basic ACL, but the basic ACL cannot distinguish different types of traffic from the same network or host, namely, it cannot complete selective filtering based on different application layer protocols or application services. Advanced ACL can receive or reject data packets based on their protocol types, source addresses, destination addresses, and port numbers in the target data packets.

When creating an ACL, users must assign a number to it, and different numbers correspond to different types of ACL, as shown in Table 1.17. Meanwhile, for the convenience of memory and recognition, users can also choose whether to set a name for each ACL when creating it. Once an ACL is created, users are no longer allowed to set a name, or modify or delete its original name.

Table 1.17 Huawei ACL classification

ACL types	Range of numbers	Basis for rules formulation
Basic ACL	2000-2999	Source address of messages
Advanced ACL	3000-3999	Source addresses, destination addresses, protocol types, port numbers, etc. of messages
Layer-2 ACL	4000-4999	Source MAC addresses, destination MAC addresses, 802.1p preferences, etc. of messages

3. Wildcard mask

The wildcard mask "wildcard-mask" is used in pairs with the IP address to indicate whether the corresponding bits in the IP address need to be checked and matched. In the wildcard mask, "1" indicates that the relevant binary bits of the corresponding IP address do not need to be matched, while "0" indicates that the relevant binary bits of the corresponding IP address need to be matched. Similar to the subnet mask, the length of the wildcard mask is 32 bits (binary), expressed in dotted decimal, for example, the wildcard mask of "0.255.255.255" indicates that the first 8 bits in the corresponding IP address need to be checked, while the last 24 bits can be ignored. See Table 1.18 for more examples of wildcard masks.

Table 1.18 Examples of wildcard masks

Test conditions	IP address	Wildcard masks
10.0.0.0/8	10.0.0.0	0.255.255.255
172.31.0.0/16	172.31.0.0	0.0.255.255
202.33.44.0/24	202.33.44.0	0.0.0.255
192.168.1.64/26	192.168.1.64	0.0.0.63
210.33.44.254/32	210.33.44.254	0.0.0.0

Related Skills

1. Planning and configuration of basic ACL

Build a topology as shown in Fig. 1.98, manage the interconnection of the entire network and implement the basic ACL to achieve such functions as follows:

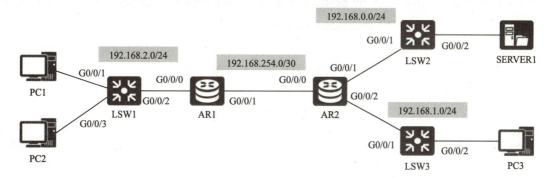

Fig. 1.98 "Basic ACL planning and configuration" topology

① Except for the host PC2 in the "192.168.2.0/24" network segment, which cannot access the server network segment, all host network segments in the network can access the server network segment.

② Only PC1 can SSH to AR1 for testing and troubleshooting.

Table 1.19 shows the IP planning of the router interface IP and PC.

Table 1.19 IP planning

Device	Interface	IP address	Gateway
AR1	G0/0/0	192.168.2.1/24	
	G0/0/1	192.168.254.1/30	
AR2	G0/0/0	192.168.254.2/30	
	G0/0/1	192.168.0.1/24	
	G0/0/2	192.168.1.1/24	
PC1	NIC	192.168.2.11/24	192.168.2.1
PC2	NIC	192.168.2.12/24	192.168.2.1
PC3	NIC	192.168.1.13/24	192.168.1.1
SERVER1	NIC	192.168.0.10/24	192.168.0.1

Use static routing to achieve interconnection and interoperability across the entire network. See Table 1.20 for the planning of static routing.

Table 1.20 Planning of static routing

Device	Target network	Target network mask/Mask length	Next hop or forwarding port
AR1	192.168.0.0	23	192.168.254.2
AR2	192.168.2.0	24	192.168.254.1

As is required, except for the host PC2 in the "192.168.2.0/24" network segment, which cannot access the server network segment, all host network segments in the network can access the server network segment. Moreover, the basic ACL can only check the source IP address in the data packet, so to achieve such a requirement, it is necessary to create a basic ACL (with a list number of 2000) on the router AR2 near the destination end (server network segment) and apply it to the outbound direction of the G0/0/1 interface on the router AR2.

As is required, only PC1 can SSH to AR1. A basic ACL needs to be created on the router AR1 and applied to the user interface view of VTY 0 4 on the router AR1.

Table 1.21 shows ACL planning.

Table 1.21 ACL planning

Device	List Number	ACL rules	Application interface and direction
AR1	2000	Allow hosts with a source address of 192.168.2.11; prohibit all	VTY 0 4 inbound
AR2	2001	Deny hosts with a source address 192.168.2.12; allow the networks 192.168.2.0/24 and 192.168.1.0/24	Interface G0/0/1 outbound of AR2

(1) Basic router configuration and static routing configuration

Refer to the following steps to configure the interface and static routing on the routers AR1 and AR2, achieving interconnection of all network segments in the network.

```
[AR1]interface g0/0/0
[AR1-GigabitEthernet0/0/0]ip address 192.168.2.1 24
[AR1-GigabitEthernet0/0/0]quit
[AR1]interface g0/0/1
[AR1-GigabitEthernet0/0/1]ip address 192.168.254.1 30
[AR1-GigabitEthernet0/0/1]quit
[AR1]ip route-static 192.168.0.0 23 192.168.254.2
[AR1]quit
<AR1>save
//Configure the interface of the router AR1 and the aggregated static routes to remote
//networks, and save the configuration
[AR2]interface g0/0/0
[AR2-GigabitEthernet0/0/0]ip address 192.168.254.2 30
[AR2-GigabitEthernet0/0/0]quit
[AR2]interface g0/0/1
[AR2-GigabitEthernet0/0/1]ip address 192.168.0.1 24
[AR2-GigabitEthernet0/0/1]quit
[AR2]interface g0/0/2
[AR2-GigabitEthernet0/0/2]ip address 192.168.1.1 24
[AR2-GigabitEthernet0/0/2]quit
```

```
[AR2]ip route-static 192.168.2.0 24 192.168.254.1
[AR2]quit
<AR2>save
//Configure the interface of the router AR2 and the static routes to remote networks,
//and save the configuration
```

(2) SSH configuration on router AR1

Refer to the following steps to configure Stelnet on the router AR1. Here follows the specific plan: Stelnet authentication uses AAA authentication, the Stelnet username is "zhangcr", the password is "Chengdu123" and the privilege level is "15".

```
[AR1]stelnet server enable            //Enable the Stelnet service
[AR1]aaa                              //Enter the AAA view
[AR1-aaa]local-user zhangcr password cipher Chengdu12
//Create a local user with the username "zhangcr",Password "Chengdu123"
[AR1-aaa]local-user zhangcr service-type ssh
//Configure user "zhangcr" to authorize SSH services
[AR1-aaa]local-user zhangcr privilege level 15
//Configure the privilege level of the user "zhangcr" to 15
[AR1-aaa]quit                         //Return to the system view
[AR1]user-interface vty 0 4           //Enter the user interface view from VTY0 to VTY4
[AR1-ui-vty0-4]protocol inbound SSH   //Configure the protocol to SSH protocol
[AR1-ui-vty0-4]authentication-mode aaa  //Configure five virtual terminals from VTY0 to VTY4
// Set the authentication mode of the session as AAA for authentication
[AR1-ui-vty0-4]quit       //Quit the user interface view of VTY0 to VTY4 and return to
                          //the user view
[AR1]rsa local-key-pair create  //Use the RSA algorithm to create a local key pair
```

(3) Network connectivity and SSH testing

Refer to the IP parameters planned in Table 1.19 to configure the IP address, subnet mask, and default gateway of all PCs and servers in the network. After the configuration is completed, use the "ipconfig" command on the DOS command line to check whether the IP configuration information is correct, and use the "ping" command to test the connectivity of the whole network. (Note: At this point, IP connectivity shall be available for all hosts in the network). Fig. 1.99 shows that the host PC1 successfully ping the server and PC3.

On PC1 and PC2, run the SecureCRT virtual terminal software and use SSH to access the IP address "192.168.2.1" of the router AR1 G0/0/0 interface.

On PC3, run the SecureCRT virtual terminal software and use SSH to access the IP address "192.168.254.1" of the router AR1 G0/0/1 interface. At this point, all hosts such as PC1, PC2, and PC3 can successfully log in to the router AR1 via SSH.

If no communication is available between the host and the server, and the host cannot log in to the router AR1 via SSH, please troubleshoot it.

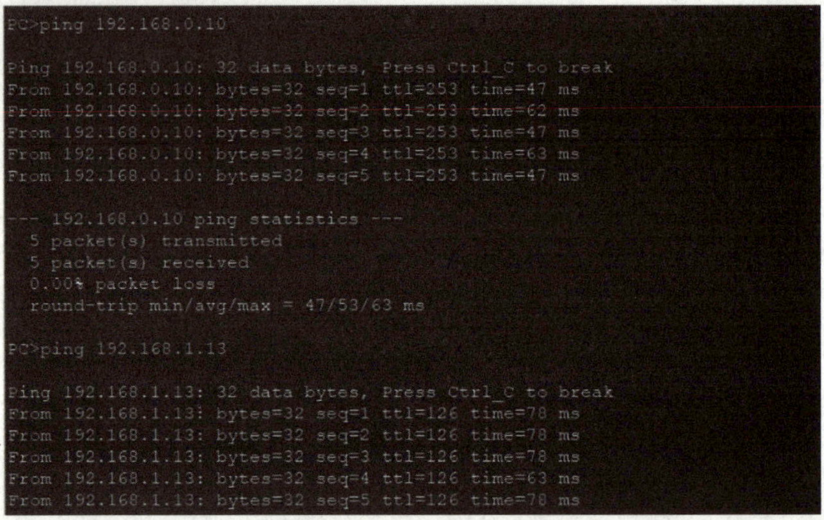

Fig. 1.99 Host PC1 successfully ping server and PC3

(4) Configuration of basic ACL

According to Table 1.21, refer to the following steps to configure the basic ACL on the router AR2, so that all host network segments in the network can access the server network segment except that the host PC2 of the network segment "192.168.2.0/24" cannot access the server network segment.

```
[AR2]acl 2001//Create a basic ACL with the number of 2001
[AR2-acl-basic-2001]rule deny source 192.168.2.12 0.0.0.0
//Configure the rule to deny the host "192.168.2.12"
[AR2-acl-basic-2001]rule permit source 192.168.2.0 0.0.0.255
//Configure the rule to permit the network "192.168.1.0/24"
[AR2-acl-basic-2001]rule permit source 192.168.1.0 0.0.0.255
//Configure the rule to permit the network "192.168.1.0/24"
[AR2-acl-basic-2001]quit
[AR2]interface g0/0/1
[AR2-GigabitEthernet0/0/1]traffic-filter outbound acl 2001
//Apply ACL 2001 to the outbound direction of the G0/0/1 interface
[AR2-GigabitEthernet0/0/1]quit
```

According to Table 1.21, refer to the following steps to configure the basic ACL on the router AR1, so that only PC1 can SSH to AR1.

```
[AR1]acl 2000     //Create a basic ACL with the number 2000
[AR1-acl-basic-2000]rule permit source 192.168.2.11 0.0.0.0
```

```
    //Configure the rule to permit the host "192.168.2.11"
[AR1-acl-basic-2000]rule deny source any         //Configure the rule to deny all traffic
[AR1]user-interface vty 0 4
[AR1-ui-vty0-4]acl 2000 inbound   //Apply ACL 2000 to the traffic from VTY 0 to VTY 4
                                  //inbound
[AR1-ui-vty0-4]quit
[AR1]
```

(5) ACL testing and troubleshooting

Use the ping utility to test the IP connectivity with the server on the DOS command line of PC1, PC2, and PC3, respectively. Fig. 1.100 shows that the hosts PC1 and PC3 can ping the server, but the host PC2 can't.

```
PC>ping 192.168.0.10

Ping 192.168.0.10: 32 data bytes, Press Ctrl_C to break
From 192.168.0.10: bytes=32 seq=1 ttl=253 time=62 ms
From 192.168.0.10: bytes=32 seq=2 ttl=253 time=47 ms

PC>ping 192.168.0.10

Ping 192.168.0.10: 32 data bytes, Press Ctrl_C to break
From 192.168.0.10: bytes=32 seq=1 ttl=254 time=62 ms
From 192.168.0.10: bytes=32 seq=2 ttl=254 time=47 ms
From 192.168.0.10: bytes=32 seq=3 ttl=254 time=47 ms

PC>ping 192.168.0.10

Ping 192.168.0.10: 32 data bytes, Press Ctrl_C to break
Request timeout!
Request timeout!
Request timeout!
Request timeout!
Request timeout!
```

Fig. 1.100 Host PC1 and PC3 can PING the server

Run the SecureCRT virtual terminal software on the hosts PC1, PC2 and PC3 respectively to test whether the router AR1 can be accessed via SSH. After ACL is implemented, only the host PC1 can successfully access the router AR1 via SSH, while other hosts cannot access the router AR1 via SSH.

If the test result is not that all host network segments in the network can access the server network segment except that the host PC2 in the "192.168.2.0/24" network segment cannot access the server network segment, please follow the steps below for troubleshooting:

Use the "display acl 2001" command on the router AR2 to display the configuration and operation status of the basic ACL 2001, as shown in Fig. 1.101.

```
<AR2>display acl 2001
Basic ACL 2001, 3 rules
Acl's step is 5
 rule 5 deny source 192.168.2.12 0 (13 matches)
 rule 10 permit source 192.168.2.0 0.0.0.255
 rule 15 permit source 192.168.1.0 0.0.0.255
```

Fig. 1.101 Display results from "display acl 2001" command

Use the "display current-configuration | begin acl" command on the router AR2 to view ACL configurations, as shown in Fig. 1.102.

```
<AR2>display current-configuration | begin acl
acl number 2001
 rule 5 deny source 192.168.2.12 0
 rule 10 permit source 192.168.2.0 0.0.0.255
 rule 15 permit source 192.168.1.0 0.0.0.255
```

Fig. 1.102 Display results from "display current-configuration | begin acl" command

If an incorrect ACL is configured, please refer to the following command to delete ACL 2001 and reconfigure a correct ACL.

```
[AR2]undo acl 2001        //Delete ACL 2001
```

Use the "display traffic-filter applied-record" command on the router AR2 to view ACL application, as shown in Fig. 1.103.

```
[AR2]display traffic-filter applied-record
--------------------------------------------------
Interface            Direction  AppliedRecord
--------------------------------------------------
GigabitEthernet0/0/1    outbound   acl 2001
```

Fig. 1.103 ACL application information via "display traffic-filter applied-record" command

If it shows that ACL 2001 is applied in a direction other than the G0/0/1 outbound, first enter the interface view wrongly configured, use the "undo traffic-filter" command to delete the applied ACL, and then enter the G0/0/1 interface to apply ACL 2001 to the G0/0/1 outbound. The following command has exampled how to delete the ACL applied to the G0/0/1 interface outbound.

```
[AR2-GigabitEthernet0/0/1]undo traffic-filter outbound
//Delete ACL applied to the G0/0/1 interface outbound
```

If the test result is not that only PC1 can SSH to AR1, please follow the steps below for troubleshooting:

Use the "display acl 2000" command on the router AR1 to display the configuration and operation status of the basic ACL 2000, as shown in Fig. 1.104.

```
<AR1>display acl 2000
Basic ACL 2000, 2 rules
Acl's step is 5
 rule 5 permit source 192.168.2.11 0 (4 matches)
 rule 10 deny (8 matches)
```

Fig. 1.104   Display results from the "display acl 2000" command

Use the "display current-configuration | begin acl" command on the router AR2 to view ACL configurations, as shown in Fig. 1.105.

```
<AR2>display current-configuration | begin acl
acl number 2001
 rule 5 deny source 192.168.2.12 0
 rule 10 permit source 192.168.2.0 0.0.0.24
 rule 15 permit source 192.168.1.0 0.0.0.24
```

Fig. 1.105   Display results from the "display current-configuration | begin acl" command

If an incorrect ACL is configured, please delete it and reconfigure a correct ACL.

In the router AR1 VTY 0 4 user interface view, use the "display this" command to view the configuration information in the VTY 0 4 user interface view and check if ACL 2000 is configured to be applied to the VTY 0 4 inbound direction, as shown in Fig. 1.106.

```
[AR1-ui-vty0-4]display this
[V200R003C00]
#
user-interface con 0
 authentication-mode password
user-interface vty 0 4
 acl 2000 inbound
 authentication-mode aaa
 protocol inbound ssh
```

Fig. 1.106   Configurations displayed via "display this" in the VTY 0 4 user interface view

If the ACL applied is incorrect in direction, use the "undo acl inbound" command to delete the ACL applied under VTY 0 4, and then apply the correct ACL to the inbound direction. The following command is used to delete the wrong ACL applied to the VTY 0 4 interface.

```
[AR1-ui-vty0-4]undo acl inbound   //Delete applied ACL inbound
[AR1-ui-vty0-4]undo acl outbound  //Delete applied ACL outbound
```

2. Planning and configuration of advanced ACL

Build a topology as shown in Fig. 1.107, completing the interconnection of the entire network and implement advanced ACL, so that all hosts in the "192.168.2.0/24" network segment can only access the WWW service of the Web server, and cannot ping the Web server, but access to others is unrestricted.

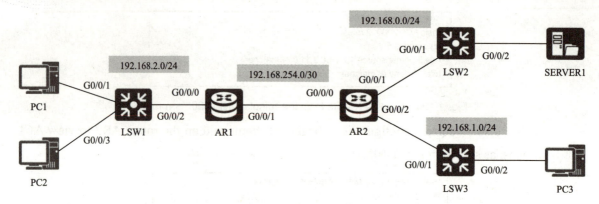

Fig. 1.107 Topology for "planning and configuration of advanced ACL"

See Table 1.22 for the IP planning of the router interface IP and PC.

Table 1.22 IP planning

Device	Interface	IP address	Gateway
AR1	G0/0/0	192.168.2.1/24	
	G0/0/1	192.168.254.1/30	
AR2	G0/0/0	192.168.254.2/30	
	G0/0/1	192.168.0.1/24	
	G0/0/2	192.168.1.1/24	
PC1	NIC	192.168.2.11/24	192.168.2.1
PC2	NIC	192.168.2.12/24	192.168.2.1
PC3	NIC	192.168.1.13/24	192.168.1.1
SERVER1	NIC	192.168.0.10/24	192.168.0.1

Use static routing to achieve interconnection and interoperability across the entire network. See Table 1.23 for the planning of static routing.

Table 1.23 Planning of static routing

Device	Target network	Target network mask/Mask length	Next hop or forward port
AR1	192.168.0.0	23	192.168.254.2
AR2	192.168.2.0	24	192.168.254.1

As is required, all hosts in the "192.168.2.0/24" network segment can only access the WWW service of the Web server, and cannot ping the Web server, but access to other resources is unrestricted. To achieve this requirement, it is necessary to create an advanced ACL (list number 3000) on the router AR1 near the source and apply it to the G0/0/0 interface inbound of the router AR1. See Table 1.24 for ACL planning.

Project 1 Enterprise Branch Internetworking Project   85

Table 1.24 ACL planning

Device	List number	ACL rule	Application interface and direction
AR1	3000	Permit the traffic with the port number www (80) and subject to the TCP protocol from the host in the source network "192.168.2.0/24" to the host "192.168.0.10"; Deny the traffic subject to the ICMP protocol from the host in the source network "192.168.2.0/24" to the host "192.168.0.10"; Permit the traffic from any source to any destination subject to the IP protocol	G0/0/0 inbound of router AR1

(1) Basic router configuration and static routing configuration

Refer to the following steps to complete interface and static routing configuration on the routers AR1 and AR2, so that interconnection can be available for all network segments in the network.

```
[AR1]interface g0/0/0
[AR1-GigabitEthernet0/0/0]ip address 192.168.2.1 24
[AR1-GigabitEthernet0/0/0]quit
[AR1]interface g0/0/1
[AR1-GigabitEthernet0/0/1]ip address 192.168.254.1 30
[AR1-GigabitEthernet0/0/1]quit
[AR1]ip route-static 192.168.0.0 23 192.168.254.2
[AR1]quit
<AR1>save
//Configure the aggregated static routing from the interface of the router AR1 to remote
//networks, and save the configuration
[AR2]interface g0/0/0
[AR2-GigabitEthernet0/0/0]ip address 192.168.254.2 30
[AR2-GigabitEthernet0/0/0]quit
[AR2]interface g0/0/1
[AR2-GigabitEthernet0/0/1]ip address 192.168.0.1 24
[AR2-GigabitEthernet0/0/1]quit
[AR2]interface g0/0/2
[AR2-GigabitEthernet0/0/2]ip address 192.168.1.1 24
[AR2-GigabitEthernet0/0/2]quit
[AR2]ip route-static 192.168.2.0 24 192.168.254.1
[AR2]quit
<AR2>save
//Configure the static routing from the interface of the router AR2 to remote networks,
//and save the configuration
```

(2) Network connectivity

Refer to the IP parameters planned in Table 1.20 to configure the IP address, subnet mask, and

default gateway of all PCs and servers in the network. After the configuration is completed, use the "ipconfig" command on the DOS command line to check whether the IP configuration information is correct, and use the "ping" command to test the connectivity of the whole network. (**Note**: At this point, IP connectivity is available for all hosts in the network). Fig. 1.108 shows that the host PC1 successfully ping the server and PC3.

If no communication is available between the host and the server, please troubleshoot it.

```
PC>ping 192.168.0.10

Ping 192.168.0.10: 32 data bytes, Press Ctrl_C to break
From 192.168.0.10: bytes=32 seq=1 ttl=253 time=47 ms
From 192.168.0.10: bytes=32 seq=2 ttl=253 time=62 ms
From 192.168.0.10: bytes=32 seq=3 ttl=253 time=47 ms
From 192.168.0.10: bytes=32 seq=4 ttl=253 time=63 ms
From 192.168.0.10: bytes=32 seq=5 ttl=253 time=47 ms

--- 192.168.0.10 ping statistics ---
  5 packet(s) transmitted
  5 packet(s) received
  0.00% packet loss
  round-trip min/avg/max = 47/53/63 ms

PC>ping 192.168.1.13

Ping 192.168.1.13: 32 data bytes, Press Ctrl_C to break
From 192.168.1.13: bytes=32 seq=1 ttl=126 time=78 ms
From 192.168.1.13: bytes=32 seq=2 ttl=126 time=78 ms
From 192.168.1.13: bytes=32 seq=3 ttl=126 time=78 ms
From 192.168.1.13: bytes=32 seq=4 ttl=126 time=63 ms
From 192.168.1.13: bytes=32 seq=5 ttl=126 time=78 ms
```

Fig. 1.108   Host PC1 successfully ping the server and PC3

(3) Configurations of advanced ACL

According to Table 1.24, refer to the following steps to configure advanced ACL on the router AR1.

```
[AR1]acl 3000//Create an advanced ACL with the number of 3000
[AR1-acl-adv-3000]rule permit tcp source 192.168.2.0 0.0.0.255 destination 192.168.0.10 0.0.0.0 destination-port eq www
//Permit the TCP protocol traffic from the source network "192.168.2.0/24" to the host
//"192.168.0.10" with the target port number of www (80)
[AR1-acl-adv-3000]rule deny icmp source 192.168.2.0 0.0.0.255 destination 192.168.0.10 0.0.0.0 icmp-type echo
//Deny the ICMP protocol traffic from the source network "192.168.2.0/24" to the host
//"192.168.0.10" with the type of echo
[AR1-acl-adv-3000]rule permit ip source any destination any
//Permit the IP protocol traffic from any source to any destination
[AR1-acl-adv-3000]quit
[AR1]interface g0/0/0
```

```
[AR1-GigabitEthernet0/0/0]traffic-filter inbound acl 3000
//Apply ACL 3000 to the port inbound
[AR1-GigabitEthernet0/0/0]quit
[AR1]
```

(4) ACL testing and troubleshooting

On any host in the network "192.168.2.0/24", open a browser, access the Web services provided by the Web server, and then use the ping utility on the DOS command line to test the IP connectivity to the Web server. Fig. 1.109 and Fig. 1.110 show the test results from the host PC1 in the network segment "192.168.2.0/24", which accesses the Web services provided by the Web server but cannot ping the Web server.

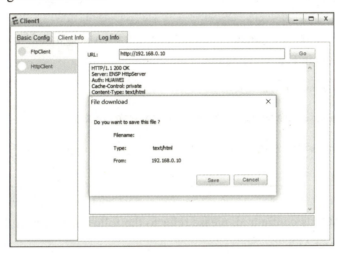

Fig. 1.109 Host PC1 can access the Web services provided by the Web server

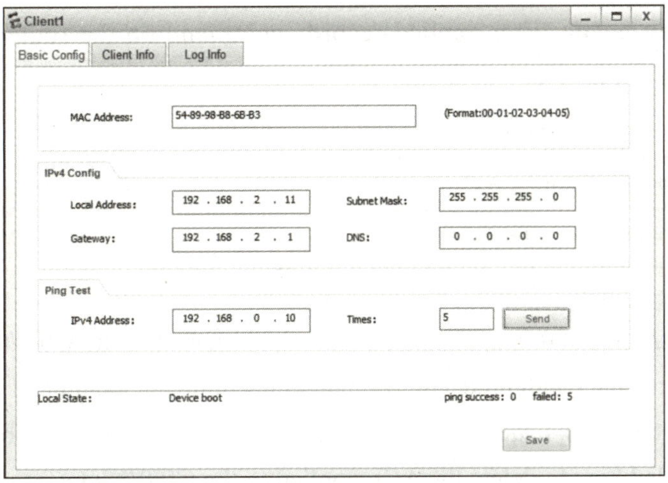

Fig. 1.110 Host PC1 cannot ping the Web server

If the test results do not meet the requirements of the subtask, please follow the steps below for troubleshooting:

Use the "display acl 3000" command on the router AR1 to display the configuration and operation status of the advanced ACL 3000, as shown in Fig. 1.111.

```
[AR1]display acl 3000
Advanced ACL 3000, 3 rules
Acl's step is 5
 rule 5 permit tcp source 192.168.2.0 0.0.0.255 destination 192.168.0.10 0 desti
nation-port eq www
 rule 10 deny icmp source 192.168.2.0 0.0.0.255 destination 192.168.0.10 0 icmp-
type echo
 rule 15 permit ip (32 matches)
```

Fig. 1.111　Display results via "display acl 3000"

On the router AR1, use the "display current-configuration | begin acl" command to view ACL configurations, as shown in Fig. 1.112.

```
[AR1]display current-configuration | begin acl
acl number 3000
 rule 5 permit tcp source 192.168.2.0 0.0.0.255 destination 192.168.0.10 0 desti
nation-port eq www
 rule 10 deny icmp source 192.168.2.0 0.0.0.255 destination 192.168.0.10 0 icmp-
type echo
 rule 15 permit ip
```

Fig. 1.112　Display results via "display current-configuration | begin acl"

If an incorrect ACL is configured, please refer to the following command to delete ACL 3000 and reconfigure a correct ACL.

```
[AR1]undo acl 3000         //Delete ACL 3000
```

Use the "display traffic-filter applied-record" command on the router AR1 to view ACL application, as shown in Fig. 1.113.

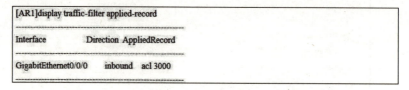

Fig. 1.113　Display results via "display traffic-filter applied-record"

If it shows that ACL 3000 is applied in a direction other than the G0/0/0 inbound, first enter the interface view wrongly configured, use the "undo traffic-filter" command to delete the applied ACL, and then enter the G0/0/1 interface to apply ACL 3000 to the G0/0/0 inbound. The following command has exampled how to delete the ACL applied to the G0/0/1 interface outbound.

```
[AR1-GigabitEthernet0/0/0]undo traffic-filter outbound
// Delete ACL applied to the interface outbound
```

Task Completion

1. Task planning

According to the task description and the networking security requirements of Project 1, as well as the network topology shown in Fig. 1.1, plan a basic ACL and a high-level ACL on the "fenbu" router as shown in Table 1.25 to satisfy the security requirements of the project.

Table 1.25 ACL planning

Device Name	S/N	ACL Rules	Application Interface and Direction
fenbu	2000	Enable hosts with the source address 172.16.128.13; Disable all traffic	Inbound direction of VTY 0 4
fenbu	3000	Disable the traffic via the TCP protocol from any source to any destination at the ports 135 to 139, 445, and 3389; Disable the traffic via the UDP protocol from any source to any destination at the ports 135 to 139, 445, and 3389; Enable the traffic via the IP protocol from any source to any destination.	Inbound and outbound directions of the interfaces g0/0/0 and s2/0/0

2. Task implementation

(1) Configuration of basic ACL

According to Table 1.25, refer to the following steps to complete the basic ACL configuration on the "fenbu" router, so that only the management department PC3 can SSH to the "fenbu" router for remote management.

```
[fenbu]acl 2000
[fenbu-acl-basic-2000]rule permit source 172.16.128.13 0.0.0.0
[fenbu-acl-basic-2000]rule deny source any
//Create a basic ACL and permit the hosts with the source address of 172.16.128.13 only
[fenbu]user-interface vty 0 4
[fenbu-ui-vty0-4]acl 2000 inbound
[fenbu-ui-vty0-4]quit
//Apply ACL 2000 to the inbound traffic from VTY 0 to VTY 4
```

(2) Configuration of advanced ACL

According to Table 1.25, refer to the following steps to complete the configuration of advanced ACL on the "fenbu" router, so that the branch routers connected to the headquarters have both outbound and inbound ACLs introduced, with the ports 135 to 139, 445, and 3389 subject to the TCP and UDP protocols disabled.

```
[fenbu]acl 3000
```

```
[fenbu-acl-adv-3000]rule deny tcp source any destination any destinationport eq 135
[fenbu-acl-adv-3000]rule deny tcp source any destination any destinationport eq 136
[fenbu-acl-adv-3000]rule deny tcp source any destination any destinationport eq 137
[fenbu-acl-adv-3000]rule deny tcp source any destination any destinationport eq 138
[fenbu-acl-adv-3000]rule deny tcp source any destination any destinationport eq 139
[fenbu-acl-adv-3000]rule deny tcp source any destination any destinationport eq 445
[fenbu-acl-adv-3000]rule deny tcp source any destination any destinationport eq 3389
[fenbu-acl-adv-3000]rule deny udp source any destination any destinationport eq 135
[fenbu-acl-adv-3000]rule deny udp source any destination any destinationport eq 136
[fenbu-acl-adv-3000]rule deny udp source any destination any destinationport eq 137
[fenbu-acl-adv-3000]rule deny udp source any destination any destinationport eq 138
[fenbu-acl-adv-3000]rule deny udp source any destination any destinationport eq 139
[fenbu-acl-adv-3000]rule deny udp source any destination any destinationport eq 445
[fenbu-acl-adv-3000]rule deny udp source any destination any destinationport eq 3389
[fenbu-acl-adv-3000]rule permit ip
[fenbu-acl-adv-3000]quit
//Configure an advanced ACL and disable the ports 135-139, 445, and 3389 subject to
//TCP and UDP protocols
[fenbu]interface g0/0/0
[fenbu-GigabitEthernet0/0/0]traffic-filter inbound acl 3000
[fenbu-GigabitEthernet0/0/0]traffic-filter outbound acl 3000
[fenbu-GigabitEthernet0/0/0]quit
[fenbu]interface s2/0/0
[fenbu-Serial2/0/0]traffic-filter inbound acl 3000
[fenbu-Serial2/0/0]traffic-filter outbound acl 3000
[fenbu-Serial2/0/0]quit
[fenbu]
//Apply ACL 3000 to the "fenbu" router inbound and outbound
```

(3) ACL testing and viewing

① Run the SecureCRT virtual terminal software on the budget department host PC2 and the management department PC3 and PC4 to test whether only the management department PC3 can access the "fenbu" router via SSH. The budget department host PC2, the management department PC4 and other hosts cannot access the "fenbu" router via SSH.

② Use the "display acl 2000" command on the "fenbu" router to check the configuration and operation of the basic ACL 2000, as shown in Fig. 1.114.

③ Use the "display acl 3000" command on the "fenbu" router to check the configuration and operation of the advanced ACL 3000, as shown in Fig. 1.115.

```
[fenbu]display acl 2000
Basic ACL 2000, 2 rules
Acl's step is 5
 rule 5 permit source 172.16.128.13 0
 rule 10 deny
[fenbu]
```

Fig. 1.114 Configuration and operation of basic ACL 2000 via the "display acl 2000" command

```
[fenbu]display acl 3000
Advanced ACL 3000, 14 rules
Acl's step is 5
 rule 5 deny tcp destination-port eq 135
 rule 10 deny tcp destination-port eq 136
 rule 15 deny tcp destination-port eq 137
 rule 20 deny tcp destination-port eq 138
 rule 25 deny tcp destination-port eq 139
 rule 30 deny tcp destination-port eq 445
 rule 35 deny tcp destination-port eq 3389
 rule 40 deny udp destination-port eq 135
 rule 45 deny udp destination-port eq 136
 rule 50 deny udp destination-port eq netbios-ns
 rule 55 deny udp destination-port eq netbios-dgm
 rule 60 deny udp destination-port eq netbios-ssn
 rule 65 deny udp destination-port eq 445
 rule 70 deny udp destination-port eq 3389
[fenbu]
```

Fig. 1.115 Configuration and operation of advanced ACL 3000 via the "display acl 3000" command

④ Use the "display current-configuration | begin acl" command on the "fenbu" router to check ACL configurations, as shown in Fig. 1.116.

```
[fenbu]display current-configuration | begin acl
acl number 2000
 rule 5 permit source 172.16.128.13 0
 rule 10 deny
#
acl number 3000
 rule 5 deny tcp destination-port eq 135
 rule 10 deny tcp destination-port eq 136
 rule 15 deny tcp destination-port eq 137
 rule 20 deny tcp destination-port eq 138
 rule 25 deny tcp destination-port eq 139
 rule 30 deny tcp destination-port eq 445
 rule 35 deny tcp destination-port eq 3389
 rule 40 deny udp destination-port eq 135
 rule 45 deny udp destination-port eq 136
 rule 50 deny udp destination-port eq netbios-ns
 rule 55 deny udp destination-port eq netbios-dgm
 rule 60 deny udp destination-port eq netbios-ssn
 rule 65 deny udp destination-port eq 445
 rule 70 deny udp destination-port eq 3389
```

Fig. 1.116 Display results from the "display current-configuration | begin acl" command

If an incorrect ACL is configured, please delete it and reconfigure a correct ACL.

⑤ Use the "display traffic-filter applied-record" command on the "fenbu" router to check the ACL application information, as shown in Fig. 1.117.

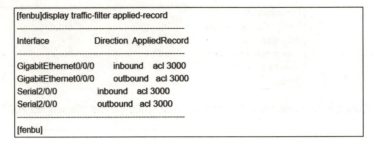

Fig. 1.117 Display results from the "display traffic-filter applied-record" command

(4) Backup of router configurations

```
<fenbu>save fenbu.cfg
//Save the current configuration and name it as "fenbu.cfg"
<fenbu>ftp 192.168.0.1
[fenbu-ftp]put fenbu.cfg
//Connect the router as an FTP client to the FTP server and upload the "fenbu.cfg"
//configuration file in the Flash to the FTP server
```

Exercises

Multiple choice

(1) The wildcard mask for 192.168.0.0/25 is (　　).

　　A. 255. 255. 255. 128　　　　　　　　B. 0. 0. 0. 127

　　C. 0. 0. 0. 1　　　　　　　　　　　　D. 255. 255. 255. 0

(2) (　　) can filter ICMP traffic.

　　A. Basic ACL　　　　　　　　　　　　B Extended ACL

(3) The basic ACL, when used for traffic filtering, shall be applied to the (　　) in the network as much as possible.

　　A. Destination　　　　　　　　　　　B. Source

(4) The advanced ACL, when used for traffic filtering, shall be applied to the (　　) in the network as much as possible.

　　A. Destination　　　　　　　　　　　B. Source

(5) Advanced ACLs are numbered within (　　).

　　A. 2000-2999　　B. 3000-2999　　C. 4000-4999　　D. 1000-1999

(6) Basic ACLs are numbered within (　　).

　　A. 2000-2999　　B. 3000-2999　　C. 4000-4999　　D. 1000-1999

Project 2

Small-sized Enterprise Internetworking Project

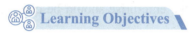
Learning Objectives

Knowledge Objectives

◎ Understand the network architecture of small-sized enterprise internetworking project.

◎ Understand the function and working process of VLAN, and understand the communication process of VLAN with single-arm routing and three-layer switch.

◎ Understand the working process of spanning tree.

◎ Understand the working process of PPP, PAP and CHAP authentication.

◎ Understand RIP working process.

◎ Understand the function, working process and classification of NAT.

Competence Objectives

◎ It is required to have the ability to independently deploy and implement a small enterprise internetworking project, including switch configuration and management, VLAN configuration, spanning tree configuration, communication of VLAN configuration, DHCP configuration, RIP configuration, PPP configuration and NAT configuration.

Quality Objectives

◎ Competent for teamwork.

◎ Competent for problem analysis and solutions.

Project Description

A small enterprise (located in a four-story building) plans to set up a network as shown in Fig. 2.1. The network adopts a two-story network architecture of access layer and core layer. There are four floor switches in the network topology diagram, in which the access switches on the first, second and third floors are connected to the core switch, which is connected to the export router AR of the enterprise, and the floor switch on the fourth floor is directly connected to the export router AR, and servers such as the enterprise portal website are directly connected to the core switch. The exit router uses a serial link to connect to the ISP router on the Internet service provider.

The enterprise has four departments: R&D, sales, after-sales and human resources. The sales department is located on the first and second floors, the after-sales department is located on the second and third floors, the R&D department is located on the third floor, and the business leaders and human resources department are located on the fourth floor. Digital cameras are installed on each floor to realize monitoring.

Enterprise network deployment requirements are as follows:

◎ The business data between different departments of the enterprise should be completely isolated.

◎ Intranets need to be interconnected.

◎ The enterprise only applies for a public address, 125.71.28.93, and the enterprise intranet is addressed by a private address. NAT is deployed on the exit router, so that all hosts in the enterprise intranet can access the Internet, and hosts on the Internet can access the enterprise external portal server.

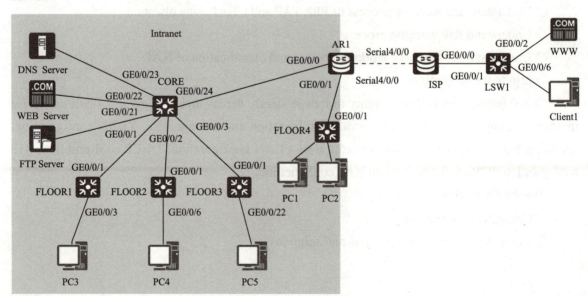

Fig. 2.1 Topology diagram of small-sized enterprise internetworking project

Basic Configuration and Management of Switches

Task Description

Set up a network topology as shown in Fig. 2.1 in the Huawei ENSP simulation environment or using Huawei's routers and switches. Complete the planning of the network IP address, test the IP configuration of the host PC and server, switch host name configuration, Console port configuration and SSH configuration, and carry out network verification and testing.

Task Objectives

- Understand the working principle of switches and the forwarding method of switches.
- The ability to initialize the configuration of switches.
- The ability to backup switch configuration files.
- The ability to clear the switch console password.

Related Knowledge

1. Working principle of switch

Ethernet switch is the most important networking equipment in Ethernet. In the Ethernet switch, it is necessary to maintain a MAC address table. This table gives information about MAC addresses of hosts connected to different ports of the switch. The contents of the MAC table can be added manually by the administrator or established by learning the source MAC address in the data frame received by the switch. Fig. 2.2 shows an example of MAC table of Huawei switch.

```
[SW1]display mac-address
MAC address table of slot 0:

MAC Address     VLAN/    PEVLAN CEVLAN Port      Type      LSP/LSR-ID
                VSI/SI                                     MAC-Tunnel

5489-981a-5f43   3        -      -      GE0/0/20  dynamic   0/-
00e0-fce7-2e5d   3        -      -      GE0/0/1   dynamic   0/-
00e0-fce7-2e5d   2        -      -      GE0/0/1   dynamic   0/-
5489-988e-307e   2        -      -      GE0/0/2   dynamic   0/-

Total matching items on slot 0 displayed = 4
```

Fig. 2.2 Example of MAC table of Huawei switch

The MAC table of a switch usually includes fields such as MAC Address, VLAN, Port and Type. Wherein, the type field indicates how the switch obtains the MAC address and the corresponding port entry. When the MAC address entry is dynamically learned by the switch, its type value is "dynamic"; When the MAC address is manually and statically specified by the

administrator, its type value is "static". The default aging time of the MAC entry dynamically learned by the switch is 300 s. If a MAC entry has not been refreshed before the aging time expires, the MAC entry will be deleted from the MAC table. MAC table entries statically configured by administrators are not affected by address aging time.

When the switch is just started, the MAC table of the switch is empty. The switch establishes a MAC table by learning the source MAC address in the received data frame, and makes a forwarding decision according to the destination MAC address in the data frame.

When a switch receives a data frame from a port, it checks the source MAC address of the data frame, and if the source MAC address does not exist in the MAC table, it adds it to the MAC table. Where VLAN is the VLAN to which the corresponding port belongs, and the port is the switch port that received the data frame. If the source MAC address exists in the MAC table, its aging time is refreshed. According to the destination MAC address in the received data frame, the switch looks up the MAC table and makes a forwarding decision according to the following rules:

① If the destination MAC address of a data frame is a multicast address or a broadcast address, the data frame is flooding, that is, the frame is forwarded to all other switch ports except the source port that received the data frame.

② If the destination MAC address of a data frame is a unicast address, but the destination MAC address does not exist in the MAC table, the data frame is also flooded.

③ If the destination MAC address of a data frame is a unicast address and the destination MAC address and the source MAC address correspond to the same port of the switch, the frame is not forwarded.

④ If the destination MAC address of a data frame is a unicast address and the destination MAC address and the source MAC address correspond to different ports of the switch, the frame is forwarded from the switch port corresponding to the destination MAC address.

2. Forwarding mode of switch

Switches forward data frames in two ways: store-and-forward and cut-through. Besides, the cut-through mode is further divided into two modes: fast-forward and fragment-free switching, as shown in Fig. 2.3.

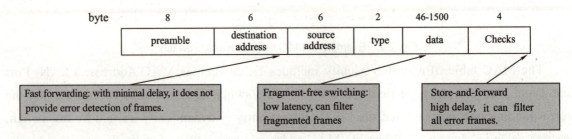

Fig. 2.3 Comparison of forwarding modes of switches

(1) Store-and-forward

Store-and-forward means that after receiving a complete data frame, the switch performs CRC(cyclic redundancy check) check, and then makes a forwarding decision according to the destination MAC address in the data frame header after confirming that the data frame is error-free.

(2) Cut-through

Cut-through means that the switch makes a forwarding decision according to the destination MAC address in the data frame only when it receives the destination MAC address of the data frame (at this time, the complete data frame is not received). Cut-through switching is further divided into two ways: fast forwarding and fragmentation-free switching. Among them, the fast forwarding mode means that as long as the switch detects the destination MAC address in the data frame, it immediately looks up the MAC table to make a forwarding decision and forward it; However, the fragmentation-free switching method requires that the received data frame must be larger than the minimum frame length (64 bytes) and then forwarded, and any data frame with a length less than 64 bytes will be discarded immediately.

3. Classification of switches

Switches have many different classification standards. Common classifications of switches are:

(1) Fixed-port switch and modular switch

The ports of fixed-port switches are fixed and cannot be expanded. The number of ports of fixed-port switches is usually 8 ports, 16 ports, 24 ports and 48 ports. The modular switch is equipped with additional open slots, and the number of ports of the switch can be expanded by inserting modules. Users can configure modules with different numbers, different rates and different interface types to meet the needs of different networks. Modular switches generally have strong fault tolerance, support redundant switching modules, and support hot-swappable dual power supplies. Modular switch has greater flexibility and scalability, but its price is much more expensive than fixed-port switch, and it is generally used in the core layer and convergence layer in large networks. Fig. 2.4 shows the shapes of Huawei fixed-port switch and modular switch.

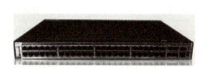
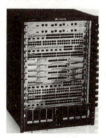

(a) Fixed-port switch   (b) Modular switch

Fig. 2.4 Shapes of Huawei fixed-port and modular switch Switch

(2) Stackable switch and non-stackable switch

Stacking technology is mainly to increase the port density of switches. When the number of ports of a single switch cannot meet the networking requirements, stacking switches can be considered. A plurality of switches stacked together become a switching device logically, and participate in data forwarding as a whole, which simplifies network networking and improves network reliability.

4. Two-layer switch and three-layer switch

According to the function of switches corresponding to OSI reference model, switches can be divided into two-layer switch, three-layer switch and four-layer switch. Two-layer switch works in the second layer of OSI reference model (namely data link layer), and forwards and filters data according to the destination MAC address of the data frame; the three-layer switch works in the third layer (i.e. the network layer) of the OSI reference model. It can not only make forwarding decisions according to the destination MAC address information of the data frame, but also according to the third layer address in the packet (such as IP address) to make forwarding decisions. The three-layer switch has the routing function, and it is usually chosen to use the three-layer switch to realize the routing between different network segments within the enterprise network.

5. Access method and command view of the switch

You can access the switch through Console port, Telnet, SSH, Web and other methods. Using the Console port to access the switch is usually used for the initial configuration and management of the switch, as well as monitoring the state of the switch and some catastrophic recovery work. Its physical connection and access method are similar to the Console port access router.

Similar to routers, switches provide a variety of different command views to satisfy users with different rights to perform different access functions. Table 2.1 shows the main command views and prompts of the switch.

Table 2.1 Main command view and prompt of the switch

Command view	Prompt example	View switching example
User view	<Huawei>	Default view after router startup
System view	[Huawei]	<Huawei>system-view
Interface view	[Huawei-GigabitEthernet0/0/0]	[Huawei]interface G0/0/0
VLAN view	[Huawei-vlan1]	[Huawei]vlan 1
User interface view	[Huawei-ui-console0]	[Huawei]user-interface console 0

Related Skills

1. Initialization configuration of switch

You should build a topology diagram as shown in Fig. 2.5. According to the topology diagram

and IP planning shown in Table 2.2, you should complete the configuration of switch host name, switch management IP address and default gateway, switch Console port, Telnet and SSH in the network, and conduct network tests to have the basic configuration ability of switches.

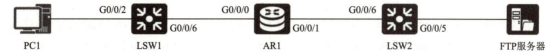

Fig. 2.5 Topology diagram of "switch basic configuration"

Table 2.2 IP planning

Device	Interface	IP address/Subnet mask	Gateway
LSW1	Vlanif1	192.168.0.1/24	192.168.0.254
LSW2	Vlanif1	192.168.1.1/24	192.168.1.254
AR1	G0/0/0	192.168.0.254/24	
	G0/0/1	192.168.1.254/24	
FTP server	NIC	192.168.1.2/24	192.168.1.254
PC1	NIC	192.168.0.2/24	192.168.0.254

(1) Basic configuration of switch

① You should connect the USB port of the host PC1 to the Console port of the switch with a USB serial port, the network card of the host PC1 to the G0/0/2 interface of the switch with a UTP direct line, and the network card of the FTP server to the G0/0/5 interface of the switch.

② On the host PC1, you should click the this PC icon on the desktop, and in the shortcut menu that pops up, you should select "Administration" → "Device Manager" → "Ports (COM and LPT)", and observe and record the port number of "USB-to-Serial Comm Port".

③ On the host PC1, you should run the SecureCRT software, select the protocol of "File" → "Quick Connection" →"Serial", the recorded port number of USB to serial port, "9600" as the baud rate value, and the data bit is 8 bits. There is no parity, and the stop bit is 1 bit. You should restore the three check boxes under Flow Control, and click the "Connect" button to enter the command view of the switch.

④ On the switch command line, with the reference to the following steps, you should complete the configuration of the host name, switch management IP address and default gateway of the switch LSW1:

```
<Huawei>sys                        //Enter the system view
[Huawei]sysname LSW1               //Configure the hostname of the switch as "LSW1"
[LSW1]interface Vlanif 1           //Create and enter the VLAN1 interface view
[LSW1-Vlanif1]ip address 192.168.0.1 24
Configure the IP address of VLAN1 interface as "192.168.0.1", The subnet mask length is "24"
```

```
[LSW1-Vlanif1]undo shutdown        //Enable the interface
[LSW1-Vlanif1]quit                 //Return to system view
[LSW1]ip route-static 0.0.0.0 0.0.0.0 192.168.0.254    //Configure the default gateway
//address as 192.168.0.254
```

⑤ On the switch command line, with the reference to the following steps, you should complete the configuration of the host name, switch management IP address and default gateway of the switch LSW2:

```
<Huawei>sys                        //Enter the system view
[Huawei]sysname LSW2               //Configure the hostname of the switch as "LSW2"
[LSW2]interface Vlanif 1           //Create and enter the VLAN1 interface view
[LSW2-Vlanif1]ip address 192.168.1.1 24  //Configure the IP address of VLAN1 interface
                                   //as "192.168.1.1",and the subnet mask length is"24"
[LSW2-Vlanif1]undo shutdown        //Enable the interface
[LSW2-Vlanif1]quit                 //Return to system view
[LSW2]ip route-static 0.0.0.0 0.0.0.0 192.168.1.254    //Configure the default gateway
                                                       //address as 192.168.1.254
```

⑥ On the switches LSW1 and LSW2, respectively, you should use the "display ip interface brief" command to check the brief information of the interface IP, and observe whether the IP address and mask of the interface are correct and whether the physical and protocol status of the interface is "UP". Fig. 2.6 is an example of interface IP brief information displayed by using the "display ip interface brief" command on switch LSW1.

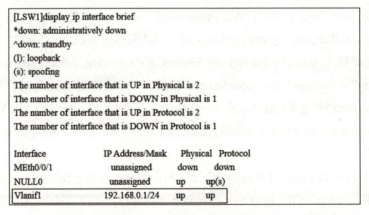

Fig. 2.6 Brief information of interface IP displayed on switch LSW1

⑦ On switches LSW1 and LSW2 respectively, you should use the "display ip routing-table" command to view the routing table information. Fig. 2.7 shows the routing table information displayed on switch LSW1.

Note: The default routing next hop address is "192.168.0.254".

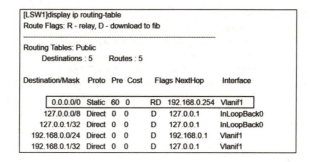

Fig. 2.7 Routing table information displayed on switch LSW1

⑧ According to the IP planning in Table 2.2, you should complete the configuration of router hostname, interface IP address and subnet mask on router AR1 with reference to the following steps:

```
<Huawei>sys
[Huawei]sysname AR1
//Configure the hostname of the router
[AR1]interface g0/0/0
[AR1-GigabitEthernet0/0/0]ip address 192.168.0.254 24
[AR1-GigabitEthernet0/0/0]quit
[AR1]
//Configure the IP address and subnet mask of G0/0/0 interface of router AR1
[AR1]interface g0/0/1
[AR1-GigabitEthernet0/0/1]ip address 192.168.1.254 24
[AR1-GigabitEthernet0/0/1]quit
[AR1]
//Configure the IP address and subnet mask of G0/0/1 interface of router AR1
```

⑨ You should complete the configuration of IP address and subnet mask between host PC1 and FTP server, and use ping utility to test the IP connectivity between host PC1, switch LSW1, switch LSW2 and FTP server. At this point, they can ping each other. Fig. 2.8 shows that switch LSW1 successfully ping the IP address of switch LSW2.

(2) Configuration of Console

The configuration steps of the switch Console are similar to those of the router Console. The tasks of configuring the control port with AAA authentication are as follows.

① Configuring legal AAA users for Console login, including configuring legal user name and password, service type of authorized users as terminal, and authority level of users.

② Entering the Console user interface view and configuring the authentication mode to AAA.

```
[LSW1]ping 192.168.1.1
  PING 192.168.1.1: 56  data bytes, press CTRL_C to break
    Reply from 192.168.1.1: bytes=56 Sequence=1 ttl=254 time=50 ms
    Reply from 192.168.1.1: bytes=56 Sequence=2 ttl=254 time=50 ms
    Reply from 192.168.1.1: bytes=56 Sequence=3 ttl=254 time=50 ms
    Reply from 192.168.1.1: bytes=56 Sequence=4 ttl=254 time=40 ms
    Reply from 192.168.1.1: bytes=56 Sequence=5 ttl=254 time=70 ms

  --- 192.168.1.1 ping statistics ---
  5 packet(s) transmitted
  5 packet(s) received
  0.00% packet loss
  round-trip min/avg/max = 40/52/70 ms
```

Fig. 2.8　Switch LSW1 successfully ping the IP address of switch LSW2

Completing the configuration of the Console on the switch LSW1 as follows:

```
[LSW1]aaa
[LSW1-aaa]local-user cdp password cipher cdp123456
[LSW1-aaa]local-user cdp service-type terminal
[LSW1-aaa]local-user cdp privilege level 15
[LSW1-aaa]quit
//Create a user for Console login, and configure the user's username as cdp, password
//as cdp123456, user's service type as terminal and privilege level as 15
[LSW1]user-interface console0
[LSW1-ui-console0]authentication-mode aaa
[LSW1-ui-console0]quit
[LSW1]quit
<LSW1>
//Configure the authentication mode of Console0 to use AAA for authentication
```

After completing the Console configuration on the switch LSW1, you should use the "quit" command to exit the user view, and the prompt "Please Press ENTER" will appear. And then you should press the Enter key, the login authentication prompt will appear and enter the legitimate Console user name cdp and the corresponding password cdp123456, and log in and enter the user view, as shown in Fig. 2.9.

```
Please Press ENTER.

Login authentication

Username:cdp
Password:
<LSW1>
```

Fig. 2.9　Login authentication test of switch LSW1 Console

(3) Configuration and testing of Telnet

The configuration of the switch Console is similar to that of the router Console. Telnet configuration on that switch include the following task:

① Enable Telnet.

② Configuring legal users for Telnet login, including entering AAA view, creating user name and password, configuring user's service type as Telnet, and configuring user's privilege level.

③ Entering the VTY user interface view, and the configuration protocol is Telnet, and the authentication mode is AAA.

Refer to the following steps to complete the configuration of Telnet on switch LSW1.

```
[LSW1]aaa
[LSW1-aaa]local-user cdptelnet password cipher cdp123456
[LSW1-aaa]local-user cdptelnet service-type telnet
[LSW1-aaa]local-user cdptelnet privilege level 15
[LSW1-aaa]quit
//Configure the user name for Telnet login, the user name is "cdptelnet", the password
//is "cdp123456", the user service type is "Telnet", and the privilege level is "15"
[LSW1]telnet server enable
//Enable Telnet service
[LSW1]user-interface vty 0 4
[LSW1-ui-vty0-4]protocol inbound telnet
[LSW1-ui-vty0-4]authentication-mode aaa
[LSW1-ui-vty0-4]quit
//The protocol for configuring VTY 0 4 is Telnet, and the authentication mode is AAA authentication
```

After completing the Telnet configuration on the switch, you should use the Telnet client software such as PUTTY or SecureCRT on the host PC1, and use the Telnet protocol to remotely log in and access the switch LSW1 (its management address is "192.168.0.1"). You can also use Telnet command to remotely log in and access switch LSW1 on router AR1 and switch LSW2. Fig. 2.10 shows the remote login to switch LSW1 using Telnet on switch LSW2.

```
<LSW2>telnet 192.168.0.1
Trying 192.168.0.1 ...
Press CTRL+K to abort
Connected to 192.168.0.1 ...

Login authentication

Username:cdptelnet
Password:
Info: The max number of VTY users is 5, and the number
    of current VTY users on line is 1.
    The current login time is 2022-11-06 10:33:32.
<LSW1>
```

Fig. 2.10 Telnet to switch LSW1 remotely on switch LSW2

(4) Configuration and testing of SSH

Not all switches support SSH. Switches must have encryption features and operating software version support to configure SSH, and different models of switches support different algorithms for generating local key pairs. Generally, configuring SSH login with local username and password on Huawei switch needs to be completed:

① Enable Stelnet service.

② Configuring a legal user for Stelnet login, including entering AAA view, creating a user name and password, configuring the user's service type as Stelnet, configuring the user's privilege level and other steps.

③ Configuring SSH users, service types and SSH authentication types.

④ Entering the VTY user interface view, with SSH as the configuration protocol and AAA as the authentication mode.

⑤ Generating a local key pair.

Corresponding to the above steps, the following is the SSH configuration and verification command taking Huawei S5700-28C-HI switch as an example:

```
[LSW1]stelnet server enable
//Enable the Stelnet service
[LSW1]aaa
[LSW1-aaa]local-user cdpssh password cipher cdp123456
[LSW1-aaa]local-user cdpssh service-type ssh
[LSW1-aaa]local-user cdpssh privilege level 15
[LSW1-aaa]quit
//Configure the user name used for SSH login. The user name is "cdpssh", the password
//is "cdp123456", the user service type is "SSH", and the privilege level is "15"
[LSW1]ssh user cdpssh
[LSW1]ssh user cdpssh service-type stelnet
[LSW1]ssh user cdpssh authentication-type password
//The user configuring SSH is cdpssh, the service type is Stelnet, and password
//authentication is used
[LSW1]user-interface vty 0 4
[LSW1-ui-vty0-4]protocol inbound ssh
[LSW1-ui-vty0-4]authentication-mode aaa
[LSW1-ui-vty0-4]quit
//Configure the protocol of VTY 0 4 as SSH and the authentication mode as AAA authentication
```

Finally, you should use the "rsa local-key-pair create" command to generate a local key pair, as shown in Fig. 2.11.

```
[LSW1]rsa local-key-pair create
The key name will be: LSW1_Host
The range of public key size is (512 ~ 2048).
NOTES: If the key modulus is greater than 512,
       it will take a few minutes.
Input the bits in the modulus[default = 512]:  ←——— Press Enter to use the default value.
Generating keys...
.........................++++++++++++
.........................++++++++++++
..................++++++++
........................................++++++++

[LSW1]
```

Fig. 2.11   Create a local key pair using RSA algorithm

After SSH configuration is completed on the switch, SSH client software such as PUTTY or SecureCRT can be used on the host PC to remotely log in and access the switch LSW1 (its management address is "192.168.0.1") using SSH protocol. You can also use the "stelnet" command on router AR1 or switch LSW2 to login to switch LSW1 remotely.

Note: Before using the "stelnet" command on a router or switch, you need to use "ssh client first-time enable" to enable the first authentication of SSH clients.

Fig. 2.12 shows the first remote access to switch LSW1 by using stelnet on switch LSW2.

```
[LSW2]ssh client first-time enable
[LSW2]stelnet 192.168.0.1
Please input the username:cdpssh
Trying 192.168.0.1 ...
Press CTRL+K to abort
Connected to 192.168.0.1 ...
The server is not authenticated. Continue to access it? [Y/N] :y
Save the server's public key? [Y/N] :y
The server's public key will be saved with the name 192.168.0.1. Please wait...
Enter password:
Info: The max number of VTY users is 5, and the number
      of current VTY users on line is 1.
      The current login time is 2022-11-06 11:56:30.
<LSW1>
```

Fig. 2.12   The example for the first time to remotely access the switch LSW1 by using stelnet on the switch LSW2

2. Management of switches

After completing the "basic configuration of switches", you should back up the configuration files of switches LSW1 and LSW2 in the network shown in Fig. 2.5 to Flash and FTP servers, and be familiar with the operation of clearing the password of switch Console.

(1) Switch file system management

① You need to use the "dir" command on switches LSW1 and LSW2 to view the list of files

in the current directory. Fig. 2.13 shows an example of viewing the file list in the current directory on switch LSW1.

```
<LSW1>dir
Directory of flash:/

  Idx  Attr    Size(Byte)  Date            Time       FileName
    0  drw-         -      Aug 06 2015  21:26:42    src
    1  drw-         -      Feb 15 2023  21:46:14    compatible

32,004 KB total (31,972 KB free)

<LSW1>
```

Fig. 2.13 Example of viewing the file list in the current directory on switch LSW1

② You need to use the "save" command on the switch LSW1 to save the current configuration and name it "LSW1", and then use the "dir" command to check whether there is a saved configuration file in Flash, as shown in Fig. 2.14.

Note: The configuration file must have the extension ".cfg" or ".zip".

```
<LSW1>save LSW1cfg.zip
Are you sure to save the configuration to flash:/LSW1cfg.zip?[Y/N]:Y
Now saving the current configuration to the slot 0.
Save the configuration successfully.
<LSW1>dir
Directory of flash:/

  Idx  Attr    Size(Byte)  Date            Time       FileName
    0  drw-         -      Aug 06 2015  21:26:42    src
    1  drw-         -      Feb 15 2023  21:46:14    compatible
    2  -rw-       452      Feb 15 2023  21:59:07    LSW1cfg.zip

32,004 KB total (31,968 KB free)

<LSW1>
```

Fig. 2.14 The sample of saving and naming the configuration file

③ You need to use the "startup saved-configuration" command to set the configuration file used for the next startup of the switch, as follows:

```
<LSW1>startup saved-configuration?            //check "startup saved-configuration" command
                                              //keywords or parameters
<LSW1>startup saved-configuration LSW1cfg.zip  //Configure the configuration file to be
                                              //used for the next startup is "LSW1cfg.zip"
<LSW1>display startup    //Display the operating system, configuration file and other
                         //information of the equipment starting this time and next time
```

④ According to the IP planning shown in Table 2.2, you should complete the IP configuration of FTP server and the construction and configuration of FTP service. Configure and input "ftp 192.168.1.2" on the switch LSW1, so that the switch can be connected to the FTP server as an FTP

client, and upload the "LSW1cfg.zip" configuration file in Flash to the FTP server, as shown in Fig. 2.15.

```
<LSW1>ftp 192.168.1.2
Trying 192.168.1.2 ...
Press CTRL+K to abort
Connected to 192.168.1.2.
220 FtpServerTry FtpD for free
User(192.168.1.2:(none)):anonymous    ← Type an anonymous account "anonymous"
331 Password required for anonymous .
Enter password:  ← Type any e-mail address as the password.
230 User anonymous logged in , proceed

[ftp]put LSW1cfg.zip
200 Port command okay.
150 Opening BINARY data connection for LSW1cfg.zip

100%
226 Transfer finished successfully. Data connection closed.
FTP: 681 byte(s) sent in 0.290 second(s) 2.34Kbyte(s)/sec.
```

Fig. 2.15 Upload the configuration file in the Flash of the switch to the FTP server

⑤ After completing the backup of files, you should enter the "reset saved-configuration" command to clear the configuration of the switch, and then enter the "reboot" command to restart the switch, as shown in Fig. 2.16.

```
<LSW1>reset saved-configuration
Warning: The action will delete the saved configuration in the device.
The configuration will be erased to reconfigure. Continue? [Y/N]:y
Warning: Now clearing the configuration in the device.
Feb 15 2023 22:17:55-08:00 LSW1 %%01CFM/4/RST_CFG(l)[0]:The user chose Y when de
ciding whether to reset the saved configuration.
Info: Succeeded in clearing the configuration in the device.
<LSW1>reboot
Info: The system is now comparing the configuration, please wait.
Warning: All the configuration will be saved to the configuration file for the n
ext startup:, Continue?[Y/N]:n
Info: If want to reboot with saving diagnostic information, input 'N' and then e
xecute 'reboot save diagnostic-information'.
System will reboot! Continue?[Y/N]:y
```

Fig. 2.16 Clearing the switch configuration and restart the switch

(2) Clearing the switch Console password

Taking Huawei switch FutureMatrix S5736-S24T4XC as an example, the steps to clear the password of the switch Console are as follows:

① You should use the control line to connect the Console with the console port of the switch to ensure the normal connection. Then, you need to power off the switch and power on the switch again, and observe the startup process of the switch. When the "Press Ctrl+B or Ctrl+E to Enter

BootLoad Menu" appears as shown in Fig. 2.17, you need to press the shortcut key 【Ctrl+B】 or 【Ctrl+E】 in time (within 3 s) to enter the Bootload menu.

```
Last reset type: Watchdog
Press Ctrl+B or Ctrl+E to enter BootLoad menu: 1
Info: The password is empty. For security purposes, change the password.
New password: ←——— 1. Set BootLoad password
Verify: ←——— 2. Confirm BootLoad password
Modify password ok.
      BootLoad Menu

  1. Boot with default mode
  2. Enter startup submenu
  3. Enter ethernet submenu
  4. Enter filesystem submenu
  5. Enter password submenu
  6. Clear password for console user
  7. Reboot
  (Press Ctrl+E to enter diag menu)

Enter your choice(1-7):6  ←——— 3. Type "6" and choose "clear password for console user"
Note: Clear password for console user? Yes or No(Y/N): Y ←——— 4. Type "Y", and confirm to clear
Clear password for console user successfully.                         password for console user
Note: Choose "1. Boot with default mode" to boot, then set a new password.

Note: If the device is restarted during startup, you need to perform this operation again.

      BootLoad Menu

  1. Boot with default mode
  2. Enter startup submenu
  3. Enter ethernet submenu
  4. Enter filesystem submenu
  5. Enter password submenu
  6. Clear password for console user
  7. Reboot
  (Press Ctrl+E to enter diag menu)

Enter your choice(1-7): 1  ←——— 5. Type "1" and continue
```

Fig. 2.17 Steps for the switch to clear the Console user password

② You should type the new BootLoad password at "New Password", enter the set BootLoad password again at "Verify".

Note: When the FutureMatrix S5736-S24T4XC of Huawei switch enters the BootLoad for the first time, it is necessary to set a new BootLoad password.

③ You should type "6" under the "BootLoader" menu and select "Clear password for Console user" to clear the console user password.

④ In the "Note: clear password for console user? Yes or No(Y/N) " input "Y",and confirm to clear the Console user password.

⑤ After clearing the Console user password successfully, You should type "1" in the next "BootLoader" menu and select "1. Boot with default mode" to start directly from the current stage. After startup, you should directly set the new password of the Console to enter the user system mode.

Task Completion

1. Task planning

According to the task description and the networking requirements of Project 2, plan the Console ports and SSH of all access and core layer switches for the enterprise intranet in the network topology as shown in Fig. 2.1 as follows:

① Console port adopts AAA authentication, user name is "cdp", password is "cdp123456", and user command line level is "15".

② SSH authentication plan is: local user authentication, user name is "cdp", password is "cdp123456" and user command line level is "15".

2. Task implementation

(1) Basic configuration of switch

① You should complete the configuration of host name, Console login authentication and SSH authentication on the access layer switch "FLOOR1". The configuration steps are as follows:

```
[Huawei]sysname FLOOR1
[FLOOR1]stelnet server enable
//Enable the Stelnet service
[FLOOR1]aaa
[FLOOR1-aaa]local-user cdp password cipher cdp123456
[FLOOR1-aaa]local-user cdp service-type terminal ssh
[FLOOR1-aaa]local-user cdp privilege level 15
[FLOOR1-aaa]quit
//Create a user for Console and SSH login, and configure the user's username as "cdp", password
//as "cdp123456", user's service types as "terminal" and "SSH", and privilege level as "15"
[FLOOR1]user-interface console 0
[FLOOR1-ui-console0]authentication-mode aaa
[FLOOR1-ui-console0]quit
//Configure the authentication mode of Console0 to use AAA for authentication
[FLOOR1]ssh user cdp
[FLOOR1]ssh user cdp service-type stelnet
[FLOOR1]ssh user cdp authentication-type password
//Configure SSH with user name "cdp" and service type "stelnet", and use password
//authentication
[FLOOR1]user-interface vty 0 4
[FLOOR1-ui-vty0-4]quit
[FLOOR1-ui-vty0-4]authentication-mode aaa
[FLOOR1-ui-vty0-4]protocol inbound ssh
//Configure VTY 0 4 with SSH protocol and AAA authentication mode
```

```
[FLOOR1]rsa local-key-pair create
//Generate a local key pair
[FLOOR1]quit
<FLOOR1>save
//Save the configuration
```

② Refer to the above steps to complete the configuration of host names, Console login authentication and SSH authentication of all switches (CORE, FLOOR2, FLOOR3 and FLOOR4) in the enterprise intranet in Figure 2.1.

(2) Console configuration verification

You should exit the user view on all switches in the intranet, press Enter to reconnect the switches through the Console, and verify whether the login authentication prompt is required. You should enter the legitimate Console user name "cdp" and the corresponding password "cdp123456" to enter the user view of the switch.

Note: Since all switches in the enterprise intranet have not finished configuring the management IP address at this time, it is necessary to wait until the "Complete Task" in Task 3 of this project to verify the SSH function configured by the switches.

Exercises

Multiple choice

(1) Two-layer switches is based on (　　) make a forwarding decision.

  A. source MAC address   B. destination MAC address

  C. source IP address   D. destination IP address

(2) (　　) command is used to display the operating system that the device starts this time and next time, configuration files and other information.

  A. Display startup   B. Display current

  C. Display interface brief   D. Display version

(3) When the destination MAC address of the data frame received by the switch is a unicast address, but the MAC address does not exist in the MAC table of the switch, the switch passes (　　) forward the data frame.

  A. discarding the data frame

  B. flooding

  C. sending out from all ports of the switch

  D. sending out from a port of the switch

(4) The following switch forwarding methods have the greatest delay is (　　).

  A. fragmentation-free exchange   B. fast forwarding

  C. store and forward   D. direct switching

Task 2 | Planning and Configuration of VLAN

Task Description

On the basis of completing Task 1 of this project, according to the network topology diagram shown in Fig. 2.1, complete the planning and configuration of VLAN on the enterprise intranet switches, and achieve the networking requirement of complete isolation of business data flows between different departments in the enterprise intranet.

Task Objectives

- Understanding how VLAN works.
- Being capable of VLAN configuration and management.
- Have the troubleshooting ability of VLAN.

Related Knowledge

1. The VLAN overview

Ethernet has experienced the development from shared Ethernet to switched Ethernet. At present, the mainstream campus network is switched Ethernet. Because the switch can divide the collision domain, the performance of the switched Ethernet network is greatly improved. However, according to the working principle of the switch, the switch cannot divide the broadcast domain, and all hosts in the local area network can receive the multicast or broadcast frame sent by any host in the switched network, which may lead to a large amount of broadcast traffic in the network and a broadcast storm phenomenon. In the network shown in Fig. 2.18, the host PC1 sends out a DHCP DISCOVER broadcast frame, which can be received by all hosts in the network.

In order to solve the problems brought by broadcast domain, VLAN(virtual local area network) technology is introduced. VLAN is a technology to form virtual workgroups or logical network segments for devices or users according to functions, departments and applications through switch software in LAN switches.

A VLAN is a broadcast domain, and nodes belonging to the same VLAN can communicate with each other directly, while nodes of different VLANs need to communicate through three-layers of equipment. VLAN can be logically divided according to users' needs without being limited by the physical location of network users, which can significantly reduce the network management overhead when users are added, deleted or moved. Using VLAN to divide logical network segments is shown in Fig. 2.19.

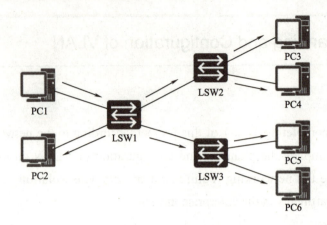

Fig. 2.18　Problems of traditional Ethernet

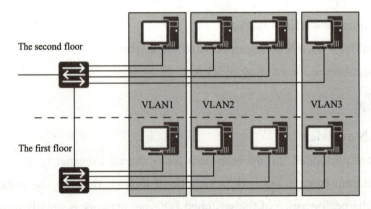

Fig. 2.19　Using VLAN to divide logical network segments

2. Types of VLAN

(1) Data VLAN

Data VLAN refers to the VLAN used to transmit all kinds of user data traffic, sometimes called "user VLAN".

(2) Voice VLAN

The communication traffic generated by IP voice terminals such as IP telephone and IP soft switch is called VoIP traffic. VoIP traffic requires high real-time transmission, so it usually needs to be set as a high priority in the network QoS policy, so that relevant network devices can transmit voice communication traffic first. Therefore, when deploying voice over IP, VoIP traffic is usually placed in an independent VLAN, which is called voice VLAN.

(3) Default VLAN

The default VLAN is also called PVID (port default VLAN ID). When the switch interface receives an untagged frame, the switch adds a tag equal to the PVID to the data frame according to the PVID, and then submits it to the switch for internal processing. When the switch interface sends a data frame, if the VID value of the Tag of this data frame is found to be the same as the PVID, the

switch will remove the Tag and send it out from this interface. Each interface has a default VLAN. By default, the default VLAN of all interfaces is VLAN1, but users can configure it according to their needs. VLAN1 cannot be deleted.

(4) Management VLAN

A management VLAN is a VLAN configured to access switch management functions. By assigning IP address, subnet mask and gateway address to the management VLAN, users can remotely manage the switch through IP network based on protocols such as HTTP, Telnet, SSH or SNMP. By default, VLAN1 is designated as the management VLAN. However, for the sake of safety, it is recommended to configure other VLANs except VLAN1 as the management VLAN.

3. Trunk protocal

If multiple VLANs are divided in a network with multiple switches, and the links between switches may carry data from multiple different VLANs, it is necessary to configure physical lines carrying data from multiple VLANs as trunk links and allow the required VLAN data to pass through. In the switching network as shown in Fig. 2.20, the link between switches LSW1 and LSW2 is a trunk link, allowing data frames of VLAN2 and VLAN3 to be transmitted on this link.

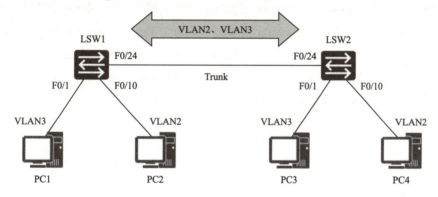

Fig. 2.20 The example of relay

Because the trunk link carries different data from multiple VLANs, there needs to be a mechanism on the switch to enable the switch to identify which VLAN the data frame on the trunk link comes from, so as to forward it correctly. Usually, a tag is inserted into the header of an Ethernet frame to identify which VLAN the data frame in the relay link belongs to. The most commonly used VLAN trunk protocol is IEEE 802.1Q, which is an open encapsulation protocol and supported by all manufacturers. The frame format of IEEE 802.1Q is shown in Fig. 2.21, which is equivalent to adding 4 bytes to the standard Ethernet frame header to become a frame with VLAN tag. Of the 4 bytes added, 2 bytes are tag protocol indentifier (TPID) and 2 bytes are tag control information (TCI). In the IEEE 802.1Q frame structure, the meaning of VLAN tag field is as follows:

8 bytes	6 bytes	6 bytes	2 bytes	2 bytes	2 bytes	46-1500 bytes	4 bytes
Preamble	Destination MAC address	Source MAC address	Marking protocol flag	Label control information	type	data	Frame check sequence

Fig. 2.21　Frame format of IEEE 802.1Q

① Tagging protocol identifier (TPID): When the length is 2 bytes and the value is "0x8100", it means an IEEE 802.1Q frame.

② Tag control information field: 2 bytes long, containing the control information of the frame. It includes 3 bit priority (PRI), 1 bit standard format indicator (CFI) and 12 bit VLAN identifier (VLAN ID). Among them, PRI is used to identify the priority of frames, mainly used for QoS. In the Ethernet environment, the value of the CFI field is 0; The VLAN ID is used to identify the VLAN to which the frame belongs.

4. Link type of switch

(1) Access interface

Access interface connects user terminal devices that can't recognize Tag, such as user host and server. The access interface must be configured with a PVID (the default PVID of the switch is VLAN1), and the VLAN tag processing process of data frames entering the access interface is shown in Fig. 2.22. When a data frame is sent from the access interface, if the VLAN ID in the data frame is the same as the PVID configured in the access interface, the label will be stripped and the unlabeled Ethernet frame will be sent from the link, as shown in Fig. 2.23.

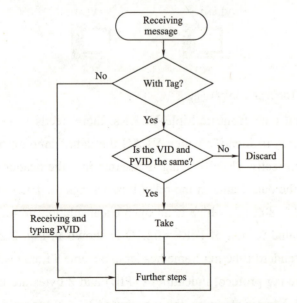
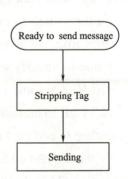

Fig. 2.22　The process of VLAN tag processing data frame entering access interface

Fig. 2.23　The process of VLAN tag processing data frame from access interface

(2) Trunk interface

Trunk interface is also called trunk interface. The port configured as trunk allows data frames of multiple VLANs to pass through, and these data frames are distinguished by 802.1Q tags, but only frames of one VLAN are allowed to be sent out from this type of interface without Tags (that is, stripped Tag). Trunk interface is usually used between switches, switches and routers. The VLAN tag processing process of data frame entering trunk interface is shown in Fig. 2.24. The VLAN tag processing process when data frames are sent from trunk interface is shown in Fig. 2.25.

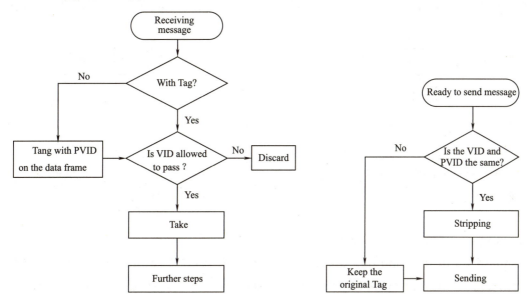

Fig. 2.24 The process of VLAN tag processing data frame entering trunk interface

Fig. 2.25 The process of VLAN tag processing of data frames sent from trunk interface

Note: VLAN data of trunk interface PVID does not carry labels on trunk links.

(3) Hybrid interface

Similar to the trunk interface, the hybrid interface also allows data frames of multiple VLAN to pass through, and these data frames are distinguished by labels. Users can flexibly specify whether the hybrid interface carries a label when sending a data frame of a certain VLAN (or some VLAN). The hybrid interface can transmit two kinds of frames: those with VLAN information and those without VLAN information. The VLAN tag processing process of data frames entering the hybrid interface is shown in Fig. 2.26. The VLAN tag processing process when the data frame is sent from the Hybrid interface is shown in Fig. 2.27.

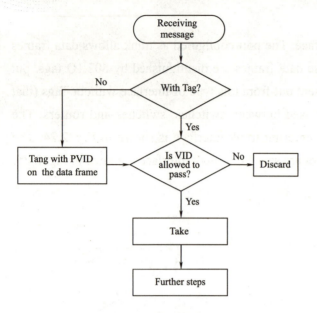
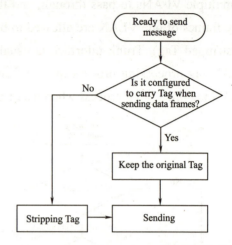

Fig. 2.26 The process of VLAN tag processing data frame entering hybrid interface

Fig. 2.27 The process of VLAN tag processing of data frames sent from hybrid interface

Related Skills

1. Planning and configuration of VLAN

You should build the topology diagram as shown in Fig. 2.28, and complete VLAN configuration and verification according to the topology diagram and VLAN planning shown in Table 2.3, in which the link between switches LSW1 and LSW2 is a trunk link, allowing VLAN2 and VLAN3 to pass, but not VLAN1 to pass.

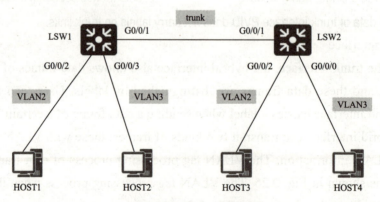

Fig. 2.28 Topology diagram of VLAN planning and configuration

See Table 2.3 for VLAN planning in switched networks.

Project 2 Small-sized Enterprise Internetworking Project

Table 2.3 VLAN planning

Device	Interface or VLAN	Description of VLAN	Interface	Explanation
LSW1	VLAN2	teacher	G0/0/2	teacher
	VLAN3	student	G0/0/3 to G0/0/20	student
LSW2	VLAN2	teacher	G0/0/2 to G0/0/9	teacher
	VLAN3	student	G0/0/10 to G0/0/20	student

The network number of VLAN2 is planned as 192.168.2.0/24, and the gateway is 192.168.2.254. The network number of VLAN3 is planned as 192.168.3.0/24, and the gateway is 192.168.3.254. See Table 2.4 for the host IP address planning.

Table 2.4 Host IP address planning

Device	IP address and mask	Default gateway
Host HOST1	192.168.2.1/24	192.168.2.254
Host HOST2	192.168.3.2/24	192.168.3.254
Host HOST3	192.168.2.3/24	192.168.2.254
Host HOST4	192.168.3.4/24	192.168.3.254

VLAN configuration generally requires three steps:

Creating VLAN on all switches of LAN according to the planned VLAN and configure VLAN description or name;

Configuring access interfaces on all switches in the local area network and specify the VLAN to which the access interfaces belong;

Configuring trunk interfaces on all switches in the local area network, which VLAN are allowed to pass through, and configure PVID of Trunk interfaces.

(1) VLAN configuration of switch LSW1

① Creating a VLAN. According to the VLAN planning in Table 2.3, two VLANs need to be created on switch LSW1, namely VLAN2 and VLAN3. Taking Huawei S5700-28C-HI switch as an example, the description steps of creating VLANs on switch LSW1 and configuring VLANs are as follows:

```
[LSW1]vlan 2                        //Create VLAN2 and enter VLAN2 view mode
[LSW1-vlan2]description teacher    //Configuration VLAN2 is described as "teacher"
[LSW1-vlan2]vlan 3                  //Create VLAN3 and enter VLAN3 view mode
[LSW1-vlan3]description student    //Configuration VLAN3 is described as "student"
[LSW1-vlan3]quit                    //Return to system view
```

② Configuring the access interface. The configuration of access interface usually needs to complete two tasks: One is to enter the interface view of access interface and configure the interface

as access interface; The second is to configure which VLAN the interface belongs to.

Taking Huawei S5700-28C-HI switch as an example, the configuration steps of configuring G0/0/2 interface of switch LSW1 as an access port and belonging to VLAN2 are as follows:

```
[LSW1]interface g0/0/2    //Enter interface view of G0/0/2 interface
[LSW1-GigabitEthernet0/0/2]port link-type access
//Configure G0/0/2 interface as access port
[LSW1-GigabitEthernet0/0/2]port default vlan 2
//Configure G0/0/2 interface to belong to VLAN2
[LSW1-GigabitEthernet0/0/2]quit                //Return to system view
```

According to the VLAN planning in Table 2.3, interfaces G0/0/3 to G0/0/20 of switch LSW1 belong to VLAN3. If each access interface is configured separately, the configuration commands are repeated and the number of configuration commands is large. Since all the interface configuration commands of G0/0/3 to G0/0/20 interfaces are the same, these interfaces can be configured by joining a port group to reduce the configuration workload.

Taking Huawei S5700-28C-HI switch as an example. Add G0/0/3 to G0/0/20 interfaces of switch LSW1 to port group 1 and configure them as access interfaces. The configuration steps belonging to VLAN3 are as follows:

```
[LSW1]port-group 1              //Create and enter the interface view of port group 1.
[LSW1-port-group-1]group-member g0/0/3 to g0/0/20
//Add interfaces G0/0/3 to G0/0/20 as members of port group 1
[LSW1-port-group-1]port link-type access   //Configure all interfaces in port group 1
                                           //as Access interfaces
[LSW1-port-group-1]port default vlan 3     //Configure all interfaces in port group 1
                                           //to belong to VLAN3
[LSW1-port-group-1]quit                    //Return to system view
```

③ Configuring the trunk interface. The configuration of trunk interface usually needs to complete three tasks: one is to enter the interface view of trunk interface and configure it as trunk interface. The second is to configure which VLAN data frames are allowed to pass through the trunk interface. Finally, specify the PVID of the trunk interface.

According to the VLAN planning in Table 2.3, interface G0/0/1 of switch LSW1 is a trunk interface, allowing VLAN2 and VLAN3 to pass, but not VLAN1. Because the trunk interface of Huawei switch allows VLAN1 to pass by default, it is necessary to disable VLAN1 to pass. The specific configuration steps are as follows:

```
[LSW1]interface g0/0/1                           //Enter the interface view of G0/0/1 interface
[LSW1-GigabitEthernet0/0/1]port link-type trunk  //Configure G0/0/1 interface as trunk interface
[LSW1-GigabitEthernet0/0/1]port trunk allow-pass vlan 2 3
```

```
//Configure G0/0/1 interface to allow VLAN2 and VLAN3 to pass through
[LSW1-GigabitEthernet0/0/1]undo port trunk allow-pass vlan 1
//Configuring G0/0/1 interface does not allow VLAN1 to pass through
[LSW1-GigabitEthernet0/0/1]quit          //Return to system view
```

Note: The default PVID of Huawei switch trunk interface is VLAN1.

(2) VLAN configuration of switch LSW2

① Creating a VLAN. According to the VLAN planning in Table 2.3, taking Huawei S5700-28C-HI switch as an example, VLAN2 and VLAN3 are created on LSW2 switch, and the description steps of configuring VLAN are as follows:

```
[LSW2]vlan 2                          //Create VLAN2 and enter VLAN2 view mode
[LSW2-vlan2]description teacher       //Configuration VLAN2 is described as "teacher"
[LSW2-vlan2]vlan 3                    //Create VLAN3 and enter VLAN3 view mode
[LSW2-vlan3]description student       //Configuration VLAN3 is described as "student"
[LSW2-vlan3]quit                      //Return to system view
```

② Configuring the access interface. According to the VLAN planning in Table 2.3, interfaces G0/0/2 to G0/0/20 of switch LSW2 are access interfaces, of which interfaces G0/0/2 to G0/0/9 belong to VLAN2, and interfaces G0/0/10 to G0/0/20 belong to VLAN3. Due to the same configuration commands for G0/0/2 to G0/0/9 interfaces, add them to a port group 1 for configuration. The G0/0/20 interface configuration commands are the same, and configure their home port group 2.

Taking Huawei S5700-28C-HI switch as an example, the steps for configuring the access interface of switch LSW2 using port groups are as follows:

```
[LSW2]port-group 1           //Create and enter the interface view of port group 1
[LSW2-port-group-1]group-member g0/0/2 to g0/0/9
//Add interfaces G0/0/2 to G0/0/9 as members of port group 1
[LSW2-port-group-1]port link-type access
//Configure all interfaces in port group 1 as access interfaces
[LSW2-port-group-1]port default vlan 2
//Configure all interfaces in port group 1 to belong to VLAN2
[LSW2-port-group-1]quit               //Return to system view
[LSW2]port-group 2
//Create and enter the interface view of port group 2
[LSW2-port-group-2]group-member g0/0/10 to g0/0/20
//Add interfaces G0/0/10 to G0/0/20 as members of port group
[LSW2-port-group-2]port link-type access
//Configure all interfaces in port group 2 as access interfaces
[LSW2-port-group-2]port default vlan 3    ///All interfaces in port group 2 belong to VLAN3
```

③ Configuring the trunk interface. According to the VLAN planning in Table 2.3, interface G0/0/1 of switch LSW2 is a trunk interface, allowing VLAN2 and VLAN3 to pass, but not VLAN1. The specific configuration steps are as follows:

```
[LSW2]interface g0/0/1    //Enter the interface view of G0/0/1 interface
[LSW2-GigabitEthernet0/0/1]port link-type trunk//Configure G0/0/1 interface as trunk interface
[LSW2-GigabitEthernet0/0/1]port trunk allow-pass vlan 2 3
//Configure G0/0/1 interface to allow VLAN2 and VLAN3 to pass through
[LSW2-GigabitEthernet0/0/1]undo port trunk allow-pass vlan 1
//Configure G0/0/1 interface does not allow VLAN1 to pass through
[LSW2-GigabitEthernet0/0/1]quit //Return to system view
```

(3) Verification and testing of VLAN

After VLAN creation, access interface configuration and trunk interface configuration are completed on the switch, you can use the "display" command to check whether the VLAN is configured correctly.

① Executing the "display vlan" command to observe whether the created VLAN information exists and is correct, and whether the interface has been added to the correct VLAN. Fig. 2.29 shows VLAN information on switch LSW1, and Fig. 2.30 shows VLAN information on switch LSW2.

```
[LSW1]display vlan
The total number of vlans is : 3
---------------------------------------------------------------
U: Up;          D: Down;         TG: Tagged;        UT: Untagged;
MP: Vlan-mapping;                ST: Vlan-stacking;
#: ProtocolTransparent-vlan;     *: Management-vlan;
---------------------------------------------------------------

VID  Type    Ports
---------------------------------------------------------------
1    common  UT:GE0/0/21(D)   GE0/0/22(D)    GE0/0/23(D)    GE0/0/24(D)
2    common  UT:GE0/0/2(U)
             TG:GE0/0/1(U)

3    common  UT:GE0/0/3(U)    GE0/0/4(D)     GE0/0/5(D)     GE0/0/6(D)
                GE0/0/7(D)    GE0/0/8(D)     GE0/0/9(D)     GE0/0/10(D)
                GE0/0/11(D)   GE0/0/12(D)    GE0/0/13(D)    GE0/0/14(D)
                GE0/0/15(D)   GE0/0/16(D)    GE0/0/17(D)    GE0/0/18(D)
                GE0/0/19(D)   GE0/0/20(D)
             TG:GE0/0/1(U)

VID  Status  Property     MAC-LRN Statistics Description
---------------------------------------------------------------
1    enable  default      enable  disable    VLAN 0001
2    enable  default      enable  disable    teacher
3    enable  default      enable  disable    studnet
```

Fig. 2.29   VLAN information on switch LSW1

② Executing the "display port vlan" command to view the switch interface VLAN information. In Fig. 2.31, the G0/0/1 interface of switch LSW1 is trunk, and only VLAN2 and VLAN3 are allowed to pass through; G0/0/2 interface is access and PVID is "2", which means it

belongs to VLAN 2; G0/0/3 to G0/0/20 interface is access and PVID is "3", which means it belongs to VLAN 3; The interface G0/0/21 to G0/0/24 is the default value, that is, the link type is hybrid and the PVID is "1", that is, it belongs to VLAN1 by default, which is consistent with the planned configuration.

```
[LSW2]display vlan
The total number of vlans is : 3
--------------------------------------------------------------------
U: Up;          D: Down;        TG: Tagged;         UT: Untagged;
MP: Vlan-mapping;               ST: Vlan-stacking;
#: ProtocolTransparent-vlan;    *: Management-vlan;
--------------------------------------------------------------------

VID  Type    Ports
--------------------------------------------------------------------
1    common  UT:GE0/0/21(D)   GE0/0/22(D)    GE0/0/23(D)    GE0/0/24(D)

2    common  UT:GE0/0/2(U)    GE0/0/3(D)     GE0/0/4(D)     GE0/0/5(D)
                GE0/0/6(D)    GE0/0/7(D)     GE0/0/8(D)     GE0/0/9(D)
             TG:GE0/0/1(U)

3    common  UT:GE0/0/10(U)   GE0/0/11(D)    GE0/0/12(D)    GE0/0/13(D)
                GE0/0/14(D)   GE0/0/15(D)    GE0/0/16(D)    GE0/0/17(D)
                GE0/0/18(D)   GE0/0/19(D)    GE0/0/20(D)
             TG:GE0/0/1(U)


VID  Status  Property    MAC-LRN  Statistics  Description
--------------------------------------------------------------------
1    enable  default     enable   disable     VLAN 0001
2    enable  default     enable   disable     teacher
3    enable  default     enable   disable     student
```

Fig. 2.30   VLAN information on switch LSW2

```
[LSW1]display port vlan
Port                      Link Type    PVID   Trunk VLAN List
--------------------------------------------------------------------
GigabitEthernet0/0/1      trunk        1      2-3
GigabitEthernet0/0/2      access       2      -
GigabitEthernet0/0/3      access       3      -
GigabitEthernet0/0/4      access       3      -
GigabitEthernet0/0/5      access       3      -
GigabitEthernet0/0/6      access       3      -
GigabitEthernet0/0/7      access       3      -
GigabitEthernet0/0/8      access       3      -
GigabitEthernet0/0/9      access       3      -
GigabitEthernet0/0/10     access       3      -
GigabitEthernet0/0/11     access       3      -
GigabitEthernet0/0/12     access       3      -
GigabitEthernet0/0/13     access       3      -
GigabitEthernet0/0/14     access       3      -
GigabitEthernet0/0/15     access       3      -
GigabitEthernet0/0/16     access       3      -
GigabitEthernet0/0/17     access       3      -
GigabitEthernet0/0/18     access       3      -
GigabitEthernet0/0/19     access       3      -
GigabitEthernet0/0/20     access       3      -
GigabitEthernet0/0/21     hybrid       1      -
GigabitEthernet0/0/22     hybrid       1      -
GigabitEthernet0/0/23     hybrid       1      -
GigabitEthernet0/0/24     hybrid       1      -
```

Fig. 2.31   Interface VLAN information of switch LSW1

In Fig. 2.32, the G0/0/1 interface of switch LSW2 is trunk, and only VLAN2 and VLAN3 are allowed to pass through; G0/0/2 to G0/0/9 interfaces are access and PVID is "2", which means VLAN 2. The interface G0/0/10 to G0/0/20 is access and the PVID is "3", which means it belongs to VLAN3, which is consistent with the planned configuration.

```
[LSW2]display port vlan
Port                    Link Type    PVID   Trunk VLAN List
--------------------------------------------------------------
GigabitEthernet0/0/1    trunk        1      2-3
GigabitEthernet0/0/2    access       2      -
GigabitEthernet0/0/3    access       2      -
GigabitEthernet0/0/4    access       2      -
GigabitEthernet0/0/5    access       2      -
GigabitEthernet0/0/6    access       2      -
GigabitEthernet0/0/7    access       2      -
GigabitEthernet0/0/8    access       2      -
GigabitEthernet0/0/9    access       2      -
GigabitEthernet0/0/10   access       3      -
GigabitEthernet0/0/11   access       3      -
GigabitEthernet0/0/12   access       3      -
GigabitEthernet0/0/13   access       3      -
GigabitEthernet0/0/14   access       3      -
GigabitEthernet0/0/15   access       3      -
GigabitEthernet0/0/16   access       3      -
GigabitEthernet0/0/17   access       3      -
GigabitEthernet0/0/18   access       3      -
GigabitEthernet0/0/19   access       3      -
GigabitEthernet0/0/20   access       3      -
GigabitEthernet0/0/21   hybrid       1      -
GigabitEthernet0/0/22   hybrid       1      -
GigabitEthernet0/0/23   hybrid       1      -
GigabitEthernet0/0/24   hybrid       1      -
```

Fig. 2.32   Port VLAN information of switch LSW2

③ In the interface view of the switch, execute the "display this" command to view the configuration information of the switch interface. The configuration information of G0/0/1 interface of switch LSW1 is shown in Fig. 2.33.

```
[LSW1-GigabitEthernet0/0/1]display this
#
interface GigabitEthernet0/0/1
 port link-type trunk
 undo port trunk allow-pass vlan 1
 port trunk allow-pass vlan 2 to 3
```

Fig. 2.33   The configuration information of G0/0/1 interface of switch LSW1

④ Refer to the parameters planned in Table 2.4 to complete the test of IP addresses, subnet masks and default gateway configurations of hosts HOST1, HOST2, HOST3 and HOST4. After the configuration is completed, you can use the "ipconfig" command at the DOS command line to check whether the IP configuration information is correct. (Note: As there is no routing device in the network topology in this task, whether the gateway is configured or not will not affect the following test.)

⑤ According to the planning of IP and VLAN, hosts in the same VLAN can communicate directly without routing equipment, while hosts in different VLANs cannot communicate directly

without routing equipment. Since HOST1 and HOST3 belong to VLAN2, HOST2 and HOST4 belong to VLAN3, in this task, theoretically, HOST1 can ping HOST3, but HOST1 cannot ping HOST2 and HOST4. Fig. 2.34 shows an example of connectivity test in which host HOST1 can ping HOST3, but cannot ping HOST2.

```
PC>ping 192.168.2.3

Ping 192.168.2.3: 32 data bytes, Press Ctrl_C to break
From 192.168.2.3: bytes=32 seq=1 ttl=128 time=47 ms
From 192.168.2.3: bytes=32 seq=2 ttl=128 time=110 ms
From 192.168.2.3: bytes=32 seq=3 ttl=128 time=93 ms
From 192.168.2.3: bytes=32 seq=4 ttl=128 time=109 ms
From 192.168.2.3: bytes=32 seq=5 ttl=128 time=78 ms

--- 192.168.2.3 ping statistics ---
 5 packet(s) transmitted
 5 packet(s) received
 0.00% packet loss
 round-trip min/avg/max = 47/87/110 ms

PC>ping 192.168.3.2

Ping 192.168.3.2: 32 data bytes, Press Ctrl_C to break
From 192.168.2.1: Destination host unreachable
From 192.168.2.1: Destination host unreachable
From 192.168.2.1: Destination host unreachable
From 192.168.2.1: Destination host unreachable
From 192.168.2.1: Destination host unreachable
```

Fig. 2.34 Host HOST1 can ping HOST3, but cannot ping connectivity test of HOST2

2. Application and configuration of hybrid interface

The topology diagram of hybrid interface application of a company is shown in Fig. 2.35. There are no routing devices in the network. The company has two departments: sales and administration. Please use hybrid interface to realize that the sales department, administrative department and server of the company need broadcast isolation, and the sales department and administrative department cannot communicate with each other, but the sales department and administrative department can access the company's server.

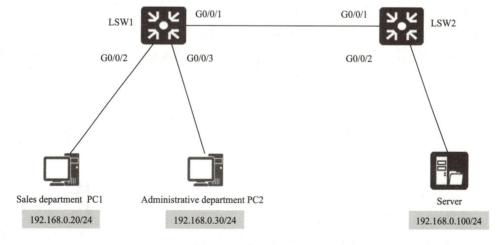

Fig. 2.35 Hybrid interface application topology diagram

According to the task requirements, the sales department, the administrative department

and the server planning belong to different VLAN to realize the requirement of broadcast isolation. Among them, the sales department belongs to VLAN20, the administrative department belongs to VLAN30, and the server belongs to VLAN100. The network number of all departments' IP addresses is "192.168.0.0/24". See Table 2.5 for the IP addresses of test hosts PC1, PC2, the server.

Table 2.5　IP address of host and server

Device	IP address and mask
Host PC1	192.168.0.20/24
Host PC2	192.168.0.30/24
Server	192.168.0.100/24

Planning the ports of switches LSW1 and LSW2 as hybrid interfaces and controlling the labeled and unlabeled VLAN can meet the task requirements. See Table 2.6 for the Hybrid interface planning of LSW1 and LSW2.

Table 2.6　Hybrid interface planning of LSW1 and LSW2

Device	Interface	Link type	PVID	List of VLAN allowed to pass
LSW1	G0/0/1	hybrid	VLAN1	Tagged VLAN：VLAN20, VLAN30, VLAN100
	G0/0/2	hybrid	VLAN20	Untagged VLAN：VLAN20, VLAN100
	G0/0/3	hybrid	VLAN30	Untagged VLAN：VLAN30, VLAN100
LSW2	G0/0/1	hybrid	VLAN1	Tagged VLAN：VLAN20, VLAN30, VLAN100
	G0/0/2	hybrid	VLAN100	Untagged VLAN：VLAN20, VLAN30, VLAN100

(1) VLAN configuration of switch LSW1

① Creating a VLAN. VLAN20, VLAN30 and VLAN100 need to be tagged through G0/0/1 interface of switch LSW1, so VLAN20, VLAN30 and VLAN100 need to be created on switch LSW1. You can create these three VLANs one by one by using the method of creating VLANs in "1. Planning and configuration of VLAN" in the related skills in this task, or you can create VLAN 20, VLAN30 and VLAN100 in batches. The steps to create VLAN20, VLAN30 and VLAN100 in batches are as follows:

```
[LSW1]vlan batch 20 30 100        //Create VLAN20, VLAN30 and VLAN100 in batches.
```

② Hybrid interface configuration. According to the hybrid interface plan shown in Table 2.6, the configuration steps of the hybrid port of switch LSW1 are as follows:

```
[LSW1]interface g0/0/1   //Enter the interface view of G0/0/1 interface
[LSW1-GigabitEthernet0/0/1]port link-type hybrid
//Configure G0/0/2 interface as a hybrid interface
[LSW1-GigabitEthernet0/0/1]port hybrid tagged vlan 20 30 100
//Configure G0/0/1 hybrid interface to add tagged VLANs as VLAN20, VLAN30 and VLAN100
```

```
[LSW1-GigabitEthernet0/0/1]quit //Return to System View
[LSW1]interface g0/0/2   //Enter the interface view of G0/0/2 interface.
[LSW1-GigabitEthernet0/0/2]port link-type hybrid
//Configure G0/0/2 interface as hybrid interface
[LSW1-GigabitEthernet0/0/2]port hybrid pvid vlan 20
//Configure the PVID of G0/0/2 port as VLAN20
[LSW1-GigabitEthernet0/0/2]port hybrid untagged vlan 20 100
//Configure G0/0/1 hybrid interface to add unlabeled VLAN as VLAN20 and VLAN100
[LSW1-GigabitEthernet0/0/2]quit //Return to system view
[LSW1]interface g0/0/3   //Enter the interface view of G0/0/3 interface
[LSW1-GigabitEthernet0/0/3]port link-type hybrid
//Configure G0/0/3 interface as hybrid interface
[LSW1-GigabitEthernet0/0/3]port hybrid pvid vlan 30
//Configure the PVID of G0/0/3 interface as VLAN30
[LSW1-GigabitEthernet0/0/3]port hybrid untagged vlan 30 100
//The unlabeled VLANs added to G0/0/3 hybrid interface are VLAN30 and VLAN100
[LSW1-GigabitEthernet0/0/3]quit //Return to system view
```

(2) VLAN configuration of switch LSW2

① Creating VLAN. VLAN20, VLAN30 and VLAN100 need to be tagged through G0/0/1 interface of switch LSW2, so VLAN20, VLAN30 and VLAN100 need to be created on switch LSW2. The steps for creating VLAN20, VLAN30 and VLAN100 in batches are as follows:

```
[LSW2]vlan batch 20 30 100      //Create VLAN20, VLAN30 and VLAN100 in batches
```

② Configuration of hybrid interface on switch LSW2. According to the hybrid interface plan shown in Table 2.6, the configuration steps of the hybrid interface of switch LSW2 are as follows:

```
[LSW2]interface g0/0/1   //Enter the interface view of G0/0/1 interface
[LSW2-GigabitEthernet0/0/1]port link-type hybrid
//Configure G0/0/2 interface as hybrid interface
[LSW2-GigabitEthernet0/0/1]port hybrid tagged vlan 20 30 100
//Configure G0/0/1 hybrid interface to add tagged VLANs as VLAN20, VLAN30 and VLAN100
[LSW2-GigabitEthernet0/0/1]quit //Return to system view
[LSW2]interface g0/0/2   //Enter the interface view of G0/0/2 interface
[LSW2-GigabitEthernet0/0/2]port link-type hybrid
//Configure G0/0/2 interface as hybrid interface
[LSW2-GigabitEthernet0/0/2]port hybrid pvid vlan 100
//Configure the PVID of G0/0/2 interface as VLAN100
[LSW2-GigabitEthernet0/0/2]port hybrid untagged vlan 20 30 100
//Configure G0/0/1 hybrid interface to add unlabeled VLANs as VLAN20, VLAN30 and VLAN100
[LSW2-GigabitEthernet0/0/2]quit //Return to system view
```

(3) Verification and testing of VLAN

After creating VLAN and configuring hybrid interface, you can use the following display command to verify whether VLAN is configured correctly.

① Execute the "display vlan" command on the switch to check the VLAN information and observe whether the created VLAN information is correct. Fig. 2.36 shows the display result of viewing VLAN information by using the "display vlan" command on switch LSW1: VLAN20 is an unlabeled VLAN added by G0/0/2 interface and a tagged VLAN added by G0/0/1 interface; VLAN30 is an unlabeled VLAN added by G0/0/3 interface and a tagged VLAN added by G0/0/1 interface; VLAN100 is an unlabeled VLAN added by G0/0/2 and G0/0/3 interfaces, and a tagged VLAN added by G0/0/1 interfaces.

Fig. 2.37 shows the display result of viewing VLAN information by using the "display vlan" command on switch LSW2: VLAN20 is an unlabeled VLAN added by G0/0/2 interface and a tagged VLAN added by G0/0/1 interface; VLAN30 is an unlabeled VLAN added by G0/0/2 interface and a tagged VLAN added by G0/0/1 interface; VLAN100 is an unlabeled VLAN added by G0/0/2 interface and a tagged VLAN added by G0/0/1 interface.

② Using the "display port vlan" command to view the PVID and tagged VLAN information of the switch hybrid interface, as shown in Fig. 2.38. The G0/0/1 interface on the switch LSW1 is a hybrid interface, and VLAN20, VLAN30 and VLAN100 are allowed for label passing.

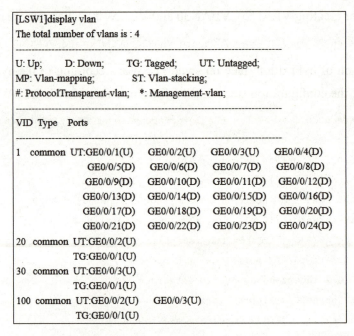

Fig. 2.36 The results of using the "display vlan" command on switch LSW1 to display

```
[LSW2]display vlan
The total number of vlans is : 4
--------------------------------------------------------------------
U: Up;       D: Down;       TG: Tagged;       UT: Untagged;
MP: Vlan-mapping;           ST: Vlan-stacking;
#: ProtocolTransparent-vlan;   *: Management-vlan;
--------------------------------------------------------------------
VID Type  Ports
--------------------------------------------------------------------
1   common UT:GE0/0/1(U)   GE0/0/2(U)    GE0/0/3(D)    GE0/0/4(D)
              GE0/0/5(D)   GE0/0/6(D)    GE0/0/7(D)    GE0/0/8(D)
              GE0/0/9(D)   GE0/0/10(D)   GE0/0/11(D)   GE0/0/12(D)
              GE0/0/13(D)  GE0/0/14(D)   GE0/0/15(D)   GE0/0/16(D)
              GE0/0/17(D)  GE0/0/18(D)   GE0/0/19(D)   GE0/0/20(D)
              GE0/0/21(D)  GE0/0/22(D)   GE0/0/23(D)   GE0/0/24(D)
20  common UT:GE0/0/2(U)
           TG:GE0/0/1(U)
30  common UT:GE0/0/2(U)
           TG:GE0/0/1(U)
100 common UT:GE0/0/2(U)
           TG:GE0/0/1(U)
```

Fig. 2.37 The results of using the "display vlan" command on switch LSW2 to display

```
[LSW1]display port vlan
Port                Link Type  PVID  Trunk VLAN List
--------------------------------------------------------------------
GigabitEthernet0/0/1  hybrid    1    20 30 100
GigabitEthernet0/0/2  hybrid    20   -
GigabitEthernet0/0/3  hybrid    30   -
GigabitEthernet0/0/4  hybrid    1    -
GigabitEthernet0/0/5  hybrid    1    -
GigabitEthernet0/0/6  hybrid    1    -
GigabitEthernet0/0/7  hybrid    1    -
GigabitEthernet0/0/8  hybrid    1    -
GigabitEthernet0/0/9  hybrid    1
```

Fig. 2.38 The results of using the "display port vlan" command on switch LSW1 to display

As shown in Fig. 2.39, the G0/0/1 interface on the switch LSW2 is a hybrid interface, and the VLANs allowed to be tagged are VLAN20, VLAN30 and VLAN100.

```
[LSW2]display port vlan
Port                Link Type  PVID  Trunk VLAN List
--------------------------------------------------------------------
GigabitEthernet0/0/1  hybrid    1    20 30 100
GigabitEthernet0/0/2  hybrid    100  -
GigabitEthernet0/0/3  hybrid    1    -
GigabitEthernet0/0/4  hybrid    1    -
GigabitEthernet0/0/5  hybrid    1    -
GigabitEthernet0/0/6  hybrid    1    -
GigabitEthernet0/0/7  hybrid    1    -
GigabitEthernet0/0/8  hybrid    1    -
GigabitEthernet0/0/9  hybrid    1    -
```

Fig. 2.39 The results of using the "display port vlan" command on switch LSW2 to display

③ In the interface view of the unlabeled hybrid interface of the switch, you can execute the "display this" command to view the configuration information of the interface. Fig. 2.40 shows the configuration of G0/0/2 interface of switch LSW1 under the interface view, using the "display this" command to view the interface.

Fig. 2.41 shows the configuration of G0/0/2 interface of switch LSW2 under the interface view to use the "display this" command to view the interface.

```
[LSW1-GigabitEthernet0/0/2]display this
#
interface GigabitEthernet0/0/2
 port hybrid pvid vlan 20
 port hybrid untagged vlan 20 100
#
return
[LSW1-GigabitEthernet0/0/2]
```

```
[LSW2-GigabitEthernet0/0/2]display this
#
interface GigabitEthernet0/0/2
 port hybrid pvid vlan 100
 port hybrid untagged vlan 20 30 100
#
return
```

Fig. 2.40　Configuration of interface G0/0/2 of switch LSW1

Fig. 2.41　Configuration of interface G0/0/2 of switch LSW2

④ You should complete the IP address and subnet mask configuration of the test hosts PC1, PC2 and the server with reference to the planned parameters in Table 2.6. After the configuration is completed, you can use the "ipconfig" command under the DOS command line to check whether their IP configuration information is correct. (**Note**: In this task, the IP address networks of the test hosts PC1 and PC2 and the server must be the same, and there is no need to configure the default gateway address.)

Fig. 2.42 shows the IP configuration information viewed by using the "ipconfig" command on the test host PC1. Pay attention to the following: whether the IP address is the planned IP address; Whether the subnet mask is the planned subnet mask.

```
PC>ipconfig

Link local IPv6 address.............: fe80::5689:98ff:fea4:33ed
IPv6 address........................: :: / 128
IPv6 gateway........................: ::
IPv4 address........................: 192.168.0.20
Subnet mask.........................: 255.255.255.0
Gateway.............................: 0.0.0.0
Physical address....................: 54-89-98-A4-33-ED
DNS server..........................:
```

Fig. 2.42　The example of using "ipconfig" command to view IP configuration information

⑤ According to the plan, the test hosts PC1 and PC2 cannot communicate with each other, but both PC1 and PC2 can access the server. Fig. 2.43 and Fig. 2.44 show the test results of connectivity between host PC1, PC2 and server. (**Note**: When ping is used for testing, if the target host turns on

the personal firewall, it is necessary to release the traffic requested by ICMP echo on the firewall before ping.)

```
PC>ping 192.168.0.100

Ping 192.168.0.100: 32 data bytes, Press Ctrl_C to break
From 192.168.0.100: bytes=32 seq=1 ttl=255 time=31 ms
From 192.168.0.100: bytes=32 seq=2 ttl=255 time=63 ms
From 192.168.0.100: bytes=32 seq=3 ttl=255 time=47 ms
From 192.168.0.100: bytes=32 seq=4 ttl=255 time=47 ms
From 192.168.0.100: bytes=32 seq=5 ttl=255 time=47 ms
```

Fig. 2.43 The example of the test host PC1 can ping the server

```
PC>ping 192.168.0.30

Ping 192.168.0.30: 32 data bytes, Press Ctrl_C to break
From 192.168.0.20: Destination host unreachable
From 192.168.0.20: Destination host unreachable
From 192.168.0.20: Destination host unreachable
From 192.168.0.20: Destination host unreachable
From 192.168.0.20: Destination host unreachable

--- 192.168.0.30 ping statistics ---
  5 packet(s) transmitted
  0 packet(s) received
  100.00% packet loss
```

Fig. 2.44 The example of the test host PC1 cannot ping test host PC2

Task Completion

1. Task planning

According to the task description and the networking requirements of Project 2, Table 2.7 gives the VLAN reference plan for the network shown in Fig. 2.1, and readers can also make their own plans according to the project requirements.

Table 2.7 A reference plan for VLAN

Device	Interface or VLAN	Description of VLAN	Interface	Explanation
FLOOR1	VLAN3	Sales	G0/0/2 to G0/0/14 interface	Sell
	VLAN4	Service	G0/0/15 to G0/0/24 interface	After service
	VLAN99	Manage		Switch management VLAN
FLOOR2	VLAN2	RD	G0/0/2 to G0/0/10 interface	Research and development
	VLAN3	Sales	G0/0/11 to G0/0/20 interface	Sell
	VLAN4	Service	G0/0/21 to G0/0/24 interface	After service
	VLAN99	Manage		Switch management VLAN

Continued

Device	Interface or VLAN	Description of VLAN	Interface	Explanation
FLOOR3	VLAN2	RD	G0/0/2 to G0/0/16 interface	Research and development
	VLAN4	Service	G0/0/17 to G0/0/24 interface	After service
	VLAN99	Manage		Switch management VLAN
FLOOR4	VLAN10	Leader	G0/0/2 to G0/0/8 interface	Business leaders
	VLAN20	HR	G0/0/9 to G0/0/24 interface	Human resources
CORE	VLAN2	RD		
	VLAN3	Sales		
	VLAN4	Service		
	VLAN99	Manage		
	VLAN100	Server	G0/0/21, G0/0/22 and G0/0/23 interface	Server VLAN
	VLAN101	linktorouter	G0/0/24 interface	Connection network segment to router
FLOOR4	VLAN10	Leader		Leader
	VLAN20	HR		Human resources

In the combination with Fig. 2.1 and Table 2.7, the links between switch CORE and switch FLOOR1, switch FLOOR2 and switch FLOOR3, and the links between router AR1 and switch FLOOR4 need to be configured as trunk links. See Table 2.8 for the planning of VLAN trunk interface.

Table 2.8 Planning of VLAN trunk interface

Name	Trunk interface	Allowing VLAN	PVID
CORE	G0/0/1	VLAN1, VLAN3, VLAN4 and VLAN99	VLAN1
	G0/0/2	VLAN1, VLAN2, VLAN3, VLAN4 and VLAN99	VLAN1
	G0/0/3	VLAN1, VLAN2, VLAN4 and VLAN99	VLAN1
FLOOR1	G0/0/1	VLAN1, VLAN3, VLAN4 and VLAN99	VLAN1
FLOOR2	G0/0/1	VLAN1, VLAN2, VLAN3, VLAN4 and VLAN99	VLAN1
FLOOR3	G0/0/1	VLAN1, VLAN2, VLAN4 and VLAN99	VLAN1
FLOOR4	G0/0/1	VLAN1, VLAN10 and VLAN20	VLAN1

Note: VLAN data of the trunk port PVID does not carry labels when the trunk link is transmitted.

2. Task implementation

(1) VLAN configuration of switch CORE

① Creating a VLAN. Taking Huawei S5700-28C-HI switch as an example to complete the VLAN configuration of the switch CORE according to Table 2.7. The steps are as follows:

```
[CORE]vlan 2                    //Create VLAN2
```

```
[CORE-vlan2]description RD        //Configuration VLAN2 is described as "RD"
[CORE-vlan2]vlan 3                //Create VLAN3
[CORE-vlan3]description Sales     //Configuration VLAN3 is described as "Sales"
[CORE-vlan3]vlan 4                //Create VLAN4
[CORE-vlan4]description Service   //Configuration VLAN4 is described as "Service"
[CORE-vlan4]vlan 99               //Create VLAN99
[CORE-vlan99]description Manage   //Configuration VLAN99 is described as "Manage"
[CORE-vlan99]vlan 100             //Create VLAN100
[CORE-vlan100]description Server  //Configuration VLAN100 is described as "Server"
[CORE-vlan99]vlan 101             //Create VLAN101
[CORE-vlan101]description linktorouter  //The description of configuring VLAN100
[CORE-vlan101]quit                //is "linktorouter"
```

② Configuration of access interface. Taking Huawei S5700-28C-HI switch as an example, the configuration steps of assigning G0/0/21, G0/0/22 and G0/0/23 interfaces of switch CORE to VLAN100 and G0/0/24 interface to VLAN 100 are as follows:

```
[CORE]interface g0/0/21  //Enter the interface view of G0/0/21 interface
[CORE-GigabitEthernet0/0/21]port link-type access    //Configure G0/0/21 interface
                                                     //link mode to access
[CORE-GigabitEthernet0/0/21]port default vlan 100
[CORE-GigabitEthernet0/0/21]quit//Configure the PVID of G0/0/21 interface as VLAN100
[CORE]interface g0/0/22  //Enter the interface view of G0/0/22 interface
[CORE-GigabitEthernet0/0/22]port link-type access    //Configure G0/0/22 interface
                                                     //link mode to access
[CORE-GigabitEthernet0/0/22]port default vlan 100
[CORE-GigabitEthernet0/0/22]quit//Configure the PVID of G0/0/22 interface as VLAN100
[CORE]interface g0/0/23  //Enter the interface view of G0/0/23 interface
[CORE-GigabitEthernet0/0/23]port link-type access    //Configure G0/0/23 interface
                                                     //link mode to access
[CORE-GigabitEthernet0/0/23]port default vlan 100
[CORE-GigabitEthernet0/0/23]quit//Configure the PVID of G0/0/23 interface as VLAN100
[CORE]interface g0/0/24  //Enter the interface view of G0/0/24 interface
[CORE-GigabitEthernet0/0/24]port link-type access    //Configure G0/0/24 interface
                                                     //link mode to access
[CORE-GigabitEthernet0/0/24]port default vlan 101    //Configure the PVID of G0/0/24
                                                     //interface as VLAN101
```

③ Configuration of trunk interface. According to the plan, G0/0/1 interface, G0/0/2 interface and G0/0/3 interface of switch CORE need to be configured as trunk interfaces, in which G0/0/1 interface allows VLAN1, VLAN3, VLAN4 and VLAN99 to pass through, G0/0/2 interface

allows VLAN1, VLAN2, VLAN3, VLAN4 and VLAN99 to pass through, and G0/0/3 interface allows VLAN1, VLAN2, VLAN4 and VLAN99 to pass through. The specific configuration steps are as follows:

```
[CORE]interface g0/0/1    //Enter the interface view of G0/0/1 interface
[CORE-GigabitEthernet0/0/1]port link-type trunk//Configure G0/0/1 interface link mode
                                              //as trunk
[CORE-GigabitEthernet0/0/1]port trunk allow-pass vlan 3 4 99
//Configure G0/0/1 interface to allow VLAN3, VLAN4 and VLAN99 to pass through
[CORE-GigabitEthernet0/0/1]quit //Return to system view
[CORE]interface g0/0/2    //Enter the interface view of G0/0/2 interface
[CORE-GigabitEthernet0/0/2]port link-type trunk //Configure G0/0/2 interface link mode as trunk
[CORE-GigabitEthernet0/0/2]port trunk allow-pass vlan 2 3 4 99
//Configure G0/0/2 interface to allow VLAN2, VLAN3, VLAN4 and VLAN99 to pass through
[CORE-GigabitEthernet0/0/2]quit //Return to system view
[CORE]interface g0/0/3    //Enter the interface view of G0/0/3 interface
[CORE-GigabitEthernet0/0/3]port link-type trunk       //Configure G0/0/3 interface link mode as trunk
[CORE-GigabitEthernet0/0/3]port trunk allow-pass vlan 2 4 99
//Configure G0/0/3 interface to allow VLAN2, VLAN4 and VLAN99 to pass through
[CORE-GigabitEthernet0/0/3]quit //Return to System View
```

Note: VLAN1, the trunk interface of Huawei switch, is allowed to pass by default.

(2) VLAN configuration of switch FLOOR1

① Creating a VLAN. Taking Huawei S5700-28C-HI switch as an example, according to Table 2.7, the steps of creating VLAN on switch FLOOR1 are as follows:

```
[FLOOR1]vlan 3                       //Create VLAN3
[FLOOR1-vlan2]description Sales      //Configuration VLAN3 is described as "Sales"
[FLOOR1-vlan3]vlan 4                 //Create VLAN4
[FLOOR1-vlan4]description Service    //Configuration VLAN4 is described as "Service"
[FLOOR1-vlan4]vlan 99                //Create VLAN99
[FLOOR1-vlan99]description Manage    //Configuration VLAN99 is described as "Manage"
```

② Configuration of access port. According to Table 2.7, interfaces G0/0/2 to G0/0/14 of switch FLOOR1 belong to VLAN3, with the same configuration, and G0/0/15 to G0/0/24 interfaces belong to VLAN4. Therefore, the configuration can be completed by adding interfaces with the same configuration to the same port group, adding G0/0/2 to G0/0/14 interfaces of switch FLOOR1 to port group 1, and adding G0/0/15 to G0/0/24 interfaces of switch FLOOR1 to port group 2. Taking Huawei S5700-28C-HI switch as an example, the configuration steps of configuring G0/0/2 to G0/0/14 interfaces of switch FLOOR1 to belong to VLAN2 and G0/0/15 to G0/0/24 interfaces to belong to VLAN4 are as follows:

```
[FLOOR1]port-group 1        //Create and enter the interface view of port group 1
[FLOOR1-port-group-1]group-member g0/0/2 to g0/0/14     //Add G0/0/2 to G0/0/14 interface
                                                        //in port group 1
[FLOOR1-port-group-1]port link-type access //Configure all interface link modes in
                                           //port group 1 to access
[FLOOR1-port-group-1]port default vlan 3   //Configure the PVID of all interfaces in
                                           //port group 1 as VLAN3
[FLOOR1-port-group-1]quit       //Return to system view
[FLOOR1]port-group 2            //Create and enter the interface view of port group 2
[FLOOR1-port-group-2]group-member g0/0/15 to g0/0/24    //Add G0/0/15 to G0/0/24 interface
                                                        //in port group 2
[FLOOR1-port-group-2]port link-type access      //Configure all interface link modes in
                                                //port group 1 to access
[FLOOR1-port-group-2]port default vlan 4        //Configure the PVID of all interfaces
                                                //in port group 1 as VLAN4
```

③ Configuration of trunk interface. According to the plan, the G0/0/1 interface of switch FLOOR1 needs to allow VLAN3, VLAN4 and VLAN99 to pass through, so it needs to be configured as a trunk interface. The specific configuration steps are as follows:

```
[FLOOR1]interface g0/0/1 //Enter the interface view of G0/0/1 interface
[FLOOR1-GigabitEthernet0/0/1]port link-type trunk       //Configure G0/0/1 interface
                                                        //link mode as trunk
[FLOOR1-GigabitEthernet0/0/1]port trunk allow-pass vlan 3 4 99
//Configure G0/0/1 interface to allow VLAN3, VLAN4 and VLAN99 to pass through
[CORE-GigabitEthernet0/0/1]quit //Return to system view
```

(3) VLAN configuration of switch FLOOR2

① Creating a VLAN. Taking Huawei S5700-28C-HI switch as an example, according to Table 2.7, the steps of creating VLAN on switch FLOOR2 are as follows:

```
[FLOOR2]vlan 2                          //Create VLAN2
[FLOOR2-vlan2]description RD            //Configuration VLAN2 is described as "RD"
[FLOOR2]vlan 3                          //Create VLAN3
[FLOOR2-vlan2]description Sales         //Configuration VLAN3 is described as "Sales"
[FLOOR2-vlan3]vlan 4                    //Create VLAN4
[FLOOR2-vlan4]description Service       //Configuration VLAN4 is described as "Service"
[FLOOR2-vlan4]vlan 99                   //Create VLAN99
[FLOOR2-vlan99]description Manage       //Configuration VLAN99 is described as "Manage"
```

② Configuration of access interface. According to Table 2.7, interfaces G0/0/2 to G0/0/10 of switch FLOOR2 belong to VLAN2, G0/0/11 to G0/0/20 interfaces belong to VLAN3, and G0/0/21

to G0/0/24 interfaces belong to VLAN4. During the configuration, G0/0/2 to G0/0/10 interfaces of switch FLOOR2 are added to port group 1, G0/0/11 to G0/0/20 interfaces are added to port group 2, and G0/0/21 to G0/0/24 interfaces are added to port group 3. The configuration steps of access interface are as follows:

```
[FLOOR2]port-group 1          //Create and enter the interface view of interface group 1
[FLOOR2-port-group-1]group-member g0/0/2 to g0/0/10    //Add G0/0/2 to G0/0/10 interface
                                                       //in port group 1
[FLOOR2-port-group-1]port link-type access   //Configure all interface link modes in
                                             //port group 1 to access
[FLOOR2-port-group-1]port default vlan 2     //Configure the PVID of all interfaces in
                                             //port group 1 as VLAN2
[FLOOR2-port-group-1]quit                    //Return to system view
[FLOOR2]port-group 2                         //Create and enter the interface view of
                                             //interface group 2
[FLOOR2-port-group-2]group-member g0/0/11 to g0/0/20   //Add G0/0/11 to G0/0/20 interface
                                                       //in port group 2
[FLOOR2-port-group-2]port link-type access   //Configure all interface link modes in
                                             //port group 2 to access
[FLOOR2-port-group-2]port default vlan 3     //Configure the PVID of all interfaces
                                             //in port group 2 as VLAN3
[FLOOR2-port-group-2]quit         //Return to system view
[FLOOR2]port-group 3           //Create and enter the port view of port group 3.
[FLOOR2-port-group-3]group-member g0/0/21 to g0/0/24   //Add G0/0/21 to G0/0/24 interface
                                                       //in port group 3
[FLOOR2-port-group-3]port link-type access   //Configure all interface link modes in
                                             //port group 3 to access
[FLOOR2-port-group-3]port default vlan 4     //Configure the PVID of all interfaces
                                             //in port group 3 as VLAN4
```

③ Configuration of trunk interface. According to the plan, the G0/0/1 interface of switch FLOOR2 needs to allow VLAN2, VLAN3, VLAN4 and VLAN99 to pass through, so it needs to be configured as a trunk interface. The specific configuration steps are as follows:

```
[FLOOR2]interface g0/0/1 //Enter the interface view of G0/0/1 interface
[FLOOR2-GigabitEthernet0/0/1]port link-type trunk      //Configure G0/0/1 interface
                                                       //link mode as trunk
[FLOOR2-GigabitEthernet0/0/1]port trunk allow-pass vlan 2 3 4 99
//Configure G0/0/1 interface to allow VLAN2, VLAN3, VLAN4 and VLAN99 to pass through
[FLOOR2-GigabitEthernet0/0/1]quit       //Return to system view
```

(4) VLAN configuration of switch FLOOR3

① Creating a VLAN. According to Table 2.7, taking Huawei S5700-28C-HI switch as an example, the steps of creating VLAN on switch FLOOR3 are as follows:

```
[FLOOR3]vlan batch 2 4 99          //Create VLAN2, VLAN4 and VLAN99 in batches
[FLOOR3]vlan 2                     //Enter the VLAN2 view
[FLOOR3-vlan2]description RD       //Configuration VLAN2 is described as "RD"
[FLOOR3-vlan2]vlan 4               //Enter the VLAN4 view
[FLOOR3-vlan4]description Service  //Configuration VLAN4 is described as "Service"
[FLOOR3-vlan4]vlan 99              //Enter the VLAN99 view
[FLOOR3-vlan99]description Manage  //Configuration VLAN99 is described as "Manage"
```

② Configuration of access interface. According to Table 2.7, interfaces G0/0/2 to G0/0/16 of switch FLOOR3 belong to VLAN2, G0/0/17 to G0/0/24 interfaces belong to VLAN4. When configuring, the interfaces G0/0/2 to G0/0/16 of the switch FLOOR3 are added to the port group 1, and the interfaces G0/0/17 to G0/0/24 are added to the port group 2. The steps for configuring the access interface are as follows:

```
[FLOOR3]port-group 1       //Create and enter the interface view of port group 1
[FLOOR3-port-group-1]group-member g0/0/2 to g0/0/16   //Add G0/0/2 to G0/0/16 interface
                                                      //in port group 1
[FLOOR3-port-group-1]port link-type access   //Configure all interface link modes in port group 1
                                             //to access
[FLOOR3-port-group-1]port default vlan 2     //Configure the PVID of all interfaces
                                             //in port group 1 as VLAN2
[FLOOR3-port-group-1]quit    //Return to system view
[FLOOR3]port-group 2         //Create and enter the interface view of port group 2
[FLOOR3-port-group-2]group-member g0/0/17 to g0/0/24  //Add G0/0/17 to G0/0/24 interface
                                                      //in port group 2
[FLOOR3-port-group-2]port link-type access   //Configure all interface link modes in port group 2
                                             //to access
[FLOOR3-port-group-2]port default vlan 4     //Configure all interfaces in port group 2
                                             //PVID is VLAN4
[FLOOR3-port-group-2]quit    //Return to system view
```

③ Configuration of trunk interface According to the plan, the G0/0/1 interface of switch FLOOR3 needs to allow VLAN2, VLAN4 and VLAN99 to pass through, so it needs to be configured as a trunk interface. The specific configuration steps are as follows:

```
[FLOOR3]interface g0/0/1  //Enter the interface view of G0/0/1 interface
[FLOOR3-GigabitEthernet0/0/1]port link-type trunk    //Configure G0/0/1 interface
                                                     //link mode as trunk
```

```
[FLOOR3-GigabitEthernet0/0/1]port trunk allow-pass vlan 2 4 99
//Configure G0/0/1 interface to allow VLAN2, VLAN4 and VLAN99 to pass through
[FLOOR3-GigabitEthernet0/0/1]quit        //Return to system view
```

(5) VLAN configuration of switch FLOOR4

① Creating a VLAN. According to Table 2.7, taking Huawei S5700-28C-HI switch as an example, the steps of creating VLAN on switch FLOOR4 are as follows:

```
[FLOOR4]vlan 10                          //Enter the VLAN10 view
[FLOOR4-vlan10]description leader        //Configuration VLAN10 is described as "leader"
[FLOOR4-vlan10]vlan 20                   //Enter VLAN20 view
[FLOOR4-vlan20]description HR            //Configuration VLAN4 is described as "HR"
```

② Configuration of access interface. According to Table 2.7, interfaces G0/0/2 to G0/0/8 of switch FLOOR4 belong to VLAN10, G0/0/9 to G0/0/24 interfaces belong to VLAN20. During configuration, the interfaces G0/0/2 to G0/0/8 of switch FLOOR4 are added to port group 1, and the interfaces G0/0/9 to G0/0/24 are added to port group 2. The configuration steps of access interface are as follows:

```
[FLOOR4]port-group 1          //Create and enter the interface view of port group 1
[FLOOR4-port-group-1]group-member g0/0/2 to g0/0/8    //Add G0/0/0 to G0/0/8 interface
                                                      //in port group 1
[FLOOR4-port-group-1]port link-type access    //Configure all interface link modes in port group 1
                                              //to access
[FLOOR4-port-group-1]port default vlan 10     //Configure the of all interfaces in port group 1
                                              //PVID is VLAN10
[FLOOR4-port-group-1]quit           //Return to system view
[FLOOR4]port-group 2                //Create and enter the interface view of port group 2
[FLOOR4-port-group-2]group-member g0/0/9 to g0/0/24   //Add G0/0/9 to G0/0/24 interface
                                                      //in port group 2
[FLOOR4-port-group-2]port link-type access //Configure all interface link modes in port group 2
                                           //to access
[FLOOR4-port-group-2]port default vlan 20  //Configure all interfaces in port group 2
                                           //PVID is VLAN20
[FLOOR4-port-group-2]quit           //Return to system view
```

③ Configuration of trunk interface. According to the plan, the G0/0/1 interface of switch FLOOR4 needs to allow VLAN10 and VLAN20 to pass through, so it needs to be configured as a trunk interface. The specific configuration steps are as follows:

```
[FLOOR4]interface g0/0/1 //Enter the interface view of G0/0/1 interface
[FLOOR4-GigabitEthernet0/0/1]port link-type trunk       //Configure G0/0/1 interface
```

```
                                                        //link mode as trunk
[FLOOR4-GigabitEthernet0/0/1]port trunk allow-pass vlan 10 20
//Configure G0/0/1 interface to allow VLAN10 and VLAN20 to pass through
[FLOOR4-GigabitEthernet0/0/1]quit       //Return to system view
```

(6) View and verification of VLAN

After VLAN creation, access interface configuration and trunk interface configuration are completed on switches CORE, FLOOR1, FLOOR2, FLOOR3 and FLOOR4, you can use the "display" command to check and verify whether the VLAN is configured correctly.

① You can use the "display vlan" command on the switches CORE, FLOOR1, FLOOR2 and FLOOR3 to check the VLAN information created and whether the VLAN added by the interface carries labels and other information. Fig. 2.45 to Fig. 2.49 show the display results after using the "display vlan" command on switches CORE, FLOOR1, FLOOR2, FLOOR3 and FLOOR4 respectively.

```
[CORE]display vlan
The total number of vlans is : 7
--------------------------------------------------------------------
U: Up;       D: Down;      TG: Tagged;      UT: Untagged;
MP: Vlan-mapping;         ST: Vlan-stacking;
#: ProtocolTransparent-vlan;   *: Management-vlan;
--------------------------------------------------------------------
VID   Type       Ports
--------------------------------------------------------------------
1     common     UT:GE0/0/1(U)    GE0/0/2(U)     GE0/0/3(U)     GE0/0/4(D)
                    GE0/0/5(D)      GE0/0/6(D)     GE0/0/7(D)     GE0/0/8(D)
                    GE0/0/9(D)      GE0/0/10(D)    GE0/0/11(D)    GE0/0/12(D)
                    GE0/0/13(D)     GE0/0/14(D)    GE0/0/15(D)    GE0/0/16(D)
                    GE0/0/17(D)     GE0/0/18(D)    GE0/0/19(D)    GE0/0/20(D)
2     common     TG:GE0/0/2(U)    GE0/0/3(U)
3     common     TG:GE0/0/1(U)    GE0/0/2(U)
4     common     TG:GE0/0/1(U)    GE0/0/2(U)    GE0/0/3(U)
99    common     TG:GE0/0/1(U)    GE0/0/2(U)    GE0/0/3(U)
100   common     UT:GE0/0/21(D)   GE0/0/22(D)   GE0/0/23(D)
101   common     UT:GE0/0/24(D)
VID   Status     Property    MAC-LRN Statistics Description
--------------------------------------------------------------------
1     enable     default     enable    disable   VLAN 0001
2     enable     default     enable    disable   RD
3     enable     default     enable    disable   Sales
4     enable     default     enable    disable   Service
99    enable     default     enable    disable   Manage
100   enable     default     enable    disable   Server
101   enable     default     enable    disable   linktorouter
```

Fig. 2.45   The result after using the "display vlan" command on the CORE

```
[FLOOR1]display vlan
The total number of vlans is : 4
--------------------------------------------------------------
U: Up;         D: Down;       TG: Tagged;      UT: Untagged;
MP: Vlan-mapping;             ST: Vlan-stacking;
#: ProtocolTransparent-vlan;  *: Management-vlan;
--------------------------------------------------------------

VID Type   Ports
--------------------------------------------------------------
1   common UT:GE0/0/1(U)

3   common UT:GE0/0/2(U)   GE0/0/3(U)   GE0/0/4(D)   GE0/0/5(D)
               GE0/0/6(D)   GE0/0/7(D)   GE0/0/8(D)   GE0/0/9(D)
               GE0/0/10(D)  GE0/0/11(D)  GE0/0/12(D)  GE0/0/13(D)
               GE0/0/14(D)
           TG:GE0/0/1(U)

4   common UT:GE0/0/15(D)  GE0/0/16(D)  GE0/0/17(D)  GE0/0/18(D)
               GE0/0/19(D)  GE0/0/20(D)  GE0/0/21(D)  GE0/0/22(D)
               GE0/0/23(D)  GE0/0/24(D)
           TG:GE0/0/1(U)

99  common TG:GE0/0/1(U)


VID Status Property    MAC-LRN Statistics Description
--------------------------------------------------------------
1   enable default     enable  disable    VLAN 0001
3   enable default     enable  disable    Sales
4   enable default     enable  disable    Service
99  enable default     enable  disable    Manage
```

Fig. 2.46   The result after using the "display vlan" command on FLOOR1

```
[FLOOR2]display vlan
The total number of vlans is : 5
--------------------------------------------------------------
U: Up;         D: Down;       TG: Tagged;      UT: Untagged;
MP: Vlan-mapping;             ST: Vlan-stacking;
#: ProtocolTransparent-vlan;  *: Management-vlan;
--------------------------------------------------------------

VID Type   Ports
--------------------------------------------------------------
1   common UT:GE0/0/1(U)

2   common UT:GE0/0/2(D)   GE0/0/3(D)   GE0/0/4(D)   GE0/0/5(D)
               GE0/0/6(U)   GE0/0/7(D)   GE0/0/8(D)   GE0/0/9(D)
               GE0/0/10(D)
           TG:GE0/0/1(U)

3   common UT:GE0/0/11(D)  GE0/0/12(D)  GE0/0/13(D)  GE0/0/14(D)
               GE0/0/15(D)  GE0/0/16(D)  GE0/0/17(D)  GE0/0/18(D)
               GE0/0/19(D)  GE0/0/20(D)
           TG:GE0/0/1(U)

4   common UT:GE0/0/21(D)  GE0/0/22(D)  GE0/0/23(D)  GE0/0/24(D)
           TG:GE0/0/1(U)

99  common TG:GE0/0/1(U)


VID Status Property    MAC-LRN Statistics Description
--------------------------------------------------------------
1   enable default     enable  disable    VLAN 0001
2   enable default     enable  disable    RD
3   enable default     enable  disable    Sales
4   enable default     enable  disable    Service
99  enable default     enable  disable    Manage
```

Fig. 2.47   The result after using the "display vlan" command on FLOOR2

```
[FLOOR3]display vlan
The total number of vlans is : 4
--------------------------------------------------------------
U: Up;       D: Down;     TG: Tagged;     UT: Untagged;
MP: Vlan-mapping;         ST: Vlan-stacking;
#: ProtocolTransparent-vlan;   *: Management-vlan;
--------------------------------------------------------------

VID  Type    Ports
--------------------------------------------------------------
1    common  UT:GE0/0/1(U)

2    common  UT:GE0/0/2(D)    GE0/0/3(D)     GE0/0/4(D)     GE0/0/5(D)
                GE0/0/6(D)    GE0/0/7(D)     GE0/0/8(D)     GE0/0/9(D)
                GE0/0/10(D)   GE0/0/11(D)    GE0/0/12(D)    GE0/0/13(D)
                GE0/0/14(D)   GE0/0/15(D)    GE0/0/16(D)
             TG:GE0/0/1(U)

4    common  UT:GE0/0/17(D)   GE0/0/18(D)    GE0/0/19(D)    GE0/0/20(D)
                GE0/0/21(D)   GE0/0/22(U)    GE0/0/23(D)    GE0/0/24(D)
             TG:GE0/0/1(U)

99   common  TG:GE0/0/1(U)

VID  Status  Property    MAC-LRN  Statistics  Description
--------------------------------------------------------------
1    enable  default     enable   disable     VLAN 0001
2    enable  default     enable   disable     RD
4    enable  default     enable   disable     Service
99   enable  default     enable   disable     Manage
```

Fig. 2.48　The result after using the "display vlan" command on FLOOR3

```
[FLOOR4]display vlan
The total number of vlans is : 3
--------------------------------------------------------------
U: Up;       D: Down;     TG: Tagged;     UT: Untagged;
MP: Vlan-mapping;         ST: Vlan-stacking;
#: ProtocolTransparent-vlan;   *: Management-vlan;
--------------------------------------------------------------

VID  Type    Ports
--------------------------------------------------------------
1    common
10   common  UT:GE0/0/2(U)    GE0/0/3(D)     GE0/0/4(D)     GE0/0/5(D)
                GE0/0/6(D)    GE0/0/7(D)     GE0/0/8(D)
             TG:GE0/0/1(U)

20   common  UT:GE0/0/9(D)    GE0/0/10(D)    GE0/0/11(D)    GE0/0/12(D)
                GE0/0/13(D)   GE0/0/14(U)    GE0/0/15(D)    GE0/0/16(D)
                GE0/0/17(D)   GE0/0/18(D)    GE0/0/19(D)    GE0/0/20(D)
                GE0/0/21(D)   GE0/0/22(D)    GE0/0/23(D)    GE0/0/24(D)
             TG:GE0/0/1(U)

VID  Status  Property    MAC-LRN  Statistics  Description
--------------------------------------------------------------
1    enable  default     enable   disable     VLAN 0001
10   enable  default     enable   disable     leader
20   enable  default     enable   disable     HR
```

Fig. 2.49　The result after using the "display vlan" command on FLOOR4

② You can use the "display port vlan" command on switches CORE, FLOOR1, FLOOR2 and FLOOR3 to view the VLAN information of the interfaces. Fig. 2.50 to Fig. 2.54 show the display results after using the "display port vlan" command on switches CORE, FLOOR1, FLOOR2, FLOOR3 and FLOOR4 respectively.

Note: Be careful to whether the link type, PVID and VLAN list allowed by Trunk of the interface are consistent with the planned configuration.

```
[CORE]display port vlan
Port                  Link Type   PVID   Trunk VLAN List
-----------------------------------------------------------------
GigabitEthernet0/0/1   trunk      1      1 3-4 99
GigabitEthernet0/0/2   trunk      1      1-4 99
GigabitEthernet0/0/3   trunk      1      1-2 4 99
GigabitEthernet0/0/4   hybrid     1      -
GigabitEthernet0/0/5   hybrid     1      -
GigabitEthernet0/0/6   hybrid     1      -
GigabitEthernet0/0/7   hybrid     1      -
GigabitEthernet0/0/8   hybrid     1      -
GigabitEthernet0/0/9   hybrid     1      -
GigabitEthernet0/0/10  hybrid     1      -
GigabitEthernet0/0/11  hybrid     1      -
GigabitEthernet0/0/12  hybrid     1      -
GigabitEthernet0/0/13  hybrid     1      -
GigabitEthernet0/0/14  hybrid     1      -
GigabitEthernet0/0/15  hybrid     1      -
GigabitEthernet0/0/16  hybrid     1      -
GigabitEthernet0/0/17  hybrid     1      -
GigabitEthernet0/0/18  hybrid     1      -
GigabitEthernet0/0/19  hybrid     1      -
GigabitEthernet0/0/20  hybrid     1      -
GigabitEthernet0/0/21  access     100    -
GigabitEthernet0/0/22  access     100    -
GigabitEthernet0/0/23  access     100    -
GigabitEthernet0/0/24  access     101    -
```

Fig. 2.50 The result after using the "display port vlan" command on the CORE

```
[FLOOR1]display port vlan
Port                  Link Type   PVID   Trunk VLAN List
-----------------------------------------------------------------
GigabitEthernet0/0/1   trunk      1      1 3-4 99
GigabitEthernet0/0/2   access     3      -
GigabitEthernet0/0/3   access     3      -
GigabitEthernet0/0/4   access     3      -
GigabitEthernet0/0/5   access     3      -
GigabitEthernet0/0/6   access     3      -
GigabitEthernet0/0/7   access     3      -
GigabitEthernet0/0/8   access     3      -
GigabitEthernet0/0/9   access     3      -
GigabitEthernet0/0/10  access     3      -
GigabitEthernet0/0/11  access     3      -
GigabitEthernet0/0/12  access     3      -
GigabitEthernet0/0/13  access     3      -
GigabitEthernet0/0/14  access     3      -
GigabitEthernet0/0/15  access     4      -
GigabitEthernet0/0/16  access     4      -
GigabitEthernet0/0/17  access     4      -
GigabitEthernet0/0/18  access     4      -
GigabitEthernet0/0/19  access     4      -
GigabitEthernet0/0/20  access     4      -
GigabitEthernet0/0/21  access     4      -
GigabitEthernet0/0/22  access     4      -
GigabitEthernet0/0/23  access     4      -
GigabitEthernet0/0/24  access     4      -
```

Fig. 2.51 The result after using the "display port vlan" command on FLOOR1

```
[FLOOR2]display port vlan
Port                    Link Type   PVID   Trunk VLAN List
--------------------------------------------------------------
GigabitEthernet0/0/1    trunk       1      1-4 99
GigabitEthernet0/0/2    access      2      -
GigabitEthernet0/0/3    access      2      -
GigabitEthernet0/0/4    access      2      -
GigabitEthernet0/0/5    access      2      -
GigabitEthernet0/0/6    access      2      -
GigabitEthernet0/0/7    access      2      -
GigabitEthernet0/0/8    access      2      -
GigabitEthernet0/0/9    access      2      -
GigabitEthernet0/0/10   access      2      -
GigabitEthernet0/0/11   access      3      -
GigabitEthernet0/0/12   access      3      -
GigabitEthernet0/0/13   access      3      -
GigabitEthernet0/0/14   access      3      -
GigabitEthernet0/0/15   access      3      -
GigabitEthernet0/0/16   access      3      -
GigabitEthernet0/0/17   access      3      -
GigabitEthernet0/0/18   access      3      -
GigabitEthernet0/0/19   access      3      -
GigabitEthernet0/0/20   access      3      -
GigabitEthernet0/0/21   access      4      -
GigabitEthernet0/0/22   access      4      -
GigabitEthernet0/0/23   access      4      -
GigabitEthernet0/0/24   access      4      -
```

Fig. 2.52　The result after using the "display port vlan" command on FLOOR2

```
[FLOOR3]display port vlan
Port                    Link Type   PVID   Trunk VLAN List
--------------------------------------------------------------
GigabitEthernet0/0/1    trunk       1      1-2 4 99
GigabitEthernet0/0/2    access      2      -
GigabitEthernet0/0/3    access      2      -
GigabitEthernet0/0/4    access      2      -
GigabitEthernet0/0/5    access      2      -
GigabitEthernet0/0/6    access      2      -
GigabitEthernet0/0/7    access      2      -
GigabitEthernet0/0/8    access      2      -
GigabitEthernet0/0/9    access      2      -
GigabitEthernet0/0/10   access      2      -
GigabitEthernet0/0/11   access      2      -
GigabitEthernet0/0/12   access      2      -
GigabitEthernet0/0/13   access      2      -
GigabitEthernet0/0/14   access      2      -
GigabitEthernet0/0/15   access      2      -
GigabitEthernet0/0/16   access      2      -
GigabitEthernet0/0/17   access      4      -
GigabitEthernet0/0/18   access      4      -
GigabitEthernet0/0/19   access      4      -
GigabitEthernet0/0/20   access      4      -
GigabitEthernet0/0/21   access      4      -
GigabitEthernet0/0/22   access      4      -
GigabitEthernet0/0/23   access      4      -
GigabitEthernet0/0/24   access      4      -
```

Fig. 2.53　The result after using the "display port vlan" command on FLOOR3

```
[FLOOR4]display port vlan
Port                    Link Type   PVID   Trunk VLAN List
--------------------------------------------------------------
GigabitEthernet0/0/1    trunk        1     10 20
GigabitEthernet0/0/2    access      10     -
GigabitEthernet0/0/3    access      10     -
GigabitEthernet0/0/4    access      10     -
GigabitEthernet0/0/5    access      10     -
GigabitEthernet0/0/6    access      10     -
GigabitEthernet0/0/7    access      10     -
GigabitEthernet0/0/8    access      10     -
GigabitEthernet0/0/9    access      20     -
GigabitEthernet0/0/10   access      20     -
GigabitEthernet0/0/11   access      20     -
GigabitEthernet0/0/12   access      20     -
GigabitEthernet0/0/13   access      20     -
GigabitEthernet0/0/14   access      20     -
GigabitEthernet0/0/15   access      20     -
GigabitEthernet0/0/16   access      20     -
GigabitEthernet0/0/17   access      20     -
GigabitEthernet0/0/18   access      20     -
GigabitEthernet0/0/19   access      20     -
GigabitEthernet0/0/20   access      20     -
GigabitEthernet0/0/21   access      20     -
GigabitEthernet0/0/22   access      20     -
GigabitEthernet0/0/23   access      20     -
GigabitEthernet0/0/24   access      20     -
```

Fig. 2.54　The result after using the "display port vlan" command on FLOOR4

③ If the VLAN information of the switch port is different from the planned configuration by using the "diplay vlan" and "display port vlan" commands, in the interface view of the switch, execute the "display this" command to view the configuration information of the switch interface. Fig. 2.55 shows an example of viewing the configuration information of G0/0/1 interface of switch CORE. Fig. 2.56 shows an example of viewing the configuration information of G0/0/3 interface of switch FLOOR1.

```
[CORE-GigabitEthernet0/0/1]display this
#
interface GigabitEthernet0/0/1
 port link-type trunk
 port trunk allow-pass vlan 3 to 4 99
```

Fig. 2.55　The example of viewing the configuration information of G0/0/1 interface of switch CORE

```
[FLOOR1-GigabitEthernet0/0/3]display this
#
interface GigabitEthernet0/0/3
 port link-type access
 port default vlan 3
#
return
```

Fig. 2.56　The example of viewing the configuration information of G0/0/3 interface of switch FLOOR1

If the switch VLAN and switch port VLAN are configured incorrectly, you need to delete the wrong VLAN in the switch system view before creating the correct VLAN. Delete the VLAN configuration of the interface under the interface view of the switch interface and reconfigure the interface VLAN.

Exercises

Multiple choice

(1) (　　) is the default VLAN.

　　A. VLAN1　　　　B. VLAN2　　　　C. VLAN99　　　　D. VLAN100

(2) (　　) command is used to view the interface VLAN information.

　　A. display vlan　　　　　　　　B. display port vlan

　　C. display interface brief　　　　D. display version

(3) The following description about the interface between trunk and access (　　) is correct.

　　A. Trunk interface can only send tagged frames

　　B. Trunk interface can only send unlabeled frames

　　C. Access interface can only send tagged frames

　　D. Access interface can only send unlabeled frames

(4) (　　) is usually configured as an access link.

　　A. Link between switches

　　B. Connecting link between router and switch in one-arm routing

　　C. Switches connect links between hosts

(5) (　　) command is used to display VLAN information of the switch.

　　A. display vlan　　　　　　　　B. display port vlan

　　C. display interface brief　　　　D. display version

Task 3 | Planning and Configuration of Communications Between the VLANs

Tasks Description

On the basis of completing the Task 1 and Task 2 of this project, according to the network topology diagram shown in Fig. 2.1, you should continue to complete the following tasks:

① You should complete the configuration of sub-interface on router AR1, and realize the communication between enterprise leaders (VLAN10) and human resources (VLAN20).

② It is required to complete the configuration of DHCP service on router AR1, and provide the service of automatically obtaining IP address for enterprise leaders (VLAN10) and human resources (VLAN20).

③ You should complete the configuration of Vlanif interface on the CORE switch, and realize the communication among sales (VLAN3), after-sales (VLAN4), access layer switch management VLAN(VLAN99), enterprise intranet server (VLAN100) and R&D (VLAN2).

④ It is required to complete the configuration of DHCP service on the CORE, and provide the service of automatically obtaining IP addresses for the hosts in the sales (VLAN3), after-sales (VLAN4) and R&D (VLAN2) departments.

⑤ You should complete IP configuration for all devices in the network that statically assign IP addresses.

Task Objectives

- Understanding the concept of logical sub-interface and master the planning and configuration of single arm routing.
- Understanding the function of three-layer switch, and master the VLAN planning and configuration of communication with the three-layer switch.

Related Knowledge

1. Single arm routing

Single arm routing is a device, which uses a physical line to connect an external router that realizes communication of VLAN with a switch in the switching network to realize communication of VLAN. Because the router has only one interface connected to the switching network, it is necessary to use Dot1q to terminate the sub-interface, which acts as the gateway of VLAN. Sub-interface is a logical interface among physical interfaces, which is created by a physical interface through specific protocols and technologies. A physical interface of a router can have multiple logical sub-interfaces. The identification of logical sub-interfaces is different from original physical interface with a mark of "."and a number behind it. For example, "G0/0/0.2" means a logical sub-interface of the physical interface "G0/0/0". Sub-interfaces, as same as physical interfaces, can also perform layer three forwarding and terminate data frames with VLAN Tag.

As shown in Fig. 2.57, router AR is used to realize the communication between VLAN2 and VLAN3 in the switching network. The G0/0/1 interface of Switch A, which is connected to router AR, is configured as Trunk, and two logical sub-interfaces are created on router AR. The VLAN terminated by the logical sub-interface G0/0/0.2 is VLAN2, and the IP address configured by this sub-interface is the gateway address of the host in VLAN2. The VLAN terminated by logical sub-interface G0/0/0.3 is VLAN3, and the IP address configured by this sub-interface is the gateway address of the host in VLAN3.

Note: The physical interface to which the logical sub-interface belongs cannot be configured with an IP address. If the physical interface to which the logical sub-interface belongs has been configured with an IP address, you need to use the "undo ip address" command to delete the IP address configured by the physical interface.

The VLAN configuration of communication using single arm routing generally requires three steps:

① Configurating the switch interface connected to the router as a trunk interface.

② Deleting the IP address of the physical interface of the router connected to the switch, and if the physical interface of the router is disabled, it needs to be enabled.

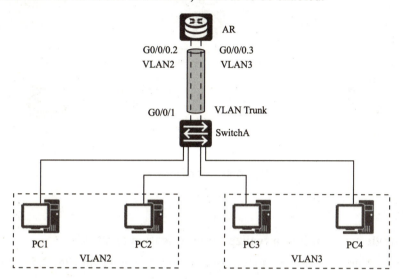

Fig. 2.57   Single arm routing topology diagram

③ Configuring VLAN, IP address and subnet mask of sub-interface termination and enable ARP(address resolution protocol) broadcast function of sub-interface for each logical sub-interface on the router according to the planning.

2. Vlanif interface

Communication between VLAN can also be realized through the Vlanif interface of the three-layer switch. The Vlanif interface is a layer three logical interface, and each VLAN corresponds to a Vlanif interface. The number of the Vlanif interface is the same as the corresponding VLAN ID (VLAN2 corresponds to Vlanif2). After configuring the IP address for the Vlanif interface, the IP address of the interface is the gateway of the host in this VLAN. As shown in Fig. 2.58, Vlanif2 interface and Vlanif3 interface are created on three-layer Switch2. The IP address of Vlanif2 interface is the gateway address of the host in VLAN2, and the IP address of Vlanif3 interface is the gateway address of the host in VLAN3. The configuration of Vlanif interface is simple, which is the most commonly used technology to realize mutual visits of VLAN.

Note: Vlanif1 interface is automatically created by the switch and cannot be deleted. The Vlanif of other VLAN is created by using the "interface Vlanif" command in the system view, and the Vlanif interface can be deleted by using the "undo interface Vlanif" command. When connecting Vlanif interface, you should pay attention to whether the corresponding VLAN already exists in the switch. If it does not exist, the Vlanif interface link protocol cannot be enabled (Up).

146　Enterprise Internetworking Technology (Bilingual)

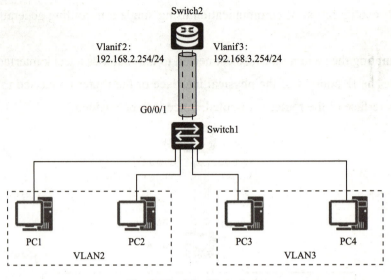

Fig. 2.58　Vlanif interface

Related Skills

1. Using single arm routing to realize communication of VLAN

You should set up the network topology diagram as shown in Fig. 2.59, complete VLAN configuration on the switch according to the VLAN plan shown in Table 2.8, and use the external router AR1 to realize communication of different VLAN by single arm routing, so as to realize interconnection between all hosts.

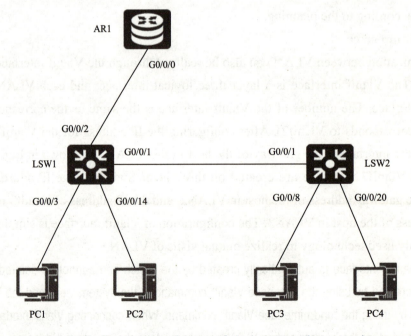

Fig. 2.59　The topology diagram of single arm routing network

The switching network is divided into three VLAN: VLAN3, VLAN4, VLAN99. Where VLAN99 is the management VLAN of the switch .Planning of VLAN is shown in Table 2.9.

Table 2.9 Planning of VLAN

Device	Number of VLAN	Description of VLAN	Interface
LSW1	VLAN3	student	G0/0/3 to G0/0/12 interface
LSW1	VLAN4	teacher	G0/0/13 to G0/0/24 interface
LSW1	VLAN99	manage	
LSW2	VLAN3	student	G0/0/2 to G0/0/12 interfaces
LSW2	VLAN4	teacher	G0/0/13 to G0/0/24 interfaces
LSW2	VLAN99	manage	

IP address planning is shown in Table 2.10, in which VLAN3 is terminated by G0/0/0.3 sub-interface of router AR1, VLAN4 is terminated by G0/0/0.4 sub-interface, and VLAN99 is terminated by G0/0/0.99 sub-interface.

Table 2.10 Planning of IP address

Device	Interface	IP address and mask	Gateway
Router AR1	G0/0/0.3	172.16.3.254/24	NA
Router AR1	G0/0/0.4	172.16.4.254/24	NA
Router AR1	G0/0/0.99	172.16.99.254/24	NA
LSW1	VLAN99	172.16.99.1/24	172.16.99.254
LSW2	VLAN99	172.16.99.2/24	172.16.99.254
PC1	NIC	172.16.3.11/24	172.16.3.254
PC2	NIC	172.16.4.12/24	172.16.4.254
PC3	NIC	172.16.3.13/24	172.16.3.254
PC4	NIC	172.16.4.14/24	172.16.4.254

(1) Configuration of VLAN on switches LSW1 and LSW2

① Refer to the following steps to complete VLAN creation and access interface configuration on switches LSW1 and LSW2.

```
[LSW1]vlan batch 3 4 99
//Create VLAN3, VLAN4 and VLAN99 in batches on switch LSW1
[LSW1]port-group 1
[LSW1-port-group-1]group-member g0/0/3 to g0/0/12
[LSW1-port-group-1]port link-type access
[LSW1-port-group-1]port default vlan 3
[LSW1-port-group-1]quit
```

```
//Create port group 1 on switch LSW1, and use interfaces G0/0/3 to G0/0/12 as members
//of port group 1, and configure all members of port group 1 as access interfaces, belonging
//to VLAN3
[LSW1]port-group 2
[LSW1-port-group-2]group-member g0/0/13 to g0/0/24
[LSW1-port-group-2]port link-type access
[LSW1-port-group-2]port default vlan 4
//Create port group 2 on the switch LSW1, and use interfaces G0/0/13 to G0/0/24 as members
//of the port group 2, and configure all members of the port group 2 as access interfaces,
//belonging to VLAN4
[LSW2]vlan batch 3 4 99
//Create VLAN3, VLAN4 and VLAN99 in batches on switch LSW2
[LSW2]port-group 1
[LSW2-port-group-1]group-member g0/0/2 to g0/0/12
[LSW2-port-group-1]port link-type access
[LSW2-port-group-1]port default vlan 3
[LSW2-port-group-1]quit
//Create port group 1 on switch LSW2, and use interfaces G0/0/2 to G0/0/12 as members
//of port group 1, and configure all members of port group 1 as access interfaces, belonging
//to VLAN3
[LSW2]port-group 2
[LSW2-port-group-2]group-member g0/0/13 to g0/0/24
[LSW2-port-group-2]port link-type access
[LSW2-port-group-2]port default vlan 4
//Create port group 2 on switch LSW2, and use interfaces G0/0/13 to G0/0/24 as members
//of port group 2, and configure all members of port group 2 as access interfaces,
//belonging to VLAN4
```

② Configuration of the trunk interface. According to the network topology diagram shown in Fig. 2.59 and VLAN planning shown in Table 2.9, the links between switch LSW1 and switch LSW2, and between switch LSW1 and router AR1 need to be configured as trunk links, allowing VLAN3, VLAN4 and VLAN99 to pass through. Take Huawei S5700-28C-HI switch as an example, configure the G0/0/1 and G0/0/2 interfaces of switch LSW1, and the G0/0/1 interface of switch LSW2 as trunk interfaces, allowing VLAN3, VLAN4 and VLAN99 to tag through, but not VLAN1. The configuration steps are as follows:

```
[LSW1]interface g0/0/1
[LSW1-GigabitEthernet0/0/1]port link-type trunk
[LSW1-GigabitEthernet0/0/1]port trunk allow-pass vlan 3 4 99
[LSW1-GigabitEthernet0/0/1]undo port trunk allow-pass vlan 1
```

```
[LSW1-GigabitEthernet0/0/1]quit
[LSW1]interface g0/0/2
[LSW1-GigabitEthernet0/0/2]port link-type trunk
[LSW1-GigabitEthernet0/0/2]port trunk allow-pass vlan 3 4 99
[LSW1-GigabitEthernet0/0/2]undo port trunk allow-pass vlan 1
[LSW1-GigabitEthernet0/0/2]quit
[LSW2]interface g0/0/1
[LSW2-GigabitEthernet0/0/1]port link-type trunk
[LSW2-GigabitEthernet0/0/1]port trunk allow-pass vlan 3 4 99
[LSW2-GigabitEthernet0/0/1]undo port trunk allow-pass vlan 1
[LSW2-GigabitEthernet0/0/1]quit
```

(2) Configuration of router sub-interface

Complete sub-interface configurations on the router AR1 as planned in Table 2.9, specifically as follows:

```
[AR1]interface g0/0/0    //Enter the interface view of G0/0/0 interface
[AR1-GigabitEthernet0/0/0]undo ip address
//Delete the IP address of G0/0/1 interface,if no IP address is configured, you can
//not configure this command
[AR1-GigabitEthernet0/0/0]quit   //Return to user view
[AR1]interface g0/0/0.3  //Enter the interface view of G0/0/0.3 subinterface
[AR1-GigabitEthernet0/0/0.3]dot1q termination vid 3
//Configure the VLAN terminated by the subinterface as VLAN3
[AR1-GigabitEthernet0/0/0.3]ip address 172.16.3.254 24
//Configure the IP address and subnet mask of the subinterface is "172.16.3.254/24"
[AR1-GigabitEthernet0/0/0.3]arp broadcast enable
//Enable the ARP broadcast function of the subinterface
[AR1-GigabitEthernet0/0/0.3]quit
[AR1]interface g0/0/0.4  //Enter the interface view of G0/0/0.4 subinterface
[AR1-GigabitEthernet0/0/0.4]dot1q termination vid 4
//Configure the VLAN terminated by the subinterface as VLAN4
[AR1-GigabitEthernet0/0/0.4]ip address 172.16.4.254 24
//Configure the IP address and subnet mask of the subinterface is "172.16.4.254/24"
[AR1-GigabitEthernet0/0/0.4]arp broadcast enable
//Enable the ARP broadcast function of the sub-port
[AR1-GigabitEthernet0/0/0.4]quit
[AR1]interface g0/0/0.99 //Enter the interface view of G0/0/0.99 subinterface
[AR1-GigabitEthernet0/0/0.99]dot1q termination vid 99
//Configure the subinterface to belong to VLAN3
[AR1-GigabitEthernet0/0/0.99]ip address 172.16.99.254 24
```

```
//Configure the IP address and subnet mask of the subinterface is "172.16.99.254/24"
[AR1-GigabitEthernet0/0/0.99]arp broadcast enable
//Enable the ARP broadcast function of the subinterface
[AR1-GigabitEthernet0/0/0.99]quit
```

(3) Single arm routing verification and network testing

① You can use the "display ip interface brief" command on router AR1 to view the IP brief information of all interfaces, and pay attention to the IP, physical and protocol status of sub-interfaces, as shown in Fig. 2.60.

```
[AR1]display ip interface brief
*down: administratively down
^down: standby
(l): loopback
(s): spoofing
The number of interface that is UP in Physical is 5
The number of interface that is DOWN in Physical is 2
The number of interface that is UP in Protocol is 4
The number of interface that is DOWN in Protocol is 3

Interface                IP Address/Mask      Physical  Protocol
GigabitEthernet0/0/0     unassigned           up        down
GigabitEthernet0/0/0.3   172.16.3.254/24      up        up
GigabitEthernet0/0/0.4   172.16.4.254/24      up        up
GigabitEthernet0/0/0.99  172.16.99.254/24     up        up
GigabitEthernet0/0/1     unassigned           down      down
GigabitEthernet0/0/2     unassigned           down      down
NULL0                    unassigned           up        up(s)
```

Fig. 2.60 The example of "display ip interface brief" command to view the IP brief information and status of all interfaces

② You can use the "display ip routing-table" command on router AR1 to view the routing table, as shown in Fig. 2.61. Pay attention to whether the three directly connected network segments "172.16.3.0/24", "172.16.4.0/24" and "172.16.99.0/24" are enlisted in the routing table.

③ You should complete the IP address, subnet mask and default gateway configuration of the test hosts PC1, PC2, PC3 and PC4 with reference to the planned parameters in Table 2.9. After the configuration is completed, use the "ipconfig /all" command at the DOS command line to check whether their IP configuration information is correct.

```
[AR1]display ip routing-table
Route Flags: R - relay, D - download to fib
------------------------------------------------------------

Routing Tables: Public
        Destinations : 13    Routes : 13

Destination/Mask      Proto   Pre  Cost  Flags  NextHop        Interface

        127.0.0.0/8   Direct  0    0     D      127.0.0.1      InLoopBack0
       127.0.0.1/32   Direct  0    0     D      127.0.0.1      InLoopBack0
 127.255.255.255/32   Direct  0    0     D      127.0.0.1      InLoopBack0
      172.16.3.0/24   Direct  0    0     D      172.16.3.254   GigabitEthernet0/0/0.3
    172.16.3.254/32   Direct  0    0     D      127.0.0.1      GigabitEthernet0/0/0.3
    172.16.3.255/32   Direct  0    0     D      127.0.0.1      GigabitEthernet0/0/0.3
      172.16.4.0/24   Direct  0    0     D      172.16.4.254   GigabitEthernet0/0/0.4
    172.16.4.254/32   Direct  0    0     D      127.0.0.1      GigabitEthernet0/0/0.4
    172.16.4.255/32   Direct  0    0     D      127.0.0.1      GigabitEthernet0/0/0.4
     172.16.99.0/24   Direct  0    0     D      172.16.99.254  GigabitEthernet0/0/0.99
   172.16.99.254/32   Direct  0    0     D      127.0.0.1      GigabitEthernet0/0/0.99
   172.16.99.255/32   Direct  0    0     D      127.0.0.1      GigabitEthernet0/0/0.99
 255.255.255.255/32   Direct  0    0     D      127.0.0.1      InLoopBack0
```

Fig. 2.61　View the routing table of router AR1

④ You should complete the configuration of management IP address and default gateway of switches LSW1 and LSW2 with reference to the planned parameters in Table 2.9. The configuration steps are as follows:

```
[LSW1]interface Vlanif 99
[LSW1-Vlanif99]ip address 172.16.99.1 24
[LSW1-Vlanif99]quit
//Configure the management IP address of switch LSW1 as "172.16.99.1/24"
[LSW1]ip route-static 0.0.0.0 0.0.0.0 172.16.99.254
//Configure the default gateway of switch LSW1 as "172.16.99.254"
[LSW2]interface Vlanif 99
[LSW2-Vlanif99]ip address 172.16.99.2 24
[LSW2-Vlanif99]quit
//Configure the management IP address of switch LSW2 as "172.16.99.2/24"
[LSW2]ip route-static 0.0.0.0 0.0.0.0 172.16.99.254
//Configure the default gateway of switch LSW2 as "172.16.99.254"
```

After completing the management IP address and default gateway configuration of switches LSW1 and LSW2, use the "display ip interface brief" command to view the IP information and status of the management VLAN interface. Fig. 2.62 shows an example of viewing IP information and status of management VLAN interface on LSW1 switch. Use the "display IP routing-table | include Static" command to check whether the default gateway is correct and effective. Fig. 2.63 shows the default gateway for viewing switch LSW1.

⑤ On any host in the network, enter the DOS command line and use the ping utility to test

the IP connectivity between other target hosts and the switch. Fig. 2.64 shows that the host PC1 can ping the IP address of the switch LSW1, which realizes the communication between different VLAN.

```
[LSW1]display ip interface brief
*down: administratively down
^down: standby
(l): loopback
(s): spoofing
The number of interface that is UP in Physical is 2
The number of interface that is DOWN in Physical is 2
The number of interface that is UP in Protocol is 2
The number of interface that is DOWN in Protocol is 2

Interface            IP Address/Mask    Physical  Protocol
MEth0/0/1            unassigned         down      down
NULL0                unassigned         up        up(s)
Vlanif1              unassigned         down      down
Vlanif99             172.16.99.1/24     up        up
```

Fig. 2.62 The example of "display ip interface brief" command to view the IP information and status of the management VLAN interface

```
[LSW1]display ip routing-table | include Static
Route Flags: R - relay, D - download to fib
------------------------------------------------------------------
Routing Tables: Public
        Destinations : 5      Routes : 5

Destination/Mask   Proto  Pre  Cost   Flags NextHop        Interface

    0.0.0.0/0    Static  60   0       RD    172.16.99.254  Vlanif99
```

Fig. 2.63 The example of viewing the default gateway of switch LSW1

```
PC>ping 172.16.99.1

Ping 172.16.99.1: 32 data bytes, Press Ctrl_C to break
From 172.16.99.1: bytes=32 seq=1 ttl=254 time=78 ms
From 172.16.99.1: bytes=32 seq=2 ttl=254 time=63 ms
From 172.16.99.1: bytes=32 seq=3 ttl=254 time=31 ms
From 172.16.99.1: bytes=32 seq=4 ttl=254 time=47 ms
From 172.16.99.1: bytes=32 seq=5 ttl=254 time=47 ms

--- 172.16.99.1 ping statistics ---
  5 packet(s) transmitted
  5 packet(s) received
  0.00% packet loss
  round-trip min/avg/max = 31/53/78 ms
```

Fig. 2.64 Host PC1 can ping the IP address of the switch LSW1

2. Using three layer switch to realize communication between VLAN

It is required to build a network topology diagram as shown in Fig. 2.65. According to the VLAN plan shown in Table 2.8, the VLAN configuration is completed on switches LSW1 and LSW2, and then the communication between VLANs is realized by using three-layer switch LSW3, so as to realize the interconnection between all hosts.

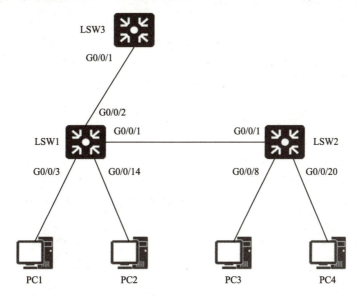

Fig. 2.65 Network topology diagram of communication between VLAN using three-layer switch

The switching network is divided into three VLANs: VLAN3, VLAN4, VLAN99, in which VLAN99 is the management VLAN of the switch. See Table 2.9 for VLAN planning.

See Table 2.11 for the planning of IP address.

Table 2.11 Planning of IP address

Device	Interface	IP address and mask	Gateway
LSW3	Vlanif3	172.16.3.254/24	
	Vlanif4	172.16.4.254/24	
	Vlanif99	172.16.99.254/24	
LSW1	Vlanif99	172.16.99.1/24	172.16.99.254
LSW2	Vlanif99	172.16.99.2/24	172.16.99.254
PC1	NIC	172.16.3.11/24	172.16.3.254
PC2	NIC	172.16.4.12/24	172.16.4.254
PC3	NIC	172.16.3.13/24	172.16.3.254
PC4	NIC	172.16.4.14/24	172.16.4.254

(1) Configuration of VLAN on switches LSW1 and LSW2

① Refer to the following steps to complete VLAN creation and access port configuration on switches LSW1 and LSW2.

```
[LSW1]vlan batch 3 4 99
//Create VLAN3, VLAN4 and VLAN99 in batches on switch LSW1
[LSW1]port-group 1
[LSW1-port-group-1]group-member g0/0/3 to g0/0/12
[LSW1-port-group-1]port link-type access
[LSW1-port-group-1]port default vlan 3
[LSW1-port-group-1]quit
//Create port group 1 on switch LSW1, and use interfaces G0/0/3 to G0/0/12 as members
//of port group 1, and configure all members of port group 1 as access interfaces,belonging
//to VLAN3
[LSW1]port-group 2
[LSW1-port-group-2]group-member g0/0/13 to g0/0/24
[LSW1-port-group-2]port link-type access
[LSW1-port-group-2]port default vlan 4
//Create port group 2 on switch LSW1, and use interfaces G0/0/13 to G0/0/24 as members
//of port group 2, and configure all members of port group 2 as access interfaces,belonging
//to VLAN4
[LSW2]vlan batch 3 4 99
//Create VLAN3, VLAN4 and VLAN99 in batches on switch LSW2.
[LSW2]port-group 1
[LSW2-port-group-1]group-member g0/0/2 to g0/0/12
[LSW2-port-group-1]port link-type access
[LSW2-port-group-1]port default vlan 3
[LSW2-port-group-1]quit
//Create port group 1 on switch LSW2, and use interfaces G0/0/2 to G0/0/12 as members
//of port group 1, and configure all members of port group 1 as access interfaces, belonging
//to VLAN3
[LSW2]port-group 2
[LSW2-port-group-2]group-member g0/0/13 to g0/0/24
[LSW2-port-group-2]port link-type access
[LSW2-port-group-2]port default vlan 4
//Create port group 2 on switch LSW2, and use interfaces G0/0/13 to G0/0/24 as members
//of port group 2, and configure all members of port group 2 as access interfaces, belonging
//to VLAN4
```

② Configuration of the trunk interface. According to the network topology diagram shown in Fig. 2.65 and VLAN planning shown in Table 2.8, the links between switch LSW1 and switch LSW2 and between switch LSW1 and switch LSW3 in the figure need to be configured as trunk links, allowing VLAN3, VLAN4 and VLAN99 to pass through. Take Huawei S5700-28C-HI switch as an example, configure the G0/0/1 and G0/0/2 interfaces of switch LSW1 as trunk interfaces, and

the G0/0/1 interface of switch LSW2 as relay ports, allowing VLAN3, VLAN4 and VLAN99 to tag through, but not VLAN1. The configuration steps are as follows:

```
[LSW1]interface g0/0/1
[LSW1-GigabitEthernet0/0/1]port link-type trunk
[LSW1-GigabitEthernet0/0/1]port trunk allow-pass vlan 3 4 99
[LSW1-GigabitEthernet0/0/1]undo port trunk allow-pass vlan 1
[LSW1-GigabitEthernet0/0/1]quit
[LSW1]interface g0/0/2
[LSW1-GigabitEthernet0/0/2]port link-type trunk
[LSW1-GigabitEthernet0/0/2]port trunk allow-pass vlan 3 4 99
[LSW1-GigabitEthernet0/0/24]undo port trunk allow-pass vlan 1
[LSW1-GigabitEthernet0/0/2]4quit
[LSW2]interface g0/0/1
[LSW2-GigabitEthernet0/0/1]port link-type trunk
[LSW2-GigabitEthernet0/0/1]port trunk allow-pass vlan 3 4 99
[LSW2-GigabitEthernet0/0/1]undo port trunk allow-pass vlan 1
[LSW2-GigabitEthernet0/0/1]quit
```

(2) Configuration of three-layer switch LSW3

① Planned VLAN need to be created in three-layer switch LSW3. The steps for creating VLAN3, VLAN4 and VLAN99 in batches are as follows:

```
[LSW3]vlan batch 3 4 99    ///Create VLAN3, VLAN4 and VLAN99 in batches on switch LSW3
```

② Configuration of the trunk interface. According to the network topology diagram shown in Fig. 2.65 and VLAN planning shown in Table 2.9, the link between LSW1 and LSW3 is a trunk link, and the G0/0/1 interface of LSW3 needs to be configured as a trunk interface, allowing VLAN3, VLAN4 and VLAN99 to pass with tags, but not VLAN1. The configuration steps are as follows:

```
[LSW3]interface g0/0/1
[LSW3-GigabitEthernet0/0/1]port link-type trunk
[LSW3-GigabitEthernet0/0/1]port trunk allow-pass vlan 3 4 99
[LSW3-GigabitEthernet0/0/1]undo port trunk allow-pass vlan 1
[LSW3-GigabitEthernet0/0/1]quit
```

③ Configuring interfaces of Vlanif2, Vlanif3 and Vlanif99 interfaces. The steps of creating Vlanif3, Vlanif4 and Vlanif99 interfaces on LSW3 and configuring the IP address and mask planned in Table 2.10 are as follows:

```
[LSW3]interface Vlanif 3  //Create the Vlanif3 interface and enter the interface view
                          //of Vlanif3 interface
```

```
[LSW3-Vlanif3]ip address 172.16.3.254 24      //Configure the IP address and subnet
                                              //mask of the interface as "172.16.3.254/24"
[LSW3-Vlanif3]quit
[LSW3]interface Vlanif 4 //Create the Vlanif4 interface and enter the interface view
                   //of Vlanif4 interface
[LSW3-Vlanif4]ip address 172.16.4.254 24      //Configure the IP address and subnet
                                              //mask of the interface as "172.16.3.254/24"
[LSW3-Vlanif4]quit
[LSW3]interface Vlanif 99       //Create the Vlanif99 interface and enter the Vlanif99
                   //interface view
[LSW3-Vlanif99]ip address 172.16.99.254 24    //Configure the IP address and subnet
                                              //mask of the interface as "172.16.99.254/24"
[LSW3-Vlanif99]quit
```

(3) Verification and network testing

① You can use the "display ip interface brief" command on the three-layer switch LSW3 to view the IP brief information of all interfaces, and you should be careful to the IP information, physical and protocol status of Vlanif3, Vlanif4 and Vlanif99 interfaces, as shown in Fig. 2.66.

```
[LSW3]display ip interface brief
*down: administratively down
^down: standby
(l): loopback
(s): spoofing
The number of interface that is UP in Physical is 4
The number of interface that is DOWN in Physical is 2
The number of interface that is UP in Protocol is 4
The number of interface that is DOWN in Protocol is 2

Interface           IP Address/Mask     Physical  Protocol
MEth0/0/1           unassigned          down      down
NULL0               unassigned          up        up(s)
Vlanif1             unassigned          down      down
Vlanif3             172.16.3.254/24     up        up
Vlanif4             172.16.4.254/24     up        up
Vlanif99            172.16.99.254/24    up        up
```

Fig. 2.66　The example of "display ip interface brief" command to view the interface IP brief information and status

② You can use the "display ip routing-table" command on the three-layer switch to view the routing table, as shown in Fig. 2.67. Be careful to whether the three directly connected network segments "172.16.3.0/24", "172.16.4.0/24" and "172.16.99.0/24" are connected or not.

```
[LSW3]display ip routing-table
Route Flags: R - relay, D - download to fib
------------------------------------------------------------------
Routing Tables: Public
        Destinations : 8      Routes : 8

Destination/Mask    Proto  Pre Cost   Flags  NextHop        Interface
        127.0.0.0/8   Direct  0   0      D    127.0.0.1      InLoopBack0
       127.0.0.1/32   Direct  0   0      D    127.0.0.1      InLoopBack0
       172.16.3.0/24  Direct  0   0      D    172.16.3.254   Vlanif3
    172.16.3.254/32   Direct  0   0      D    127.0.0.1      Vlanif3
       172.16.4.0/24  Direct  0   0      D    172.16.4.254   Vlanif4
    172.16.4.254/32   Direct  0   0      D    127.0.0.1      Vlanif4
      172.16.99.0/24  Direct  0   0      D    172.16.99.254  Vlanif99
   172.16.99.254/32   Direct  0   0      D    127.0.0.1      Vlanif99
```

Fig. 2.67 The routing table of three-layer switch LSW3

③ You can refer to the parameters planned in Table 2.10 to complete the IP address, subnet mask and default gateway configuration of the test hosts PC1, PC2, PC3 and PC4. After configuration, you can use the "ipconfig /all" command at the DOS command line to check whether their IP configuration information is correct.

④ For completing the configuration of management IP address and default gateway of switches LSW1 and LSW2 with reference to the planned parameters in Table 2.10, the configuration steps are as follows:

```
[LSW1]interface Vlanif 99
[LSW1-Vlanif99]ip address 172.16.99.1 24
[LSW1-Vlanif99]quit
//Configure the management IP address of switch LSW1 as "172.16.99.1/24"
[LSW1]ip route-static 0.0.0.0 0.0.0.0 172.16.99.254
//Configure the default gateway of switch LSW1 as "172.16.99.254"
[LSW2]interface Vlanif 99
[LSW2-Vlanif99]ip address 172.16.99.2 24
[LSW2-Vlanif99]quit
//Configure the management IP address of switch LSW2 as "172.16.99.2/24"
[LSW2]ip route-static 0.0.0.0 0.0.0.0 172.16.99.254
//Configure the default gateway of switch LSW2 as "172.16.99.254"
```

After completing the management IP address and default gateway configuration of switches LSW1 and LSW2, you can use the "display ip interface brief" command to view the IP information and status of the management VLAN interface. Fig. 2.68 shows an example of viewing IP information and status of management VLAN interface on switch LSW1. Then you can use the "display IP routing-table | include Static" command to check whether the default gateway is correct and effective shown as Fig. 2.69(viewing the default gateway on switch LSW1).

```
[LSW1]display ip interface brief
*down: administratively down
^down: standby
(l): loopback
(s): spoofing
The number of interface that is UP in Physical is 2
The number of interface that is DOWN in Physical is 2
The number of interface that is UP in Protocol is 2
The number of interface that is DOWN in Protocol is 2

Interface            IP Address/Mask    Physical  Protocol
MEth0/0/1            unassigned         down      down
NULL0                unassigned         up        up(s)
Vlanif1              unassigned         down      down
Vlanif99             172.16.99.1/24     up        up
```

Fig. 2.68 Example of viewing IP information and status of management VLAN interface on switch LSW1

```
[LSW1]display ip routing-table | include Static
Route Flags: R - relay, D - download to fib
------------------------------------------------
Routing Tables: Public
        Destinations : 5    Routes : 5

Destination/Mask  Proto   Pre  Cost   Flags NextHop       Interface

    0.0.0.0/0    Static   60   0        RD  172.16.99.254 Vlanif99
```

Fig. 2.69 The default gateway on switch LSW1

⑤ On any host in the network, you can enter the DOS command line and use the ping utility to test the IP connectivity between other target hosts and the switch. Fig. 2.70 shows that the host PC1 can ping the IP address of the switch LSW1, which realizes the communication between different VLAN.

```
PC>ping 172.16.99.1

Ping 172.16.99.1: 32 data bytes, Press Ctrl_C to break
From 172.16.99.1: bytes=32 seq=1 ttl=254 time=78 ms
From 172.16.99.1: bytes=32 seq=2 ttl=254 time=63 ms
From 172.16.99.1: bytes=32 seq=3 ttl=254 time=31 ms
From 172.16.99.1: bytes=32 seq=4 ttl=254 time=47 ms
From 172.16.99.1: bytes=32 seq=5 ttl=254 time=47 ms

--- 172.16.99.1 ping statistics ---
  5 packet(s) transmitted
  5 packet(s) received
  0.00% packet loss
  round-trip min/avg/max = 31/53/78 ms
```

Fig. 2.70 Host PC1 can ping the IP address of switch LSW1

Task Completion

1. Task planning

According to the task description and the networking requirements of Project 2, Table 2.12 gives the IP address reference plan for communication between VLANs. The VLAN terminated by the G0/0/0.10 sub-interface of router AR1 is VLAN10, and the VLAN terminated by the G0/0/0.20 sub-interface is VLAN20.

Table 2.12 Planning of IP address

Device	Interface	IP address and mask	Gateway
Router AR1	G0/0/0	172.16.254.1/30	NA
	G0/0/1.10	172.16.10.254/24	NA
	G0/0/1.20	172.16.20.254/24	NA
	S4/0/0	125.71.28.93/24	NA
Router ISP	S4/0/0	125.71.28.1/24	NA
	G0/0/0	210.33.44.254/24	NA
CORE	Vlanif2	172.16.2.254/24	NA
	Vlanif3	172.16.3.254/24	NA
	Vlanif4	172.16.4.254/24	NA
	Vlanif99	172.16.99.254/24	NA
	Vlanif100	172.16.100.254/24	NA
	Vlanif101	172.16.254.2/30	NA
FLOOR1	Vlanif99	172.16.99.1/24	172.16.99.254
FLOOR2	Vlanif99	172.16.99.2/24	172.16.99.254
FLOOR3	Vlanif99	172.16.99.3/24	172.16.99.254
Web server	NIC	172.16.100.1/24	172.16.100.254
FTP server	NIC	172.16.100.2/24	172.16.100.254
DNS server	NIC	172.16.100.3/24	172.16.100.254
WWW	NIC	210.33.44.1/24	210.33.44.254
CLIENT	NIC	210.33.44.2/24	210.33.44.254

See Table 2.13 for the planning of DHCP address pool on router AR1 and switch CORE

Table 2.13 Planning of DHCP address pool

Device	Address pool	IP address and mask	Gateway	DNS address	Address lease period	Exclusion address
AR1	vlan10	172.16.10.0/24	172.16.10.254	172.16.100.3	8 h	172.16.10.50 -172.16.10.70
	vlan20	172.16.20.0/24	172.16.20.254	172.16.100.3	8 h	

Continued

Device	Address pool	IP address and mask	Gateway	DNS address	Address lease period	Exclusion address
CORE	vlan2	172.16.2.0/24	172.16.2.254	172.16.100.3	8 h	
	vlan3	172.16.3.0/24	172.16.3.254	172.16.100.3	8 h	
	vlan4	172.16.4.0/24	172.16.4.254	172.16.100.3	8 h	

2. Task implementation

(1) Configuration of router AR1

① According to the IP address planning shown in Table 2.12, the steps to configure the sub-interface and interface IP of router AR1 are as follows:

```
[AR1]interface g0/0/1
[AR1-GigabitEthernet0/0/1]undo ip address
//Delete the IP address of G0/0/0 interface
[AR1-GigabitEthernet0/0/1]quit
[AR1]interface g0/0/1.10
[AR1-GigabitEthernet0/0/1.10]dot1q termination vid 10
[AR1-GigabitEthernet0/0/1.10]ip address 172.16.10.254 24
[AR1-GigabitEthernet0/0/1.10]arp broadcast enable
[AR1-GigabitEthernet0/0/1.10]quit
//Create sub-interface G0/0/1.10, configure the VLAN terminated by sub-interface as
//VLAN10, and the IP address and subnet mask as "172.16.10.254/24", and enable the
//ARP broadcast function of sub-interface
[AR1]interface g0/0/1.20
[AR1-GigabitEthernet0/0/1.20]dot1q termination vid 20
[AR1-GigabitEthernet0/0/1.20]ip address 172.16.20.254 24
[AR1-GigabitEthernet0/0/1.20]arp broadcast enable
[AR1-GigabitEthernet0/0/1.20]quit
//Create sub-interface G0/0/1.20, configure the VLAN terminated by sub-interface as
//VLAN20, IP address and subnet mask as "172.16. 20.254/24", and enable the ARP
//broadcast function of subinterface
[AR1]interface g0/0/0
[AR1-GigabitEthernet0/0/0]ip address 172.16.254.1 30
[AR1-GigabitEthernet0/0/0]quit
//Configure the IP address and subnet mask of G0/0/0 interface as "172.16.254.1/30"
[AR1]interface s4/0/0
[AR1-Serial4/0/0]ip address  125.71.28.93 24
[AR1-Serial4/0/0]quit
//Configure the IP address and subnet mask of S4/0/0 interface as "125.71.28.93/24"
```

② Configuration of DHCP. According to Table 2.13, the steps of configuring DHCP on router AR1 are as follows:

```
[AR1]dhcp enable
//Enable DHCP service
[AR1]ip pool vlan10
[AR1-ip-pool-vlan10]network 172.16.10.0 mask 24
[AR1-ip-pool-vlan10]gateway-list 172.16.10.254
[AR1-ip-pool-vlan10]dns-list 172.16.100.3
[AR1-ip-pool-vlan10]excluded-ip-address 172.16.10.50 172.16.10.70
[AR1-ip-pool-vlan10]lease day 0 hour 8
[AR1-ip-pool-vlan10]quit
//Configure DHCP address pool "vlan10" with network number "172.16.10.0/24"; The gate-
//way is "172.16.10.254"; The DNS address is "172.16.100.3", the address lease period
//is 8 h, and the exclusion address is "172.16.10.50-172.16.10.70"
[AR1]interface g0/0/1.10
[AR1-GigabitEthernet0/0/1.10]dhcp select global
[AR1-GigabitEthernet0/0/1.10]quit
//Turn on the DHCP Server function of G0/0/1.10 sub-interface with global address pool
[AR1]ip pool vlan20
[AR1-ip-pool-vlan20]network 172.16.20.0 mask 24
[AR1-ip-pool-vlan20]gateway-list 172.16.20.254
[AR1-ip-pool-vlan20]dns-list 172.16.100.3
[AR1-ip-pool-vlan20]lease day 0 hour 8
[AR1-ip-pool-vlan20]quit
//Configure DHCP address pool "vlan20" with network number "172.16.20.0/24"; The gate-
//way is "172.16.20.254"; The DNS is "172.16.100.3" and the address lease period is 8 h
[AR1]interface g0/0/1.20
[AR1-GigabitEthernet0/0/1.20]dhcp select global
[AR1-GigabitEthernet0/0/1.20]quit
//Turn on the DHCP Server function of G0/0/1.20 sub-interface with global address pool
```

(2) Configuration of switch CORE

① The steps of creating Vlanif2, Vlanif3, Vlanif4, Vlanif99, Vlanif100 and Vlanif101 interfaces on the switch CORE and configuring the IP address and mask planned in Table 2.12 are as follows:

```
[CORE]interface Vlanif 2
[CORE-Vlanif2]ip address 172.16.2.254 24
[CORE-Vlanif2]quit
//Create and enter the Vlanif2 interface, and configure the IP address and subnet mask
```

```
//of the Vlanif2 interface as "172.16.2.254/24"
[CORE]interface Vlanif 3
[CORE-Vlanif3]ip address 172.16.3.254 24
[CORE-Vlanif3]quit
//Create and enter the Vlanif3 interface, and configure the IP address and subnet mask
//of the Vlanif3 interface as "172.16.3.254/24"
[CORE]interface Vlanif 4
[CORE-Vlanif4]ip address 172.16.4.254 24
[CORE-Vlanif4]quit
//Create and enter the Vlanif4 interface, and configure the IP address and subnet mask
//of the Vlanif4 interface as "172.16.4.254/24"
[CORE]interface Vlanif 99
[CORE-Vlanif99]ip address 172.16.99.254 24
[CORE-Vlanif99]quit
//Create and enter the Vlanif99 interface, and configure the IP address and subnet mask
//of the Vlanif99 interface as "172.16.99.254/24"
[CORE]interface Vlanif 100
[CORE-Vlanif100]ip address 172.16.100.254 24
[CORE-Vlanif100]quit
//Create and enter the Vlanif100 interface, and configure the IP address and subnet mask
//of the Vlanif100 interface as "172.16.100.254/24"
[CORE]interface Vlanif 101
[CORE-Vlanif101]ip address 172.16.254.2 30
[CORE-Vlanif101]quit
//Create and enter the Vlanif101 interface, and configure the IP address and subnet mask
//of the Vlanif101 interface as "172.16.254.2/30"
```

② Configuration of DHCP. According to Table 2.13, the steps of configuring DHCP on the switch CORE are as follows:

```
[CORE]dhcp enable
//Enable DHCP service
[CORE]ip pool vlan2
[CORE-ip-pool-vlan2]network 172.16.2.0 mask 24
[CORE-ip-pool-vlan2]gateway-list 172.16.2.254
[CORE-ip-pool-vlan2]dns-list 172.16.100.3
[CORE-ip-pool-vlan2]lease day 0 hour 8
[CORE-ip-pool-vlan2]quit
//Configure DHCP address pool "vlan2" with network number "172.16.2.0/24"; The gateway
//is "172.16.2.254"; The DNS address is "172.16.100.3", the address lease time is 8 h
[CORE]interface Vlanif 2
```

```
[CORE-Vlanif2]dhcp select global
[CORE-Vlanif2]quit
//Turn on the DHCP Server function of Vlanif 2 interface with global address pool
[CORE]ip pool vlan3
[CORE-ip-pool-vlan3]network 172.16.3.0 mask 24
[CORE-ip-pool-vlan3]gateway-list 172.16.3.254
[CORE-ip-pool-vlan3]dns-list 172.16.100.3
[CORE-ip-pool-vlan3]lease day 0 hour 8
[CORE-ip-pool-vlan3]quit
//Configure DHCP address pool "vlan3" with network number "172.16.3.0/24"; The gateway
//is "172.16.3.254"; The DNS is "172.16.100.3" and the address lease time is 8 h
[CORE]interface Vlanif 3
[CORE-Vlanif3]dhcp select global
[CORE-Vlanif3]quit
//Turn on the DHCP Server function of Vlanif 3 interface with global address pool
[CORE]ip pool vlan4
[CORE-ip-pool-vlan4]network 172.16.4.0 mask 24
[CORE-ip-pool-vlan4]gateway-list 172.16.4.254
[CORE-ip-pool-vlan4]dns-list 172.16.100.3
[CORE-ip-pool-vlan4]lease day 0 hour 8
[CORE-ip-pool-vlan4]quit
//Configure DHCP address pool "vlan4" with network number "172.16.4.0/24"; The gateway
//is "172.16.4.254"; The DNS is "172.16.100.3" and the address lease time is 8 h
[CORE]interface Vlanif 4
[CORE-Vlanif4]dhcp select global
[CORE-Vlanif4]quit
//Turn on the DHCP Server function of Vlanif 4 interface with global address pool
```

(3) Configuration of management IP address of access layer switch

According to Table 2.12, with the reference of the following steps on the access layer switches FLOOR1, FLOOR2 and FLOOR3, you can complete the configuration of management IP and default gateway.

```
[FLOOR1]interface Vlanif 99
[FLOOR1-Vlanif99]ip address 172.16.99.1 24
[FLOOR1]quit
[FLOOR1]ip route-static 0.0.0.0 0.0.0.0 172.16.99.254
//Configure the management IP address of switch FLOOR1 as "172.16.99.1/24" and the default
//gateway as "172.16.99.254"
[FLOOR2]interface Vlanif 99
[FLOOR2-Vlanif99]ip address 172.16.99.2 24
```

```
    [FLOOR2]quit
    [FLOOR2]ip route-static 0.0.0.0 0.0.0.0 172.16.99.254
    //Configure the management IP address of switch FLOOR2 as "172.16.99.2/24" and the default
    //gateway as "172.16.99.254"
    [FLOOR3]interface Vlanif 99
    [FLOOR3-Vlanif99]ip address 172.16.99.3 24
    [FLOOR3]quit
    [FLOOR3]ip route-static 0.0.0.0 0.0.0.0 172.16.99.254
    //Configure the management IP address of switch FLOOR3 as "172.16.99.3/24" and the default
    //gateway as "172.16.99.254"
```

(4) Verification and network testing

After completing the task configuration, theoretically, the test hosts of enterprise leaders (VLAN10), human resources (VLAN20), R&D (VLAN2), sales (VLAN3) and after-sales (VLAN4) departments can obtain IP addresses from DHCP servers. The test hosts (PC 1 and PC 2) of enterprise leader (VLAN 10) and human resources (VLAN 20) ping each other, the test hosts (PC3、PC4、PC5), intranet server (VLAN 100), FLOOR1 management address, FLOOR2 management address and FLOOR3 management address for R&D (VLAN 2), sales (VLAN 3) and after-sales (VLAN 4) can all be ping.

① You are advised to use the "display ip interface brief" command to view the IP brief information and status of interfaces, and pay attention to the IP addresses and subnet masks of interfaces and sub-interfaces configured by the router, and the status of physical interfaces and link protocols, as shown in Fig. 2.71.

② It is required to enter the router AR1 sub-interface and check the configuration of the sub-interface. Fig. 2.72 shows the configuration of router AR1 sub-interface G0/0/1.10.

③ It will be obtained automatically to configure the IP addresses and DNS addresses of the test hosts PC1 and PC2, then you can use the"ipconfig"command at the command line to check the obtained IP information, and use the ping command to test the IP connectivity between PC1 and PC2. Fig. 2.73 shows an example of viewing IP information on PC1 and ping the IP address of PC2.

④ You are advised to use the "display ip pool name vlan10 used" command on the router to check the used addresses in the address pool "vlan10", as shown in Fig. 2.74. The figure shows that "172.16.10.253" has been dynamically assigned to the host with the MAC address of "5489- 9872- 427a" (that is, the test host PC1).

```
[AR1]display ip interface brief
*down: administratively down
^down: standby
(l): loopback
(s): spoofing
The number of interface that is UP in Physical is 6
The number of interface that is DOWN in Physical is 2
The number of interface that is UP in Protocol is 5
The number of interface that is DOWN in Protocol is 3

Interface                IP Address/Mask      Physical   Protocol
GigabitEthernet0/0/0     172.16.254.1/30      up         up
GigabitEthernet0/0/1     unassigned           up         down
GigabitEthernet0/0/1.10  172.16.10.254/24     up         up
GigabitEthernet0/0/1.20  172.16.20.254/24     up         up
GigabitEthernet0/0/2     unassigned           down       down
NULL0                    unassigned           up         up(s)
Serial4/0/0              125.71.28.93/24      up         up
Serial4/0/1              unassigned           down       down
```

Fig. 2.71 The example of the "display ip interface brief" command to view the interface IP brief information and status.

```
[AR1-GigabitEthernet0/0/1.10]display this
[V200R003C00]
#
interface GigabitEthernet0/0/1.10
 dot1q termination vid 10
 ip address 172.16.10.254 255.255.255.0
 arp broadcast enable
 dhcp select global
#
return
[AR1-GigabitEthernet0/0/1.10]
```

Fig. 2.72 Viewing the configuration of router AR1 sub-interface G0/0/1.10

```
PC>ipconfig

Link local IPv6 address..............: fe80::5689:98ff:fe72:427a
IPv6 address.........................: :: / 128
IPv6 gateway.........................: ::
IPv4 address.........................: 172.16.10.253
Subnet mask..........................: 255.255.255.0
Gateway..............................: 172.16.10.254
Physical address.....................: 54-89-98-72-42-7A
DNS server...........................: 172.16.100.3

PC>ping 172.16.20.253

Ping 172.16.20.253: 32 data bytes, Press Ctrl_C to break
From 172.16.20.253: bytes=32 seq=1 ttl=127 time=78 ms
From 172.16.20.253: bytes=32 seq=2 ttl=127 time=78 ms
From 172.16.20.253: bytes=32 seq=3 ttl=127 time=78 ms
From 172.16.20.253: bytes=32 seq=4 ttl=127 time=47 ms
```

Fig. 2.73 The example of viewing IP information on PC1 and ping the IP address of PC2

```
[AR1]display ip pool name vlan10 used
  Pool-name      : vlan10
  Pool-No        : 0
  Lease          : 0 Days 8 Hours 0 Minutes
  Domain-name    : -
  DNS-server0    : 172.16.100.3
  NBNS-server0   : -
  Netbios-type   : -
  Position       : Local        Status       : Unlocked
  Gateway-0      : 172.16.10.254
  Mask           : 255.255.255.0
  VPN instance   : --
  -----------------------------------------------------------------
    Start       End      Total  Used  Idle(Expired)  Conflict  Disable
  -----------------------------------------------------------------
    172.16.10.1  172.16.10.254  253    1      231(0)       0       21
  -----------------------------------------------------------------

  Network section :
  -----------------------------------------------------------------
    Index     IP             MAC           Lease   Status
  -----------------------------------------------------------------
    252    172.16.10.253   5489-9872-427a   163    Used
  -----------------------------------------------------------------
```

Fig. 2.74   Viewing the used addresses in the address pool "vlan10"

⑤ You are advised to use the "display ip interface brief" command on the switch CORE to view the IP brief information of the interface, and pay attention to the IP address and subnet mask of the Vlanif interface, and the status of the physical interface and link protocol, as shown in Fig. 2.75.

```
[CORE]display ip interface brief
*down: administratively down
^down: standby
(l): loopback
(s): spoofing
The number of interface that is UP in Physical is 8
The number of interface that is DOWN in Physical is 1
The number of interface that is UP in Protocol is 7
The number of interface that is DOWN in Protocol is 2

Interface              IP Address/Mask      Physical  Protocol
MEth0/0/1              unassigned           down      down
NULL0                  unassigned           up        up(s)
Vlanif1                unassigned           up        down
Vlanif2                172.16.2.254/24      up        up
Vlanif3                172.16.3.254/24      up        up
Vlanif4                172.16.4.254/24      up        up
Vlanif99               172.16.99.254/24     up        up
Vlanif100              172.16.100.254/24    up        up
Vlanif101              172.16.254.2/30      up        up
```

Fig. 2.75   The example of the "display ip interface brief" command to view the interface IP brief information and status

⑥ It will be obtained automatically to configure the IP addresses and DNS addresses of the test hosts PC3, PC4 and PC5, and then you can use the "ipconfig" command at the command line to see if they have obtained the IP information correctly. Fig. 2.76 shows the IP address information obtained by the test host PC3.

```
PC>ipconfig

Link local IPv6 address............: fe80::5689:98ff:fe3b:5afd
IPv6 address.......................: :: / 128
IPv6 gateway.......................: ::
IPv4 address.......................: 172.16.3.253
Subnet mask........................: 255.255.255.0
Gateway............................: 172.16.3.254
Physical address...................: 54-89-98-3B-5A-FD
DNS server.........................: 172.16.100.3
```

Fig. 2.76 Test the IP address information obtained by the host PC3

⑦ After the testing host PC3 for obtaining the IP address automatically, you should use the "display ip pool name vlan3 used" command on the switch CORE to check the used addresses in the address pool vlan3, as shown in Fig. 2.77. Fig. 2.77 shows that "172.16.3.253" has been dynamically assigned to the host with the MAC address of "5489-983b-5afd" (i.e. the test host PC3). Similarly, after the testing host PC4 for obtaining the IP address automatically, you should use the "display ip pool name vlan2 used" command to check the used addresses in the address pool vlan2; after the testing host PC5 for obtaining the IP address automatically, you should use the "display ip pool name vlan4 used" command to check the used addresses in the address pool "vlan4".

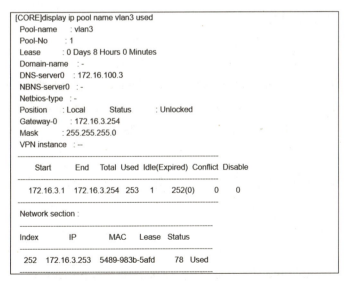

Fig. 2.77 Viewing the used addresses of the address pool "vlan3"

⑧ It is advisable for you to set the IP addresses, subnet masks, default gateways and DNS addresses of DNS servers, Web servers and FTP servers, and use the "ipconfig" command at the DOS command line to check whether the set IP information is effective or not, according to Table 2.12.

⑨ You also need to use ping command to test the IP connectivity between PC4, PC5, FLOOR1, FLOOR2, FLOOR3 and the server, based on the PC3. Fig. 2.78 shows that the host PC3 successfully ping the switch FLOOR2.

Fig. 2.78 Host PC3 successfully ping switch FLOOR2.

Exercises

Multiple choice

(1) (　　) is the sub-interface of router G0/0/0 interface.
  A. G0/0/0. 1 B. G0/0/0-1 C. G0/0/1. 0 D. G0/0/1-1

(2) When configuring a router's sub-interface for single arm routing, (　　) cannot be configured.
  A. VLAN terminated by sub-interface
  B. the IP address and subnet mask of the sub-interface
  C. the ARP broadcast function of the sub-interface
  D. an IP address for the physical interface to which the sub-interface belongs

(3) The management address of the access layer switch is configured in (　　) interface.
  A. Vlanif1 B. Vlanif99 C. Vlanif100 D. Managing VLAN

Task 4　Planning and Configuration of Spanning Tree

Task Description

On the basis of completing the Task 1 to Task 3 of this project, deploy the spanning tree protocol on the enterprise intranet switch according to the network topology shown in Fig. 2.1 to

avoid two-level loops in the enterprise intranet.

Task Objectives

- Understanding the basic concept and working principle of STP.
- Mastering the configuration of STP.
- Mastering the configuration of rapid spanning tree.

Related Knowledge

1. The concept of spanning tree

In the network, if the links between network devices are not redundant, there will be a single point of failure. For example, in the network shown in Fig. 2.79, if the link between switch LSW1 and switch LSW2 fails, host PC1 and host PC2 will no longer have network connectivity.

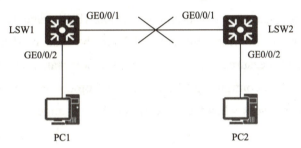

Fig. 2.79 The example of non-redundant network topology

In order to eliminate network interruption caused by single point failure, redundant links can be used to enhance network availability. Networks with redundant paths and redundant devices will have longer network uptime. In the network with redundant links shown in Fig. 2.80, when the link between G0/0/1 interface of switch LSW1 and G0/0/1 interface of switch LSW2 fails, the traffic from PC1 to PC2 will still be communicated through the redundant link shown by the arrow in Fig. 2.80, for ensuring the reliability of the network.

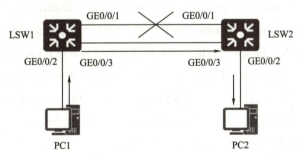

Fig. 2.80 The example of redundant network topology

Link redundancy is a very important part of switching network design. The redundancy design can effectively ensure the availability of the network, according to the working principle of the

switch, but it will cause the trouble problems about second layer loop, broadcast storm, unstable MAC address table of the switch and repeated unicast frame.

In order to solve the loop problem in the switching network, IEEE proposed spanning tree protocol (STP), which is a broad concept, including STP, RSTP and MSTP. It means that you need to break the two layer data loop by blocking the traffic of some ports of the switch. Spanning tree protocol has many types or variants, and its development has experienced the following three generations:

(1) STP

In 1990, IEEE released the first STP technology in the form of IEEE 802.1D standard. STP is a single spanning tree protocol— all VLAN in the switching network share the same spanning tree instance, which can not reach the goal of balancing load of two layer traffic.

(2) RSTP(rapid spanning tree protocol)

RSTP is also a single spanning tree protocol, and the IEEE standard is 802.1W, which is an upgraded version of STP and is compatible with STP. On the basis of IEEE 802.1D standard, the improvement is made, and measures such as reducing the port status, increasing the port role and changing the sending mode of configuring BPDU (bridge protocol data unit) are taken to realize the faster convergence of the network when the network topology changes.

(3) MSTP

Due to the concept of spanning tree was put forward earlier than the concept of VLAN, VLAN was not considered in the implementation of STP. RSTP only improves the convergence mechanism of STP, and also does not take VLAN into account. Therefore, both STP and RSTP belong to the single spanning tree protocol, all VLANs in the switching network share the same spanning tree, which may lead to the waste of network bandwidth.

MSTP(multiple spanning tree protocol) defined by IEEE 802.1S standard calculates multiple spanning trees based on spanning tree instances. In an MSTP domain, multiple VLANs can be mapped to the same spanning tree instance, but one VLAN can only be mapped to one spanning tree instance. The load balance of links can be achieved by configuring multiple spanning tree instances on the switch. As shown in Fig. 2.81, switches LSW1, LSW2 and LSW3 belong to the same MST region, in which VLAN1 and VLAN2 are mapped to multiple spanning tree instance (MSTI)1, VLAN3 and VLAN4 are mapped to instance 2, the root of which is LSW1, and the G0/0/1 interface of switch LSW3 is blocked, while the root of instance 2 is LSW2. In this way, the data of VLAN1 and VLAN2 are forwarded from G0/0/2 interface of LSW3, while the data of VLAN3 and VLAN4 are forwarded from G0/0/1 interface of LSW3, thus realizing load balancing.

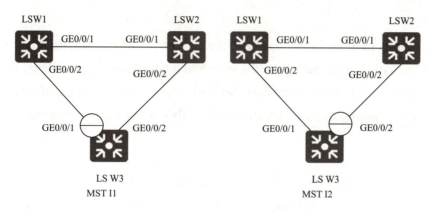

Fig. 2.81 The example of MSTP

2. Concepts of STP

(1) BPDU

Bridge protocol data unit (BPDU) is a message frame exchanged between bridges running STP. STP determines the topology of the network by transmitting BPDU messages between devices, and completes the calculation of spanning tree. The BPDU packet is encapsulated in an Ethernet data frame, and the destination MAC is the multicast address "01-80-C2-00-00-00". BPDU message of STP protocol is divided into configuration BPDU and topology change notification BPDU (TCN BPDU). Configure BPDU to calculate spanning tree and maintain spanning tree topology, while topology change notification BPDU is used to notify related devices when network topology changes.

(2) BID

BID (bridge ID) is used to uniquely identify a bridge or switch in STP. The bridge ID consists of two parts, with a total length of 8 bytes, of which the upper 16 bits are the bridge priority and the lower 48 bits are the MAC address of the bridge, as shown in Fig. 2.82. The default value of the bridge priority is 32 768, and the network administrator can modify the priority of the bridge manually. Because the MAC address of the bridge is unique, the bridge ID is also unique in the network. When STP executes spanning tree algorithm, it selects the root of spanning tree according to the bridge ID in the network. The switch with the smallest bridge ID is selected as the root bridge of the spanning tree.

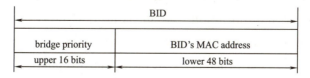

Fig. 2.82 Composition of BID

(3) Path cost

Path cost is used to measure the quality of the path between switches. Root path cost refers to the sum of all link costs on the path from the switch to the root bridge. By default, the port cost is determined by the running speed of the port. Network administrators can modify the cost value of a switch port by configuring its port cost, so as to flexibly control the spanning tree path to the root bridge.

3. Roles of root and port

There are two special bridges in the spanning tree: root bridge and designated bridge. The switch with the smallest bridge ID in the spanning tree is selected as the root bridge of the spanning tree. The designated bridge is a separate bridge responsible for data forwarding on the physical segment. In the process of spanning tree work, bridge/switch ports mainly have the following port roles.

(1) Root port

The root port refers to the port closest to the root bridge on a non-root bridge. Each non-root bridge has only one root port. As shown in Fig. 2.83, the G0/0/1 interface of the non-root switch LSW2 is the root port, and the G0/0/2 interface of the non-root switch LSW3 is the root port.

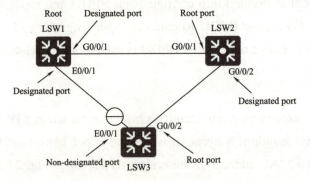

Fig. 2.83　Example of STP port roles

(2) Designated port

Each network segment in the switching network needs to specify a port for forwarding the data of that network segment, and this port is the designated port. The designated port is a port that can forward traffic in STP except the root port. All ports of the root bridge are designated ports. In Fig. 2.91, switch LSW1 is the root bridge, so all its ports (G0/0/1 and E0/0/1 interfaces) are designated ports. For the network segment where the switch LSW2 is connected with the switch LSW3, the G0/0/2 interface of the switch LSW2 is the designated port.

(3) Non-designated port

Ports that are set to blocking state due to the need of STP to prevent two layer loop are called non-designated ports. In Fig. 2.83, the E0/0/1 interface of switch LSW3 is an unspecified port,

which cannot forward data frames because it is blocked.

The root port and the designated port are finally in the forwarding state, but the non-designated switch will eventually become blocked. In STP, five port states are defined: disabled, blocking, listening, learning, forwarding, as shown in Table 2.14.

Table 2.14 STP port status

Port status	Explanation
Disabled	When the port status is Down, BPDU messages will not be processed and user traffic will not be forwarded
Listening	In the transitional state, spanning tree calculation is started, and the port can receive and send BPDU, but does not forward user traffic
Learning	In the transitional state, a loop-free MAC address forwarding table is established, and user traffic is not forwarded
Forwarding	Ports can receive and send BPDU and also forward user traffic. Only the root port or the specified port can enter the Forwarding state
Blocking	The port only receives and processes BPDU, and does not forward user traffic

4. The working process of STP

The working process of STP mainly includes three tasks: one is to elect the root bridge; second, select root ports for all non-root bridges/switches; the third is to select the designated port for each network segment. The root bridge ID, root path cost, sending device BID and sending port PID (port ID) in the BPDU report are mainly used by STP to select the root bridge, root port and designated port.

(1) Elected root bridge

In the initialization state, all switches in the network have not received the configuration BPDU sent by other switches, so they will assume that they are the root bridge in the network, and the switches will immediately start sending the configuration BPDU based on themselves as soon as they complete the startup process. When a switch receives a configuration BPDU from a neighbor switch, it compares the root ID in the received BPDU with the root ID of the switch, and sets the value of the root ID field in the BPDU to a root ID with a small value. Finally, the switch with the minimum BID is selected as the minimum BID of the root bridge.

(2) Selection of the root port and designated port

The root port and the designated port make decisions according to four conditions: minimum root path cost, minimum sender BID, minimum sender port ID and minimum receiver port ID. The specific process is as follows:

① Comparing the root path cost of each port on the non-root switch to the root bridge, the port with the minimum path cost value is selected as the root port.

② If there are multiple ports on the non-root switch with the same root path cost to the root bridge and the minimum, compare the BIDs in the configured BPDU received by the ports, and the port that receives the BPDU with the minimum transmitted BID is the root port.

③ If there are multiple ports on the non-root switch with the same and minimum root path cost to the root bridge, and at the same time have the same minimum sending BID, the sender port ID is compared, and the bridge port with the smallest port ID is selected as the root port.

④ If there are multiple ports on the non-root switch with the same minimum root path cost to the root bridge and the same minimum sending BID and minimum sender port ID, the port with the smallest receiver port ID is selected as the root port by comparing the port ID of the receiving BPDU.

(3) Selection of the designated port

The selection of designated port is similar to the selection of root port, and the decision is also made according to four conditions: minimum root path cost, minimum sender BID, minimum sender port ID and minimum receiver port ID.

Related Skills

1. Planning and configuration of STP

It is required to build the network topology diagram as shown in Fig. 2.84, and complete the VLAN configuration according to the IP and VLAN planning in Table 2.15. STP is deployed on switches LSW1, LSW2 and LSW3, so LSW1 is the root bridge and LSW2 is the backup root bridge, and the G0/0/2 interface of switch LSW3 is blocked.

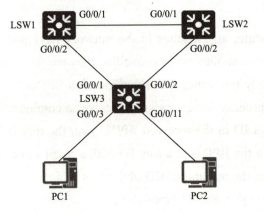

Fig. 2.84 The Network topology diagram of STP configuration

Table 2.15 Planning of IP and VLAN

Device	Interface or VLAN	Description of VLAN	Two-layer/Three-layer planning	Explanation
LSW1	VLAN2	teacher		Teacher
	VLAN3	student		Student
LSW2	VLAN2	teacher		Teacher
	VLAN3	student		Student

Continued

Device	Interface or VLAN	Description of VLAN	Two-layer/Three-layer planning	Explanation
LSW3	VLAN2	teacher	G0/0/3 to G0/0/11 interface	Teacher
	VLAN3	student	G0/0/12 to G0/0/24 interface	Student
PC1			192.168.2.11/24	
PC2			192.168.2.12/24	

(1) Planning of STP

① All switch spanning trees use STP protocol.

② The spanning tree priority value of switch LSW1 is 4096, which is the root of STP.

③ The spanning tree priority value of switch LSW2 is 8192, which is the backup root of STP.

④ The spanning tree priority value of switch LSW3 is the default value.

(2) VLAN configuration

Taking Huawei S5700-28C-HI switch as an example, the steps for completing VLAN configuration shown in Table 2.15 are as follows:

```
//Create VLAN2 and VLAN3 on LSW1, and configure G0/0/1 and G0/0/2 interfaces as trunk interfaces.
[LSW1]vlan 2
[LSW1-vlan2]description teacher
[LSW1-vlan2]vlan 3
[LSW1-vlan3]description student
[LSW1]port-group 1
[LSW1-port-group-1]group-member g0/0/1 to g0/0/2
[LSW1-port-group-1]port link-type trunk
[LSW1-port-group-1]port trunk allow-pass vlan 2 3
//Create VLAN2 and VLAN3 on LSW2, and configure G0/0/1 and G0/0/2 interfaces as trunk interfaces.
[LSW2]vlan 2
[LSW2-vlan2]description teacher
[LSW2-vlan2]vlan 3
[LSW2-vlan3]description student
[LSW2]port-group 1
[LSW2-port-group-1]group-member g0/0/1 to g0/0/2
[LSW2-port-group-1]port link-type trunk
[LSW2-port-group-1]port trunk allow-pass vlan 2 3
//Create VLAN2 and VLAN3 on LSW3, configure G0/0/1 and G0/0/2 interfaces as trunk interfaces,
//configure G0/0/3 to G0/0/11 interfaces as VLAN2, and G0/0/12 to G0/0/24 interfaces
//as VLAN3
[LSW3]vlan 2
[LSW3-vlan2]description teacher
```

```
[LSW3-vlan2]vlan 3
[LSW3-vlan3]description student
[LSW3]port-group 1
[LSW3-port-group-1]group-member g0/0/1 to g0/0/2
[LSW3-port-group-1]port link-type trunk
[LSW3-port-group-1]port trunk allow-pass vlan 2 3
[LSW3-port-group-1]quit
[LSW3]port-group 2
[LSW3-port-group-2]group-member g0/0/3 to g0/0/11
[LSW3-port-group-2]port link-type access
[LSW3-port-group-2]port default vlan 2
[LSW3-port-group-2]quit
[LSW3]port-group 3
[LSW3-port-group-3]group-member g0/0/12 to g0/0/24
[LSW3-port-group-3]port link-type access
[LSW3-port-group-3]port default vlan 3
```

(3) Configuration of STP

Configuring STP on a switch usually requires:

① Enabling STP.

② Configuring the switch spanning tree mode to STP.

③ Configuring the spanning tree priority of the switch according to the plan to make it the root bridge or the backup root bridge.

Taking Huawei S5700-28C-HI switch as an example, the steps to complete STP configuration on switches LSW1, LSW2 and LSW3 are as follows:

```
[LSW1]stp enable              //Enable STP
[LSW1]stp mode stp            //Configure spanning tree mode as STP
[LSW1]stp priority 4096       //Configure LSW1 spanning tree priority to 4096
[LSW2]stp enable              //Enable STP
[LSW2]stp mode stp            //Configure spanning tree mode as STP
[LSW2]stp priority 8192       //Configure lsw1 spanning tree priority to 8192
[LSW3]stp enable              //Enable STP
[LSW3]stp mode stp            //Configure spanning tree mode as STP
```

(4) Review and verification of STP

After completing STP configuration on switches LSW1, LSW2 and LSW3, you should check and verify STP after STP calculation is stable.

① To execute the "display stp" command on LSW1, LSW2 and LSW3 for viewing the spanning tree information. Fig. 2.85 shows the result of executing the "display stp" command on

switch LSW1. The results show that the spanning tree mode of switch LSW1 is STP, the BID of switch LSW1 is "4096.4c1f-cc7c-26e4", the BID of root bridge is "4096.4c1f-cc7c-26e4", and the ID of root bridge is the same as that of switch LSW1. Therefore, switch LSW1 is the root of spanning tree, and the selection result is consistent with the planning result.

```
[LSW1]display stp
-------[CIST Global Info][Mode STP]-------
CIST Bridge         :4096 .4c1f-cc7c-26e4
Config Times        :Hello 2s MaxAge 20s FwDly 15s MaxHop 20
Active Times        :Hello 2s MaxAge 20s FwDly 15s MaxHop 20
CIST Root/ERPC      :4096 .4c1f-cc7c-26e4 / 0
CIST RegRoot/IRPC   :4096 .4c1f-cc7c-26e4 / 0
CIST RootPortId     :0.0
BPDU-Protection     :Disabled
TC or TCN received  :16
TC count per hello  :0
STP Converge Mode   :Normal
Time since last TC  :0 days 12h:48m:14s
Number of TC        :10
Last TC occurred    :GigabitEthernet0/0/1
----[Port1(GigabitEthernet0/0/1)][FORWARDING]----
Port Protocol       :Enabled
Port Role           :Designated Port
Port Priority       :128
Port Cost(Dot1T )   :Config=auto / Active=20000
Designated Bridge/Port :4096.4c1f-cc7c-26e4 / 128.1
Port Edged          :Config=default / Active=disabled
Point-to-point      :Config=auto / Active=true
Transit Limit       :147 packets/hello-time
Protection Type     :None
---- More ----
```

Fig. 2.85 The result of executing "display stp" command on switch LSW1

② To execute the "display stp brief " command on LSW1, LSW2 and LSW3 for viewing the brief information of the spanning tree, and paying attention to the switch port role and port status. As shown in Fig. 2.86, all port roles of switch LSW1 are designated ports (DESI), and the spanning tree status of ports is FORWARDING, so the switch is the root of the tree. As shown in Fig. 2.87, the GigabitEthernet0/0/1 port of switch LSW2 has the role of ROOT and the state of FORWARDING. The role of GigabitEthernet0/0/2 port is designated port (DESI) and its status is FORWARDING. As shown in Fig. 2.88, the GigabitEthernet0/0/1 port of switch LSW3 has the role of ROOT and the state is FORWARDING. GigabitEthernet0/0/3 and GigabitEthernet0/0/11 ports are designated ports (DESI) and the state is FORWARDING. GigabitEthernet0/0/2 has the role of other port (ALTE) and the state is DISCARDING.

```
[LSW1]display stp brief
MSTID  Port                  Role  STP State    Protection
  0    GigabitEthernet0/0/1  DESI  FORWARDING   NONE
  0    GigabitEthernet0/0/2  DESI  FORWARDING   NONE
```

Fig. 2.86 The result of executing the "display stp brief " command on the switch LSW1

```
[LSW2]display stp brief
MSTID  Port                        Role  STP State   Protection
  0    GigabitEthernet0/0/1        ROOT  FORWARDING  NONE
  0    GigabitEthernet0/0/2        DESI. FORWARDING  NONE
```

Fig. 2.87 The result of executing the "display stp brief" command on the switch LSW2

```
[LSW3]display stp brief
MSTID  Port                        Role  STP State   Protection
  0    GigabitEthernet0/0/1        ROOT  FORWARDING  NONE
  0    GigabitEthernet0/0/2        ALTE  DISCARDING  NONE
  0    GigabitEthernet0/0/3        DESI  FORWARDING  NONE
  0    GigabitEthernet0/0/11       DESI  FORWARDING  NONE
```

Fig. 2.88 The result of executing the "display stp brief" command on the switch LSW3

③ You should turn off the test host PC1 in the network topology, then turn it on again, and observe the signal light of interface G0/0/3 of switch LSW3 connected to PC1. After that, you should execute the "display stp brief" command on the switch LSW3, and keep observing the Role and STP State of G0/0/3 interface.

2. Planning and configuration of RSTP

It is required to build the network topology diagram as shown in Fig. 2.84, and complete the VLAN configuration according to the IP and VLAN planning in Table 2.13. RSTP is deployed on switches LSW1, LSW2 and LSW3, so LSW1 is the root bridge, LSW2 is the backup root bridge, the G0/0/2 interface of switch LSW3 is blocked, and the port connected to PC is the edge port of spanning tree, and immediately enters the forwarding state, configuring the protection function of spanning tree to protect equipment and links. For example, the configuration the root protection function at the designated port of the root bridge device is needed.

(1) Planning of RSTP

① All switch spanning trees use RSTP protocol.

② Switch LSW1 is a tree root.

③ Switch LSW2 is the backup root bridge.

④ G0/0/1 and G0/0/2 interfaces of switch LSW1 are configured with root protection function.

⑤ G0/0/3 and G0/0/24 interfaces of switch LSW3 are configured as edge ports, and BPDU protection is enabled.

(2) VLAN configuration

You should refer to the steps of "VLAN configuration" in "STP planning and configuration" to complete VLAN configuration.

(3) Configuration of RSTP

According to the plan, to complete the configuration of RSTP, the following configuration tasks need to be completed:

① Enabling STP.
② Configuring the switch spanning tree mode to RSTP.
③ Configuring switch LSW1 as the root bridge and enabling root protection.
④ Configuring switch LSW2 as backup root bridge.
⑤ Configuring G0/0/3 to G0/0/24 interfaces of switch LSW3 as edge ports, and enabling BPDU protection.

Taking Huawei S5700-28C-HI switch as an example, the steps to complete RSTP configuration on switches LSW1, LSW2 and LSW3 are as follows:

```
[LSW1]stp enable              //Enable STP
[LSW1]stp mode rstp           //Configure spanning tree mode as RSTP
[LSW1]stp root primary        //Configure LSW1 as the root bridge of the spanning tree
[LSW1]interface g0/0/1
[LSW1-GigabitEthernet0/0/1]stp root-protection //Enable root protection
[LSW1-GigabitEthernet0/0/1]quit
[LSW1]interface g0/0/2
[LSW1-GigabitEthernet0/0/2]stp root-protection //Enable root protection
[LSW1-GigabitEthernet0/0/2]quit
[LSW2]stp enable              //Enable STP
[LSW2]stp mode rstp           //Configure spanning tree mode as RSTP
[LSW2]stp root secondary      //Configure LSW1 as the backup root bridge of spanning tree
[LSW3]stp enable              //Enable STP
[LSW3]stp mode rstp           //Configure spanning tree mode as RSTP
[LSW3]port-group 2
[LSW3-port-group-2]group-member g0/0/3 to g0/0/24
[LSW3-port-group-2]stp edged-port enable  //Configured as an edge port
[LSW3-port-group-2]quit
[LSW3]stp bpdu-protection     //Configure the BPDU protection function of the switch
```

(4) Viewing and verification of RSTP

After RSTP configuration is completed on switches LSW1, LSW2 and LSW3, RSTP is checked and verified after RSTP calculation is stable.

① To execute the "display stp" command on LSW1, LSW2 and LSW3 for viewing the spanning tree information. Fig. 2.89 shows the result of executing the "display stp" command on switch LSW1. The results show that the spanning tree mode of switch LSW1 is RSTP, the BID of switch LSW1 is "0.4c1f-cc7c-26e4", the BID of root bridge is "0.4c1f-cc7c-26e4", and the ID of root bridge is the same as that of switch LSW1. Therefore, switch LSW1 is the root of spanning tree, and the selection result is consistent with the planning result.

```
[LSW1]display stp
-------[CIST Global Info][Mode RSTP]-------
CIST Bridge          :0    .4c1f-cc7c-26e4
Config Times         :Hello 2s MaxAge 20s FwDly 15s MaxHop 20
Active Times         :Hello 2s MaxAge 20s FwDly 15s MaxHop 20
CIST Root/ERPC       :0    .4c1f-cc7c-26e4 / 0
CIST RegRoot/IRPC    :0    .4c1f-cc7c-26e4 / 0
CIST RootPortId      :0.0
BPDU-Protection      :Disabled
CIST Root Type       :Primary root
TC or TCN received   :9
TC count per hello   :0
STP Converge Mode    :Normal
Time since last TC   :0 days 2h:45m:16s
Number of TC         :9
Last TC occurred     :GigabitEthernet0/0/2
----[Port1(GigabitEthernet0/0/1)][FORWARDING]----
Port Protocol        :Enabled
Port Role            :Designated Port
Port Priority        :128
Port Cost(Dot1T)     :Config=auto / Active=20000
Designated Bridge/Port :0.4c1f-cc7c-26e4 / 128.1
Port Edged           :Config=default / Active=disabled
Point-to-point       :Config=auto / Active=true
Transit Limit        :147 packets/hello-time
---- More ----
```

Fig. 2.89 The result of executing "display stp" command on switch LSW1

② To execute the "display stp brief" command on LSW1, LSW2 and LSW3 for viewing the brief information of the spanning tree, and pay attention to the switch port role and port status. As shown in Fig. 2.90, all port roles of switch LSW1 are designated ports (DESI), and the spanning tree status of ports is FORWARDING, so the switch is the root of the tree. As shown in Fig. 2.91, the GigabitEthernet0/0/1 port of switch LSW2 has the role of ROOT and the state of FORWARDING. The role of GigabitEthernet0/0/2 port is designated port (DESI) and the state is FORWARDING. As shown in Fig. 2.92, the GigabitEthernet0/0/1 port of switch LSW3 has the role of ROOT and the state is FORWARDING. GigabitEthernet0/0/3 and GigabitEthernet0/0/11 ports are designated ports (DESI) and the states is FORWARDING. GigabitEthernet 0/0/2 has the role of other port (ALTE) and the state is DISCARDING.

```
[LSW1]display stp brief
 MSTID  Port                Role  STP State   Protection
   0    GigabitEthernet0/0/1  DESI  FORWARDING  NONE
   0    GigabitEthernet0/0/2  DESI  FORWARDING  NONE
```

Fig. 2.90 The result of executing the "display stp brief" command on the switch LSW1

```
[LSW2]display stp brief
 MSTID  Port                Role  STP State   Protection
   0    GigabitEthernet0/0/1  ROOT  FORWARDING  NONE
   0    GigabitEthernet0/0/2  DESI  FORWARDING  NONE
```

Fig. 2.91 The result of executing the "display stp brief" command on the switch LSW2

```
[LSW3]display stp brief
MSTID  Port                  Role  STP State    Protection
    0  GigabitEthernet0/0/1  ROOT  FORWARDING   NONE
    0  GigabitEthernet0/0/2  ALTE  DISCARDING   NONE
    0  GigabitEthernet0/0/3  DESI  FORWARDING   BPDU
    0  GigabitEthernet0/0/11 DESI  FORWARDING   BPDU
```

Fig. 2.92 The result of executing the "display stp brief" command on the switch LSW3

③ You should turn off the test host PC1 in the network topology, then turn it on again, and observe the signal light of interface G0/0/3 of switch LSW3 connected to PC1. After that, you should execute the "display stp brief" command on switch LSW3, and observe the Role and STP State of G0/0/3 interface. Because the G0/0/3 interface of LSW3 is configured as an edge port, it immediately enters the forwarding state.

3. Planning and configuration of MSTP

It is required to build the network topology diagram as shown in Fig. 2.93, and complete the VLAN configuration according to the IP and VLAN planning in Table 2.16. MSTP is deployed on switches LSW1, LSW2 and LSW3, so the root bridge of VLAN1 and VLAN2 is LSW1, LSW2 is the backup root bridge, and the G0/0/2 interface of switch LSW3 is blocked. The root bridge of VLAN3 and VLAN4 is LSW2, LSW1 is the backup root bridge, and the G0/0/1 interface of switch LSW3 is blocked. The ports connected to the PC are configured as spanning tree edge ports.

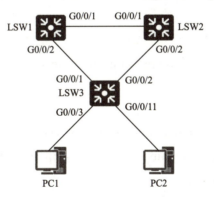

Fig. 2.93 The Network topology diagram of MSTP configuration

Table 2.16 Planning of IP and VLAN

Device	Interface or VLAN	Description of VLAN	Two-layer/Three-layer planning	Explanation
LSW1	VLAN2	teacher		Teacher
	VLAN3	student		Student
	VLAN4	jwc		Dean's office
LSW2	VLAN2	teacher		Teacher
	VLAN3	student		Student
	VLAN4	jwc		Dean's office

Continued

Device	Interface or VLAN	Description of VLAN	Two-layer/Three-layer planning	Explanation
LSW3	VLAN2	teacher	G0/0/3 to G0/0/10 ports	Teacher
	VLAN3	student	G0/0/11 to G0/0/19 ports	Student
	VLAN4	jwc	G0/0/20 to G0/0/24 ports	
PC1			192.168.2.11/24	
PC2			192.168.3.12/24	

(1) Planning of MSTP

① All switch spanning trees use MSTP protocol.

② MSTP domain name is cdp, with revision level of 1. Instance 1 includes VLAN1 and VLAN2, and instance 2 includes VLAN3 and VLAN 4.

③ Switch LSW1 is the primary root of instances 0 and 1 and the backup root of instance 2;

④ Switch LSW2 is the backup root of instances 0 and 1 and the primary root of instance 2;

⑤ Interface G0/0/3 to G0/0/24 of switch LSW3 are configured as edge ports, and BPDU protection is enabled.

(2) VLAN configuration

You should refer to the steps of "VLAN configuration" in "STP planning and configuration" to complete VLAN configuration.

(3) Configuration of MSTP

According to the plan, the following configuration tasks are required to complete MSTP configuration:

① Enabling STP.

② Configuring the switch spanning tree mode to MSTP.

③ Configuring MSTP domain name, revision, mapping relationship between instance and VLAN, and activate MSTP domain.

④ Configuring switch LSW1 as that root bridge of instance 1 and the backup root bridge of instance 2.

⑤ Configuring switch LSW2 as the backup root bridge of instance 1 and the root bridge of instance 2.

⑥ Configuring G0/0/3 to G0/0/24 interfaces of switch LSW3 as edge ports, and enable BPDU protection.

Taking Huawei S5700-28C-HI switch as an example, the steps to complete MSTP configuration on switches LSW1, LSW2 and LSW3 are as follows:

```
[LSW1]stp enable                    //Enable STP
[LSW1]stp mode mstp                 //Configure the spanning tree mode to MSTP
```

```
[LSW1]stp region-configuration          //Enter the MSTP domain view
[LSW1-mst-region]region-name cdp        //Configure the MSTP domain name as cdp
[LSW1-mst-region]revision-level 1       //Configuration revision level is 1
[LSW1-mst-region]instance 1 vlan 1 2    //Configuration instance 1 includes VLAN1 and VLAN2
[LSW1-mst-region]instance 2 vlan 3 4    //Configuration instance 2 includes VLAN3 and VLAN4
[LSW1-mst-region]active region-configuration  //Activate MSTP domain
[LSW1]stp instance 0 root primary       //Configure LSW1 as the root bridge of instance 0
[LSW1]stp instance 1 root primary       //configure LSW1 as the root bridge of instance 1
[LSW1]stp instance 2 root secondary     //Configure LSW1 as the backup root bridge for
                                        //instance 2
[LSW2]stp enable                        //Enable STP
[LSW2]stp mode mstp                     //Configure the spanning tree mode to MSTP
[LSW2]stp region-configuration          //Enter the MSTP domain view
[LSW2-mst-region]region-name cdp        //Configure the MSTP domain name as cdp
[LSW2-mst-region]revision-level 1       //Configuration revision level is 1
[LSW2-mst-region]instance 1 vlan 1 2    //Configuration instance 1 includes VLAN1 and VLAN2
[LSW2-mst-region]instance 2 vlan 3 4    //Configuration instance 2 includes VLAN3 and VLAN4
[LSW2-mst-region]active region-configuration    //Activate MSTP domain
[LSW2]stp instance 0 root secondary     //Configure LSW2 as the backup root bridge for
                                        //instance 0
[LSW2]stp instance 1 root secondary'    //Configure LSW1 as the backup root bridge for
                                        //instance 1
[LSW2]stp instance 2 root primary       //configure LSW1 as the root bridge of instance 2
[LSW3]stp enable                        //Enable STP
[LSW3]stp mode mstp                     //Configure the spanning tree mode to MSTP
[LSW3]stp region-configuration          //Enter the MSTP domain view
[LSW3-mst-region]region-name cdp        //Configure the MSTP domain name as cdp
[LSW3-mst-region]revision-level 1       //Configuration revision level is 1
[LSW3-mst-region]instance 1 vlan 1 2    //Configuration instance 1 includes VLAN1 and VLAN2
[LSW3-mst-region]instance 2 vlan 3 4    //Configuration instance 2 includes VLAN3 and VLAN4
[LSW3-mst-region]active region-configuration  //Activate MSTP domain
[LSW3]port-group 5
[LSW3-port-group-2]group-member g0/0/3 to g0/0/24
[LSW3-port-group-2]stp edged-port enable    //Configure G0/0/3 to G0/0/24 as edge ports
[LSW3-port-group-2]quit
[LSW3]stp bpdu-protection               //Configure the BPDU protection function of the switch
```

(4) Viewing and verification of MSTP

After MSTP configuration is completed on switches LSW1, LSW2 and LSW3, check and verify MSTP after MSTP calculation is stable.

① To execute the "display stp region-configuration" command on LSW1, LSW2 and LSW3 for viewing the MSTP domain configuration information. Fig. 2.94 shows the configuration information of MSTP domain on LSW1. You should pay more attention to observe whether the spanning tree domain name, revision level and VLAN contained in the instance are consistent with the planned configuration. Instance 0 in Fig. 2.94 is the default instance, and all VLAN that specify which instance belong to are not configured to belong to instance 0.

```
[LSW1]display stp region-configuration
Oper configuration
   Format selector   :0
   Region name       :cdp
   Revision level    :1

   Instance   VLANs Mapped
      0       5 to 4094
      1       1 to 2
      2       3 to 4
```

Fig. 2.94   The configuration information of MSTP domain on LSW1

② To execute the "display stp brief" command on LSW1 for viewing the brief information of spanning tree, and pay attention to the switch port role and port status. As shown in Fig. 2.95, all port roles of spanning tree instances 0 and 1 of switch LSW1 are designated ports (DESI), and the spanning tree state is FORWARDING. The switch is the root of instances 0 and 1. The GigabitEthernet0/0/1 port of spanning tree instance 2 of switch LSW1 is the root port and the GigabitEthernet0/0/2 port is the designated port, all of which are in FORWARDING state.

```
[LSW1]display stp brief
 MSTID  Port                 Role   STP State    Protection
   0    GigabitEthernet0/0/1  DESI   FORWARDING   NONE
   0    GigabitEthernet0/0/2  DESI   FORWARDING   NONE
   1    GigabitEthernet0/0/1  DESI   FORWARDING   NONE
   1    GigabitEthernet0/0/2  DESI   FORWARDING   NONE
   2    GigabitEthernet0/0/1  ROOT   FORWARDING   NONE
   2    GigabitEthernet0/0/2  DESI   FORWARDING   NONE
```

Fig. 2.95   The result of executing the "display stp brief" command on the switch LSW1

③ To execute the "display stp brief" command on LSW2 for viewing the spanning tree summary information. As shown in Fig. 2.96, all port roles of spanning tree instance 2 of switch LSW2 are designated ports (DESI), and the spanning tree state is FORWARDING, and the switch is the root of instance 2. The GigabitEthernet0/0/1 port of spanning tree instance 0 and instance 1 of switch LSW2 is the root port, and the GigabitEthernet0/0/2 port is the designated port, both of which are in FORWARDING state.

```
[LSW2]display stp brief
MSTID  Port                 Role    STP State    Protection
  0    GigabitEthernet0/0/1  ROOT   FORWARDING    NONE
  0    GigabitEthernet0/0/2  DESI   FORWARDING    NONE
  1    GigabitEthernet0/0/1  ROOT   FORWARDING    NONE
  1    GigabitEthernet0/0/2  DESI   FORWARDING    NONE
  2    GigabitEthernet0/0/1  DESI   FORWARDING    NONE
  2    GigabitEthernet0/0/2  DESI   FORWARDING    NONE
```

Fig. 2.96 The result of executing the "display stp brief" command on the switch LSW2

④ To execute the "display stp brief" command on LSW3 for viewing the spanning tree summary information. As shown in Fig. 2.97, the GigabitEthernet0/0/2 port of the spanning tree instance 0 and instance 1 of the switch LSW3 is in the state of DISCARDING (ALTE), while the GigabitEthernet0/0/1 port is in the state of DISCARDING (ALTE) for instance 2.

```
[LSW3]display stp brief
MSTID  Port                   Role    STP State     Protection
  0    GigabitEthernet0/0/1   ROOT    FORWARDING    NONE
  0    GigabitEthernet0/0/2   ALTE    DISCARDING    NONE
  0    GigabitEthernet0/0/3   DESI    FORWARDING    BPDU
  0    GigabitEthernet0/0/11  DESI    FORWARDING    BPDU
  1    GigabitEthernet0/0/1   ROOT    FORWARDING    NONE
  1    GigabitEthernet0/0/2   ALTE    DISCARDING    NONE
  1    GigabitEthernet0/0/3   DESI    FORWARDING    BPDU
  2    GigabitEthernet0/0/1   ALTE    DISCARDING    NONE
  2    GigabitEthernet0/0/2   ROOT    FORWARDING    NONE
  2    GigabitEthernet0/0/11  DESI    FORWARDING    BPDU
```

Fig. 2.97 The result of executing the "display stp brief" command on the switch LSW3

Task Completion

1. Task planning

According to the task description and the network topology shown in Fig. 2.1, the MSTP protocol is adopted in the enterprise intranet to break the two-level loops. The planning of MSTP is as follows:

① The spanning tree of all switches in the enterprise intranet uses MSTP protocol.

② Switches CORE, FLOOR1, FLOOR2 and FLOOR3 belong to the same MSTP domain. The MSTP domain name is cdp, and the revision level is 1. All VLAN are mapped to default instance 0, and the root bridge of instance 0 is switch CORE.

③ The switch FLOOR4 belongs to another MSTP domain, the domain name is leader, the revision level is 1, all VLAN are mapped to the default instance 0, and the root bridge of the instance 0 is switch FLOOR 4.

④ Enabling the root protection function at all designated ports of the core switch CORE.

⑤ The G0/0/24 interface of router AR1 connected with the core switch CORE does not participate in the spanning tree.

⑥ The ports of all access layer switches connected to the host are configured as edge ports,

and BPDU protection is enabled.

2. Task implementation

(1) Switch spanning tree configuration in MSTP domain "cdp"

① According to the planning of spanning tree, the configuration steps of CORE spanning tree of core switch are as follows:

```
[CORE]stp enable
[CORE]stp mode mstp
//Enable spanning tree and configure spanning tree mode as MSTP
[CORE]stp region-configuration
[CORE-mst-region]region-name cdp
[CORE-mst-region]revision-level 1
[CORE-mst-region]active region-configuration
[CORE-mst-region]quit
[CORE]
//Configure the MSTP domain and activate the domain. The MSTP domain name is "cdp",
//and the revision level is 1. All VLAN are mapped to the default instance 0
[CORE]stp instance 1 root primary
//Configure the switch CORE as the root bridge of instance 0 in cdp domain
[CORE]interface g0/0/24
[CORE-GigabitEthernet0/0/24]stp disable
[CORE-GigabitEthernet0/0/24]quit
[CORE]
//The G0/0/24 interface of the switch CORE does not participate in the spanning tree election
[CORE]port-group 1
[CORE-port-group-1]group-member g0/0/21 to g0/0/24
[CORE-port-group-1]stp edged-port enable
[CORE-port-group-1]quit
[CORE]stp bpdu-protection
//Configure G0/0/21 to G0/0/24 interfaces of the switch CORE connection server as edge
//ports, and enable BPDU protection
[CORE]port-group 2
[CORE-port-group-2]group-member g0/0/1 to g0/0/3
[CORE-port-group-2]stp root-protection
//Switch CORE interfaces G0/0/1 to G0/0/3 enable root protection
```

② According to the planning of spanning tree, the configuration steps of switch FLOOR1 spanning tree are as follows:

```
[FLOOR1]stp enable
[FLOOR1]stp mode mstp
```

```
[FLOOR1]// Enable spanning tree and configure spanning tree mode as MSTP
[FLOOR1]stp region-configuration
[FLOOR1-mst-region]region-name cdp
[FLOOR1-mst-region]revision-level 1
[FLOOR1-mst-region]active region-configuration
[FLOOR1-mst-region]quit
//Configure the MSTP domain and activate the domain. The MSTP domain name is "cdp",
//and the revision level is 1. All VLAN are mapped to the default instance 0
[FLOOR1]port-group 5
[FLOOR1-port-group-5]group-member g0/0/2 to g0/0/24
[FLOOR1-port-group-5]stp edged-port enable
[FLOOR1-port-group-5]quit
[FLOOR1]stp bpdu-protection
//Configure G0/0/2 to G0/0/24 interfaces of switch FLOOR1 as edge ports, and enable
//BPDU protection
```

③ According to the planning of spanning tree, the configuration steps of switch FLOOR2 spanning tree are as follows:

```
[FLOOR2]stp enable
[FLOOR2]stp mode mstp
[FLOOR2]
//Enable spanning tree and configure spanning tree mode as MSTP
[FLOOR2]stp region-configuration
[FLOOR2-mst-region]region-name cdp
[FLOOR2-mst-region]revision-level 1
[FLOOR2-mst-region]active region-configuration
[FLOOR2-mst-region]quit
//Configure the MSTP domain and activate the domain. The MSTP domain name is "cdp",
//and the revision level is 1. All VLAN are mapped to the default instance 0
[FLOOR2]port-group 5
[FLOOR2-port-group-5]group-member g0/0/2 to g0/0/24
[FLOOR2-port-group-5]stp edged-port enable
[FLOOR2-port-group-5]quit
[FLOOR2]stp bpdu-protection
//Configure G0/0/2 to G0/0/24 interfaces of switch FLOOR2 as edge ports, and enable
//BPDU protection
```

④ According to the planning of spanning tree, the configuration steps of switch FLOOR3 spanning tree are as follows:

```
[FLOOR3]stp enable
```

```
[FLOOR3]stp mode mstp
[FLOOR3]
//Enable spanning tree and configure spanning tree mode as MSTP
[FLOOR3]stp region-configuration
[FLOOR3-mst-region]region-name cdp
[FLOOR3-mst-region]revision-level 1
[FLOOR3-mst-region]active region-configuration
[FLOOR3-mst-region]quit
//Configure the MSTP domain and activate the domain. The MSTP domain name is "cdp",
//and the revision level is 1. All VLAN are mapped to the default instance 0
[FLOOR3]port-group 5
[FLOOR3-port-group-5]group-member g0/0/2 to g0/0/24
[FLOOR3-port-group-5]stp edged-port enable
[FLOOR3-port-group-5]quit
[FLOOR3]stp bpdu-protection
//Configure G0/0/2 to G0/0/24 interfaces of switch FLOOR3 as edge ports, and enable
//BPDU protection
```

(2) Switch spanning tree configuration in MSTP domain "leader"

According to the planning of spanning tree, the configuration steps of switch FLOOR4 spanning tree are as follows:

```
[FLOOR4]stp enable
[FLOOR4]stp mode mstp
[FLOOR4]
//Enable spanning tree and configure spanning tree mode as MSTP
[FLOOR4]stp region-configuration
[FLOOR4-mst-region]region-name leader
[FLOOR4-mst-region]revision-level 1
[FLOOR2-mst-region]active region-configuration
[FLOOR2-mst-region]quit
//Configure the MSTP domain and activate the domain. The MSTP domain name is "leader"
//and the revision level is 1. All VLAN are mapped to the default instance 0
[FLOOR4]stp instance 0 root primary
//Configure switch FLOOR4 as the root bridge of instance 0 in the leader domain
[FLOOR4]port-group 5
[FLOOR4-port-group-5]group-member g0/0/2 to g0/0/24
[FLOOR4-port-group-5]stp edged-port enable
[FLOOR4-port-group-5]quit
[FLOOR4]stp bpdu-protection
```

```
//Configure G0/0/2 to G0/0/24 interfaces of switch FLOOR4 as edge ports, and enable
//BPDU protection
```

(3) Verification and network testing

After MSTP configuration is completed on all switches in the enterprise intranet, you should check and verify MSTP after MSTP calculation is stable.

① To execute the "display stp region-configuration" command on switches CORE, FLOOR1, FLOOR2 and FLOOR3 for viewing the MSTP domain configuration information. Fig. 2.98 shows the result of MSTP domain configuration of switch CORE spanning tree.

Note: According to the plan, the MSTP domain information (domain name, revision level, mapping relationship between instance and VLAN) of switches CORE, FLOOR1, FLOOR2 and FLOOR3 need to be exactly the same.

```
[CORE]display stp region-configuration
Oper configuration
  Format selector   :0
  Region name       :cdp
  Revision level    :1

  Instance   VLANs Mapped
     0       1 to 4094
```

Fig. 2.98   The result of switch CORE spanning tree MSTP domain configuration

② To execute the "display stp brief" command on switches CORE, FLOOR1, FLOOR2 and FLOOR3 for viewing the brief information of spanning tree, and the results are shown in Fig. 2.99 to Fig. 2.102. It is required to observe the switch port role and port status.

Note: According to the plan, the CORE is the root, and all ports in its instance 0 are designated ports, all in forwarding state. The ports of the access layer switches FLOOR1, FLOOR2 and FLOOR3 connected to the CORE are root ports, and even the ports of the host are designated ports, which are all in the forwarding state.

```
[CORE]display stp brief
MSTID  Port                    Role    STP State    Protection
  0    GigabitEthernet0/0/1    DESI    FORWARDING   NONE
  0    GigabitEthernet0/0/2    DESI    FORWARDING   NONE
  0    GigabitEthernet0/0/3    DESI    FORWARDING   NONE
  0    GigabitEthernet0/0/21   DESI    FORWARDING   BPDU
  0    GigabitEthernet0/0/22   DESI    FORWARDING   BPDU
  0    GigabitEthernet0/0/23   DESI    FORWARDING   BPDU
```

Fig. 2.99   The result of executing the "display stp brief" command on the switch CORE

```
[FLOOR1]display stp brief
MSTID  Port                    Role    STP State    Protection
  0    GigabitEthernet0/0/1    ROOT    FORWARDING   NONE
  0    GigabitEthernet0/0/3    DESI    FORWARDING   BPDU
```

Fig. 2.100   The result of executing the "display stp brief" command on switch FLOOR1

```
[FLOOR2]display stp brief
MSTID  Port                    Role    STP State    Protection
  0    GigabitEthernet0/0/1    ROOT    FORWARDING   NONE
  0    GigabitEthernet0/0/6    DESI    FORWARDING   BPDU
```

Fig. 2.101 The result of executing the "display stp brief" command on switch FLOOR2

```
[FLOOR3]display stp brief
MSTID  Port                     Role    STP State    Protection
  0    GigabitEthernet0/0/1     ROOT    FORWARDING   NONE
  0    GigabitEthernet0/0/22    DESI    FORWARDING   BPDU
```

Fig. 2.102 The result of executing the "display stp brief" command on switch FLOOR3

③ To execute the "display stp region-configuration" command on switch FLOOR4 for viewing the MSTP domain configuration information, as shown in Fig. 2.103.

```
[FLOOR4]display stp region-configuration
Oper configuration
  Format selector   :0
  Region name       :leader
  Revision level    :1

  Instance   VLANs Mapped
     0       1 to 4094
```

Fig. 2.103 The result of executing the "display stp region-configuration" command on switch FLOOR4

④ To execute "display stp brief" command on switch FLOOR4 for viewing the brief information of spanning tree, and the result is shown in Fig. 2.104. It is required to observe the switch port role and port status. According to the plan, FLOOR4 is the root of the leader domain, and all ports in its instance 0 are designated ports, all of which are in forwarding state.

```
[FLOOR4]display stp brief
MSTID  Port                     Role    STP State    Protection
  0    GigabitEthernet0/0/1     DESI    FORWARDING   NONE
  0    GigabitEthernet0/0/2     DESI    FORWARDING   BPDU
  0    GigabitEthernet0/0/14    DESI    FORWARDING   BPDU
```

Fig. 2.104 The result of executing the "display stp brief" command on switch FLOOR4

Exercises

Multiple choice

(1) When STP selects the root bridge, (   ) will be chosen as the root.

　　A. the switch with the smallest MAC address

　　B. switch with the largest MAC address

　　C. switch with minimum BID

　　D. switch with the largest BID

(2) (   ) can learn the source MAC address of a data frame into the MAC address table, but don't forward the data frame.

A. Forwarding   B. Learning
C. Blocking   D. Listening

(3) In STP, (   ) will not send configuration BPDU.

A. forwarding   B. learning
C. blocking   D. listening

(4) The function of configuring the switch port with the "stp edged-port enable" command in the switch is (   ).

A. to eliminate quickly the two-layer loop

B. that the command is configured on trunk interfaces connected to other switches to reduce STP convergence time

C. that if the command is configured on the access port, the access interface will immediately enter the forwarding state from the blocking state

(5) (   ) is the official standard of link aggregation technology.

A. IEEE 802.1Q   B. IEEE 802.3AD   C. IEEE 802.1W   D. IEEE 802.1D

(6) (   ) is the official standard of MSTP.

A. IEEE 802.1S   B. IEEE 802.3AD   C. IEEE 802.1W   D. IEEE 802.1D

Task 5 Planning and Configuration of PPP

Task Description

On the basis of completing the Task 1 to Task 4 of this project, the PPP authentication function is used to plan and configure the bidirectional identity authentication between the enterprise boundary router AR1 and the operator router ISP in the network shown in Fig. 2.1.

Task Objectives

- Understanding the working process of PPP.
- Understanding the working process of PAP and master the configuration of PAP.
- Understanding the working process of CHAP and master the configuration of CHAP.

Related Knowledge

1. Overview of PPP

PPP(point-to-point protocol) is a common WAN data link layer protocol, which is developed on the basis of SLIP (serial line internet protocol) protocol and can be used on synchronous and asynchronous private lines, asynchronous dial-up links and synchronous dial-up links.

PPP is a protocol cluster composed of a series of protocols, and the hierarchical structure of PPP protocols is shown in Fig. 2.105. PPP provides LCP(link control protocol) for negotiation of various link layer parameters, such as maximum receiving unit and authentication mode, and NCP(network control protocol), such as IPCP(IP control protocol), for negotiation of network layer parameters, which better supports network layer protocols.

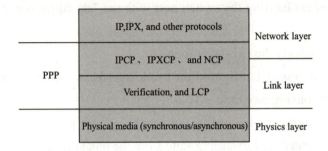

Fig. 2.105　The Hierarchical structure of PPP protocol

2. Establishment of PPP session

The establishment of PPP session is divided into three stages: link layer negotiation, authentication negotiation (optional) and network layer negotiation, as shown in Fig. 2.106.

(1) Link layer negotiation

Negotiate link parameters through LCP message, such as establishing link layer connection.

(2) Authentication negotiation (optional)

Link authentication is carried out through the authentication mode negotiated in the link establishment stage.

(3) Network layer negotiation

Select and configure a network layer protocol through NCP negotiation and negotiate network layer parameters.

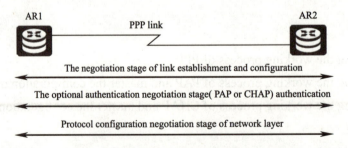

Fig. 2.106　The process of PPP session establishment

3. PPP authentication

PPP provides PAP and CHAP.

(1) PAP

PAP authentication protocol is a two-way handshake authentication protocol, and the password is sent in clear text on the link, as shown in Fig. 2.107. The authentication process is as follows:

① The consumer sends the configured user name and password information to the authenticator in clear text.

② After receiving the user name and password information sent by the provider, the authenticator should checks whether the user name and password information match according to the locally configured user name and password database, and if they match, the authenticator should send an ACK message to inform the provider that it has passed the authentication and is allowed to enter the next stage; If not, the authenticator should send a NAK message to inform the provider that it's failed.

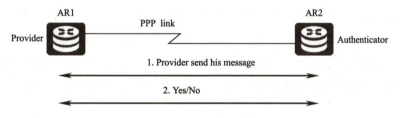

Fig. 2.107 The process of PPP PAP authentication

(2) CHAP

CHAP authentication uses three-way handshake authentication, and the negotiation message is encrypted before being transmitted on the link, as shown in Fig. 2.108. The authentication process is as follows:

① The authenticator initiates an authentication request, and the authenticator sends a Challenge message to the provider, which contains a Random number and an ID.

② After receiving the Challenge message, the provider performs an encryption operation to obtain a summary message, and then encapsulates the summary message and the CHAP user name configured on the port in a Response message and sends it back to the authenticator.

③ After receiving the Response message sent by the provider, the authenticator looks up the corresponding password information locally according to the user name in it, and after obtaining the password information, performs an encryption operation in the same way as that of the provider. Then, the digest information obtained by encryption operation is compared with the digest information encapsulated in the Response message. Therefore, if they are the same, the authentication is successful, and vice versa.

194 Enterprise Internetworking Technology (Bilingual)

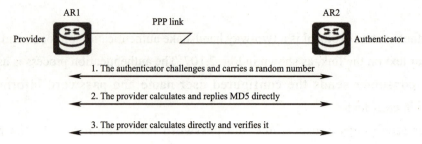

Fig. 2.108 The process of PPP CHAP authentication

When using CHAP authentication method, the provider's password is encrypted before transmission, which greatly improves the security.

Related Skills

1. Planning and configuration of PPP PAP local authentication

It is required to build the network topology diagram as shown in Fig. 2.109, configure the serial ports of routers AR1 and AR2, and use PPP PAP for one-way identity authentication, in which router AR1 is the authenticator and router AR2 is the provider, and test and check.

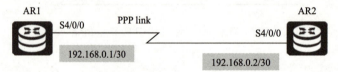

Fig. 2.109 The network topology diagram of PPP PAP

PPP PAP planning is as follows;

① AR1 is the authenticator, the legal user name is AR2, and the password is cdp123456.

② AR2 is the provider.

(1) Observing the default data link layer protocol of Huawei serial port

You should use the "display interface s4/0/0" command on routers AR1 and AR2 respectively to display the IP information about interface S4/0/0 and pay attention to the information shown in Fig. 2.110.

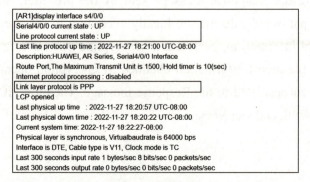

Fig. 2.110 Default encapsulation protocol for serial port of Huawei router

(2) Configuration of authenticator router AR1

The authenticator configured as PPP PAP on AR1 needs to complete two tasks:

① Creating a legal PPP account and password for the remote authenticated router.

② Entering S4/0/0 interface, configuring IP address and encapsulation protocol as PPP, and enabling PAP authentication.

The steps of configuration are as follows:

```
[AR1]aaa
[AR1-aaa]local-user AR2 password cipher cdp123456
[AR1-aaa]local-user AR2 service-type ppp
[AR1-aaa]quit
//Create a legal PPP account and password, and the account service type is PPP
[AR1]interface s4/0/0
[AR1-Serial4/0/0]ip address 192.168.0.1 30
[AR1-Serial4/0/0]link-protocol ppp
[AR1-Serial4/0/0]ppp authentication-mode pap
[AR1-Serial4/0/0]quit
//Configure S4/0/0 interface protocol as PPP protocol, and enable PAP authentication
```

(3) Configuration of authenticated router AR2

The provider configured as PPP on AR2 only needs to enter the serial port, configure the encapsulation protocol as PPP (the default protocol of serial port of Huawei router is PPP), and send the user name and password of PPP PAP (the user name and password must match the legal user name and password configured by the authenticator). The configuration steps are as follows:

```
[AR2]interface s4/0/0
[AR2-Serial4/0/0]ip address 192.168.0.2 30
[AR2-Serial4/0/0]link-protocol ppp
[AR2-Serial4/0/0]ppp pap local-user AR2 password cipher cdp123456
[AR2-Serial4/0/0]
```

(4) Verification and testing

① You should disable the S4/0/0 interface on the router AR1 or AR2, and then enable the interface, and then execute the "display interface serial 4/0/0" command to view the configuration information of the interface. The physical and link protocol states of the interface are "UP" and the LCP and IPCP of PPP are "opened", indicating that the PPP negotiation of the link has been successful, as shown in Fig. 2.111.

② Routers AR1 and AR2 can ping each other. As shown in Fig. 2.112, router AR1 can ping the IP address of the S4/0/0 interface of AR2.

```
[AR2-Serial4/0/0]shutdown        //Disable the interface
```

```
[AR2-Serial4/0/0]undo shutdown   //Enable the interface
```

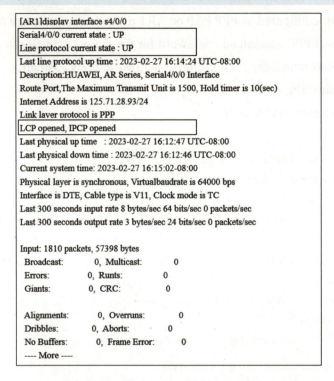

Fig. 2.111   The result of PPP negotiation

```
[AR1]ping 192.168.0.2
 PING 192.168.0.2: 56  data bytes, press CTRL_C to break
   Reply from 192.168.0.2: bytes=56 Sequence=1 ttl=255 time=30 ms
   Reply from 192.168.0.2: bytes=56 Sequence=2 ttl=255 time=20 ms
   Reply from 192.168.0.2: bytes=56 Sequence=3 ttl=255 time=20 ms
   Reply from 192.168.0.2: bytes=56 Sequence=4 ttl=255 time=30 ms
   Reply from 192.168.0.2: bytes=56 Sequence=5 ttl=255 time=20 ms
```

Fig. 2.112   The router AR1 ping the S4/0/0 interface of router AR2

2. Planning and configuration of local authentication of PPP CHAP

It is required to set up the network topology diagram as shown in Fig. 2.113, configure the serial ports of routers AR1 and AR2, and use PPP CHAP for one-way identity authentication, in which router AR1 is the authenticator and router AR2 is the provider, and test and check.

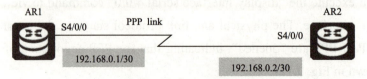

Fig. 2.113   The topology diagram of PPP CHAP network

PPP CHAP planning is as follows;

① AR1 is the authenticator, the legal user name is "AR2" and the password is "cdp123456".

② AR2 is the provider.

(1) Configuration of authenticator router AR1

The authenticator configured as PPP CHAP on AR1 needs to complete two tasks:

① Creating a legal PPP account and password for the remote authenticated router.

② Entering S4/0/0 interface, configuring IP address and encapsulation protocol as PPP, and enabling CHAP authentication.

The steps of configuration are as follows:

```
[AR1]aaa
[AR1-aaa]local-user AR2 password cipher cdp123456
[AR1-aaa]local-user AR2 service-type ppp
[AR1-aaa]quit
//Create a legal PPP account and password, and the account service type is PPP
[AR1]interface s4/0/0
[AR1-Serial4/0/0]ip address 192.168.0.1 30
[AR1-Serial4/0/0]link-protocol ppp
[AR1-Serial4/0/0]ppp authentication-mode chap
[AR1-Serial4/0/0]quit
//Configure S4/0/0 interface protocol as PPP protocol, and enable CPAP authentication
```

(2) Configuration of authenticated router AR2

The provider configured as PPP on AR2 only needs to enter the serial port, configure the encapsulation protocol as PPP, and send the username and password of PPP CHAP (the username and password must match the legal username and password configured by the authenticator). The steps of configuration are as follows:

```
[AR2]interface s4/0/0
[AR2-Serial4/0/0]ip address 192.168.0.2 30
[AR2-Serial4/0/0]link-protocol ppp
[AR2-Serial4/0/0]ppp chap user AR2
[AR2-Serial4/0/0]ppp chap password cipher cdp123456
[AR2-Serial4/0/0]
```

(3) Verification and testing

① Disable the S4/0/0 interface on the router AR1 or AR2, and then enable the interface, and then execute the "display interface serial 4/0/0" command to check the configuration information of the interface. The physical and link protocol states of the interface are "UP" and the LCP and IPCP of PPP are "opened", indicating that the PPP negotiation of the link has been successful, as shown

in Fig. 2.114.

```
[AR2-Serial4/0/0]shutdown          //Disable the interface
[AR2-Serial4/0/0]undo shutdown     //Enable the interface
```

```
[AR1]display interface s4/0/0
Serial4/0/0 current state : UP
Line protocol current state : UP
Last line protocol up time : 2023-03-30 14:47:28 UTC-08:00
Description:HUAWEI, AR Series, Serial4/0/0 Interface
Route Port,The Maximum Transmit Unit is 1500, Hold timer is 10(sec)
Internet Address is 192.168.0.1/30
Link layer protocol is PPP
LCP opened, IPCP opened
Last physical up time   : 2023-03-30 14:47:25 UTC-08:00
Last physical down time : 2023-03-30 14:47:20 UTC-08:00
Current system time: 2023-03-30 14:47:36-08:00
Physical layer is synchronous, Virtualbaudrate is 64000 bps
Interface is DTE, Cable type is V11, Clock mode is TC
Last 300 seconds input rate 6 bytes/sec 48 bits/sec 0 packets/sec
Last 300 seconds output rate 2 bytes/sec 16 bits/sec 0 packets/sec
```

Fig. 2.114　The result of PPP negotiation

② Routers AR1 and AR2 can ping each other. As shown in Fig. 2.115, router AR1 ping the S4/0/0 interface of router AR2.

```
[AR1]ping 192.168.0.2
  PING 192.168.0.2: 56  data bytes, press CTRL_C to break
    Reply from 192.168.0.2: bytes=56 Sequence=1 ttl=255 time=30 ms
    Reply from 192.168.0.2: bytes=56 Sequence=2 ttl=255 time=20 ms
    Reply from 192.168.0.2: bytes=56 Sequence=3 ttl=255 time=20 ms
    Reply from 192.168.0.2: bytes=56 Sequence=4 ttl=255 time=30 ms
    Reply from 192.168.0.2: bytes=56 Sequence=5 ttl=255 time=20 ms
```

Fig. 2.115　The router AR1 ping the S4/0/0 interface of router AR2

Task Completion

1. Task planning

According to the task description and the topology shown in Fig. 2.1, the enterprise boundary router AR1 and the operator router ISP use PPP CHAP to achieve bidirectional authentication subject to the password "cdp123456". See Table 2.12 for the IP planning.

2. Task implementation

(1) Configuration of PPP CHAP local authentication

① According to the planning of PPP CHAP, the configuration steps of router AR1 are as follows:

```
[AR1]aaa
[AR1-aaa]local-user ISP password cipher cdp123456
[AR1-aaa]local-user ISP service-type ppp
[AR1-aaa]quit
//Create a legal PPP account and password, and the account service type is PPP
[AR1]interface s4/0/0
[AR1-Serial4/0/0]link-protocol ppp
[AR1-Serial4/0/0]ip address 125.71.28.93 24
[AR1-Serial4/0/0]ppp authentication-mode chap
[AR1-Serial4/0/0]ppp chap user AR1
[AR1-Serial4/0/0]ppp chap password cipher cdp123456
[AR1-Serial4/0/0]quit
//Configure S4/0/0 interface protocol as PPP protocol, and enable CPAP authentication. The
//user name for CHAP authentication is "AR1" and the authentication password is "cdp123456"
```

② According to the planning of PPP CHAP, the configuration steps of router ISP are as follows:

```
[ISP]aaa
[ISP-aaa]local-user AR1 password cipher cdp123456
[ISP-aaa]local-user AR1 service-type ppp
[ISP-aaa]quit
//Create a legal PPP account and password, and the account service type is PPP
[ISP]interface s4/0/0
[ISP-Serial4/0/0]link-protocol ppp
[ISP-Serial4/0/0]ip address 125.71.28.1 24
[ISP-Serial4/0/0]ppp authentication-mode chap
[ISP-Serial4/0/0]ppp chap user ISP
[ISP-Serial4/0/0]ppp chap password cipher cdp123456
[ISP-Serial4/0/0]quit
//Configure S4/0/0 interface protocol as PPP protocol, and enable CPAP authentication. The
//user name for CHAP authentication is "ISP" and the authentication password is "cdp123456"
```

(2) Verification and testing

① You should disable the interface configured with PPP CHAP authentication on the router, then enable the interface, and execute the "display interface serial 4/0/0" command to view the configuration information of the interface. The physical and link protocol states of the interface are "UP", and the LCP and IPCP of PPP are "opened", indicating that the PPP negotiation of the link has been successful, as shown in Fig. 2.116.

```
[AR1-Serial4/0/0]shutdown        //Disable the interface
[AR1-Serial4/0/0]undo shutdown   //Enable the interface
```

```
[AR1]display interface s4/0/0
Serial4/0/0 current state : UP
Line protocol current state : UP
Last line protocol up time : 2023-02-27 16:14:24 UTC-08:00
Description:HUAWEI, AR Series, Serial4/0/0 Interface
Route Port,The Maximum Transmit Unit is 1500, Hold timer is 10(sec)
Internet Address is 125.71.28.93/24
Link layer protocol is PPP
LCP opened, IPCP opened
Last physical up time   : 2023-02-27 16:12:47 UTC-08:00
Last physical down time : 2023-02-27 16:12:46 UTC-08:00
Current system time: 2023-02-27 16:15:02-08:00
Physical layer is synchronous, Virtualbaudrate is 64000 bps
Interface is DTE, Cable type is V11, Clock mode is TC
Last 300 seconds input rate 8 bytes/sec 64 bits/sec 0 packets/sec
Last 300 seconds output rate 3 bytes/sec 24 bits/sec 0 packets/sec

Input: 1810 packets, 57398 bytes
  Broadcast:        0,  Multicast:      0
  Errors:           0,  Runts:          0
  Giants:           0,  CRC:            0

  Alignments:       0,  Overruns:       0
  Dribbles:         0,  Aborts:         0
  No Buffers:       0,  Frame Error:    0
---- More ----
```

Fig. 2.116   The result of PPP negotiation

② Routers AR1 and ISP can ping each other. As shown in Fig. 2.117, router AR1 ping the S4/0/0 interface of router ISP.

```
[AR1]ping 125.71.28.1
 PING 125.71.28.1: 56  data bytes, press CTRL_C to break
   Reply from 125.71.28.1: bytes=56 Sequence=1 ttl=255 time=40 ms
   Reply from 125.71.28.1: bytes=56 Sequence=2 ttl=255 time=30 ms
   Reply from 125.71.28.1: bytes=56 Sequence=3 ttl=255 time=30 ms
   Reply from 125.71.28.1: bytes=56 Sequence=4 ttl=255 time=20 ms
   Reply from 125.71.28.1: bytes=56 Sequence=5 ttl=255 time=20 ms

 --- 125.71.28.1 ping statistics ---
   5 packet(s) transmitted
   5 packet(s) received
   0.00% packet loss
   round-trip min/avg/max = 20/28/40 ms
```

Fig. 2.117   The Router AR1 ping the S4/0/0 interface of router ISP

Exercises

Multiple choice

(1) PPP PAP authentication adopts (   ) handshake.

    A. one   B. two   C. three   D. four

(2) PPP CHAP authentication adopts (　　) handshake.
  A. one   B. two   C. three   D. four

(3) (　　) is not a part of PPP protocol.
  A. LCP   B. NCP   C. CHAP   D. IP

Task 6   Planning and Configuration of RIP

Task Description

On the basis of completing the Task 1 to Task 5 of this project, in the network shown in Fig. 2.1, complete the routing planning and configuration tasks of using the default and dynamic routing protocol RIP to achieve interconnection in the enterprise intranet and enable the enterprise intranet access to the internet.

Tasks Objectives

- Understanding the difference between dynamic routing and static routing.
- Understanding the working principle and characteristics of RIPv1.
- Mastering the configuration, viewing and troubleshooting of RIPv1.
- Understanding the working principle and characteristics of RIPv2.
- Mastering the configuration, viewing and troubleshooting of RIPv2.

Related Knowledge

1. Dynamic routing and routing protocol

Static routing is selected by the network administrator and manually configured, which has less overhead but is not flexible. When the scale of network interconnection increases, it will be very complicated to generate and maintain a routing table manually by static routing, and it can not adapt to the changes of network state in time. Therefore, when the scale of network interconnection is large or there are many unstable factors in the network, dynamic routing is usually used to generate and maintain routing tables. Running the routing protocol on the router and configuring the corresponding routing protocol can ensure that the router automatically generates and maintains routing table information.

There are many classification methods for routing protocols:

① It can be divided into exterior gateway protocol (EGP) and interior gateway protocol (IGP) according to the different scope of application.

② According to different routing algorithms, routing protocols are divided into distance vector

routing protocol and link state routing protocol.

IPv4 routing protocols include: RIPv1, RIPv2, OSPF, IS-IS and BGP,etc., and IPv6 routing protocols include RIPng, OSPFv3 and BGPv4,etc., as shown in Fig. 2.118.

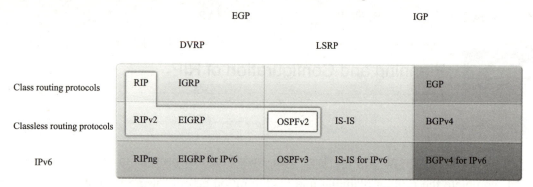

Fig. 2.118 The classification of routing protocols

The routing table dynamically constructed by routing protocol can better adapt to the changes of network state, such as the changes of network topology or network traffic, and also reduce the workload of manually generating and maintaining routing tables. However, the price paid for this is the resource consumption for routing information update, including network bandwidth, router CPU and storage resources.

2. RIP protocol

RIP is short for routing information protocol, which is an interior gateway protocol. RIP is a protocol based on distance vector (D-V) algorithm, which uses hop count as a metric to measure the distance to the target network. The number of hops refers to the number of routers that need to pass to reach the target network. RIP exchanges routing information through UDP messages, and the port number used is 520.

Although the implementation and configuration of RIP are very simple, the distance vector algorithm used in this protocol may form a routing loop in the network. Therefore, RIP introduces some mechanisms to reduce or prevent the occurrence of routing loops, such as horizontal division, suppressing update timers, specifying the maximum number of hops, and poisoning routing. The maximum hop count of RIP is set at 15. When the destination network exceeds 15 hops, RIP considers the destination network unreachable. Routing information is exchanged between neighboring routers running RIP protocol by means of periodic update.

There are two versions of RIP, namely RIPv1 and RIPv2. RIPv1 uses broadcast to send routing update information, and the protocol standard of RIPv1 was put forward before VLSM and CIDR. Therefore, RIPv1 does not support VLSM and CIDR, and cannot be enabled in the interconnection environment of VLSM and CIDR. The subnet mask information is not included in the routing

update information of RIPv1. RIPv2 inherits all the functions of RIPv1, and improves it on this basis. The main contents of the improvement include support for VLSM and authentication. RIPv2 uses multicast to send routing update information, and the multicast address is 224.0.0.9.

Related Skills

1. Planning and configuration of RIPv2 based on router

It is required to build the network topology diagram and IP address assignment as shown in Fig. 2.119, and use RIPv2 routing protocol on routers AR1, AR2 and AR3 to realize the interconnection of networks, and check and test the routes.

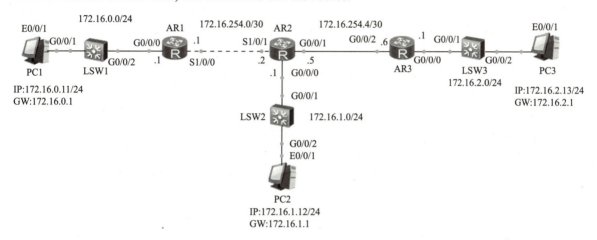

Fig. 2.119 The topology diagram of RIPv2 network based on router

See Table 2.17 for the planning of RIPv2.

Table 2.17 Planning of RIPv2

Device	RIP number	Networks participating in RIP updates	Silent interface	Auto-summary
AR1	1	172.16.0.0	G0/0/0	Yes
AR2	1	172.16.0.0	G0/0/0	Yes
AR3	1	172.16.0.0	G0/0/0	Yes

(1) Configuration of router interface

① Configuring the interfaces on the routers AR1, AR2 and AR3 according to the IP planned in Fig. 2.119, and the steps are as follows:

```
[AR1]interface g0/0/0
[AR1-GigabitEthernet0/0/0]ip address 172.16.0.1 24
[AR1-GigabitEthernet0/0/0]quit
//Configure the IP address and mask of G0/0/0 interface of router AR1 as
//"172.16.0.1/24"
```

```
[AR1]interface s1/0/0
[AR1-Serial1/0/0]ip address 172.16.254.1 30
[AR1-Serial1/0/0]quit
//Configure the IP address and mask of the S1/0/0 interface of router AR1 as
//"172.16.254.1/30"
[AR2]interface g0/0/0
[AR2-GigabitEthernet0/0/0]ip address 172.16.1.1 24
[AR2-GigabitEthernet0/0/0]quit
//Configure the IP address and mask of the G0/0/0 interface of router AR2 as
//"172.16.1.1/24"
[AR2]interface s1/0/1
[AR2-Serial1/0/1]ip address 172.16.254.2 30
[AR2-Serial1/0/1]quit
//Configure the IP address and mask of the S1/0/1 interface of router AR2 as
//"172.16.254.2/30"
[AR2]interface g0/0/1
[AR2-GigabitEthernet0/0/1]ip address 172.16.254.5 30
[AR2-GigabitEthernet0/0/1]quit
//Configure the IP address and mask of the G0/0/1 interface of router AR2 as
//"172.16.254.5/30"
[AR3]interface g0/0/0
[AR3-GigabitEthernet0/0/0]ip address 172.16.2.1 24
[AR3-GigabitEthernet0/0/0]quit
//Configure the IP address and mask of G0/0/0 interface of router AR3 as
//"172.16.2.1/24"
[AR3]interface g0/0/2
[AR3-GigabitEthernet0/0/2]ip address 172.16.254.6 30
[AR3-GigabitEthernet0/0/2]quit
//Configure the IP address and mask of G0/0/2 interface of router AR3 as
//"172.16.254.6/30"
```

② You should use the "display ip interface brief" command on the routers AR1, AR2 and AR3 to check whether the IP address of the router interface is correct and whether the physical and protocol states of the interfaces are "UP".

(2) Configuration of RIPv2

The configuration of RIPv2 needs to complete the following tasks:

① Creating and run RIP process, and specify RIP process number.

② The specified version of RIP is "2".

③ You should use the "network" command to specify which networks participate in RIP routing updates.

The "network" command has two meanings: on the one hand, it is used to specify which directly connected networks on this machine participate in RIP updates; on the other hand, it is used to specify which interfaces can send and receive RIP updates.

④ Configuring the silent interface (optional configuration). The ports configured as silent interfaces can only receive RIP updates without sending RIP updates.

⑤ Configuring to disable automatic summarization (optional configuration).

The steps of configuration are as follows:

```
[AR1]rip1                          //Enable RIP protocol, and the process number is 1
[AR1-rip-1]version 2               //Specify RIP version as RIPv2
[AR1-rip-1]network 172.16.0.0      //The network "172.16.0.0" participates in the routing
                                   //update of RIPv2. Note: RIP is equipped
[AR1-rip-1]silent-interface g0/0/0 //Configure G0/0/0 interface as silent interface
[AR1-rip-1]undo summary            //Disable automatic summarization
[AR2]rip 1                         //Enable RIP protocol, and the process number is 1
[AR2-rip-1]version 2               //Specify RIP version as RIPv2
[AR2-rip-1]network 172.16.0.0      //The network "172.16.0.0" participates in the routing
                                   //update of RIPv2. Note:The network number specified
                                   //by RIP is classful.
[AR2-rip-1]silent-interface g0/0/0 //Configure G0/0/0 interface as silent interface
[AR2-rip-1]undo summary            //Disable automatic summarization
[AR3]rip 1                         //Enable RIP protocol, and the process number is 1
[AR3-rip-1]version 2               //Specify RIP version as RIPv2
[AR3-rip-1]network 172.16.0.0      //The network "172.16.0.0" participates in the routing
                                   //update of RIPv2. Note:The network number specified
                                   //by RIP is classful.
[AR3-rip-1]silent-interface g0/0/0 //Configure G0/0/0 interface as silent interface
[AR3-rip-1]undo summary            //Disable automatic summarization
```

(3) Verification and testing

① To execute the command "display IP routing-table | include RIP" on routers AR1, AR2 and AR3 respectively for viewing the routing entries with RIP protocol in the router routing table. Fig. 2.120, Fig. 2.121 and Fig. 2.122 are the result after executing the "display IP routing-table | include rip" command on routers AR1, AR2 and AR3 respectively.

Note: "RIP" in a routing entry means that the routing entry has been learned through RIP protocol. The default routing priority value of RIP is "100" and the Cost value is the number of hops to the target network.

② You should configure the IP address, subnet mask and default gateway on the test hosts PC1, PC2 and PC3, according to the IP planned in Fig. 2.119, and then verify the correctness of the

IP address, subnet mask and default gateway by using the "ipconfig /all" command on the command line. After that, you can use ping command to test the connectivity of the whole network. Fig. 2.123 shows that the test host PC1 successfully ping PC2 and PC3.

```
[AR1]display ip routing-table | include RIP
Route Flags: R - relay, D - download to fib
------------------------------------------------------------
Routing Tables: Public
         Destinations : 14    Routes : 14

Destination/Mask  Proto  Pre  Cost   Flags NextHop       Interface

172.16.1.0/25      RIP    100   1      D   172.16.254.2  Serial1/0/0
172.16.2.0/24      RIP    100   2      D   172.16.254.2  Serial1/0/0
172.16.254.4/30    RIP    100   1      D   172.16.254.2  Serial1/0/0
```

Fig. 2.120 The result of executing the command "display IP routing-table | include rip" on AR1

```
[AR2]display ip routing-table | include RIP
Route Flags: R - relay, D - download to fib
------------------------------------------------------------
Routing Tables: Public
         Destinations : 16    Routes : 16

Destination/Mask  Proto  Pre  Cost   Flags NextHop       Interface

172.16.0.0/24     RIP    100   1      D   172.16.254.1  Serial1/0/1
172.16.2.0/24     RIP    100   1      D   172.16.254.6  GigabitEthernet0/0/1
```

Fig. 2.121 The result of executing the command "display IP routing-table | include rip" on AR2

```
[AR3]display ip routing-table | include RIP
Route Flags: R - relay, D - download to fib
------------------------------------------------------------
Routing Tables: Public
         Destinations : 13    Routes : 13

Destination/Mask  Proto  Pre  Cost   Flags NextHop       Interface

172.16.0.0/24     RIP    100   2      D   172.16.254.5  GigabitEthernet0/0/2
172.16.1.0/25     RIP    100   1      D   172.16.254.5  GigabitEthernet0/0/2
172.16.254.0/30   RIP    100   1      D   172.16.254.5  GigabitEthernet0/0/2
```

Fig. 2.122 The result of executing the command "display IP routing-table | include rip" on AR3

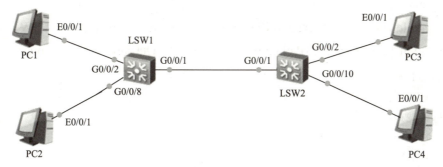

Fig. 2.123 The result of testing host PC1 successfully ping PC2 and PC3

2. Planning and configuration of RIPv2 based on three-layer switch

It is required to build the network topology diagram as shown in Fig. 2.124 (Note: switches LSW1 and LSW2 are three-layer switches supporting RIP protocol). See Table 2.18 for VLAN planning, and Table 2.19 for IP address allocation. RIPv2 routing protocol is used to realize network interconnection, and the routing is checked and tested.

Fig. 2.124 RIPv2 topology diagram based on three-layer switch

Table 2.18 Planning of VLAN

Device	Interface or VLAN	Description of VLAN	Interface	Explanation
LSW1	VLAN2	student	G0/0/2	
	VLAN3	teacher	G0/0/8	
	VLAN100	linktoLSW2	G0/0/1	Used to connect to LSW2
LSW2	VLAN20	jwc	G0/0/2 interface	
	VLAN30	cw	G0/0/10 interface	
	VLAN100	linktoLSW1	G0/0/1	Used to connect to LSW1

Table 2.19 Planning of IP address

Device	Interface	IP address and mask	Gateway
LSW1	Vlanif2	192.168.0.1/25	
	Vlanif3	192.168.0.129/25	
	Vlanif100	192.168.254.1/30	
LSW2	Vlanif20	192.168.1.1/25	
	Vlanif30	192.168.1.129/25	
	Vlanif100	192.168.254.2/30	
PC1	NIC	192.168.0.2/25	192.168.0.1
PC2	NIC	192.168.0.130/25	192.168.0.129
PC3	NIC	192.168.1.2/25	192.168.1.1
PC4	NIC	192.168.1.130/25	192.168.1.129

See Table 2.20 for the planning of RIPv2.

Table 2.20 Planning of RIPv2

Device	RIP number	Networks participating in RIP updates	Silent interface	Auto-summary
LSW1	1	192.168.0.0, 192.168.254.0	VLAN2, VLAN3	be
LSW2	1	192.168.1.0, 192.168.254.0	VLAN20, VLAN30	be

(1) Configuration of VLAN and IP

① You should complete VLAN and IP configuration on three-layer switch LSW1 according to VLAN planning in Table 2.18 and IP planning in Table 2.19. The steps are as follows:

```
[LSW1]vlan 2
[LSW1-vlan2]description student
[LSW1-vlan2]vlan 3
[LSW1-vlan3]description teacher
[LSW1-vlan3]vlan 100
[LSW1-vlan100]description linktoLSW2
[LSW1-vlan100]quit
[LSW1]interface g0/0/2
[LSW1-GigabitEthernet0/0/2]port link-type access
[LSW1-GigabitEthernet0/0/2]port default vlan 2
[LSW1-GigabitEthernet0/0/2]quit
[LSW1]interface g0/0/8
[LSW1-GigabitEthernet0/0/8]port link-type access
[LSW1-GigabitEthernet0/0/8]port default vlan 3
[LSW1-GigabitEthernet0/0/8]quit
[LSW1]interface g0/0/1
[LSW1-GigabitEthernet0/0/1]port link-type access
```

```
[LSW1-GigabitEthernet0/0/1]port default vlan 100
[LSW1-GigabitEthernet0/0/1]quit
[LSW1]interface Vlanif 2
[LSW1-Vlanif2]ip address 192.168.0.1 25
[LSW1-Vlanif2]quit
[LSW1]interface Vlanif 3
[LSW1-Vlanif3]ip address 192.168.0.129 25
[LSW1-Vlanif3]quit
[LSW1]interface Vlanif 100
[LSW1-Vlanif100]ip address 192.168.254.1 30
[LSW1-Vlanif100]quit
```

② You should complete VLAN and IP configuration on three-layer switch LSW2 according to VLAN planning in Table 2.18 and IP planning in Table 2.19. The steps are as follows:

```
[LSW2]vlan 20
[LSW2-vlan20]description jwc
[LSW2-vlan20]vlan 30
[LSW2-vlan30]description cw
[LSW2-vlan30]vlan 100
[LSW2-vlan100]description linktoLSW1
[LSW2-vlan100]quit
[LSW2]interface g0/0/2
[LSW2-GigabitEthernet0/0/2]port link-type access
[LSW2-GigabitEthernet0/0/2]port default vlan 20 [LSW2-GigabitEthernet0/0/2]quit
[LSW2]interface g0/0/10
[LSW2-GigabitEthernet0/0/10]port link-type access
[LSW2-GigabitEthernet0/0/10]port default vlan 30
[LSW2-GigabitEthernet0/0/10]quit
[LSW2]interface g0/0/1
[LSW2-GigabitEthernet0/0/1]port link-type access
[LSW2-GigabitEthernet0/0/1]port default vlan 100
[LSW2-GigabitEthernet0/0/1]quit
[LSW2]interface Vlanif 20
[LSW2-Vlanif20]ip address 192.168.1.1 25
[LSW2-Vlanif20]quit
[LSW2]interface Vlanif 30
[LSW2-Vlanif30]ip address 192.168.1.129 25
[LSW2-Vlanif30]quit
[LSW2]interface Vlanif 100
[LSW2-Vlanif100]ip address 192.168.254.2 30
```

```
[LSW2-Vlanif100]quit
```

③ You should use the "display ip interface brief" command on the three-layer switches LSW1 and LSW2 to check whether the Vlanif interface IP address and subnet mask of the three-layer switches are correct and whether the physical and protocol states of the interfaces are "UP".

(2) Configuration of RIPv2

The RIP configuration steps of LSW1 and LSW2 are as follows:

```
[LSW1]rip 1                              //Enable RIP protocol, and the process number is 1
[LSW1-rip-1]version 2                    //Specify RIP version as RIPv2
[LSW1-rip-1]network 192.168.0.0          //The network "192.168.0.0" participates in the
                                         //routing update of RIPV2. Note:Subnets "192.168.0.0/25"
                                         //and "192.168.0.128/25" The classful network numbers
                                         //of are all "192.168.0.0"
[LSW1-rip-1]network 192.168.254.0        //The network "192.168.254.0" participates in
                                         //the routing update of RIPv2
[LSW1-rip-1]silent-interface Vlanif 2    //Configure Vlanif2 interface as silent interface
[LSW1-rip-1]silent-interface Vlanif 3    //Configure Vlanif3 interface as silent interface
[LSW1-rip-1]undo summary                 //Disable automatic summarization
[LSW1-rip-1]quit
[LSW2]rip 1                              //Enable RIP protocol, and the process number is 1
[LSW2-rip-1]version 2    //Specify RIP version as RIPv2
[LSW2-rip-1]network 192.168.1.0 //The network "192.168.1.0" participates in the rout
//ing update of RIPv2. Note:Subnets "192.168.1.0/25" and "192.168.1.128/25" The
//classful network numbers of are all "192.168.1.0"
[LSW2-rip-1]network 192.168.254.0        //The network "192.168.254.0" participates in
                                         //the routing update of RIPv2
[LSW2-rip-1]silent-interface Vlanif 20   //Configure Vlanif20 interface as silent interface
[LSW2-rip-1]silent-interface Vlanif 30   //Configure Vlanif30 interface as silent interface
[LSW2-rip-1]undo summary                 //Disable automatic summarization
[LSW2-rip-1]quit
```

(3) Verification and testing

① Executing the "display ip routing-table" command on routers AR1, AR2 and AR3 respectively to display the routing table. Fig. 2.125 and Fig. 2.126 show the result after executing the "display ip routing-table" command on three-layer switches LSW1 and LSW2, respectively. You should observe the Cost value in the routing entry with the routing protocol "RIP". And you can the next hop and forwarding.

② Configuring the IP address, subnet mask and default gateway on the test hosts PC1, PC2, PC3 and PC4 according to the IP planned in Table 2.19. Then you should use the "ipconfig /all"

command on the command line to verify the correctness of the IP address, subnet mask and default gateway, and use the ping command to test the connectivity of the whole network. Fig. 2.127 shows that the test host PC1 successfully ping PC2 and PC3.

```
[LSW1]display ip routing-table
Route Flags: R - relay, D - download to fib
-----------------------------------------------------------------
Routing Tables: Public
        Destinations : 10    Routes : 10

Destination/Mask   Proto   Pre  Cost    Flags NextHop      Interface

        127.0.0.0/8   Direct   0    0      D   127.0.0.1      InLoopBack0
        127.0.0.1/32  Direct   0    0      D   127.0.0.1      InLoopBack0
      192.168.0.0/25  Direct   0    0      D   192.168.0.1    Vlanif2
      192.168.0.1/32  Direct   0    0      D   127.0.0.1      Vlanif2
    192.168.0.128/25  Direct   0    0      D   192.168.0.129  Vlanif3
    192.168.0.129/32  Direct   0    0      D   127.0.0.1      Vlanif3
      192.168.1.0/25  RIP    100    1      D   192.168.254.2  Vlanif100
    192.168.1.128/25  RIP    100    1      D   192.168.254.2  Vlanif100
    192.168.254.0/30  Direct   0    0      D   192.168.254.1  Vlanif100
    192.168.254.1/32  Direct   0    0      D   127.0.0.1      Vlanif100
```

Fig. 2.125   The result of executing the "display ip routing-table" command on LSW1

```
[LSW2]display ip routing-table
Route Flags: R - relay, D - download to fib
-----------------------------------------------------------------
Routing Tables: Public
        Destinations : 10    Routes : 10

Destination/Mask   Proto   Pre  Cost    Flags NextHop      Interface

        127.0.0.0/8   Direct   0    0      D   127.0.0.1      InLoopBack0
        127.0.0.1/32  Direct   0    0      D   127.0.0.1      InLoopBack0
      192.168.0.0/25  RIP    100    1      D   192.168.254.1  Vlanif100
    192.168.0.128/25  RIP    100    1      D   192.168.254.1  Vlanif100
      192.168.1.0/25  Direct   0    0      D   192.168.1.1    Vlanif20
      192.168.1.1/32  Direct   0    0      D   127.0.0.1      Vlanif20
    192.168.1.128/25  Direct   0    0      D   192.168.1.129  Vlanif30
    192.168.1.129/32  Direct   0    0      D   127.0.0.1      Vlanif30
    192.168.254.0/30  Direct   0    0      D   192.168.254.2  Vlanif100
    192.168.254.2/32  Direct   0    0      D   127.0.0.1      Vlanif100
```

Fig. 2.126   The result of executing the "display ip routing-table" command on LSW2

```
PC>ping 192.168.0.130

Ping 192.168.0.130: 32 data bytes, Press Ctrl_C to break
From 192.168.0.130: bytes=32 seq=1 ttl=127 time=78 ms
From 192.168.0.130: bytes=32 seq=2 ttl=127 time=63 ms
From 192.168.0.130: bytes=32 seq=3 ttl=127 time=62 ms
From 192.168.0.130: bytes=32 seq=4 ttl=127 time=47 ms
From 192.168.0.130: bytes=32 seq=5 ttl=127 time=47 ms

--- 192.168.0.130 ping statistics ---
  5 packet(s) transmitted
  5 packet(s) received
  0.00% packet loss
  round-trip min/avg/max = 47/59/78 ms

PC>ping 192.168.1.2

Ping 192.168.1.2: 32 data bytes, Press Ctrl_C to break
From 192.168.1.2: bytes=32 seq=1 ttl=126 time=140 ms
From 192.168.1.2: bytes=32 seq=2 ttl=126 time=125 ms
From 192.168.1.2: bytes=32 seq=3 ttl=126 time=110 ms
From 192.168.1.2: bytes=32 seq=4 ttl=126 time=94 ms
From 192.168.1.2: bytes=32 seq=5 ttl=126 time=125 ms

--- 192.168.1.2 ping statistics ---
```

Fig. 2.127   The result of testing host PC1 successfully ping PC2 and PC3

Task Completion

1. Task planning

Based on the task description, network topology shown in Fig. 2.1, IP planning in Table 2.12, and networking requirements of Project 2, the specific routing requirements for the current task are as follows:

① Configuring the default route to the Internet on the enterprise border router AR1.

② RIPv2 protocol is used in the enterprise to realize the interconnection and intercommunication of all network segments in the intranet.

③ Introducing the default route to the Internet into RIPv2 routing protocol on AR1.

④ Realizing the optimization of routing.

The routing plan is as follows:

① Configuring the default route to the Internet on router AR1, and the forwarding exit is S4/0/0.

② RIPv2 is configured on the router AR1 and the three-layer switch CORE to realize the interconnection of all network segments in the enterprise internal network, and the RIP process number is 1.

③ Router AR1 has the category network number "172.16.0.0" to participate in the routing update of RIPv2. There is a category network number "172.16.0.0" on the CORE of three-layer switch to participate in the routing update of RIPv2.

Note: the classed network numbers of the networks 172.16.254.1/30, 172.16.10.254/24 and 172.16.20.254/24 directly connected to the internal network are "172.16.0.0"; The networks directly connected by the three-layer switch CORE are 172.16.2.254/24, 172.16.3.254/24, 172.16.4.254/24, 172.16.99.254/24 and 172.16.100.

④ Introducing the default route into RIP on router AR1.

⑤ G0/0/1.10 and G0/0/1.20 of AR1 are configured as silent interfaces, and VLAN2, VLAN3, VLAN4, VLAN99 and VLAN100 of CORE are configured as silent interfaces.

2. Task implementation

(1) Configuration of routing on AR1

```
[AR1]ip route-static 0.0.0.0 0.0.0.0 s4/0/0
//Configure the default route to the Internet, and the forwarding exit is S4/0/0
[AR1]rip 1
[AR1-rip-1]version 2
[AR1-rip-1]network 172.16.0.0
[AR1-rip-1]silent-interface g0/0/1.10
[AR1-rip-1]silent-interface g0/0/1.20 [AR1-rip-1]undo summary
```

```
//The RIP protocol is enabled, the process number is 1, the RIP version is RIPv2, the
//network "172.16.0.0" participates in the routing update of RIPv2, G0/0/1.10 and
//G0/0/1.20 are silent interfaces, and the automatic summarization function is disabled
[AR1-rip-1]default-route-originate
//Introduce default route into RIP routing protocol
```

(2) Configuration of Routing on three-layer switch CORE

```
[CORE]rip 1
[CORE-rip-1]version 2
[CORE-rip-1]network 172.16.0.0
[CORE-rip-1]silent-interface Vlanif 2
[CORE-rip-1]silent-interface Vlanif 3
[CORE-rip-1]silent-interface Vlanif 4
[CORE-rip-1]silent-interface Vlanif 99
[CORE-rip-1]silent-interface Vlanif 100
[CORE-rip-1]quit
//The RIP protocol is enabled, the process number is 1, the RIP version is RIPv2, the net
//work "172.16.0.0" participates in the routing update of RIPv2, and Vlanif2, Vlanif3,
//Vlanif4, Vlanif99 and Vlanif100 are silent interfaces, and the automatic summarization
//function is disabled.
```

(3) Verification and testing

① To execute the "display ip routing-table" command on router AR1 and three-layer switch CORE respectively for viewing the routing table, and observe the default routing of the routing table and the routing entry of the remote target network learned through RIPv2.

② Executing the "display IP routing-table | include Static" command on router AR1 to display the static routing entries in the routing table, as shown in Fig. 2.128. Then you can execute the command "display IP routing-table | include RIP" to display all the routing entries learned through RIP in the routing table, as shown in Fig. 2.129.

```
[AR1]display ip routing-table | include Static
Route Flags: R - relay, D - download to fib
--------------------------------------------------
Routing Tables: Public
        Destinations : 23    Routes : 23

Destination/Mask   Proto   Pre  Cost    Flags  NextHop        Interface

0.0.0.0/0          Static  60   0       D      125.71.28.93   Serial4/0/0
```

Fig. 2.128　The static route entry in the routing table of router AR1

```
[AR1]display ip routing-table | include RIP
Route Flags: R - relay, D - download to fib
------------------------------------------------------------
Routing Tables: Public
     Destinations : 23     Routes : 23

Destination/Mask   Proto  Pre  Cost    Flags NextHop        Interface

   172.16.2.0/24    RIP   100   1       D   172.16.254.2   GigabitEthernet0/0/0
   172.16.3.0/24    RIP   100   1       D   172.16.254.2   GigabitEthernet0/0/0
   172.16.4.0/24    RIP   100   1       D   172.16.254.2   GigabitEthernet0/0/0
  172.16.99.0/24    RIP   100   1       D   172.16.254.2   GigabitEthernet0/0/0
 172.16.100.0/24    RIP   100   1       D   172.16.254.2   GigabitEthernet0/0/0
```

Fig. 2.129 The routing entries learned by RIP in the routing table on router AR1

③ You can use the command "display IP routing-table | include RIP" on the CORE of the three-layer switch to display all the routing entries in the routing table learned through RIP, as shown in Fig. 2.130.

Note: The default routing entries displayed in the routing table on CORE are learned through RIP.

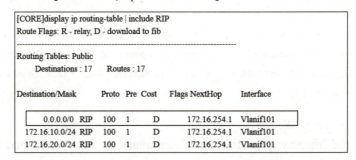

Fig. 2.130 The routing entries learned by RIP in the routing table on CORE

Exercises

Multiple choice

(1) RIP protocol sends an update by default (   ).
   A. 20 s   B. 30 s   C. 60 s   D. 90 s

(2) When RIPv1 sends an update, its destination IP address is (   ).
   A. 255. 255. 255. 255   B. 255. 255. 255. 0
   C. 224. 0. 0. 9   D. 224. 0. 0. 10

(3) When RIPv2 sends an update, its destination IP address is (   ).
   A. 255. 255. 255. 255   B. 255. 255. 255. 0
   C. 224. 0. 0. 9   D. 224. 0. 0. 10

(4) For Huawei's VRP platform, the default routing priority value of RIP protocol is (   ).
   A. 10   B. 60   C. 100   D. 120

(5) RIP protocol calculates the best path to the target network according to (   ).

    A. cost   B. delay   C. hop counts   D. load

Task 7 Planning and Configuration of NAT

Task Description

On the basis of completing the Task 1 to Task 6 of this project, complete the NAT planning and configurations in the network shown in Fig. 2.1, and achieve the networking requirements of Project 2 that "the hosts in the enterprise intranet can have access to the Internet, and the Internet can have access to the enterprise Web server for external Web services".

Task Objectives

- Understanding the role of private IP addresses.
- Understanding the application scenarios and working processes of NAT SERVER and EASY NAT.
- Mastering the configuration of NAT SERVER and EASY NAT.

Related Knowledge

1. The working principle of NAT

Private addresses specified in RFC1918 are reserved for private or internal network use, including 10.0.0.0/8, 172.16.0.0./12 and 192.168.0.0/16. Private addresses can only be used to form an internal IP network, but hosts using private addresses cannot communicate with IP hosts using public addresses on the external Internet. NAT is generally deployed on network exit devices, such as routers or firewalls.

Network address translation (NAT) is a technology used to convert one address into another. The typical application is to convert a private address into a public IP address that can be routed on the public network, so that hosts using private addresses can access the Internet. The equipment that provides NAT function generally runs on the boundary of stub area. When a host in stub area wants to transmit data to an external host, it sends the data packet to NAT equipment first. The NAT process looks at the received IP packet header and, if appropriate, replaces the internal private address in the source address field with a local global address (usually a public address). When the external target host sends a response packet, the NAT process will receive it and replace the local global address in the "target address" address field with the original private address by looking at the network address translation table. Fig. 2.131 shows an example of NAT working process.

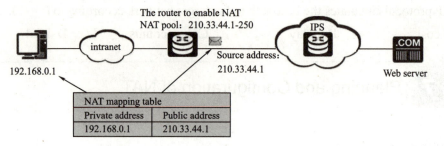

Fig. 2.131 An example of NAT working process

2. Types of NAT

(1) Static NAT

Static NAT refers to the one-to-one conversion between the internal local address and the internal global address by the administrator through manual configuration, as shown in Fig. 2.132. The static NAT mapping will remain unchanged unless the administrator reconfigures the static NAT. Static NAT supports two-way access: The host of the internal network can access the external network through NAT; The host of the external network can access the host of the internal network through the converted global address (usually public address).

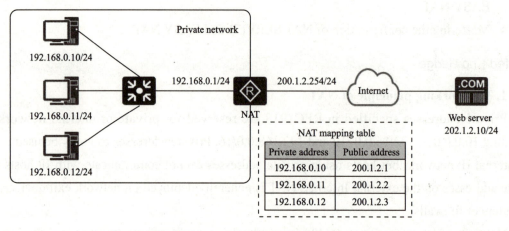

Fig. 2.132 Examples of static NAT in working process

(2) Dynamic NAT

Dynamic NAT dynamically realizes one-to-one mapping between internal local addresses and local global addresses in a first-come-first-served manner by defining a local global address pool. First, you need to define a local global address pool on the NAT device. When the data of the internal host passes through the NAT device, dynamic NAT selects a local global address from the internal global address pool that is not occupied by other hosts for address translation, and writes the dynamically obtained address translation or mapping record into the network address translation table. Once there are not enough addresses in the local global address pool for translation, NAT translation will fail, as shown in Fig. 2.133.

Dynamic NAT will not convert the port number when selecting the address in the address pool for address translation, that is, No-PAT (no-port address translation), and the mapping relationship between public address and private address is still 1:1, which cannot improve the utilization rate of public address.

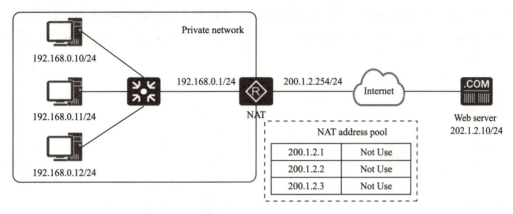

Fig. 2.133 An example of dynamic NAT working process

(3) NAPT

NAPT (network address and port translation) not only converts IP addresses, but also converts port numbers when selecting addresses from the address pool for address translation. As shown in Fig. 2.134, NAPT can realize that a public address corresponds to multiple private addresses at the same time by means of ports. Because NAPT converts IP address and transport layer port at the same time, it can map different private addresses (different private addresses and different source ports) to the same public address (same public address and different source ports), which effectively improves the utilization rate of public addresses.

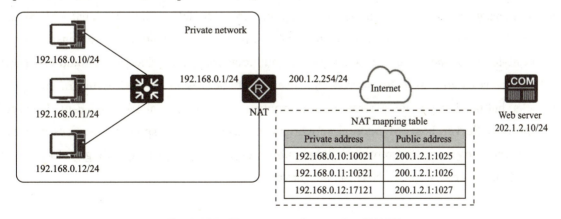

Fig. 2.134 The converted example of NAPT

(4) Easy IP

Easy IP is implemented in the same way as NAPT. When converting, the IP address is

converted and the transport layer port is also converted. However, Easy IP does not have the concept of address pool, and it uses the interface IP address of NAT device exit as the public address after address translation.

Easy IP is usually suitable for scenarios that do not have a fixed public IP address. For example, the private network exit with the address obtained through DHCP and PPPoE dialing can be directly converted by using the obtained dynamic address. Fig. 2.135 shows the converted example of Easy IP.

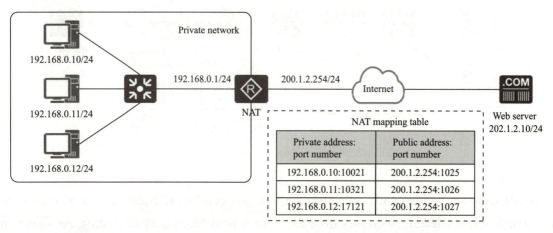

Fig. 2.135 The converted example of Easy IP

(5) NAT Server

NAT Server is very similar to static NAT, and it is also a fixed one-to-one conversion mapping generated by the administrator's manual configuration. The NAT Server translation mapping will remain unchanged unless deleted or reconfigured by the administrator. However, unlike static address translation, NAT Server translation mapping not only defines internal local address and internal global address, but also specifies the protocol and port information of the transport layer. It is a one-to-one mapping relationship between [public address: port] and [private address: port]. NAT Server is usually used for servers that use private addresses for addressing in enterprise networks, but need to provide external access services, such as external Web and E-mail servers. Through NAT Server, hosts of external networks can access the corresponding services provided by these servers by using the converted global public address and the corresponding protocol ports. Fig. 2.136 shows an example of NAT Server conversion.

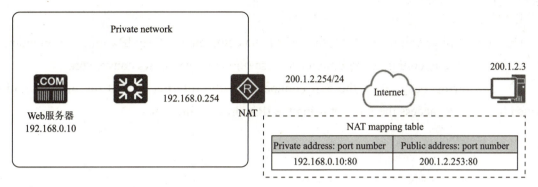

Fig. 2.136 The converted example of NAT Server

Task Completion

1. Task planning

According to the task description, network topology shown in Fig. 2.1, IP planning in Table 2.12, and networking requirements in Project 2, Easy IP and NAT Server can be configured on the exit router AR1 to achieve the network requirements that the enterprise intranet hosts can have simultaneous access to the Internet and the test hosts on the Internet can have access to the external WWW and DNS services provided by the enterprise. See Table 2.21 for the planning of NAT Server.

Table 2.21 Planning of NAT Server

Server	Protocol	Internal address	Internal port number	Interface	Global port number
Web	TCP	172.16.100.1/24	80	AR1 S4/0/0 current interface	80
DNS	TCP and UDP	172.16.100.3/24	53	AR1 S4/0/0 current interface	53

Planning of Easy IP:

The ACL list number is 2000, and allow to perform Easy IP conversion of network segment addresses including 172.16.2.0/24, 172.16.3.0/24, 172.16.4.0/24, 172.16.99.0/24, 172.16.100.0/24, and 172.16.20.0/24.

2. Task implementation

(1) Configuration of NAT Server

On the enterprise exit router AR1, you should complete the configuration according to the NAT Server planned in Table 2.21. The steps are as follows:

```
[AR1]interface s4/0/0
[AR1-Serial4/0/0]nat server protocol tcp global current-interface 80 inside 172.16.100.1 80
[AR1-Serial4/0/0]nat server protocol tcp global current-interface 53 inside 172.16.100.3 53
[AR1-Serial4/0/0]nat server protocol udp global current-interface 53 inside 172.16.100.3 53
```

(2) Easy IP configuration

On the enterprise exit router AR1, the tasks to be completed in Easy IP configuration include:

① Using ACL to configure the intranet address segment that allows conversion.

② Configuring Easy IP on the exit connected to the Internet so that the intranet address defined by ACL can access the Internet using the IP address of the exit.

```
[AR1]acl 2000
[AR1-acl-basic-2000]rule permit source 172.16.3.0 0.0.0.255
[AR1-acl-basic-2000]rule permit source 172.16.4.0 0.0.0.255
[AR1-acl-basic-2000]rule permit source 172.16.99.0 0.0.0.255
[AR1-acl-basic-2000]rule permit source 172.16.100.0 0.0.0.255
[AR1-acl-basic-2000]rule permit source 172.16.10.0 0.0.0.255
[AR1-acl-basic-2000]rule permit source 172.16.20.0 0.0.0.255 [AR1-acl-basic-2000]quit
//Configure the intranet address segments that allow Easy IP conversion. The ACL list
//number is 2000. The intranet address segments that allow Easy IP conversion include:
//172.16.2.0/24, 172.16.3.0/24, 172.16.4.0/24 and 172.16.99.0/24, 172.16.100.0/24,
//172.16.10.0/24 and 172.16.20.0/24
[AR1]interface s4/0/0
[AR1-Serial4/0/0]nat outbound 2000
//Configure Easy IP, so that hosts with network segment addresses defined by ACL2000
//can access the Internet through the public IP address of S4/0/0 interface
```

(3) Viewing and verification of NAT

After completing the configuration of NAT, you can use the display command to view and verify NAT.

① To execute the "display nat server" command for viewing NAT server information. Fig. 2.137 shows an example of viewing NAT server information on router AR1.

```
[AR1]display nat server
Nat Server Information:
Interface : Serial4/0/0
  Global IP/Port   : current-interface/80(www) (Real IP : 125.71.28.93)
  Inside IP/Port   : 172.16.100.1/80(www)
  Protocol : 6(tcp)
  VPN instance-name : ----
  Acl number      : ----
  Description : ----

  Global IP/Port   : current-interface/53(domain) (Real IP : 125.71.28.93)
  Inside IP/Port   : 172.16.100.3/53(domain)
  Protocol : 6(tcp)
  VPN instance-name : ----
  Acl number      : ----
  Description : ----

  Global IP/Port   : current-interface/53(dns) (Real IP : 125.71.28.93)
  Inside IP/Port   : 172.16.100.3/53(dns)
  Protocol : 17(udp)
  VPN instance-name : ----
  Acl number      : ----
  Description : ----

Total :  3
```

Fig. 2.137 The example of viewing NAT server information on router AR1

② To execute the "display this" command under the S4/0/0 interface view of router AR1 for viewing the configuration under this interface, and pay attention to whether the configuration information of NAT is consistent with the plan, as shown in Fig. 2.138.

```
[AR1-Serial4/0/0]display this
[V200R003C00]
#
interface Serial4/0/0
 link-protocol ppp
 ppp authentication-mode chap
 ppp chap user AR1
 ppp chap password cipher %$%$WJ(-3^`dfCH0^ZT!~%d&,+Jl%$%$
 ip address 125.71.28.93 255.255.255.0
 nat server protocol tcp global current-interface www inside 172.16.100.1 www
 nat server protocol tcp global current-interface domain inside 172.16.100.3 dom
ain
 nat server protocol udp global current-interface dns inside 172.16.100.3 dns
 nat outbound 2000
#
return
```

Fig. 2.138 The configuration under the S4/0/0 interface of router AR1

③ On the Internet test host Client1, you should open a browser and enter the converted public address of the enterprise Web server (that is, the IP address of the S4/0/0 interface of router AR1) in the Web address to access the Web servers provided by the enterprise Web server, as shown in Fig. 2.139.

④ On multiple hosts in the intranet, you can access any test host on the Internet by using the uninterrupted ping command at the same time. Fig. 2.140 shows that PC3 and PC4 in the enterprise intranet ping Client1 on the Internet at the same time.

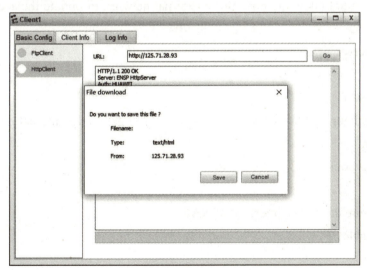

Fig. 2.139 The example of test host on the Internet accessing the Web Servers provided by the enterprise Web server

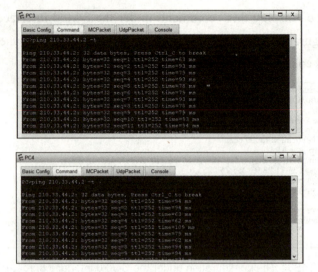

Fig. 2.140 The example of PC3 and PC4 ping the test host Client1 on the Internet

Exercises

Multiple choice

(1) (   ) can make multiple hosts in the enterprise intranet share the IP address of NAT device exit to access the Internet.

      A. Static NAT   B. Dynamic NAT   C. Easy IP   D. NAPT

(2) (   ) can not allow the host of the external network to initiate the access to the host of the internal network.

      A. NAT server

      B. NAPT

      C. static NAT

(3) When configuring Easy IP, you can use (   ) to define the internal network address for address translation.

      A. address pool

      B. ACL

      C. Configure the device interface of NAT

      D. NAT table

Project 3
Medium-sized Enterprise Internetworking Project

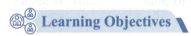
Learning Objectives

Knowledge Objectives

◎ Understand the network architecture of medium-sized enterprise internetworking project.

◎ Understand the three-layer network architecture of access layer, aggregation layer, and core layer.

◎ Understand the working process of link aggregation.

◎ Understand the concept and working process of OSPF.

Competence Objectives

◎ Competent for independent deployment and implementation of a medium-sized enterprise internetworking project, including the configuration of IP and VLAN, router switch, link aggregation, spanning tree, communication between VLANs, OSPF and NAT, etc.

Quality Objectives

◎ Competent for teamwork.

◎ Competent for problem analysis and solutions.

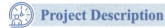
Project Description

A medium-sized enterprise plans to establish a network as shown in Fig. 3.1, subject to a three-layer network architecture consisting of access layer, aggregation layer and core layer. The administrative building accommodates such departments as the Security Department, Students Affairs Department, Finance Department, and Personnel Department. The teaching building has

classrooms and offices; the training building has software training rooms and network training rooms. Each building is equipped with digital cameras for monitoring. The enterprise has applied for five public addresses ranging from 125.71.28.65 to 125.71.28.69. Now it is necessary to plan and deploy the enterprise network so that interconnectivity can be achieved between various network segments within the enterprise, its internal hosts can access the Internet, and hosts on the Internet can access the external WWW and DNS services provided by the enterprise.

Enterprise network deployment (requirements) are as follows:

◎ The business data between different departments of the enterprise shall be completely isolated.

◎ The administrator can log in to all switches within the enterprise remotely via SSH.

◎ The link between the convergence layer switch SXL-AGG and the access layer switch SXL-FLOOR2 in the training building can use the link aggregation technology to improve the link bandwidth and increase the high reliability of the link.

◎ The link between the convergence layer switch SXL-AGG and the core switch CORE in the training building can use the link aggregation technology to improve the link bandwidth and increase the high reliability of the link.

◎ The corporate intranet needs to be interconnected.

◎ The corporate intranet adopts private addressing, with NAPT and NAT SERVER deployed on the outbound router, so that all hosts in the corporate intranet can access the Internet and the hosts on the Internet can access the external network services of the enterprise.

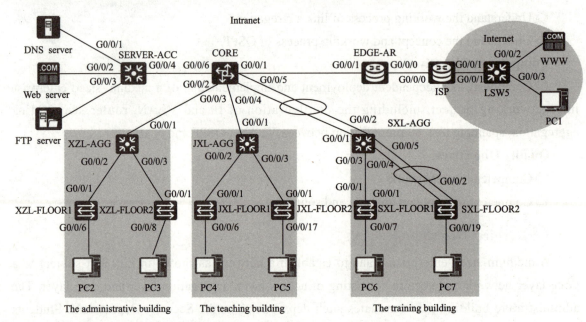

Fig. 3.1　Network topology of medium-sized enterprise internetworking project

Task 1 Planning and Configuration of IP, VLAN and NAT

Task Description

Set up a network topology as shown in Fig.3.1 in the Huawei ENSP simulation environment or using Huawei's routers and switches. According to the networking requirements of Project 3, complete the planning and configurations of the network IP, VLAN, NAPT, and NAT Server, and carry out network verification and testing.

Task Objectives

- Competent to deploy and implement VLANs for medium-sized enterprise internetworking projects.
- Competent for IP deployment and implementation in medium-sized enterprise internetworking projects.
- Master the methods for Easy IP planning and configuration in medium-sized internetworking projects.
- Master the methods for NAT Server configuration in medium-sized internetworking projects.

Related Knowledge

1. Layered network design in medium-sized enterprises

When building a network for the needs of medium-sized enterprises, a layered design model is usually adopted in order to make the established network easier to manage, more flexible, and more scalable. The layered design model, also known as the structured design model, divides the entire network structure into several layers by means of a clear division of the network functions, making the design of each layer relatively simple. In large and medium-sized networks, the layered design model usually adopts a three-layer structure, namely the access layer, aggregation layer, and core layer, as shown in Fig. 3.2.

(1) Access layer

The access layer mainly helps users or workgroups for network access. The access layer of the enterprise network usually completes such tasks as VLAN, spanning tree and IP address management configuration or link aggregation.

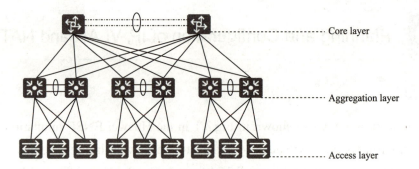

Fig. 3.2   Three-layer structure

(2) Aggregation layer

The aggregation layer, also known as the distribution layer, is the boundary point between the access layer and the core layer. The aggregation layer provides policy-based connection management to achieve controllable transmission from the access layer to the core layer. The aggregation layer mainly implements such functions as routing between VLANs.

(3) Core layer

As the high-speed backbone of an enterprise network, the core layer mainly provides high-speed optimized bandwidth transmission between nodes that communicate with each other. Therefore, the core layer needs to maintain high availability and redundancy, avoiding such functions as access to control lists and packet filtering.

Note: In actual network design, not all networks must have all such three layers. When a network is small in scale or limited in functions, the three-layer design can often be simplified into a two-layer design (access layer + core layer), or even a single-layer design.

2. Switches in a layered switch network

Corresponding to the layered switching network mentioned above, the switches in the enterprise network are divided into access layer switches, aggregation layer switches and core layer switches. Taking the enterprise network topology shown in Fig. 3.1 as an example, the switch CORE in the figure is the core layer switch, the switches XZL-AGG, JXL-AGG and SXL-AGG are the aggregation layer switches, and the switches such as XZL-FLOOR1 and XZL-FLOOR2 are the access layer switches.

① The access layer switches usually should have such functions as VLAN and certain security, and usually a two-layer switch design can be used.

② The aggregation layer switches should achieve communication between VLANs and be competent for routing, so a three-layer switch design needs to be selected. The aggregation layer switches need to provide routing function, so the link between the aggregation layer switch and the core layer switch is usually configured as a three-layer link. The ports of most three-layer switches

in the market can be directly configured as routing ports, such as H3C, Cisco, and other three-layer switches. Moreover, the physical port of the three-layer link can also be added to a separate VLAN, and the IP address of the three-layer switch is configured on the VLAN port. Usually, the spanning tree function is often disabled on this physical port.

③ The core layer switch is mainly used for fast forwarding of data packets on the backbone link, so it is also necessary to choose a three-layer switch, with the requirements for switching and forwarding performance higher than those of the aggregation layer switches. The basic configuration is the routing protocol on the core layer switch. Furthermore, if the enterprise-level server of the enterprise network is connected to the core switch, then the core layer switch also needs to create VLANs, assign the ports of the switch to the corresponding VLANs, and configure communication between VLANs.

Task Completion

1. Task planning

① The IP and VLAN planning shown in Table 3.1 is designed based on the principle of a three-layer network design model. Please complete the VLAN configuration according to Fig. 3.1 and Table 3.1. Configure the IP addresses on the access layer switch, aggregation layer switch, core layer switch, outbound router EDGE-AR, operator router ISP, server, and test host PC1 on the Internet, and configure DHCP services on the aggregation layer switch to provide IP address services for the end access devices on that floor. See Table 3.2 for the planning of DHCP services.

② According to Fig. 3.1 and Table 3.3, configure NAT Server and NAPT on the outbound router EDGE-AR to ensure that the intranet host of the enterprise can access the Internet simultaneously, and the test host on the Internet can access the external WWW and DNS services provided by the enterprise.

Note: The management IP address of the access layer switch is statically assigned, and the IP address of the monitoring device is also statically assigned, so it is unnecessary to configure DHCP services for it.

Table 3.1   Planning of IP and VLAN

Device	Interface or VLAN	Description of VLAN	Two or three-layer planning	Instructions
XZL-FLOOR1	VLAN2	baoan	Interface G0/0/2 to G0/0/12	Security department
	VLAN3	student	Interface G0/0/13 to G0/0/18	Students affairs department
	VLAN90	manage	10.0.90.1/24	Switch management VLAN
	VLAN91	monitor	Interface G0/0/19 to G0/0/24	Network monitoring

Continued

Device	Interface or VLAN	Description of VLAN	Two or three-layer planning	Instructions
XZL-FLOOR2	VLAN4	caiwu	Interface G0/0/2 to G0/0/12	Finance department
	VLAN5	hr	Interface G0/0/13 to G0/0/18	Personnel department
	VLAN90	manage	10.0.90.2/24	Switch management VLAN
	VLAN91	monitor	Interface G0/0/19 to G0/0/24	Monitoring
XZL-AGG	VLAN2	baoan	10.0.2.254/24	
	VLAN3	student	10.0.3.254/24	
	VLAN4	caiwu	10.0.4.254/24	
	VLAN5	hr	10.0.5.254/24	
	VLAN90	manage	10.0.90.254/24	
	VLAN91	monitor	10.0.91.254/24	
	VLAN199	linktocore	10.254.254.2/30	Link network segment VLAN and IP address to CORE
JXL-FLOOR1	VLAN12	jiaoshi	Interface G0/0/2 to G0/0/15	Classroom
	VLAN13	office	Interface G0/0/16 to G0/0/18	Office
	VLAN92	manage	10.1.92.1/24	Switch management VLAN
	VLAN93	monitor	Interface G0/0/19 to G0/0/24	Network monitoring
JXL-FLOOR2	VLAN12	jiaoshi	Interface G0/0/2 to G0/0/15	Classroom
	VLAN13	office	Interface G0/0/16 to G0/0/18	Office
	VLAN92	manage	10.1.92.2/24	Switch management VLAN
	VLAN93	monitor	Interface G0/0/19 to G0/0/24	Network monitoring
JXL-AGG	VLAN12	jiaoshi	10.1.12.254/24	
	VLAN13	office	10.1.13.254/24	
	VLAN92	manage	10.1.92.254/24	
	VLAN93	monitor	10.1.93.254/24	
	VLAN199	linktocore	10.254.254.6/30	Link network segment VLAN and IP address to CORE
SXL-FLOOR1	VLAN22	rj-sxs	Interface G0/0/2 to G0/0/10	Software training room
	VLAN23	wl-sxs	Interface G0/0/11 to G0/0/18	Network training room
	VLAN94	manage	10.2.94.1/24	Switch management VLAN
	VLAN95	monitor	Interface G0/0/19 to G0/0/24	Network monitoring
SXL-FLOOR2	VLAN22	rj-sxs	Interface G0/0/3 to G0/0/10	Software training room
	VLAN23	wl-sxs	Interface G0/0/11 to G0/0/18	Network training room
	VLAN94	manage	10.2.94.2/24	Switch management VLAN
	VLAN95	monitor	Interface G0/0/19 to G0/0/24	Network monitoring

Continued

Device	Interface or VLAN	Description of VLAN	Two or three-layer planning	Instructions
SXL-AGG	VLAN22	rj-sxs	10.2.22.254/24	
	VLAN23	wl-sxs	10.2.23.254/24	
	VLAN94	manage	10.2.94.254/24	
	VLAN95	monitor	10.2.95.254/24	
	VLAN199	linktocore	10.254.254.10/30	Link network segment VLAN to switch CORE
CORE	VLAN199	linkto-sxl-agg	10.254.254.9/30	Link network segment VLAN to switch SXL-AGG
	VLAN200	linkto-jxl-agg	10.254.254.5/30	Link network segment VLAN to switch JXL-AGG
	VLAN201	linkto-xzl-agg	10.254.254.1/30	Link network segment VLAN to switch XZL-AGG
	VLAN202	SERVER	10.254.253.254/24	Server network segment
	VLAN203	linkto-edge-ar	10.254.254.13/30	Link network segment VLAN and IP address to CORE
EDGE-AR	G0/0/1	NA	10.254.254.14/30	
	G0/0/0	NA	125.71.28.65/24	
SXL-AGG	VLAN22	rj-sxs	10.2.22.254/24	
ISP	G0/0/1	NA	202.1.1.254/24	
	G0/0/0	NA	125.71.28.1/24	
Web	Fa0	NA	10.254.253.1/24	
DNS	Fa0	NA	10.254.253.2/24	
FTP	Fa0	NA	10.254.253.3/24	
WWW	Fa0	NA	202.1.1.1/24	
PC1	Fa0	NA	202.1.1.10/24	

Table 3.2 Planning of DHCP Services

Device	DHCP address pool name	IP address network number	Gateway address	DNS address	Lease term
XZL-AGG	VLAN2	10.0.2.0/24	10.0.2.254	10.254.253.2	8 h
	VLAN3	10.0.3.0/24	10.0.3.254	10.254.253.2	8 h
	VLAN4	10.0.4.0/24	10.0.4.254	10.254.253.2	8 h
	VLAN5	10.0.5.0/24	10.0.5.254	10.254.253.2	8 h
JXL-AGG	VLAN12	10.1.12.0/24	10.1.12.254	10.254.253.2	8 h
	VLAN13	10.1.13.0/24	10.1.13.254	10.254.253.2	8 h
SXL-AGG	VLAN22	10.2.22.0/24	10.2.22.254	10.254.253.2	8 h
	VLAN23	10.2.23.0/24	10.2.23.254	10.254.253.2	8 h

Table 3.3 Planning of NAT Server

Server	Protocol	Internal address	Internal port number	External address	Global port number
Web	TCP	10.254.253.1/24	80	125.71.28.66	80
DNS	TCP and UDP	10.254.253.2/24	53	125.71.28.66	53

Note:

① The aggregation layer switch of each building realizes communication between the VLANs of the building, namely, the aggregation layer switch XZL-AGG realizes communication between administrative building VLANs, JXL-AGG realizes communication between teaching building VLANs, and SXL-AGG realizes communication between training building VLANs.

② A three-layer link shall be adopted between the aggregation layer switch and the core switch CORE in each building.

NAPT planning:

① The ACL list number is 2000, allowing for Easy IP conversion of the addresses in such network segments as 10.0.2.0/24, 10.0.3.0/24, 10.0.4.0/24, 10.0.5.0/24, 10.1.12.0/24, 10.1.13.0/24, 10.2.22.0/24 and 10.2.23.0/24.

② The converted public address pool ID is "1" and the address pool ranges from 125.71.28.67 to 125.71.28.69.

2. Task implementation

(1) Configuration of access layer switch VLAN

According to the plan in Table 3.1, the access layer switch needs to complete such VLAN configuration tasks as follows: create VLANs and configure VLAN descriptions; configure the link to the aggregation layer switch as the trunk link; configure the access link.

① Complete VLAN configurations on the access layer switch XZL-FLOOR1 as follows:

```
[XZL-FLOOR1]vlan 2
[XZL-FLOOR1-vlan2]name baoan
[XZL-FLOOR1-vlan2]description baoan
[XZL-FLOOR1-vlan2]vlan 3
[XZL-FLOOR1-vlan3]description student
[XZL-FLOOR1-vlan3]vlan 90
[XZL-FLOOR1-vlan90]description manage
[XZL-FLOOR1-vlan90]vlan 91
[XZL-FLOOR1-vlan91]description monitor
[XZL-FLOOR1-vlan91]quit
//Create VLAN2, VLAN3, VLAN90 and VLAN91; configure VLAN descriptions
[XZL-FLOOR1]port-group 1
[XZL-FLOOR1-port-group-1]group-member g0/0/2 to g0/0/12
```

```
[XZL-FLOOR1-port-group-1]port link-type access
[XZL-FLOOR1-port-group-1]port default vlan 2
//Configure the interfaces from G0/0/2 to G0/0/12 as access interfaces, under VLAN2
[XZL-FLOOR1]port-group 2
[XZL-FLOOR1-port-group-2]group-member g0/0/13 to g0/0/18
[XZL-FLOOR1-port-group-2]port link-type access
[XZL-FLOOR1-port-group-2]port default vlan 3
[XZL-FLOOR1-port-group-2]quit
//Configure the interfaces from G0/0/13 to G0/0/18 as access interfaces, under VLAN3
[XZL-FLOOR1]port-group 3
[XZL-FLOOR1-port-group-3]group-member g0/0/19 to g0/0/24
[XZL-FLOOR1-port-group-3]port link-type access
[XZL-FLOOR1-port-group-3]port default vlan 91
//Configure the interfaces from G0/0/19 to G0/0/24 as access interfaces, under VLAN91
[XZL-FLOOR1]interface g0/0/1
[XZL-FLOOR1-GigabitEthernet0/0/1]port link-type trunk
[XZL-FLOOR1-GigabitEthernet0/0/1]port trunk allow-pass vlan 2 3 90 91
[XZL-FLOOR1-GigabitEthernet0/0/1]quit
//Configure the G0/0/1 interface as the trunk interface, allowing VLAN2, VLAN3, VLAN90
//and VLAN91 access
```

② Complete VLAN configurations on the access layer switch XZL-FLOOR2 as follows:

```
[XZL-FLOOR2]vlan 4
[XZL-FLOOR2-vlan4]description caiwu
[XZL-FLOOR2-vlan4]vlan 5
[XZL-FLOOR2-vlan5]description HR
[XZL-FLOOR2-vlan5]vlan 90
[XZL-FLOOR2-vlan90]description manage
[XZL-FLOOR2-vlan90]vlan 91
[XZL-FLOOR2-vlan91]description monitor
[XZL-FLOOR2-vlan91]quit
[XZL-FLOOR2]
//Create VLAN4, VLAN5, VLAN90 and VLAN91, configure VLAN descriptions
[XZL-FLOOR2]port-group 1
[XZL-FLOOR2-port-group-1]group-member g0/0/2 to g0/0/12
[XZL-FLOOR2-port-group-1]port link-type access
[XZL-FLOOR2-port-group-1]port default vlan 4
//Configure the interfaces from G0/0/2 to G0/0/12 as access interfaces, under VLAN4.
[XZL-FLOOR2]port-group 2
[XZL-FLOOR2-port-group-2]group-member g0/0/13 to g0/0/18
```

```
[XZL-FLOOR2-port-group-2]port link-type access
[XZL-FLOOR2-port-group-2]port default vlan 5
[XZL-FLOOR2-port-group-2]quit
//Configure the interfaces from G0/0/13 to G0/0/18 as access interfaces, under VLAN5
[XZL-FLOOR2]port-group 3
[XZL-FLOOR2-port-group-3]group-member g0/0/19 to g0/0/24
[XZL-FLOOR2-port-group-3]port link-type access
[XZL-FLOOR2-port-group-3]port default vlan 91
[XZL-FLOOR2-port-group-3]quit
//Configure the interfaces from G0/0/19 to G0/0/24 as access interfaces, under VLAN91
[XZL-FLOOR2]interface g0/0/1
[XZL-FLOOR2-GigabitEthernet0/0/1]port link-type trunk
[XZL-FLOOR2-GigabitEthernet0/0/1]port trunk allow-pass vlan 4 5 90 91
[XZL-FLOOR2-GigabitEthernet0/0/1]quit
//Configure the G0/0/1 interface as the trunk interface, allowing VLAN4, VLAN5, VLAN90
//and VLAN91 access
```

③ Complete VLAN configurations on the access layer switch JXL-FLOOR1 as follows:

```
[JXL-FLOOR1]vlan 12
[JXL-FLOOR1-vlan12]name jiaoshi
[JXL-FLOOR1-vlan12]description jiaoshi
[JXL-FLOOR1-vlan12]vlan 13
[JXL-FLOOR1-vlan13]description office
[JXL-FLOOR1-vlan13]vlan 92
[JXL-FLOOR1-vlan92]description manage
[JXL-FLOOR1-vlan92]vlan 93
[JXL-FLOOR1-vlan93]description monitor
[JXL-FLOOR1-vlan93]quit
//Create VLAN12, VLAN13, VLAN92 and VLAN93, configure VLAN descriptions
[JXL-FLOOR1]port-group 1
[JXL-FLOOR1-port-group-1]group-member g0/0/2 to g0/0/15
[JXL-FLOOR1-port-group-1]port link-type access
[JXL-FLOOR1-port-group-1]port default vlan 12
//Configure the interfaces from G0/0/2 to G0/0/15 as access interfaces, under VLAN12
[JXL-FLOOR1]port-group 2
[JXL-FLOOR1-port-group-2]group-member g0/0/16 to g0/0/18
[JXL-FLOOR1-port-group-2]port link-type access
[JXL-FLOOR1-port-group-2]port default vlan 13
[JXL-FLOOR1-port-group-2]quit
//Configure the interfaces from G0/0/16 to G0/0/18 as access interfaces, under VLAN13
[JXL-FLOOR1]port-group 3
```

```
[JXL-FLOOR1-port-group-3]group-member g0/0/19 to g0/0/24
[JXL-FLOOR1-port-group-3]port link-type access
[JXL-FLOOR1-port-group-3]port default vlan 93
[JXL-FLOOR1-port-group-3]quit
//Configure the interfaces from G0/0/19 to G0/0/24 as access interfaces, under VLAN93
[JXL-FLOOR1]interface g0/0/1
[JXL-FLOOR1-GigabitEthernet0/0/1]port link-type trunk
[JXL-FLOOR1-GigabitEthernet0/0/1]port trunk allow-pass vlan 12 13 92 93
[JXL-FLOOR1-GigabitEthernet0/0/1]quit
//Configure the G0/0/1 interface as the trunk interface, allowing VLAN12, VLAN13,
//VLAN92 and VLAN93 access
```

④ **Complete VLAN configurations on the access layer switch JXL-FLOOR2 as follows:**

```
[JXL-FLOOR2]vlan 12
[JXL-FLOOR2-vlan12]description jiaoshi
[JXL-FLOOR2-vlan12]vlan 13
[JXL-FLOOR2-vlan13]description office
[JXL-FLOOR2-vlan13]vlan 92
[JXL-FLOOR2-vlan92]description manage
[JXL-FLOOR2-vlan92]vlan 93
[JXL-FLOOR2-vlan93]description monitor
[JXL-FLOOR2-vlan93]quit
//Create VLAN12, VLAN13, VLAN92 and VLAN93, configure VLAN descriptions
[JXL-FLOOR2]port-group 1
[JXL-FLOOR2-port-group-1]group-member g0/0/2 to g0/0/15
[JXL-FLOOR2-port-group-1]port link-type access
[JXL-FLOOR2-port-group-1]port default vlan 12
[JXL-FLOOR2-port-group-1]quit
//Configure the interfaces from G0/0/2 to G0/0/15 as access interfaces, under VLAN12
[JXL-FLOOR2]port-group 2
[JXL-FLOOR2-port-group-2]group-member g0/0/16 to g0/0/18
[JXL-FLOOR2-port-group-2]port link-type access
[JXL-FLOOR2-port-group-2]port default vlan 13
[JXL-FLOOR2-port-group-2]quit
//Configure the interfaces from G0/0/16 to G0/0/18 as access interfaces, under VLAN13
[JXL-FLOOR2]port-group 3
[JXL-FLOOR2-port-group-3]group-member g0/0/19 to g0/0/24
[JXL-FLOOR2-port-group-3]port link-type access
[JXL-FLOOR2-port-group-3]port default vlan 93
[JXL-FLOOR2-port-group-3]quit
```

```
//Configure the interfaces from G0/0/19 to G0/0/24 as access interfaces, under VLAN93
[JXL-FLOOR2]interface g0/0/1
[JXL-FLOOR2-GigabitEthernet0/0/1]port link-type trunk
[JXL-FLOOR2-GigabitEthernet0/0/1]port trunk allow-pass vlan 12 13 92 93
[JXL-FLOOR2-GigabitEthernet0/0/1]quit
//Configure the G0/0/1 interface as the trunk interface, allowing VLAN12, VLAN13,
//VLAN92 and VLAN93 access
```

⑤ Complete VLAN configurations on the access layer switch SXL-FLOOR1 as follows:

```
[SXL-FLOOR1]vlan 22
[SXL-FLOOR1-vlan22]description rj-sxs
[SXL-FLOOR1-vlan22]vlan 23
[SXL-FLOOR1-vlan23]description wl-sxs
[SXL-FLOOR1-vlan23]vlan 94
[SXL-FLOOR1-vlan94]description manage
[SXL-FLOOR1-vlan94]vlan 95
[SXL-FLOOR1-vlan95]description monitor
[SXL-FLOOR1-vlan95]quit
[SXL-FLOOR1]
//Create VLAN22, VLAN23, VLAN94 and VLAN95; configure VLAN descriptions
[SXL-FLOOR1]port-group 1
[SXL-FLOOR1-port-group-1]group-member g0/0/2 to g0/0/10
[SXL-FLOOR1-port-group-1]port link-type access
[SXL-FLOOR1-port-group-1]port default vlan 22
[SXL-FLOOR1-port-group-1]quit
//Configure the interfaces from G0/0/2 to G0/0/10 as access interfaces, under VLAN22
[SXL-FLOOR1]port-group 2
[SXL-FLOOR1-port-group-2]group-member g0/0/11 to g0/0/18
[SXL-FLOOR1-port-group-2]port link-type access
[SXL-FLOOR1-port-group-2]port default vlan 23
[SXL-FLOOR1-port-group-2]quit
//Configure the interfaces from G0/0/11 to G0/0/18 as access interfaces, under VLAN23
[SXL-FLOOR1]port-group 3
[SXL-FLOOR1-port-group-3]group-member g0/0/19 to g0/0/24
[SXL-FLOOR1-port-group-3]port link-type access
[SXL-FLOOR1-port-group-3]port default vlan 95
[SXL-FLOOR1-port-group-3]quit
//Configure the interfaces from G0/0/19 to G0/0/24 as access interfaces, under VLAN95
[SXL-FLOOR1]interface g0/0/1
[SXL-FLOOR1-GigabitEthernet0/0/1]port link-type trunk
```

```
[SXL-FLOOR1-GigabitEthernet0/0/1]port trunk allow-pass vlan 22 23 94 95
[SXL-FLOOR1-GigabitEthernet0/0/1]quit
//Configure the G0/0/1 interface as the trunk interface, allowing VLAN22, VLAN23,
//VLAN94 and VLAN95 access
```

⑥ Complete VLAN configurations on the access layer switch SXL-FLOOR2 as follows:

```
[SXL-FLOOR2]vlan 22
[SXL-FLOOR2-vlan22]description rj-sxs
[SXL-FLOOR2-vlan22]vlan 23
[SXL-FLOOR2-vlan23]description wl-sxs
[SXL-FLOOR2-vlan23]vlan 94
[SXL-FLOOR2-vlan94]description manage
[SXL-FLOOR2-vlan94]vlan 95
[SXL-FLOOR2-vlan95]description monitor
[SXL-FLOOR2-vlan95]quit
[SXL-FLOOR2]
//Create VLAN22, VLAN23, VLAN94 and VLAN95, configure VLAN descriptions
[SXL-FLOOR2]port-group 1
[SXL-FLOOR2-port-group-1]group-member g0/0/3 to g0/0/10
[SXL-FLOOR2-port-group-1]port link-type access
[SXL-FLOOR2-port-group-1]port default vlan 22
[SXL-FLOOR2-port-group-1]quit
//Configure the interfaces from G0/0/3 to G0/0/10 as access interfaces, under VLAN22
[SXL-FLOOR2]port-group 2
[SXL-FLOOR2-port-group-2]group-member g0/0/11 to g0/0/18
[SXL-FLOOR2-port-group-2]port link-type access
[SXL-FLOOR2-port-group-2]port default vlan 23
[SXL-FLOOR2-port-group-2]quit
//Configure the interfaces from G0/0/11 to G0/0/18 as access interfaces, under VLAN23
[SXL-FLOOR2]port-group 3
[SXL-FLOOR2-port-group-3]group-member g0/0/19 to g0/0/24
[SXL-FLOOR2-port-group-3]port link-type access
[SXL-FLOOR2-port-group-3]port default vlan 95
[SXL-FLOOR2-port-group-3]quit
//Configure the interfaces from G0/0/19 to G0/0/24 as access interfaces, under VLAN95
```

Note: The two links between the access layer switch SXL-FLOOR2 and the aggregation layer switch SXL-AGG of the training building use link aggregation technology to improve the link bandwidth and high reliability. Therefore, VLAN will temporarily not be configured for the interfaces (G0/0/1 and G0/0/2) on the two links of the training building access layer switch SXL-FLOOR2.

(2) Configuration of aggregation layer switch VLAN

The aggregation layer switch of each building needs to achieve communication between different VLANs in the building. According to the plan in Table 3.1, the aggregation layer switch needs to complete such VLAN configuration tasks as follows: create VLANs and configure VLAN descriptions; configure the link to the access layer switch as the trunk link; configure the link to CORE as the access link.

① Complete VLAN configurations on the aggregate layer switch XZL-AGG as follows:

```
[XZL-AGG]vlan 2
[XZL-AGG-vlan2]description baoan
[XZL-AGG-vlan2]vlan 3
[XZL-AGG-vlan3]description student
[XZL-AGG-vlan3]vlan 4
[XZL-AGG-vlan4]description caiwu
[XZL-AGG-vlan4]vlan 5
[XZL-AGG-vlan5]description hr
[XZL-AGG-vlan5]vlan 90
[XZL-AGG-vlan99]description manage
[XZL-AGG-vlan99]vlan 91
[XZL-AGG-vlan91]description monitor
[XZL-AGG-vlan91]vlan 199
[XZL-AGG-vlan199]description linktocore
[XZL-AGG-vlan199]quit
[XZL-AGG]
//Create VLAN2, VLAN3, VLAN4, VLAN5, VLAN90, VLAN91 and VLAN99, configure VLAN
//descriptions
[XZL-AGG]interface g0/0/2
[XZL-AGG-GigabitEthernet0/0/2]port link-type trunk
[XZL-AGG-GigabitEthernet0/0/2]port trunk allow-pass vlan 2 3 90 91
[XZL-AGG-GigabitEthernet0/0/2]quit
//Configure the G0/0/2 interface as the trunk interface, allowing VLAN2, VLAN3, VLAN90
//and VLAN91 access
[XZL-AGG]interface g0/0/3
[XZL-AGG-GigabitEthernet0/0/3]port link-type trunk
[XZL-AGG-GigabitEthernet0/0/3]port trunk allow-pass vlan 4 5 90 91
[XZL-AGG-GigabitEthernet0/0/3]quit
//Configure the G0/0/3 interface as the trunk interface, allowing VLAN4, VLAN5, VLAN90
//and VLAN91 access
[XZL-AGG]interface g0/0/1
[XZL-AGG-GigabitEthernet0/0/1]port link-type access
```

```
[XZL-AGG-GigabitEthernet0/0/1]port default vlan 199
[XZL-AGG-GigabitEthernet0/0/1]quit
//Configure the G0/0/1 interface as the access port, under VLAN199
```

② Complete VLAN configurations on the aggregate layer switch JXL-AGG as follows:

```
[JXL-AGG]vlan 12
[JXL-AGG-vlan12]description jiaoshi
[JXL-AGG-vlan12]vlan 13
[JXL-AGG-vlan13]description office
[JXL-AGG-vlan13]vlan 92
[JXL-AGG-vlan92]description manage
[JXL-AGG-vlan92]vlan 93
[JXL-AGG-vlan93]description monitor
[JXL-AGG-vlan93]vlan 199
[JXL-AGG-vlan199]description linktocore
[JXL-AGG-vlan199]quit
//Create VLAN12, VLAN13, VLAN92, VLAN93 and VLAN199, configure VLAN descriptions
[JXL-AGG]port-group 1
[JXL-AGG-port-group-1]group-member g0/0/2 to g0/0/3
[JXL-AGG-port-group-1]port link-type trunk
[JXL-AGG-port-group-1]port trunk allow-pass vlan 12 13 92 93
[JXL-AGG-port-group-1]quit
//Configure the interfaces G0/0/2 and G0/0/3 as the trunk interface, allowing VLAN12,
//VLAN13, VLAN92 and VLAN93 access
[JXL-AGG]interface g0/0/1
[JXL-AGG-GigabitEthernet0/0/1]port link-type access
[JXL-AGG-GigabitEthernet0/0/1]port default vlan 199
[JXL-AGG-GigabitEthernet0/0/1]quit
//Configure the G0/0/1 interface as the access interface, under VLAN199
```

③ Complete VLAN configurations on the aggregate layer switch SXL-AGG as follows:

```
[SXL-AGG]vlan 22
[SXL-AGG-vlan22]description rj-sxs
[SXL-AGG-vlan22]vlan 23
[SXL-AGG-vlan23]description wl-sxs
[SXL-AGG-vlan23]vlan 94
[SXL-AGG-vlan94]description manage
[SXL-AGG-vlan94]vlan 95
[SXL-AGG-vlan95]description monitor
[SXL-AGG-vlan95]vlan 199
[SXL-AGG-vlan199]description linktocore
```

```
[SXL-AGG-vlan199]quit
//Create VLAN22, VLAN23, VLAN94, VLAN95 and VLAN199, configure VLAN descriptions
[SXL-AGG]interface g0/0/3
[SXL-AGG-GigabitEthernet0/0/3]port link-type trunk
[SXL-AGG-GigabitEthernet0/0/3]port trunk allow-pass vlan 22 23 94 95
[SXL-AGG-GigabitEthernet0/0/3]quit
//Configure the G0/0/3 interface as the trunk interface, allowing VLAN22, VLAN23,
//VLAN94 and VLAN95 access
```

Note: The two links connecting the aggregation layer switch SXL-AGG and the core layer switch CORE of the training building, and the two links connecting the aggregation layer switch SXL-AGG and the access layer switch SXL-FLOOR2 of the training building should use link aggregation technology to improve link bandwidth and high reliability. Therefore, VLAN will temporarily not be configured for the interfaces (G0/0/1, G0/0/2, G0/0/4 and G0/0/5) on the four links of the training building aggregation layer switch SXL-AGG.

(3) Configuration of core layer switch VLAN

According to the plan in Table 3.1, the core layer switch needs to complete such VLAN configuration tasks as follows: create VLAN and configure VLAN descriptions; configure the link to the aggregation layer switch as the access link; configure the link to the server access switch SERVER-ACC as the access link. The steps are detailed as follows:

```
[CORE]vlan 199
[CORE-vlan199]description linkto-sxl-agg
[CORE-vlan199]vlan 200
[CORE-vlan200]description linkto-jxl-agg
[CORE-vlan200]vlan 201
[CORE-vlan201]description linkto-xzl-agg
[CORE-vlan201]vlan 202
[CORE-vlan202]description server
[CORE-vlan202]vlan 203
[CORE-vlan203]description linkto-edge-ar
[CORE-vlan203]quit
//Create VLAN199, VLAN200, VLAN201, VLAN202 and VLAN203, and configure VLAN descriptions
[CORE]interface g0/0/1
[CORE-GigabitEthernet0/0/1]port link-type access
[CORE-GigabitEthernet0/0/1]port default vlan 203
[CORE-GigabitEthernet0/0/1]quit
//Configure the G0/0/1 interface as the access interface, under VLAN203. Note: This
//interface is connected to the outbound router EDGE-AR
[CORE]interface g0/0/2
```

```
[CORE-GigabitEthernet0/0/2]port link-type access
[CORE-GigabitEthernet0/0/2]port default vlan 201
[CORE-GigabitEthernet0/0/2]quit
//Configure the G0/0/2 interface as the access interface, under VLAN201. Note: This
//interface is connected to the aggregation layer switch XZL-AGG in the administrative building
[CORE]interface g0/0/3
[CORE-GigabitEthernet0/0/3]port link-type access
[CORE-GigabitEthernet0/0/3]port default vlan 200
[CORE-GigabitEthernet0/0/3]quit
//Configure the G0/0/3 interface as the access interface, under VLAN200. Note: This
//interface is connected to the aggregation layer switch JXL-AGG in the teaching building
[CORE]interface g0/0/6
[CORE-GigabitEthernet0/0/6]port link-type access
[CORE-GigabitEthernet0/0/6]port default vlan 202
[CORE-GigabitEthernet0/0/6]quit
//Configure the G0/0/6 interface as the access interface, under VLAN202. Note: This
//interface is connected to the server access layer switch SERVER-ACC
```

(4) VLAN verification

Execute the "display vlan" command and "display port vlan" command on all switches, and see whether the VLAN information displayed on switches and the port VLAN information are consistent with the plan in Table 3.1. Fig. 3.3 shows the display results of the port VLAN on the core layer switch CORE. Fig. 3.4 shows the display results of the port VLAN on the aggregation layer switch XZL-AGG. Fig. 3.5 shows the display results of the port VLAN on the access layer switch XZL-FLOOR1. Check whether the results displayed in the highlighted boxes are consistent with the plan in Table 3.1.

(5) Configuration of access layer switch management IP

The access layer switch needs to complete such management IP configuration tasks as follows: create management VLANs and configure the IP address and mask; configure the default gateway address of the access layer switch.

① Complete the configuration of the management IP on the access layer switch XZL-FLOOR1, as follows:

```
[XZL-FLOOR1]interface Vlanif 90
[XZL-FLOOR1-Vlanif90]ip address 10.0.90.1 24
[XZL-FLOOR1-Vlanif90]quit
[XZL-FLOOR1]ip route-static 0.0.0.0 0.0.0.0 10.0.90.254
```

//Configure the management IP address of the switch XZL-FLOOR1 as 10.0.90.1/24,
//subject to the default gateway of 10.0.90.254

```
[CORE]display port vlan
Port              Link Type   PVID   Trunk VLAN List
--------------------------------------------------------
GigabitEthernet0/0/1    access    203    -
GigabitEthernet0/0/2    access    201    -
GigabitEthernet0/0/3    access    200    -
GigabitEthernet0/0/4    hybrid    1      -
GigabitEthernet0/0/5    hybrid    1      -
GigabitEthernet0/0/6    access    202    -
GigabitEthernet0/0/7    hybrid    1      -
GigabitEthernet0/0/8    hybrid    1      -
GigabitEthernet0/0/9    hybrid    1      -
GigabitEthernet0/0/10   hybrid    1      -
GigabitEthernet0/0/11   hybrid    1      -
GigabitEthernet0/0/12   hybrid    1      -
GigabitEthernet0/0/13   hybrid    1      -
GigabitEthernet0/0/14   hybrid    1      -
GigabitEthernet0/0/15   hybrid    1      -
```

Fig. 3.3　Display results of the port VLAN on the core layer switch CORE

```
[XZL-AGG]display port vlan
Port              Link Type   PVID   Trunk VLAN List
--------------------------------------------------------
GigabitEthernet0/0/1    access    199    -
GigabitEthernet0/0/2    trunk     1      1-3 90-91
GigabitEthernet0/0/3    trunk     1      1 4-5 90-91
GigabitEthernet0/0/4    hybrid    1      -
GigabitEthernet0/0/5    hybrid    1      -
GigabitEthernet0/0/6    hybrid    1      -
GigabitEthernet0/0/7    hybrid    1      -
GigabitEthernet0/0/8    hybrid    1      -
GigabitEthernet0/0/9    hybrid    1      -
GigabitEthernet0/0/10   hybrid    1      -
GigabitEthernet0/0/11   hybrid    1      -
GigabitEthernet0/0/12   hybrid    1      -
GigabitEthernet0/0/13   hybrid    1      -
GigabitEthernet0/0/14   hybrid    1      -
GigabitEthernet0/0/15   hybrid    1      -
```

Fig. 3.4　Display results of the port VLAN on the aggregation layer switch XZL-AGG

```
[XZL-FLOOR1]display vlan
The total number of vlans is : 5
----------------------------------------------------------------
U: Up;       D: Down;      TG: Tagged;     UT: Untagged;
MP: Vlan-mapping;          ST: Vlan-stacking;
#: ProtocolTransparent-vlan;   *: Management-vlan;
----------------------------------------------------------------
VID  Type    Ports
----------------------------------------------------------------
1    common  UT:GE0/0/1(U)
2    common  UT:GE0/0/2(D)    GE0/0/3(D)    GE0/0/4(D)    GE0/0/5(D)
             GE0/0/6(U)    GE0/0/7(D)    GE0/0/8(D)    GE0/0/9(D)
             GE0/0/10(D)   GE0/0/11(D)   GE0/0/12(D)
             TG:GE0/0/1(U)
3    common  UT:GE0/0/13(D)   GE0/0/14(D)   GE0/0/15(D)   GE0/0/16(D)
             GE0/0/17(D)   GE0/0/18(D)
             TG:GE0/0/1(U)
90   common  TG:GE0/0/1(U)
91   common  UT:GE0/0/19(D)   GE0/0/20(D)   GE0/0/21(D)   GE0/0/22(D)
             GE0/0/23(D)   GE0/0/24(D)
             TG:GE0/0/1(U)
VID  Status  Property    MAC-LRN Statistics Description
----------------------------------------------------------------
1    enable  default     enable  disable   VLAN 0001
2    enable  default     enable  disable   baoan
3    enable  default     enable  disable   student
90   enable  default     enable  disable   manage
91   enable  default     enable  disable   monitor
```

Fig. 3.5 Display results of the port VLAN on the access layer switch XZL-FLOOR1

② Complete the configuration of the management IP on the access layer switch XZL-FLOOR2, as follows:

```
[XZL-FLOOR2]interface Vlanif 90
[XZL-FLOOR2-Vlanif90]ip address 10.0.90.2 24
[XZL-FLOOR2-Vlanif90]quit
[XZL-FLOOR2]ip route-static 0.0.0.0 0.0.0.0 10.0.90.254
//Configure the management IP address of the switch XZL-FLOOR2 as 10.0.90.2/24,
//subject to the default gateway of 10.0.90.254
```

③ Complete the configuration of the management IP on the access layer switch JXL-FLOOR1, as follows:

```
[JXL-FLOOR1]interface Vlanif 92
[JXL-FLOOR1-Vlanif92]ip address 10.1.92.1 24
[JXL-FLOOR1-Vlanif92]quit
[JXL-FLOOR1]ip route-static 0.0.0.0 0.0.0.0 10.1.92.254
//Configure the management IP address of the switch JXL-FLOOR1 as 10.1.92.1/24,
//subject to the default gateway of 10.0.92.254
```

④ Complete the configuration of the management IP on the access layer switch JXL-FLOOR2, as follows:

```
[JXL-FLOOR2]interface Vlanif 92
[JXL-FLOOR2-Vlanif92]ip address 10.1.92.2 24
[JXL-FLOOR2-Vlanif92]quit
[JXL-FLOOR2]ip route-static 0.0.0.0 0.0.0.0 10.1.92.254
//Configure the management IP address of the switch JXL-FLOOR2 as 10.1.92.2/24,
//subject to the default gateway of 10.0.92.254
```

⑤ Complete the configuration of the management IP on the access layer switch SXL-FLOOR1, as follows:

```
[SXL-FLOOR1]interface Vlanif 94
[SXL-FLOOR1-Vlanif94]ip address 10.2.94.1 24
[SXL-FLOOR1-Vlanif94]quit
[SXL-FLOOR1]ip route-static 0.0.0.0 0.0.0.0 10.2.94.254
[SXL-FLOOR1]quit
//Configure the management IP address of the switch SXL-FLOOR1 as 10.1.94.1/24,
//subject to the default gateway of 10.0.94.254
```

⑥ Complete the configuration of the management IP on the access layer switch SXL-FLOOR2, as follows:

```
[SXL-FLOOR2]interface Vlanif 94
[SXL-FLOOR2-Vlanif94]ip address 10.2.94.2 24
[SXL-FLOOR2-Vlanif94]quit
[SXL-FLOOR2]ip route-static 0.0.0.0 0.0.0.0 10.2.94.254
//Configure the management IP address of the switch SXL-FLOOR2 as 10.1.94.2/24,
//subject to the default gateway of 10.0.94.254
```

(6) Configuration of aggregation layer switch IP address and DHCP service

The aggregation layer switch needs to dynamically allocate IP addresses for terminal devices in the building, so it is necessary to configure the interface IP address as planned in Table 3.1, and also the DHCP service.

① Configure the IP and DHCP service as planned in Table 3.1 on the aggregation layer switch XZL-AGG, as follows:

```
[XZL-AGG]dhcp enable
//Enable DHCP service
[XZL-AGG]ip pool vlan2
[XZL-AGG-ip-pool-vlan2]network 10.0.2.0 mask 24
[XZL-AGG-ip-pool-vlan2]gateway-list 10.0.2.254
```

```
[XZL-AGG-ip-pool-vlan2]dns-list 10.254.253.2
[XZL-AGG-ip-pool-vlan2]lease day 0 hour 8
[XZL-AGG-ip-pool-vlan2]quit
//Configure the DHCP address pool (named vlan2) for VLAN2, as well as the network
//number 10.0.2.0/24, gateway address 10.0.2.254, DNS server address 10.254.253.2,
//and address lease period of 8 hours
[XZL-AGG]ip pool vlan3
[XZL-AGG-ip-pool-vlan3]network 10.0.3.0 mask 24
[XZL-AGG-ip-pool-vlan3]gateway-list 10.0.3.254
[XZL-AGG-ip-pool-vlan3]dns-list 10.254.253.2
[XZL-AGG-ip-pool-vlan3]lease day 0 hour 8
[XZL-AGG-ip-pool-vlan3]quit
//Configure the DHCP address pool (named vlan3) for VLAN3, as well as the network
//number 10.0.3.0/24, gateway address 10.0.3.254, DNS server address 10.254.253.2,
//and address lease period of 8 hours
[XZL-AGG]ip pool vlan4
[XZL-AGG-ip-pool-vlan4]network 10.0.4.0 mask 24
[XZL-AGG-ip-pool-vlan4]gateway-list 10.0.4.254
[XZL-AGG-ip-pool-vlan4]dns-list 10.254.253.2
[XZL-AGG-ip-pool-vlan4]lease day 0 hour 8
[XZL-AGG-ip-pool-vlan4]quit
//Configure the DHCP address pool (named vlan4) for VLAN4, as well as the network
//number 10.0.4.0/24, gateway address 10.0.4.254, DNS server address 10.254.253.2,
//and address lease period of 8 hours
[XZL-AGG]ip pool vlan5
[XZL-AGG-ip-pool-vlan5]network 10.0.5.0 mask 24
[XZL-AGG-ip-pool-vlan5]gateway-list 10.0.5.254
[XZL-AGG-ip-pool-vlan5]dns-list 10.254.253.2
[XZL-AGG-ip-pool-vlan5]lease day 0 hour 8
[XZL-AGG-ip-pool-vlan5]quit
//Configure the DHCP address pool (named vlan5) for VLAN5, as well as the network
//number 10.0.5.0/24, gateway address 10.0.5.254, DNS server address 10.254.253.2,
//and address lease period of 8 hours
[XZL-AGG]interface Vlanif 2
[XZL-AGG-Vlanif2]ip address 10.0.2.254 24
[XZL-AGG-Vlanif2]dhcp select global
[XZL-AGG-Vlanif2]quit
[XZL-AGG]interface Vlanif 3
[XZL-AGG-Vlanif3]ip address 10.0.3.254 24
[XZL-AGG-Vlanif3]dhcp select global
```

```
[XZL-AGG-Vlanif3]quit
[XZL-AGG]interface Vlanif 4
[XZL-AGG-Vlanif4]ip address 10.0.4.254 24
[XZL-AGG-Vlanif4]dhcp select global
[XZL-AGG-Vlanif4]quit
[XZL-AGG]interface Vlanif 5
[XZL-AGG-Vlanif5]ip address 10.0.5.254 24
[XZL-AGG-Vlanif5]dhcp select global
[XZL-AGG-Vlanif5]quit
[XZL-AGG]interface Vlanif 90
[XZL-AGG-Vlanif91]ip address 10.0.9.254 24
[XZL-AGG-Vlanif91]quit
[XZL-AGG]interface Vlanif 91
[XZL-AGG-Vlanif91]ip address 10.0.91.254 24
[XZL-AGG-Vlanif91]quit
[XZL-AGG]interface Vlanif 199
[XZL-AGG-Vlanif199]ip address 10.254.254.2 30
[XZL-AGG-Vlanif199]quit
//Configure the IP addresses of the Vlanif2, Vlanif3, Vlanif4, Vlanif5, Vlanif90, Vlanif91,
//and Vlanif199 interfaces of the switch SXL-AGG, and enable the Vlanif2, Vlanif3, Vlanif4,
//and Vlanif5 interfaces to use the DHCP Server function of the global address pool
```

② Configure the IP and DHCP service as planned in Table 3.1 on the aggregation layer switch JXL-AGG, as follows:

```
[JXL-AGG]dhcp enable
//Enable DHCP service
[JXL-AGG]ip pool vlan12
[JXL-AGG-ip-pool-vlan12]network 10.1.12.0 mask 24
[JXL-AGG-ip-pool-vlan12]gateway-list 10.1.12.254
[JXL-AGG-ip-pool-vlan12]dns-list 10.254.253.2
[JXL-AGG-ip-pool-vlan12]lease day 0 hour 8
[JXL-AGG-ip-pool-vlan12]quit
//Configure the DHCP address pool (named vlan12) for VLAN12, as well as the network
//number 10.1.12.0/24, gateway address 10.1.12.254, DNS server address 10.254.253.2,
//and address lease period of 8 hours
[JXL-AGG]ip pool vlan13
[JXL-AGG-ip-pool-vlan13]network 10.1.13.0 mask 24
[JXL-AGG-ip-pool-vlan13]gateway-list 10.1.13.254
[JXL-AGG-ip-pool-vlan13]dns-list 10.254.253.2
[JXL-AGG-ip-pool-vlan13]lease day 0 hour 8
```

```
[JXL-AGG-ip-pool-vlan13]quit
//Configure the DHCP address pool (named vlan13) for VLAN13, as well as the network
//number 10.1.13.0/24, gateway address 10.1.13.254, DNS server address 10.254.253.2,
//and address lease period of 8 hours
[JXL-AGG]interface Vlanif 12
[JXL-AGG-Vlanif12]ip address 10.1.12.254 24
[JXL-AGG-Vlanif12]dhcp select global
[JXL-AGG-Vlanif12]quit
[JXL-AGG]interface Vlanif 13
[JXL-AGG-Vlanif13]ip address 10.1.13.254 24
[JXL-AGG-Vlanif13]dhcp select global
[JXL-AGG-Vlanif13]quit
[JXL-AGG]interface Vlanif 92
[JXL-AGG-Vlanif92]ip address 10.1.92.254 24
[JXL-AGG-Vlanif92]quit
[JXL-AGG]interface Vlanif 93
[JXL-AGG-Vlanif93]ip address 10.1.93.254 24
[JXL-AGG-Vlanif93]quit
[JXL-AGG]interface Vlanif 199
[JXL-AGG-Vlanif199]ip add
[JXL-AGG-Vlanif199]ip address 10.254.254.6 30
[JXL-AGG-Vlanif199]quit
//Configure the IP addresses of the Vlanif12, Vlanif13, Vlanif92, Vlanif93 and Vlanif199
//interfaces of the switch JXL-AGG, and enable the Vlanif12 and Vlanif13 interfaces
//to use the DHCP Server function of the global address pool
```

③ Configure the IP and DHCP service as planned in Table 3.1 on the aggregation layer switch SXL-AGG, as follows:

```
[SXL-AGG]dhcp enable
//Enable DHCP service
[SXL-AGG]ip pool vlan22
[SXL-AGG-ip-pool-vlan22]network 10.2.22.0 mask 24
[SXL-AGG-ip-pool-vlan22]gateway-list 10.2.22.254
[SXL-AGG-ip-pool-vlan22]dns-list 10.254.253.2
[SXL-AGG-ip-pool-vlan22]lease day 0 hour 8
[SXL-AGG-ip-pool-vlan22]quit
//Configure the DHCP address pool (named vlan22) for VLAN22, as well as the network
//number 10.2.22.0/24, gateway address 10.2.22.254, DNS server address 10.254.253.2,
//and address lease period of 8 hours
[SXL-AGG]ip pool vlan23
```

```
[SXL-AGG-ip-pool-vlan23]network 10.2.23.0 mask 24
[SXL-AGG-ip-pool-vlan23]gateway-list 10.2.23.254
[SXL-AGG-ip-pool-vlan23]dns-list 10.254.253.2
[SXL-AGG-ip-pool-vlan23]lease day 0 hour 8
[SXL-AGG-ip-pool-vlan23]quit
//Configure the DHCP address pool (named vlan23) for VLAN23, as well as the network
//number 10.2.23.0/24, gateway address 10.2.23.254, DNS server address 10.254.253.2,
//and address lease period of 8 hours
[SXL-AGG]interface Vlanif 22
[SXL-AGG-Vlanif22]ip address 10.2.22.254 24
[SXL-AGG-Vlanif22]dhcp select global
[SXL-AGG-Vlanif22]quit
[SXL-AGG]interface Vlanif 23
[SXL-AGG-Vlanif23]ip address 10.2.23.254 24
[SXL-AGG-Vlanif23]dhcp select global
[SXL-AGG-Vlanif23]quit
[SXL-AGG]interface Vlanif 94
[SXL-AGG-Vlanif94]ip address 10.2.94.254 24
[SXL-AGG-Vlanif94]quit
[SXL-AGG]interface Vlanif 95
[SXL-AGG-Vlanif95]ip address 10.2.95.254 24
[SXL-AGG-Vlanif95]quit
[SXL-AGG]interface Vlanif 199
[SXL-AGG-Vlanif199]ip address 10.254.254.10 30
[SXL-AGG-Vlanif199]quit
//Configure the IP addresses of the Vlanif22, Vlanif23, Vlanif94, Vlanif95 and
//Vlanif199 interfaces of the switch SXL-AGG, and enable the Vlanif22 and Vlanif23
//interfaces to use the DHCP Server function of the global address pool
```

(7) Configuration of core layer switch IP address

According to the IP plan in Table 3.1, configure the IP addresses of the Vlanif199, Vlanif200, Vlanif 201, Vlanif 202 and Vlanif 203 interfaces of the core layer switch CORE, as follows:

```
[CORE]interface Vlanif 199
[CORE-Vlanif199]ip address 10.254.254.9 30
[CORE-Vlanif199]quit
[CORE]interface Vlanif 200
[CORE-Vlanif200]ip address 10.254.254.5 30
[CORE-Vlanif200]quit
[CORE]interface Vlanif 201
[CORE-Vlanif201]ip address 10.254.254.1 30
```

```
[CORE-Vlanif201]quit
[CORE]interface Vlanif 202
[CORE-Vlanif202]ip address 10.254.253.254 24
[CORE-Vlanif202]quit
[CORE]interface Vlanif 203
[CORE-Vlanif203]ip address 10.254.254.13 30
[CORE-Vlanif203]quit
```

(8) IP configuration for outbound router and operator router interfaces

① According to the IP plan in Table 3.1, configure the interface IP address of the outbound router EDGE-AR as follows:

```
[EDGE-AR]interface g0/0/1
[EDGE-AR-GigabitEthernet0/0/1]ip address 10.254.254.14 30
[EDGE-AR-GigabitEthernet0/0/1]undo shutdown
[EDGE-AR-GigabitEthernet0/0/1]quit
[EDGE-AR]interface g0/0/0
[EDGE-AR-GigabitEthernet0/0/0]ip address 125.71.28.65 24
[EDGE-AR-GigabitEthernet0/0/0]undo shutdown
[EDGE-AR-GigabitEthernet0/0/0]quit
```

② According to the IP plan in Table 3.1, configure the interface IP address of the operating router ISP as follows:

```
[ISP]interface g0/0/1
[ISP-GigabitEthernet0/0/1]ip address 202.1.1.254 24
[ISP-GigabitEthernet0/0/1]undo shutdown
[ISP-GigabitEthernet0/0/1]quit
[ISP]interface g0/0/0
[ISP-GigabitEthernet0/0/0]ip address 125.71.28.1 24
[ISP-GigabitEthernet0/0/0]undo shutdown
[ISP-GigabitEthernet0/0/0]quit
```

(9) Configuration of NAT Server

On the enterprise outbound router EDGE-AR, complete the configuration according to the NAT Server planned in Table 3.3, as follows:

```
[EDGE-AR]interface g0/0/0
[EDGE-AR-GigabitEthernet0/0/0]nat server protocol tcp global 125.71.28.66 www inside 10.254.253.1 www
[EDGE-AR-GigabitEthernet0/0/0]nat server protocol tcp global 125.71.28.66 53 inside 10.254.253.2 53
[EDGE-AR-GigabitEthernet0/0/0]nat server protocol udp global 125.71.28.66 53 inside 10.254.253.2 53
```

(10) NAPT configuration

On the enterprise outbound router EDGE-AR, the NAPT configuration shall complete such tasks as follows: use ACL to configure the intranet address segments that allow conversion; configure the NAT public address pool.

```
[EDGE-AR]acl 2000
[EDGE-AR-acl-basic-2000]rule permit source 10.0.2.0 0.0.0.255
[EDGE-AR-acl-basic-2000]rule permit source 10.0.3.0 0.0.0.255
[EDGE-AR-acl-basic-2000]rule permit source 10.0.4.0 0.0.0.255
[EDGE-AR-acl-basic-2000]rule permit source 10.0.5.0 0.0.0.255
[EDGE-AR-acl-basic-2000]rule permit source 10.1.12.0 0.0.0.255
[EDGE-AR-acl-basic-2000]rule permit source 10.1.13.0 0.0.0.255
[EDGE-AR-acl-basic-2000]rule permit source 10.2.22.0 0.0.0.255
[EDGE-AR-acl-basic-2000]rule permit source 10.2.23.0 0.0.0.255
[EDGE-AR-acl-basic-2000]rule permit source 10.0.90.0 0.0.0.255
[EDGE-AR-acl-basic-2000]rule permit source 10.0.91.0 0.0.0.255
[EDGE-AR-acl-basic-2000]rule permit source 10.1.92.0 0.0.0.255
[EDGE-AR-acl-basic-2000]rule permit source 10.1.93.0 0.0.0.255
[EDGE-AR-acl-basic-2000]rule permit source 10.2.94.0 0.0.0.255
[EDGE-AR-acl-basic-2000]rule permit source 10.2.95.0 0.0.0.255
[EDGE-AR-acl-basic-2000]rule permit source 10.254.253.0 0.0.0.255
[EDGE-AR-acl-basic-2000]rule permit source 10.254.254.0 0.0.0.255
//Configure the intranet address segments that allow NAT conversion, and set the
//ACL list number as 2000. The intranet address segments that allow NAT conversion
//include: 10.0.2.0/24, 10.0.3.0/24, 10.0.4.0/24, 10.0.5.0/24, 10.1.12.0/24,
//10.1.13.0/24, 10.2.22.0/24, 10.2.23.0/24, 10.0.90.0/24, 10.0.91.0/24, 10.1.92.0/24,
//10.1.93.0/24, 10.2.94.0/24, 10.2.95.0/24, 10.254.253.0/24 and 10.254.254.0/24
[EDGE-AR]nat address-group 1 125.71.28.67 125.71.28.69
//Configure the NAT public address pool subject to the address pool number 1 and
//address pool range 125.71.28.67-125.71.28.69
[EDGE-AR-GigabitEthernet0/0/0]nat outbound 2000 address-group 1
//Associate ACL under the outbound interface G0/0/0 for NAPT conversion with the
//address pool
```

(11) IP address planning for servers and PC1

According to the IP address planning in Table 3.1, the IP addresses are statically allocated to the corporate intranet server, the Internet server and the test host PC1. Please configure the IP address, subnet mask, default gateway and DNS server IP addresses as planned, and use the ipconfig command in the command line to check whether the configured IP information takes effect.

(12) Verification and testing

After the IP addresses and DHCP services are configured, carry out verification as follows:

① Use the "display ip interface brief" command to display the interface IP brief information on all access layer switches, aggregation layer switches, core layer switches, and routers in the building. Check if the management VLAN port, management IP address, physical and protocol states are UP. Fig. 3.6 shows the display results via the "display ip interface brief" command on the access layer switch XZL-FLOOR1. Fig. 3.7 shows the display results via the "display ip interface brief" command on the aggregation layer switch XZL-AGG. Fig. 3.8 shows the display results via the "display ip interface brief" command on the core layer switch CORE.

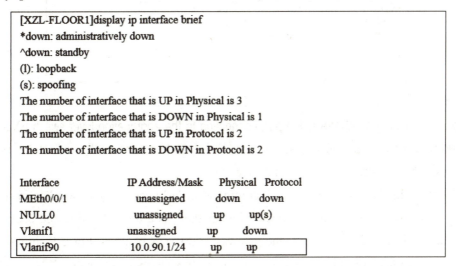

Fig. 3.6 Display results via the "display ip interface brief" command on the access layer switch XZL-FLOOR1

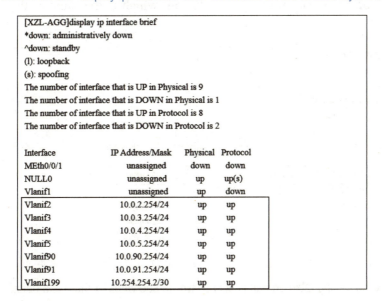

Fig. 3.7 Display results via the "display ip interface brief" command on the aggregation layer switch XZL-AGG

```
[CORE]display ip interface brief
*down: administratively down
^down: standby
(l): loopback
(s): spoofing
The number of interface that is UP in Physical is 6
The number of interface that is DOWN in Physical is 2
The number of interface that is UP in Protocol is 5
The number of interface that is DOWN in Protocol is 3

Interface        IP Address/Mask      Physical   Protocol
MEth0/0/1        unassigned           down       down
NULL0            unassigned           up         up(s)
Vlanif1          unassigned           up         down
Vlanif199        10.254.254.9/30      down       down
Vlanif200        10.254.254.5/30      up         up
Vlanif201        10.254.254.1/30      up         up
Vlanif202        10.254.253.254/24    up         up
Vlanif203        10.254.254.13/30     up         up
```

Fig. 3.8　Display results via the "display ip interface brief" command on the core layer switch CORE

Note: The interface VLAN199 is the port connecting to the aggregation layer switch SXL-AGG in the training building. So far, no physical interface under VLAN199 has been configured, so the physical and protocol states of the interface Vlanif199 are both down.

② Use the "display ip routing table" command on all access layer switches in the building to display the routing table contents, check whether there is a default route, and see whether the next hop address of the default route is the gateway address as planned. Fig. 3.9 shows the display results via the "display ip routing-table" command on the access layer switch XZL-FLOOR1. The next hop address of the default route is 10.0.90.254, which is the same as the planned gateway address.

```
[XZL-FLOOR1]display ip routing-table
Route Flags: R - relay, D - download to fib
------------------------------------------------------------
Routing Tables: Public
         Destinations : 5     Routes : 5

Destination/Mask    Proto   Pre  Cost   Flags  NextHop        Interface

0.0.0.0/0           Static  60   0      RD     10.0.90.254    Vlanif90
10.0.90.0/24        Direct  0    0      D      10.0.90.1      Vlanif90
10.0.90.1/32        Direct  0    0      D      127.0.0.1      Vlanif90
127.0.0.0/8         Direct  0    0      D      127.0.0.1      InLoopBack0
127.0.0.1/32        Direct  0    0      D      127.0.0.1      InLoopBack0
```

Fig. 3.9　Display results via the "display ip routing-table" command on the access layer switch XZL-FLOOR1

Note: Check the routing table on each access layer to see whether there is a default route and whether

the next hop address is the planned gateway address.

③ Use the ping command on all access layer switches in the building to test the connectivity of the default gateway. Fig. 3.10 shows that the access layer switch XZL-FLOOR1 successfully ping its default gateway configured.

```
[XZL-FLOOR1]ping 10.0.90.254
 PING 10.0.90.254: 56  data bytes, press CTRL_C to break
   Reply from 10.0.90.254: bytes=56 Sequence=1 ttl=255 time=10 ms
   Reply from 10.0.90.254: bytes=56 Sequence=2 ttl=255 time=20 ms
   Reply from 10.0.90.254: bytes=56 Sequence=3 ttl=255 time=50 ms
   Reply from 10.0.90.254: bytes=56 Sequence=4 ttl=255 time=30 ms
   Reply from 10.0.90.254: bytes=56 Sequence=5 ttl=255 time=50 ms

 --- 10.0.90.254 ping statistics ---
   5 packet(s) transmitted
   5 packet(s) received
   0.00% packet loss
   round-trip min/avg/max = 10/32/50 ms
```

Fig. 3.10 The access layer switch XZL-FLOOR1 successfully ping its default gateway configured

④ Connect a test host to any port on all access layer switches in the building, configure its IP address and DNS server address to be automatically obtained, use the ipconfig command on the test host to check if the IP address is obtained normally, and then use the ping command to test the IP connectivity to the default gateway. Fig. 3.11 shows that the test host PC2 has successfully obtained the IP address, subnet mask and default gateway from the DHCP server, and successfully ping its default gateway address.

```
PC>ipconfig

Link local IPv6 address............: fe80::5689:98ff:febc:7cb1
IPv6 address.......................: :: / 128
IPv6 gateway.......................: ::
IPv4 address.......................: 10.0.2.253
Subnet mask........................: 255.255.255.0
Gateway............................: 10.0.2.254
Physical address...................: 54-89-98-BC-7C-B1
DNS server.........................: 10.254.253.2

PC>ping 10.0.2.254

Ping 10.0.2.254: 32 data bytes, Press Ctrl_C to break
From 10.0.2.254: bytes=32 seq=1 ttl=255 time=47 ms
From 10.0.2.254: bytes=32 seq=2 ttl=255 time=47 ms
From 10.0.2.254: bytes=32 seq=3 ttl=255 time=31 ms
From 10.0.2.254: bytes=32 seq=4 ttl=255 time=47 ms
From 10.0.2.254: bytes=32 seq=5 ttl=255 time=31 ms
```

Fig. 3.11 The test host PC2 has successfully obtained the IP address from the DHCP server

Use the following display command for NAT view and verification.

① Use the "display nat server" command to view NAT Server information. Fig. 3.12 shows an

example of viewing the NAT server information on the router EDGE-AR.

② In the G0/0/0 interface view of the router EDGE-AR, use the "display this" command to view the configuration under the interface. Take care to check whether the NAT configuration information is consistent with the plan, as shown in Fig. 3.13.

```
[EDGE-AR]display nat server
 Nat Server Information:
 Interface : GigabitEthernet0/0/0
   Global IP/Port   : 125.71.28.66/80(www)
   Inside IP/Port   : 10.254.253.1/80(www)
   Protocol : 6(tcp)
   VPN instance-name  : ----
   Acl number        : ----
   Description : ----
   Global IP/Port   : 125.71.88.66/53(domain)
   Inside IP/Port   : 10.254.253.2/53(domain)
   Protocol : 6(tcp)
   VPN instance-name  : ----
   Acl number        : ----
   Description : ----
   Global IP/Port   : 125.71.88.66/53(dns)
   Inside IP/Port   : 10.254.253.2/53(dns)
   Protocol : 17(udp)
   VPN instance-name  : ----
   Acl number        : ----
   Description : ----

 Total :  3
```

Fig. 3.12　An example of viewing the NAT Server information

```
[EDGE-AR-GigabitEthernet0/0/0]display this
[V200R003C00]
#
interface GigabitEthernet0/0/0
 ip address 125.71.28.65 255.255.255.0
 nat server protocol tcp global 125.71.28.66 www inside 10.254.253.1 www
 nat server protocol tcp global 125.71.88.66 domain inside 10.254.253.2 domain
 nat server protocol udp global 125.71.88.66 dns inside 10.254.253.2 dns
 nat outbound 2000 address-group 1 return
```

Fig. 3.13　Configuration of router EDGE-AR under G0/0/0 interface

Exercises

Multiple choice

(1) The following statement (　　) about the layered design of enterprise networks is incorrect.

　　A. The core layer provides high-speed optimized bandwidth transmission between nodes that communicate with each other

　　B. The aggregation layer provides high-speed optimized bandwidth transmission between nodes that communicate with each other

C. Redundant fiber links are usually used between core layer devices, as well as between core layer devices and aggregation layer devices

D. The access layer network is used to connect terminal devices to the network

(2) The following (   ) is usually obtained automatically via DHCP.

A. router interface IP address

B. management IP address of access layer switch

C. server IP address

D. laptop computers for ordinary users

(3) Access layer switches typically do not complete (   ).

A. communication configurations between VLANs

B. VLAN configurations

C. configurations of management IP address

D. configurations of spanning tree

Task 2 Planning and Configuration of High Reliability Links

Task Description

On the basis of completing Task 1 of this project, plan and configure the link aggregation and MSTP in the network shown in Fig. 3.1, so as to achieve the networking requirements that the link bandwidth and reliability can be increased and the two-level loops can be avoided.

Task Objectives

- Understand the application of link aggregation.
- Master the link aggregation configuration, view and troubleshooting in manual mode.
- Master the link aggregation configuration, view and troubleshooting in LACP mode.
- Competent for deployment and implementation of link aggregation and MSTP.

Related Knowledge

1. Basic principle of link aggregation

The link aggregation technology, also known link bonding, can help bond multiple physical links between two devices to form a logical link, also known as aggregation link. The link aggregation technology, subject to the formal standard IEEE 802.3ad developed by IEEE 802 Committee, is applicable to such Ethernet technologies as 1000 Mbit/s and 10 Gbit/s. The link aggregation technology can have the data traffic shared among all physical links constituting the

aggregation link, so as to effectively improve the bandwidth of the network access. In addition, the various physical links that form an aggregation link are dynamically backed up to each other. Only if there is a functioning physical link, the entire logical link will not fail, so the reliability of the network can be effectively increased. In the link aggregation example shown in Fig. 3.14, the three physical links between the switches LSW1 and LSW2 are aggregated into one logical link.

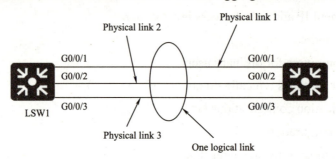

Fig. 3.14   An example of link aggregation technology

In the enterprise network, multiple physical links are usually deployed between the core layer switch and the aggregation layer switch, between the aggregation layer switch and the access layer switch, and between the server access layer switch and the server, and link aggregation is used to improve the link bandwidth between switches and between switches and servers for ensuring its reliability.

Note: Huawei's link aggregation is Eth-Trunk.

2. Link aggregation modes

Links can be aggregated in manual and LACP modes.

(1) Manual mode

In manual mode, the Eth-Trunk creation and the addition of member interfaces are manually configured. Normally, all links are active, that is, they all can forward data. If some active link fails, Eth-Trunk can automatically share the traffic equally among the remaining active links. In manual mode, it must ensure that the peer member interfaces corresponding to all member interfaces in the local aggregation link are added from the same device, which can only be confirmed manually by the administrator. The device can only determine whether the peer interface is working properly based on the physical layer status. Usually, the manual mode shall be used when there is only one device at both ends of the aggregation link that does not support the LACP protocol.

(2) LACP mode

The LACP mode is a link aggregation mode using LACP protocol. The link aggregation devices interact with each other through LACPDU (link aggregation control protocol data unit), and negotiate to ensure that the peer is a member interface of the same device and the same aggregation

interface. The LACPDU message contains such fields as device priority, MAC address, interface priority, and interface number. In LACP mode, the active end is determined by the system LACP priority (the default priority is 32,768, and the lower the priority value, the higher the priority), and the other end (passive end) follows the active end to select an active interface. The LACP mode supports the maximum number of active interfaces to be configured. When the number of member interfaces exceeds the maximum number of active interfaces, the preferred link aggregation member interface will be selected as the active interface by comparing the interface priority and interface number, and the remaining member interfaces are inactive interfaces (backup ports). Switches will only send and receive data frames from active interfaces. When active interfaces fail, inactive interfaces will be elected as active interfaces. As shown in Fig. 3.15, the link aggregation between the switches LSW1 and LSW2 has three member interfaces, but only two interfaces are active at most. Under normal circumstances, data frames are forwarded from the active interfaces 1 and 2, and the interface 3 is an inactive interface. When some active link (such as port 1) fails, port 3 will be selected as the active interface.

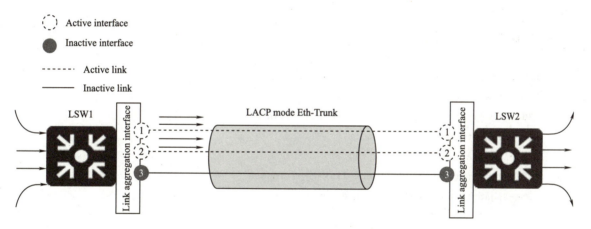

Fig. 3.15   An example of maximum active interfaces in link aggregation

Related Skills

1. Planning and configuration of manual mode link aggregation

Construct a network topology as shown in Fig. 3.16. Configure link aggregation between the switches LSW1 and LSW2 manually, and let the three links between LSW1 and LSW2 kept forwarding data at the same time, so as to make full use of the bandwidth of the three links for verification and test of link aggregation.

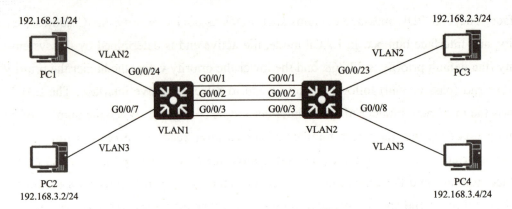

Fig. 3.16 Network topology for manual mode link aggregation

(1) Environment construction and VLAN configurations

① Construct a network topology as shown in Fig. 3.16, and configure the host IP according to the IP planning in the figure.

② On the switch LSW1, complete VLAN configurations according to the VLAN planned in Fig. 3.16, as follows:

```
[LSW1]vlan 2
[LSW1-vlan2]vlan 3
[LSW1]interface g0/0/24
[LSW1-GigabitEthernet0/0/24]port link-type access
[LSW1-GigabitEthernet0/0/24]port default vlan 2
[LSW1-GigabitEthernet0/0/24]quit
[LSW1]interface g0/0/7
[LSW1-GigabitEthernet0/0/7]port link-type access
[LSW1-GigabitEthernet0/0/7]port default vlan 3
[LSW1-GigabitEthernet0/0/7]quit
```

③ On the switch LSW2, complete VLAN configurations according to the VLAN planned in Fig. 3.16, as follows:

```
[LSW2]vlan 2
[LSW2-vlan2]vlan 3
[LSW2-vlan3]quit
[LSW2]interface g0/0/23
[LSW2-GigabitEthernet0/0/23]port link-type access
[LSW2-GigabitEthernet0/0/23]port default vlan 2
[LSW2-GigabitEthernet0/0/23]quit
[LSW2]interface g0/0/8
[LSW2-GigabitEthernet0/0/8]port link-type access
[LSW2-GigabitEthernet0/0/8]port default vlan 3
```

```
[LSW2-GigabitEthernet0/0/8]quit
```

Note: The member interfaces of the link aggregation group of Huawei switches cannot have any VLAN configured. Therefore, no VLAN access link can be configured for the member interfaces of the switch (LSW1 and LSW2) access links. The VLAN information must be configured in the link aggregation group Eth-Trunk.

(2) Configuration of manual mode link aggregation

The tasks are as follows for configurations of manual mode link aggregation:

① Create a link aggregation group Eth-Trunk.

② Configure the link aggregation mode as manual mode.

③ Add the member interface to the link aggregation group.

④ Enter the link aggregation group and configure Layer 2 or Layer 3 attributes for it (optional).

⑤ Configure the load-balance mode (optional).

Follow the steps below for configuration:

```
    [LSW1]interface Eth-Trunk 1       //Create a link aggregation group Eth-Trunk1 and en-
ter the Eth-Trunk interface view
    [LSW1-Eth-Trunk1]mode manual load-balance
    //Configure the link aggregation mode as manual load sharing. Note: Eth-Trunk works
    //subject to manual load sharing by default. This command can be left unconfigured
    [LSW1-Eth-Trunk1]trunkport g0/0/1      //Add the interface G0/0/1 to Eth-Trunk1
    [LSW1-Eth-Trunk1]trunkport g0/0/2      //Add the interface G0/0/2 to Eth-Trunk1
    [LSW1-Eth-Trunk1]trunkport g0/0/3      //Add the interface G0/0/3 to Eth-Trunk1
    [LSW1-Eth-Trunk1]port link-type trunk  //Configure the link aggregation group
                                           //Eth-Trunk1 as the trunk interface.
    [LSW1-Eth-Trunk1]port trunk allow-pass vlan 2 3  //Configure the link aggregation
                                           //group Eth-Trunk1 to allow VLAN 2, VLAN 3
    [LSW1-Eth-Trunk1]quit
    [LSW2]interface Eth-Trunk 1       //Create a link aggregation group Eth-Trunk1 and enter
                                      //Eth-Trunk interface view
    [LSW2-Eth-Trunk1]trunkport g0/0/1      //Add the interface G0/0/1 to Eth-Trunk1
    [LSW2-Eth-Trunk1]trunkport g0/0/2      //Add the interface G0/0/2 to Eth-Trunk1
    [LSW2-Eth-Trunk1]trunkport g0/0/3      //Add the interface G0/0/3 to Eth-Trunk1
    [LSW2-Eth-Trunk1]port link-type trunk  //Configure the link aggregation group
                                           //Eth-Trunk1 as the trunk interface
    [LSW2-Eth-Trunk1]port trunk allow-pass vlan 2 3  //Configure the link aggregation
                                           //group Eth-Trunk1 to allow VLAN 2, VLAN 3
    [LSW2-Eth-Trunk1]quit
```

Note: You can use the trunkport command in the Eth-Trunk port view to add member interfaces in the link aggregation group, or use the "eth-trunk" command in the interface view of the member interface to add it

to the link aggregation group Eth-Trunk. When adding member interfaces to Eth-Trunk, attention shall be paid to such problems as follows:

① One Ethernet interface can only be added to one Eth-Trunk interface.

② The Eth-Trunk interface cannot be nested, namely, the member interface of the Eth-Trunk interface cannot be an Eth-Trunk interface.

③ If a local device uses Eth-Trunk, the peer interface directly connected to the member interface must also be bonded as an Eth-Trunk interface for normal communication between the two ends.

④ The number, rate and duplex mode must be consistent for physical interfaces connected at both ends of the Eth-Trunk link.

(3) Verification and testing

① Execute the "display eth-trunk" command on the switches LSW1 and LSW2 respectively to see the status of the Eth-Trunk interface, as shown in Fig. 3.17 and Fig. 3.18.

```
[LSW1]display eth-trunk 1            NORMAL represents the     Load sharing mode, which can be configured
Eth-Trunk1's state information is:   manual mode               using the load-balance command
WorkingMode: NORMAL         Hash arithmetic: According to SIP-XOR-DIP
Least Active-linknumber: 1  Max Bandwidth-affected-linknumber: 8
Operate status: up          Number Of Up Port In Trunk: 3  ← There three member interfaces are
                                                             Up in the Eth-Trunk interface
--------------------------------------------------------------
PortName               Status      Weight  ← Weight of member interface
GigabitEthernet0/0/1   Up          1
GigabitEthernet0/0/2   Up          1
GigabitEthernet0/0/3   Up          1
                       ↑
          Up indicates that the interface is normally enabled
```

Fig. 3.17 Eth-Trunk1 interface status of switch LSW1

```
[LSW2]display eth-trunk 1
Eth-Trunk1's state information is:
WorkingMode: NORMAL         Hash arithmetic: According to SIP-XOR-DIP
Least Active-linknumber: 1  Max Bandwidth-affected-linknumber: 8
Operate status: up          Number Of Up Port In Trunk: 3
--------------------------------------------------------------
PortName               Status      Weight
GigabitEthernet0/0/1   Up          1
GigabitEthernet0/0/2   Up          1
GigabitEthernet0/0/3   Up          1
```

Fig. 3.18 Eth-Trunk1 interface status of switch LSW2

② Execute the "display port vlan" command on the switches LSW1 and LSW2, respectively. Fig. 3.19 shows the display results on LSW1, and it can be seen that the link type of the Ether-Trunk1 interface is trunk, allowing VLAN1 to VLAN3.

```
<LSW1>display port vlan
Port              Link Type  PVID  Trunk VLAN List
--------------------------------------------------------
Eth-Trunk1        trunk      1     1-3
GigabitEthernet0/0/1   hybrid   0    -
GigabitEthernet0/0/2   hybrid   0    -
GigabitEthernet0/0/3   hybrid   0    -
GigabitEthernet0/0/4   hybrid   1    -
GigabitEthernet0/0/5   hybrid   1    -
GigabitEthernet0/0/6   hybrid   1    -
GigabitEthernet0/0/7   access   3    -
GigabitEthernet0/0/8   hybrid   1    -
GigabitEthernet0/0/9   hybrid   1    -
GigabitEthernet0/0/10  hybrid   1    -
GigabitEthernet0/0/11  hybrid   1    -
```

Fig. 3.19　Port VLAN information of switch LSW1

2. Planning and configuration of LACP mode link aggregation

Construct a network topology as shown in Fig. 3.16. Configure link aggregation between the switches LSW1 and LSW2 in LACP mode and set the maximum number of active links as 2, so that two of the three links between LSW1 and LSW2 are active and one is inactive. Then, verify and test the link aggregation.

(1) Environment construction and VLAN configurations

① Construct a network topology as shown in Fig. 3.16, and configure the host IP according to the IP planning in the figure.

② Refer to "Planning and configuration of manual mode link aggregation" and complete the VLAN configuration on the switch LSW1 according to the VLAN planned in Fig. 3.16.

Note: Do not configure the VLAN access link for the physical ports connecting the switches LSW1 and LSW2.

(2) Configuration of LACP mode link aggregation

The tasks are as follows for configuring link aggregation in LACP mode:

① Configure system LACP priority (optional).

② Create a link aggregation group.

③ Configure the link aggregation mode to LACP mode.

④ Enable LACP priority preemption function (optional).

⑤ Configure the maximum number of active links (optional).

⑥ Configure the load-balance mode (optional).

⑦ Add the member interface to the link aggregation group.

⑧ Configure two-layer or three-layer attributes for the link aggregation port (optional);

The steps are as follows:

```
[LSW1]lacp priority 20000            //Configure the LACP priority of the switch LSW1 to
                                     //20000
[LSW1]interface Eth-Trunk 1          //Create a link aggregation group Eth-Trunk1 and
                                     //enter the Eth-Trunk interface view
[LSW1-Eth-Trunk1]mode lacp-static    //Configure the link aggregation mode to LACP mode
[LSW1-Eth-Trunk1]lacp preempt enable //Enable the LACP priority preemption function
[LSW1-Eth-Trunk1]max active-linknumber 2  //Configure the maximum number of active links
                                     //to 2
[LSW1-Eth-Trunk1]load-balance dst-ip //Configure load balance by destination IP
[LSW1-Eth-Trunk1]trunkport g0/0/1    //Add the interface G0/0/1 to Eth-Trunk1
[LSW1-Eth-Trunk1]trunkport g0/0/2    //Add the interface G0/0/2 to Eth-Trunk1
[LSW1-Eth-Trunk1]trunkport g0/0/3    //Add the interface G0/0/3 to Eth-Trunk1
[LSW1-Eth-Trunk1]port link-type trunk
[LSW1-Eth-Trunk1]port trunk allow-pass vlan 2 3
[LSW1-Eth-Trunk1]quit
//Configure the Eth-Trunk1 port as the trunk interface, allowing VLAN2 and VLAN3 access
[LSW2]interface Eth-Trunk 1          //Create a link aggregation group Eth-Trunk1
                                     //and enter the Eth-Trunk interface view
[LSW2-Eth-Trunk1]mode lacp-static    //Configure the link aggregation mode to LACP mode
[LSW2-Eth-Trunk1]lacp preempt enable //Enable the LACP priority preemption function
[LSW2-Eth-Trunk1]load-balance dst-ip //Configure the maximum number of active links to 2
[LSW2-Eth-Trunk1]trunkport g0/0/1    //Add the interface G0/0/1 to Eth-Trunk1
[LSW2-Eth-Trunk1]trunkport g0/0/2    //Add the interface G0/0/2 to Eth-Trunk1
[LSW2-Eth-Trunk1]trunkport g0/0/3    //Add the interface G0/0/3 to Eth-Trunk1
[LSW2-Eth-Trunk1]port link-type trunk
[LSW2-Eth-Trunk1]port trunk allow-pass vlan 2 3
[LSW2-Eth-Trunk1]quit
//Configure the Eth-Trunk1 port as the trunk interface, allowing VLAN2 and VLAN3 access
```

(3) Verification and testing

① Execute the "display eth-trunk" command on the switches LSW1 and LSW2 to check the status of the Eth-Trunk interface. Fig. 3.20 shows the status of the Eth-Trunk 1 interface on the switch LSW1. Fig. 3.21 shows the status of the Eth-Trunk 1 interface on the switch LSW2.

```
[LSW1]display eth-trunk 1
Eth-Trunk1's state information is:    It indicates LACP mode
Local:      The LACP priority is 20000           Use the destination IP for load sharing
LAG ID: 1              WorkingMode: STATIC
Preempt Delay Time: 30      Hash arithmetic: According to DIP
System Priority: 20000    System ID: 4c1f-ccd4-7dc2
Least Active-linknumber: 1  Max Active-linknumber: 2
Operate status: up        Number Of Up Port In Trunk: 2
--------------------------------------------------------------------
                    "Selected" indicates that the interface is an active one
ActorPortName       Status  PortType PortPri PortNo PortKey PortState Weight
GigabitEthernet0/0/1 Selected 1GE    32768   2      305     10111100  1
GigabitEthernet0/0/2 Selected 1GE    32768   3      305     10111100  1
GigabitEthernet0/0/3 Unselect 1GE    32768   4      305     10100000  1
Partner:
                    "Unselected" indicates that the interface is an inactive one
--------------------------------------------------------------------
ActorPortName       SysPri  SystemID        PortPri PortNo PortKey PortState
GigabitEthernet0/0/1 32768  4c1f-cc1e-578f  32768   2      305     10111100
GigabitEthernet0/0/2 32768  4c1f-cc1e-578f  32768   3      305     10111100
GigabitEthernet0/0/3 32768  4c1f-cc1e-578f  32768   4      305     10110000
```

Fig. 3.20 Status of Eth-Trunk 1 interface on switch LSW1

```
[LSW2]display eth-trunk
Eth-Trunk1's state information is:
Local:
LAG ID: 1              WorkingMode: STATIC
Preempt Delay Time: 30    Hash arithmetic: According to DIP
System Priority: 32768    System ID: 4c1f-cc1e-578f
Least Active-linknumber: 1  Max Active-linknumber: 8
Operate status: up        Number Of Up Port In Trunk: 2
--------------------------------------------------------------------
ActorPortName       Status  PortType PortPri PortNo PortKey PortState Weight
GigabitEthernet0/0/1 Selected 1GE    32768   2      305     10111100  1
GigabitEthernet0/0/2 Selected 1GE    32768   3      305     10111100  1
GigabitEthernet0/0/3 Unselect 1GE    32768   4      305     10110000  1

Partner:
--------------------------------------------------------------------
ActorPortName       SysPri  SystemID        PortPri PortNo PortKey PortState
GigabitEthernet0/0/1 20000  4c1f-ccd4-7dc2  32768   2      305     10111100
GigabitEthernet0/0/2 20000  4c1f-ccd4-7dc2  32768   3      305     10111100
GigabitEthernet0/0/3 20000  4c1f-ccd4-7dc2  32768   4      305     10100000
```

Fig. 3.21 Status of Eth-Trunk 1 interface on switch LSW2

② Execute the "display port vlan" command on the switches LSW1 and LSW2, respectively. Fig. 3.22 shows the port VLAN information of the switch LSW1. It can be seen that the link type is trunk for the Eth-Trunk1 interface, allowing VLAN 1 to VLAN 3.

③ Remove the active link between the switches LSW1 and LSW2 (such as the interface G0/0/1 link) or disable the interface G0/0/0 to simulate a link failure, and then use the "display eth-trunk" command on the switches LSW1 and LSW2 to check the status of the Eth-Trunk interface.

Fig. 3.23 shows the display results on LSW1. It can be seen that the original backup link (G0/0/3) interface has been selected as the active interface, while the interface G0/0/1 is an inactive interface.

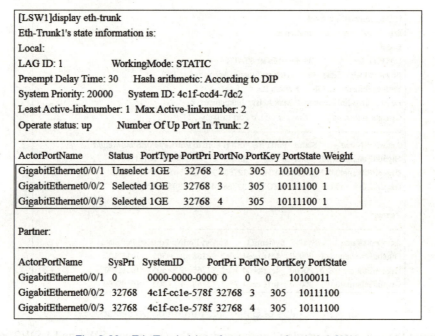

Fig. 3.22 VLAN information of switch LSW1 Port

Fig. 3.23 Eth-Trunk 1 interface status of switch LSW1

Task Completion

1. Task planning

According to the task description, network topology shown in Fig. 3.1, and networking requirements of Project 3, use the LACP mode to plan and configure the link aggregation as shown

in Table 3.4 for the link between the access layer switch SXL-FLOOR2 and the aggregation layer switch SXL-AGG, and the link between the aggregation layer switch SXL-AGG and the core layer switch CORE.

Table 3.4 Planning of LACP link aggregation

Device	Aggregation port ID	Member interface	System LACP priority	Maximum number of active links	Description
CORE	1	G0/0/4, G0/0/5	Default	Default	linkto-sxl-agg
SXL-AGG	1	G0/0/1, G0/0/2	20000	Default	linkto-core
	2	G0/0/4, G0/0/5	20000	Default	linkto-sxl-floor2
SXL-FLOOR2	1	G0/0/1, G0/0/2	Default	Default	linkto-sxl-agg

MSTP is planned as follows:

① The switch spanning trees all use the MSTP protocol.

② Set the MSTP domain name and revision level of the administrative building switch as xzl and 1 respectively. Instance 1 includes VLAN2, VLAN3, VLAN4, VLAN5, VLAN90 and VLAN91; The aggregation layer switch XZL-AGG is the main root of instance 0 and instance 1, and the port of the access layer switch to the terminal is configured as a portfast, with BPDU protection enabled.

③ Set the MSTP domain name and revision level of the teaching building switch as jxl and 1 respectively. Instance 1 includes VLAN12, VLAN13, VLAN92 and VLAN93; The aggregation layer switch JXL-AGG is the main root of instance 0 and instance 1, and the port of the access layer switch to the terminal is configured as a portfast, with BPDU protection enabled.

④ Set the MSTP domain name and revision level of the training building switch as sxl and 1 respectively. Instance 1 includes VLAN22, VLAN23, VLAN94 and VLAN95; The aggregation layer switch SXL-AGG is the main root of instance 0 and instance 1, and the port of the access layer switch to the terminal is configured as a portfast, with BPDU protection enabled.

⑤ Set the MSTP domain names and revision levels as server and 1 respectively for the core layer switch CORE and server access layer switch SERVER-ACC. Instance 1 includes VLAN202; The core layer switch CORE is the main root of instance 0 and instance 1, and the port of the server access layer switch to the terminal is configured as a portfast, with BPDU protection enabled.

⑥ The STP function is disabled on the link port between the core layer switch and the aggregation layer switch, and also on the port where the core layer switch connects to the edge router.

2. Task implementation

(1) Configuration of link aggregation

① On the core layer switch CORE, configure the link aggregation port Eth-Trunk1 according to the link aggregation plan in Table 3.4 and the VLAN plan in Table 3.1 (the aggregation port Eth-

Trunk1 is in the ACCESS mode and belongs to VLAN199). The steps are as follows:

```
[CORE] interface Eth-Trunk 1
[CORE-Eth-Trunk1]mode lacp-static
[CORE-Eth-Trunk1]trunkport g0/0/4
[CORE-Eth-Trunk1]trunkport g0/0/5
[CORE-Eth-Trunk1]port link-type access
[CORE-Eth-Trunk1]port default vlan 199
[CORE-Eth-Trunk1]description linkto-sxl-agg
[CORE-Eth-Trunk1]quit
//Create the aggregation port Eth-Trunk1. The aggregation mode is LACP. The member
//interfaces are G0/0/4 and G0/0/5. The link type is ACCESS. It belongs to VLAN199
//and is described as "linkto-sxl-agg"
```

② On the aggregation layer switch SXL-AGG, configure the link aggregation ports Eth-Trunk1 and Eth-Trunk2 according to the link aggregation plan in Table 3.4 and the VLAN plan in Table 3.1 (The aggregation port Eth-Trunk1 is in ACCESS mode, subject to VLAN199; The aggregation port Eth-Trunk2 is in TRUNK mode, allowing VLAN22, VLAN23, VLAN94 and VLAN95 access). The steps are as follows:

```
[SXL-AGG]lacp priority 20000
//Configure the LACP priority to 20000
[SXL-AGG]interface Eth-Trunk 1
[SXL-AGG-Eth-Trunk1]mode lacp-static
[SXL-AGG-Eth-Trunk1]trunkport g0/0/1
[SXL-AGG-Eth-Trunk1]trunkport g0/0/2
[SXL-AGG-Eth-Trunk1]description linkto-core
[SXL-AGG-Eth-Trunk1]port link-type access
[SXL-AGG-Eth-Trunk1]port default vlan 199
[SXL-AGG-Eth-Trunk1]quit
//Create the aggregation port Eth-Trunk1. The aggregation mode is LACP. The member
//interfaces are G0/0/1 and G0/0/2. The link type is ACCESS. It belongs to VLAN199
//and is described as "linkto-core"
[SXL-AGG]interface Eth-Trunk 2
[SXL-AGG-Eth-Trunk2]mode lacp-static
[SXL-AGG-Eth-Trunk2]trunkport g0/0/4
[SXL-AGG-Eth-Trunk2]trunkport g0/0/5
[SXL-AGG-Eth-Trunk2]port link-type trunk
[SXL-AGG-Eth-Trunk2]port trunk allow-pass vlan 22 23 94 95
[SXL-AGG-Eth-Trunk2]description linkto-sxl-floor2
[SXL-AGG-Eth-Trunk2]quit
//Create the aggregation port Eth-Trunk2. The aggregation mode is LACP. The member
```

```
//interfaces are G0/0/4 and G0/0/5. The link type is TRUNK. It allows VLAN22, VLAN23,
//VLAN94 and VLAN95 and is described as "linkto-sxl-floor2"
```

③ On the access layer switch SXL-FLOOR2, configure the link aggregation port Eth-Trunk1 according to the link aggregation plan in Table 3.4 and the VLAN plan in Table 3.1 (The aggregation port Eth-Trunk1 is in TRUNK mode, allowing VLAN22, VLAN23, VLAN94 and VLAN95 access). The steps are as follows:

```
[SXL-FLOOR2]interface Eth-Trunk 1
[SXL-FLOOR2-Eth-Trunk1]mode lacp-static
[SXL-FLOOR2-Eth-Trunk1]trunkport g0/0/1
[SXL-FLOOR2-Eth-Trunk1]trunkport g0/0/2
[SXL-FLOOR2-Eth-Trunk1]port link-type trunk
[SXL-FLOOR2-Eth-Trunk1]port trunk allow-pass vlan 22 23 94 95
[SXL-FLOOR2-Eth-Trunk1]description linkto-sxl-agg
[SXL-FLOOR2-Eth-Trunk1]quit
//Create the aggregation port Eth-Trunk1. The aggregation mode is LACP. The member
//interfaces are G0/0/4 and G0/0/5. The link type is TRUNK. It allows VLAN22, VLAN23,
//VLAN94 and VLAN95 and is described as "linkto-sxl-agg"
```

(2) Verification of link aggregation

① After the link aggregation is configured, use the "display eth-trunk" command on the CORE, SXL-AGG and SXL-FLOOR2 switches to view the link aggregation Eth-Trunk interface information. Fig. 3.24 shows the Eth-Trunk1 interface information displayed on the switch SXL-AGG. Fig. 3.25 shows the Eth-Trunk1 interface information displayed on the switch SXL FLOOR2.

```
[SXL-AGG]display eth-trunk 1
Eth-Trunk1's state information is:
Local:
LAG ID: 1              WorkingMode: STATIC
Preempt Delay: Disabled    Hash arithmetic: According to SIP-XOR-DIP
System Priority: 20000    System ID: 4c1f-cc95-3bd2
Least Active-linknumber: 1  Max Active-linknumber: 8
Operate status: up     Number Of Up Port In Trunk: 2
--------------------------------------------------------------------
ActorPortName      Status  PortType PortPri PortNo PortKey PortState Weight
GigabitEthernet0/0/1  Selected 1GE    32768   2      305    10111100  1
GigabitEthernet0/0/2  Selected 1GE    32768   3      305    10111100  1

Partner:
--------------------------------------------------------------------
ActorPortName       SysPri  SystemID       PortPri PortNo PortKey PortState
GigabitEthernet0/0/1 32768  4c1f-cc89-64b2  32768   5     305    10111100
GigabitEthernet0/0/2 32768  4c1f-cc89-64b2  32768   6     305    10111100
```

Fig. 3.24　Eth-Trunk 1 interface information displayed on switch SXL-AGG

```
[SXL-FLOOR2]display eth-trunk 1
Eth-Trunk1's state information is:
Local:
LAG ID: 1                WorkingMode: STATIC
Preempt Delay: Disabled  Hash arithmetic: According to SIP-XOR-DIP
System Priority: 32768   System ID: 4c1f-cc49-5291
Least Active-linknumber: 1  Max Active-linknumber: 8
Operate status: up       Number Of Up Port In Trunk: 2
--------------------------------------------------------------
ActorPortName       Status    PortType PortPri PortNo PortKey PortState Weight
GigabitEthernet0/0/1 Selected 1GE      32768   2      305     10111100  1
GigabitEthernet0/0/2 Selected 1GE      32768   3      305     10111100  1

Partner:
--------------------------------------------------------------
ActorPortName       SysPri  SystemID        PortPri PortNo PortKey PortState
GigabitEthernet0/0/1 20000  4c1f-cc95-3bd2  32768   5      561     10111100
GigabitEthernet0/0/2 20000  4c1f-cc95-3bd2  32768   6      561     10111100
```

Fig. 3.25 Eth-Trunk 1 interface information displayed on switch SXL-FLOOR2

② Use the "display ip interface brief" command to display the interface IP brief information on the switches CORE and SXL-AGG. Fig. 3.26 shows the display results via the "display ip interface brief" command on the aggregation layer switch SXL-AGG, and Fig. 3.27 shows the display results via the "display ip interface brief" command on the core layer switch CORE.

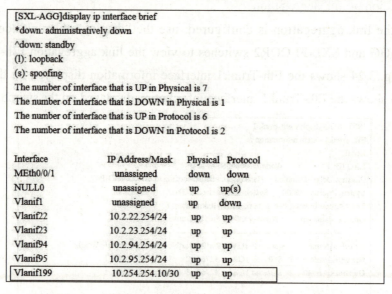

```
[SXL-AGG]display ip interface brief
*down: administratively down
^down: standby
(l): loopback
(s): spoofing
The number of interface that is UP in Physical is 7
The number of interface that is DOWN in Physical is 1
The number of interface that is UP in Protocol is 6
The number of interface that is DOWN in Protocol is 2

Interface        IP Address/Mask      Physical  Protocol
MEth0/0/1        unassigned           down      down
NULL0            unassigned           up        up(s)
Vlanif1          unassigned           up        down
Vlanif22         10.2.22.254/24       up        up
Vlanif23         10.2.23.254/24       up        up
Vlanif94         10.2.94.254/24       up        up
Vlanif95         10.2.95.254/24       up        up
Vlanif199        10.254.254.10/30     up        up
```

Fig. 3.26 Display results via the "display ip interface brief" command on the aggregation layer switch SXL-AGG

```
[CORE]display ip interface brief
*down: administratively down
^down: standby
(l): loopback
(s): spoofing
The number of interface that is UP in Physical is 6
The number of interface that is DOWN in Physical is 2
The number of interface that is UP in Protocol is 6
The number of interface that is DOWN in Protocol is 2

Interface         IP Address/Mask    Physical  Protocol
MEth0/0/1         unassigned         down      down
NULL0             unassigned         up        up(s)
Vlanif1           unassigned         down      down
Vlanif199         10.254.254.9/30    up        up
Vlanif200         10.254.254.5/30    up        up
Vlanif201         10.254.254.1/30    up        up
Vlanif202         10.254.253.254/24  up        up
Vlanif203         10.254.254.13/30   up        up
```

Fig. 3.27 Display results via the "display ip interface brief" command on the core layer switch CORE

(3) Configuration of MSTP

① Complete MSTP configurations on the switch in the administrative building according to the MSTP plan. Set the MSTP domain names and revision levels of all switches as xzl and 1 respectively. Instance 1 includes VLAN2, VLAN3, VLAN4, VLAN5, VLAN90 and VLAN91; The aggregation layer switch XZL-AGG is the main root of instance 0 and instance 1, and the interface G0/0/1 needs to have the spanning tree function disabled; The interface (G0/0/2 to G0/0/24) of the access layer switches XZL-FLOOR1 and XZL-FLOOR2 connected to the terminal devices are spanning tree portfasts, with BPDU protection disabled. The steps are as follows for configuring the administrative building switch MSTP:

```
[XZL-AGG]stp enable
[XZL-AGG]stp mode mstp
//Enable the spanning tree, subject to the mode MSTP
[XZL-AGG]stp region-configuration
[XZL-AGG-mst-region]region-name xzl
[XZL-AGG-mst-region]revision-level 1
[XZL-AGG-mst-region]instance 1 vlan 2 3 4 5 90 91
[XZL-AGG-mst-region]active region-configuration
//Configure the MSTP domain name, revision level, and VLAN corresponding to instance
//1, and activate the MSTP domain
[XZL-AGG]interface g0/0/1
[XZL-AGG-GigabitEthernet0/0/1]stp disable
//Disable the spanning tree function on the interface G0/0/1 of the switch XZL-AGG
[XZL-AGG]stp instance 0 root primary
```

```
[XZL-AGG]stp instance 1 root primary
//Configure the switch XZL-AGG as the tree root of instance 0 and instance 1.
[XZL-FLOOR1]stp enable
[XZL-FLOOR1]stp mode mstp
//Enable the spanning tree, subject to the mode MSTP
[XZL-FLOOR1]stp region-configuration
[XZL-FLOOR1-mst-region]region-name xzl
[XZL-FLOOR1-mst-region]revision-level 1
[XZL-FLOOR1-mst-region]instance 1 vlan 2 3 4 5 90 91
[XZL-FLOOR1-mst-region]active region-configuration
[XZL-FLOOR1-mst-region]quit
//Configure the MSTP domain name, revision level, and VLAN corresponding to instance
//1, and activate the MSTP domain
[XZL-FLOOR1]port-group 4
[XZL-FLOOR1-port-group-4]group-member g0/0/2 to g0/0/24
[XZL-FLOOR1-port-group-4]stp edged-port enable
[XZL-FLOOR1-port-group-4]quit
[XZL-FLOOR1]stp bpdu-protection
//Configure the interfaces (G0/0/2 to G0/0/24) of the switch XZL-FLOOR1 as the
//spanning tree portfasts with BPDU protection enabled
[XZL-FLOOR2]stp enable
[XZL-FLOOR2]stp mode mstp
//Enable the spanning tree, subject to the mode MSTP
[XZL-FLOOR2]stp region-configuration
[XZL-FLOOR2-mst-region]region-name xzl
[XZL-FLOOR2-mst-region]revision-level 1
[XZL-FLOOR2-mst-region]instance 1 vlan 2 3 4 5 90 91
[XZL-FLOOR2-mst-region]active region-configuration
[XZL-FLOOR2-mst-region]quit
//Configure the MSTP domain name, revision level, and VLAN corresponding to instance
//1, and activate the MSTP domain
[XZL-FLOOR2]port-group 4
[XZL-FLOOR2-port-group-4]group-member g0/0/2 to g0/0/24
[XZL-FLOOR2-port-group-4]stp edged-port enable
[XZL-FLOOR2-port-group-4]quit
[XZL-FLOOR2]stp bpdu-protection
//Configure the interfaces (G0/0/2 to G0/0/24) of the switch XZL-FLOOR2 as the spanning
//tree portfasts with BPDU protection enabled
```

② Complete MSTP configurations on the switch in the teaching building according to

the MSTP plan. Set the MSTP domain names and revision levels of all switches as jxl and 1 respectively. Instance 1 includes VLAN12, VLAN13, VLAN92 and VLAN93; The aggregation layer switch JXL-AGG is the main root of instance 0 and instance 1, and the interface G0/0/1 needs to have the spanning tree function disabled; the interfaces (G0/0/2 to G0/0/24) of the access layer switches JXL-FLOOR1 and JXL-FLOOR2 connected to the terminal devices are spanning tree portfasts, with BPDU protection enabled. The steps are as follows for configuring the teaching building switch MSTP:

```
[JXL-AGG]stp enable
[JXL-AGG]stp mode mstp
//Enable the spanning tree, subject to the mode MSTP
[JXL-AGG]stp region-configuration
[JXL-AGG-mst-region]region-name jxl
[JXL-AGG-mst-region]revision-level 1
[JXL-AGG-mst-region]instance 1 vlan 12 13 92 93
[JXL-AGG-mst-region]active region-configuration
[JXL-AGG-mst-region]quit
//Configure the MSTP domain name, revision level, and VLAN corresponding to instance
//1, and activate the MSTP domain
[JXL-AGG]stp instance 0 root primary
[JXL-AGG]stp instance 1 root primary
//Configure the switch JXL-AGG as the tree root of instance 0 and instance 1
[JXL-AGG]interface g0/0/1
[JXL-AGG-GigabitEthernet0/0/1]stp disable
[JXL-AGG-GigabitEthernet0/0/1]quit
//Disable the spanning tree function on the G0/0/1 interface of the switch JXL-AGG
[JXL-FLOOR1]stp enable
[JXL-FLOOR1]stp mode mstp
//Enable the spanning tree, subject to the mode MSTP
[JXL-FLOOR1]stp region-configuration
[JXL-FLOOR1-mst-region]region-name jxl
[JXL-FLOOR1-mst-region]revision-level 1
[JXL-FLOOR1-mst-region]instance 1 vlan 12 13 92 93
[JXL-FLOOR1-mst-region]active region-configuration
[JXL-FLOOR1-mst-region]quit
//Configure the MSTP domain name, revision level, and VLAN corresponding to instance
//1, and activate the MSTP domain
[JXL-FLOOR1]port-group 4
[JXL-FLOOR1-port-group-4]group-member g0/0/2 to g0/0/24
[JXL-FLOOR1-port-group-4]stp edged-port enable
```

```
[JXL-FLOOR1]stp bpdu-protection
// Configure MSTP of the access layer switch JXL-FLOOR1 in the teaching building
[JXL-FLOOR2]stp enable
[JXL-FLOOR2]stp mode mstp
//Enable the spanning tree, subject to the mode MSTP
[JXL-FLOOR2]stp region-configuration
[JXL-FLOOR2-mst-region]region-name jxl
[JXL-FLOOR2-mst-region]revision-level 1
[JXL-FLOOR2-mst-region]instance 1 vlan 12 13 92 93
[JXL-FLOOR2-mst-region]active region-configuration
[JXL-FLOOR2-mst-region]quit
//Configure the MSTP domain name, revision level, and VLAN corresponding to instance
//1, and activate the MSTP domain
[JXL-FLOOR2]port-group 4
[JXL-FLOOR2-port-group-4]group-member g0/0/2 to g0/0/24
[JXL-FLOOR2-port-group-4]stp edged-port enable
[JXL-FLOOR2]stp bpdu-protection
//Configure the interfaces (G0/0/2 to G0/0/24) of the switch JXL-FLOOR2 as the spanning
//tree portfasts with BPDU protection enabled
```

③ Complete MSTP configurations on the switch in the training building according to the MSTP plan. Set the MSTP domain names and revision levels of all switches as sxl and 1 respectively. Instance 1 includes VLAN22, VLAN23, VLAN94 and VLAN95; The aggregation layer switch SXL-AGG is the main root of instance 0 and instance 1, and its Eth-Trunk 1 port shall have the spanning tree function disabled; The ports of the access layer switches SXL-FLOOR1 and JXL-FLOOR2 connected to the terminal devices (G0/0/2 to G0/0/24 interfaces of SXL-FLOOR1, and G0/0/3 to G0/0/24 interfaces of SXL-FLOOR2) are spanning tree portfasts, with BPDU protection enabled. The steps are as follows for switch MSTP configuration in the training building:

```
[SXL-AGG]stp enable
[SXL-AGG]stp mode mstp
//Enable the spanning tree, subject to the mode MSTP
[SXL-AGG]stp region-configuration
[SXL-AGG-mst-region]region-name sxl
[SXL-AGG-mst-region]revision-level 1
[SXL-AGG-mst-region]instance 1 vlan 22 23 94 95
[SXL-AGG-mst-region]active region-configuration
[SXL-AGG-mst-region]quit
//Configure the MSTP domain name, revision level, and VLAN corresponding to instance
//1, and activate the MSTP domain
```

```
[SXL-AGG]stp instance 0 root primary
[SXL-AGG]stp instance 1 root primary
//Configure the switch SXL-AGG as the tree root of instance 0 and instance 1
[SXL-AGG]interface Eth-Trunk 1
[SXL-AGG-Eth-Trunk1]stp disable
[SXL-AGG-Eth-Trunk1]quit
//Disable the spanning tree function on the Eth-Trunk 1 port of the switch SXL-AGG
[SXL-FLOOR1]stp enable
[SXL-FLOOR1]stp mode mstp
//Enable the spanning tree, subject to the mode MSTP
[SXL-FLOOR1]stp region-configuration
[SXL-FLOOR1-mst-region]region-name sxl
[SXL-FLOOR1-mst-region]revision-level 1
[SXL-FLOOR1-mst-region]instance 1 vlan 22 23 94 95
[SXL-FLOOR1-mst-region]active region-configuration
[SXL-FLOOR1-mst-region]quit
//Configure the MSTP domain name, revision level, and VLAN corresponding to instance
//1, and activate the MSTP domain
[SXL-FLOOR1]port-group 4
[SXL-FLOOR1-port-group-4]group-member g0/0/2 to g0/0/24
[SXL-FLOOR1-port-group-4]stp edged-port enable
[SXL-FLOOR1]stp bpdu-protection
//Configure the interfaces (G0/0/2 to G0/0/24) of the switch SXL-FLOOR1 as the spanning
//tree portfasts with BPDU protection enabled
[SXL-FLOOR2]stp enable
[SXL-FLOOR2]stp mode mstp
//Enable the spanning tree, subject to the mode MSTP
[SXL-FLOOR2]stp region-configuration
[SXL-FLOOR2-mst-region]region-name sxl
[SXL-FLOOR2-mst-region]revision-level 1
[SXL-FLOOR2-mst-region]instance 1 vlan 22 23 94 95
[SXL-FLOOR2-mst-region]active region-configuration
[SXL-FLOOR2-mst-region]quit
//Configure the MSTP domain name, revision level, and VLAN corresponding to Instance
//1, and activate the MSTP domain.
[SXL-FLOOR2]port-group 4
[SXL-FLOOR2-port-group-4]group-member g0/0/3 to g0/0/24
[SXL-FLOOR2-port-group-4]stp edged-port enable
[SXL-FLOOR2-port-group-4]quit
[SXL-FLOOR2]stp bpdu-protection
```

```
//Configure the interfaces (G0/0/3 to G0/0/24) of the switch SXL-FLOOR2 as the spanning
//tree portfasts with BPDU protection enabled.
```

④ Complete MSTP configurations on the core layer switch CORE and server access layer switch SERVER-ACC according to the MSTP plan. Set the MSTP domain names and revision levels of the switches CORE and SERVER-ACC as server and 1 respectively. Instance 1 includes VLAN202; The core layer switch CORE is the main root of instance 0 and instance 1, and its interfaces (Eth-Trunk 1, G0/0/1, G0/0/2 and G0/0/3) have the spanning tree function disabled (these ports are connected to the connection network segment ports of the aggregation layer switch and edge router EDGE-AR); the interfaces (G0/0/1 to G0/0/3 interfaces) with the server access layer switch SERVER-ACC connected to the server act as spanning tree portfasts with BPDU protection enabled. The steps are as follows for MSTP configurations of the core switch and server access layer switch:

```
[CORE]stp enable
[CORE]stp mode mstp
//Enable the spanning tree, subject to the mode MSTP
[CORE]stp region-configuration
[CORE-mst-region]region-name server
[CORE-mst-region]revision-level 1
[CORE-mst-region]instance 1 vlan 202
[CORE-mst-region]active region-configuration
[CORE-mst-region]quit
//Configure the MSTP domain name, revision level, and VLAN corresponding to instance
//1, and activate the MSTP domain
[CORE]stp instance 0 root primary
[CORE]stp instance 1 root primary
//Configure the switch CORE as the tree root of Instance 0 and Instance 1.
[CORE]port-group 1
[CORE-port-group-1]group-member g0/0/1 to g0/0/3 Eth-Trunk 1
[CORE-port-group-1]stp disable
[CORE-port-group-1]quit
//Disable the spanning tree function on the interfaces (G0/0/1 to G0/0/3 and Eth-Trunk 1)
//of the switch CORE
[SERVER-ACC]stp enable
[SERVER-ACC]stp mode mstp
//Enable the spanning tree, subject to the mode MSTP
[SERVER-ACC]stp region-configuration
[SERVER-ACC-mst-region]region-name server
[SERVER-ACC-mst-region]revision-level 1
```

```
[SERVER-ACC-mst-region]instance 1 vlan 202
[SERVER-ACC-mst-region]active region-configuration
[SERVER-ACC-mst-region]quit
//Configure the MSTP domain name, revision level, and VLAN corresponding to instance
//1, and activate the MSTP domain
[SERVER-ACC]port-group 1
[SERVER-ACC-port-group-1]group-member g0/0/1 to g0/0/3
[SERVER-ACC-port-group-1]stp edged-port enable
[SERVER-ACC-port-group-1]quit
[SERVER-ACC]stp bpdu-protection
//Configure the ports (G0/0/1 to G0/0/3) of the switch SERVER-ACC as the spanning tree
//portfasts with BPDU protection enabled
```

(4) Verification and testing

① Execute the "display stp region-configuration" command on all switches respectively to view the configuration information of the MSTP domain. The switch MSTP domain shall be configured identically in the same building. Fig. 3.28 shows the MSTP domain configuration on the aggregation layer switch SXL-AGG of the training building. Fig. 3.29 shows the MSTP domain configuration on the access layer switch SXL-FLOOR2 of the training building.

```
[SXL-AGG]display stp region-configuration
Oper configuration
  Format selector   :0
  Region name       :sxl
  Revision level    :1

  Instance   VLANs Mapped
     0       1 to 21, 24 to 93, 96 to 4094
     1       22 to 23, 94 to 95
```

Fig. 3.28 MSTP domain configuration on the aggregation layer switch SXL-AGG in the training building

```
[SXL-FLOOR2]display stp region-configuration
Oper configuration
  Format selector   :0
  Region name       :sxl
  Revision level    :1

  Instance   VLANs Mapped
     0       1 to 21, 24 to 93, 96 to 4094
     1       22 to 23, 94 to 95
```

Fig. 3.29 MSTP domain configuration on the access layer switch SXL-FLOOR2 in the training building

② Execute the "display stp brief" command on all switches respectively to view the brief information of the spanning tree. Fig. 3.30 shows the brief information of the spanning tree on the

aggregation layer switch SXL-AGG of the training building. Fig. 3.31 shows the brief information of the spanning tree on the access layer switch SXL-FLOOR2 of the training building.

```
[SXL-AGG]display stp brief
 MSTID  Port                Role  STP State   Protection
   0    GigabitEthernet0/0/3  DESI  FORWARDING  NONE
   0    Eth-Trunk2            DESI  FORWARDING  NONE
   1    GigabitEthernet0/0/3  DESI  FORWARDING  NONE
   1    Eth-Trunk2            DESI  FORWARDING  NONE
```

Fig. 3.30　Spanning tree brief information on the aggregation layer switch SXL-AGG of the training building

```
[SXL-FLOOR2]display stp brief
 MSTID  Port                 Role  STP State   Protection
   0    GigabitEthernet0/0/19  DESI  FORWARDING  BPDU
   0    Eth-Trunk1             ROOT  FORWARDING  NONE
   1    GigabitEthernet0/0/19  DESI  FORWARDING  BPDU
   1    Eth-Trunk1             ROOT  FORWARDING  NONE
```

Fig. 3.31　Spanning tree brief information on the access layer switch SXL-FLOOR2 of the training building

Exercises

Multiple choice

(1) The following (　　) is used to view the status of the Eth-Trunk interface.

　　A. display eth-trunk　　　　　　　　B. display current-configuration

　　C. display this　　　　　　　　　　　D. display ip interface brief

(2) When the LACP mode is used for link aggregation, the default system priority of Huawei switches is (　　).

　　A. 0　　　　　　B. 64　　　　　　C. 4096　　　　　　D. 32768

(3) Fig. 3.17 shows the output information of the Eth-Trunk 1 interface on the switch LSW1. If you want to delete Eth-Trunk 1, the correct command is (　　).

　　A. interface GigabitEthernet 0/0/1

　　　　undo eth-trunk

　　　　quit

　　　　undo interface Eth-Trunk

　　B. interface GigabitEthernet 0/0/1

　　　　undo eth-trunk

　　　　quit

　　　　interface GigabitEthernet 0/0/2

　　　　undo eth- trunk

　　　　quit

interface GigabitEthernet 0/0/3

undo eth- trunk

quit

undo interface Eth-Trunk 1

C. undo interface Eth-Trunk 1

D. inter GigabitEthernet 0/0/1

undo eth trunk

quit undo interface Eth-Trunk 1

Planning and Configuration of OSPF

Task Description

On the basis of completing the Task 1 and Task 2 of this project, in the network shown in Fig. 3.1, complete the routing planning and configuration tasks of using default and dynamic routing protocol OSPF to achieve interconnection in the enterprise intranet and enable the enterprise intranet to have access to the Internet.

Task Objectives

- Understand the basic concepts of OSPF.
- Understand the working process of OSPF.
- Master the configuration of single domain OSPF.
- Master the verification and troubleshooting of single domain OSPF.

Related Knowledge

1. Overview of OSPF

OSPF (open shortest path first) is a link state routing protocol based on open standards.

OSPF adopts a link state routing algorithm, where each OSFP router uses the HELLO protocol to identify neighboring routers, establish communications with them and inform all other OSPF routers in the same area of its own link state and received link state information through flooding. When the OSPF routers located in the same area have a complete link state database, that is, after a unified network topology is obtained, each OSPF router will take the local router as the root and use the SPF (shortest path first) algorithm to calculate the shortest path to each destination network and then fill the IP routing table with the best path to each network based on the SPF tree, as shown in Fig. 3.32.

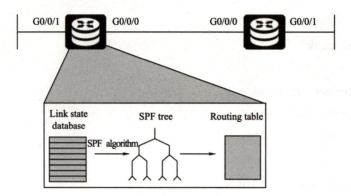

Fig. 3.32 OSPF via SPF algorithm

2. OSPF router working process

The working process of OSPF routers can be roughly divided into three steps:

Firstly, each OSPF router generates an LSA (link state advertisement) that describes its own interface state.

Secondly, OSPF routers achieve synchronization of LSDB (link state database) by exchanging LSAs, and all OSPF routers in the same area have the same LSDB.

Finally, the OSPF router calculates the route via SPF based on LSDB and inserts it into the routing table.

The working process of OSPF includes such stages as neighbor discovery, routing information exchange, route calculation and route maintenance. During these stages, each OSPF router needs to maintain the three basic data structures related to OSPF implementation, namely the adjacencies database, link-state database and route table.

① Adjacencies database: Used to store all neighboring routers that have established two-way communications with the router.

② LSDB also known as topological database, is used to store the link state information about all other routers in the OSPF network. The database displays the topology of the entire network. When the network converges, all OSPF routers in the same OSPF region will have the same link-state database.

③ Route table: The routes generated by running the SPF algorithm on the link-state database are inserted into the route table.

3. Important concepts related to OSPF

The working process of OSPF involves such concepts as follows:

(1) Link

It refers to a network communication channel composed of circuits and transmission paths.

(2) Link-state

It refers to the state of the link between two routers or two router interfaces, as well as the connection between a router and its neighboring routers.

(3) Cost

It refers to the metric value assigned to the OSPF link. By default, the cost is calculated based on the bandwidth on the interface, namely, "10^8/interface bandwidth". Administrators can manually configure the cost value for the link.

(4) SPF algorithm

The SPF algorithm is also known as the shortest-path-first algorithm (Dijkstra algorithm), in which each OSPF router uses itself as the root node to calculate the shortest path from the root node to each network. When there are multiple reachable network paths, the path with the minimum cost accumulation value is considered the shortest path.

(5) Network type

The OSPF interface automatically recognizes three types of networks, namely BMA (broadcast multiaccess), NBMA (nonbroadcast multiaccess) and P2P (point to point) networks.

(6) DR (designated router) and BDR (backup designated router)

In a multiaccess (such as Ethernet) OSPF network environment, as there can be multiple OSPF routers, a lot of additional cost will incur when each router establishes an adjacent relationship with all other routers. Therefore, in a multiaccess network, one OSPF router is selected as DR, and another OSFP router as BDR. Both DR and BDR establish an adjacent relationship with all other routers in the network. DR sends LSA packets (i.e. Class-2 LSA) to other networks regarding all local routers. BDR is used as a backup for DR to improve network fault tolerance. But BDR is not responsible for sending routing update information to other routers, nor does it send network LSA. When DR fails, BDR takes over the role of DR, and then the network will elect another BDR.

Note: DR or BDR will not be elected in point-to-point networks.

Principle for selecting DR and BDR in BMA and NBMA networks: The OSPF router with the highest priority is DR, and the OSPF router with the second highest priority is BDR. If the priorities are the same, the one with the highest router ID is DR, and the one with the second highest router ID is BDR.

(7) OSPF router ID

The OSPF router ID is used to uniquely identify each router in OSPF. A router ID is actually an IP address. The ROUTER-ID of the router can be configured via commands or automatically selected.

Related Skills

1. Planning and configuration of single domain OSPF based on routers

Construct a network topology as shown in Fig. 3.33, and use the single domain OSPF routing protocol to achieve network interconnection for routing view and testing.

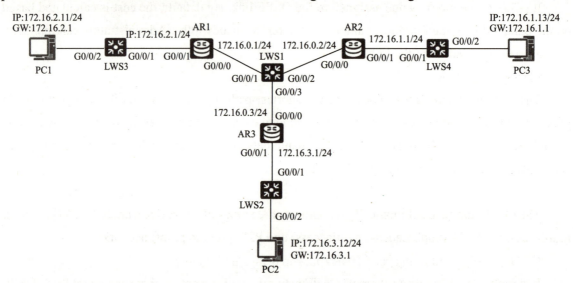

Fig. 3.33 Single domain OSPF network topology

(1) Planning of OSPF

① Use single domain OSPF with the region number 0.

② Set ROUTER-ID as 1.1.1.1, 2.2.2.2 and 3.3.3.3 for AR1, AR2 and AR3 respectively.

③ Configure the G0/0/1 interfaces of AR1, AR2, and AR3 as silent interfaces, forwarding no routing update.

④ Select the priority values as 10 and 5 respectively for AR1 and AR2 regarding G0/0/0 interface DR.

(2) Configuration of router interfaces

Configure the interfaces on the routers AR1, AR2 and AR3 based on the IP planned in Fig. 3.33, as follows:

```
[AR1]interface g0/0/0
[AR1-GigabitEthernet0/0/0]ip address 172.16.0.1 24
[AR1-GigabitEthernet0/0/0]quit
//Configure the G0/0/0 interface IP address and mask of the router AR1 to be "172.16.0.1/24"
[AR1]interface g0/0/1
[AR1-GigabitEthernet0/0/1]ip address 172.16.2.1 24
[AR1-GigabitEthernet0/0/1]quit
//Configure the G0/0/1 interface IP address and mask of the router AR1 to be
//"172.16.2.1/24"
```

```
[AR2]interface g0/0/0
[AR2-GigabitEthernet0/0/0]ip address 172.16.0.2 24
[AR2-GigabitEthernet0/0/0]quit
//Configure the G0/0/0 interface IP address and mask of the router AR2 to be
//"172.16.0.1/24"
[AR2]interface g0/0/1
[AR2-GigabitEthernet0/0/1]ip address 172.16.1.1 24
[AR2-GigabitEthernet0/0/1]quit
//Configure the G0/0/1 interface IP address and mask of the router AR2 to be
//"172.16.1.1/24"
[AR3]interface g0/0/0
[AR3-GigabitEthernet0/0/0]ip address 172.16.0.3 24
[AR3-GigabitEthernet0/0/0]quit
//Configure the G0/0/0 interface IP address and mask of the router AR3 to be
//"172.16.0.3/24"
[AR3]interface g0/0/1
[AR3-GigabitEthernet0/0/1]ip address 172.16.3.1 24
[AR3-GigabitEthernet0/0/1]quit
//Configure the G0/0/1 interface IP address and mask of the router AR3 to be
//"172.16.3.1/24"
[AR3]interface LoopBack 0
[AR3-LoopBack0]ip address 3.3.3.3 32
[AR3-LoopBack0]quit
//Configure the LoopBack0 interface IP address and mask of the router AR3 to be
//"3.3.3.3/32"
```

(3) Configuration of single domain OSPF

Such tasks as follows shall be completed for configuration of single domain OSPF:

① Modify the priority value of the router interface when selecting DR (optional).

② Create and run the OSPF process, specify the OSPF process number, and configure ROUTER-ID of the router. Set the default OSPF process number as 1, and configure ROUTER-ID of the router (optional).

③ Create and enter the OSPF area.

④ Specify the interface for running OSPF.

⑤ Configure the router interface as a silent interface (optional).

The steps for configuration are as follows:

```
[AR1]interface g0/0/0
[AR1-GigabitEthernet0/0/0]ospf-dr-priority 10   //Set the priority value of the G0/0/0
```

```
                                              //interface when selecting DR as 10
                                              //(1 by default)
[AR1]ospf 1 router-id 1.1.1.1         //Create and run the OSPF process 1, and
                                      //configure the router's ROUTER-ID as "1.1.1.1"
[AR1-ospf-1]area 0                    //Create and enter the OSPF area 0
[AR1-ospf-1-area-0.0.0.0]network 172.16.2.0 0.0.0.255
//Let the network 172.16.2.0/24 interface participate in the OSPF process and run in
//the area 0
[AR1-ospf-1-area-0.0.0.0]network 172.16.0.0 0.0.0.255
// Let the network 172.16.0.0/24 interface participate in the OSPF process and run in
// the area 0
[AR1-ospf-1-area-0.0.0.0]quit
[AR1-ospf-1]silent-interface g0/0/1   //Configure the interface as a silent interface
[AR2]interface g0/0/0
[AR2-GigabitEthernet0/0/0]ospf dr-priority 5    //Set the G0/0/0 interface priority as 5
                                                //when selecting DR
[AR2]ospf 1 router-id 2.2.2.2         //Create and run the OSPF process 1, and configure
                                      //the router ROUTER-ID as "2.2.2.2"
[AR2-ospf-1]area 0.0.0.0              //Create and enter the OSPF area 0
[AR2-ospf-1-area-0.0.0.0]network 172.16.1.0 0.0.0.255
// Let the network 172.16.1.0/24 interface participate in the OSPF process and run in
// the area 0
[AR2-ospf-1-area-0.0.0.0]network 172.16.0.0 0.0.0.255
// Let the network 172.16.0.0/24 interface participate in the OSPF process and run in
// the area 0
[AR2-ospf-1-area-0.0.0.0]quit
[AR2-ospf-1]silent-interface g0/0/1   //Configure the interface as a silent interface
[AR3]ospf 1 router-id 3.3.3.3    //Create and run the OSPF process 1, and configure
                                 //the router ROUTER-ID as "3.3.3.3"
[AR3-ospf-1]area 0         //Create and enter the OSPF area 0
[AR3-ospf-1-area-0.0.0.0]network 172.16.3.0 0.0.0.255
// Let the network 172.16.3.0/24 interface participate in the OSPF process and run in
// the area 0
[AR3-ospf-1-area-0.0.0.0]network 172.16.0.0 0.0.0.255
// Let the network 172.16.0.0/24 interface participate in the OSPF process and run in
// the area 0
[AR3-ospf-1-area-0.0.0.0]quit
[AR3-ospf-1]silent-interface g0/0/1   //Configure the interface as a silent interface
```

(4) Verification and testing

① Execute the "display ospf peer brief " command on AR1, AR2 and AR3 respectively to view the OSPF neighbor table, as shown in Fig. 3.34 to Fig. 3.36. Pay attention to the neighbor ID value of each router and check if the status is FULL.

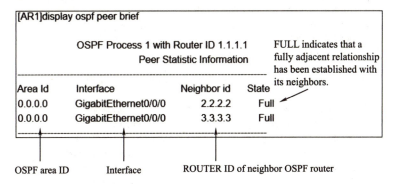

Fig. 3.34   Display results after the "display ospf peer brief " command is executed on AR1

```
[AR2]display ospf peer brief

        OSPF Process 1 with Router ID 2.2.2.2
                Peer Statistic Information
-----------------------------------------------------------
Area Id     Interface              Neighbor id    State
0.0.0.0     GigabitEthernet0/0/0   1.1.1.1        Full
0.0.0.0     GigabitEthernet0/0/0   3.3.3.3        Full
-----------------------------------------------------------
```

Fig. 3.35   Display results after the "display ospf peer brief " command is executed on AR2

```
[AR3]display ospf peer brief

        OSPF Process 1 with Router ID 3.3.3.3
                Peer Statistic Information
-----------------------------------------------------------
Area Id     Interface              Neighbor id    State
0.0.0.0     GigabitEthernet0/0/0   1.1.1.1        Full
0.0.0.0     GigabitEthernet0/0/0   2.2.2.2        Full
-----------------------------------------------------------
```

Fig. 3.36   Display results after the "display ospf peer brief " command is executed on AR3

② Execute the "display ip routing-table" command on AR1, AR2 and AR3 to view the routing table of the router. Fig. 3.37 shows the routing table information on AR1.

③ Execute the "display ip routing-table | include OSPF" command on AR1, AR2 and AR3 to view the routing entries containing OSPF in the routing table. The display results are shown in Fig. 3.38 to Fig. 3.40.

④ Configure the IP address, subnet mask and default gateway of the test hosts PC1, PC2 and PC3 according to the parameters planned in Fig. 3.33. After the configuration is completed, use the "ipconfig" command on the DOS command line to check whether the IP configuration information

is correct. On the DOS command line of one host, use the ping utility to test the IP connectivity to the other two hosts. Fig. 3.41 shows the successful ping of the test host PC1 to the test hosts PC2 and PC3.

```
[AR1]display ip routing-table
Route Flags: R - relay, D - download to fib
--------------------------------------------------------------------------------
Routing Tables: Public
        Destinations : 12      Routes : 12

Destination/Mask    Proto   Pre  Cost     Flags NextHop        Interface

       127.0.0.0/8  Direct  0    0          D   127.0.0.1      InLoopBack0
      127.0.0.1/32  Direct  0    0          D   127.0.0.1      InLoopBack0
127.255.255.255/32  Direct  0    0          D   127.0.0.1      InLoopBack0
      172.16.0.0/24 Direct  0    0          D   172.16.0.1     GigabitEthernet0/0/0
     172.16.0.1/32  Direct  0    0          D   127.0.0.1      GigabitEthernet0/0/0
   172.16.0.255/32  Direct  0    0          D   127.0.0.1      GigabitEthernet0/0/0
      172.16.1.0/24 OSPF    10   2          D   172.16.0.2     GigabitEthernet0/0/0
      172.16.2.0/24 Direct  0    0          D   172.16.2.1     GigabitEthernet0/0/1
     172.16.2.1/32  Direct  0    0          D   127.0.0.1      GigabitEthernet0/0/1
   172.16.2.255/32  Direct  0    0          D   127.0.0.1      GigabitEthernet0/0/1
      172.16.3.0/24 OSPF    10   2          D   172.16.0.3     GigabitEthernet0/0/0
255.255.255.255/32  Direct  0    0          D   127.0.0.1      InLoopBack0
```

Fig. 3.37　Routing table information on AR1

```
[AR1]display ip routing-table | include OSPF
Route Flags: R - relay, D - download to fib
--------------------------------------------------------------------------------
Routing Tables: Public
        Destinations : 12      Routes : 12

Destination/Mask   Proto  Pre  Cost     Flags NextHop      Interface

172.16.1.0/24   OSPF   10   2          D   172.16.0.2    GigabitEthernet0/0/0
172.16.3.0/24   OSPF   10   2          D   172.16.0.3    GigabitEthernet0/0/0
```

Fig. 3.38　Display results after executing the "display ip routing-table | include OSPF" command on AR1

```
[AR2]display ip routing-table | include OSPF
Route Flags: R - relay, D - download to fib
--------------------------------------------------------------------------------
Routing Tables: Public
        Destinations : 12      Routes : 12

Destination/Mask   Proto  Pre  Cost     Flags NextHop      Interface

172.16.2.0/24   OSPF   10   2          D   172.16.0.1    GigabitEthernet0/0/0
172.16.3.0/24   OSPF   10   2          D   172.16.0.3    GigabitEthernet0/0/0
```

Fig. 3.39　Display results after executing the "display ip routing-table | include OSPF" command on AR2

```
[AR3]display ip routing-table | include OSPF
Route Flags: R - relay, D - download to fib
------------------------------------------------
Routing Tables: Public
       Destinations : 13    Routes : 13

Destination/Mask   Proto  Pre  Cost    Flags  NextHop      Interface

172.16.1.0/24  OSPF   10   2        D    172.16.0.2   GigabitEthernet0/0/0
172.16.2.0/24  OSPF   10   2        D    172.16.0.1   GigabitEthernet0/0/0
```

Fig. 3.40 Display results after executing the "display ip routing-table | include OSPF" command on AR3

```
PC>ping 172.16.3.12

Ping 172.16.3.12: 32 data bytes, Press Ctrl_C to break
From 172.16.3.12: bytes=32 seq=1 ttl=126 time=78 ms
From 172.16.3.12: bytes=32 seq=2 ttl=126 time=94 ms
From 172.16.3.12: bytes=32 seq=3 ttl=126 time=125 ms
From 172.16.3.12: bytes=32 seq=4 ttl=126 time=125 ms
From 172.16.3.12: bytes=32 seq=5 ttl=126 time=109 ms

--- 172.16.3.12 ping statistics ---
 5 packet(s) transmitted
 5 packet(s) received
 0.00% packet loss
 round-trip min/avg/max = 78/106/125 ms

PC>ping 172.16.1.13

Ping 172.16.1.13: 32 data bytes, Press Ctrl_C to break
From 172.16.1.13: bytes=32 seq=1 ttl=126 time=110 ms
From 172.16.1.13: bytes=32 seq=2 ttl=126 time=78 ms
From 172.16.1.13: bytes=32 seq=3 ttl=126 time=78 ms
From 172.16.1.13: bytes=32 seq=4 ttl=126 time=79 ms
From 172.16.1.13: bytes=32 seq=5 ttl=126 time=109 ms
```

Fig. 3.41 Test host PC1 successfully ping through test hosts PC2 and PC3

2. OSPF planning and configuration based on three-layer switches

Construct a network topology as shown in Fig. 3.42. The switches CORE, AGG1 and AGG2 in the diagram are all three-layer switches, and the three-layer connection network segments are located between CORE and AGG1, between CORE and AGG2. The IP addresses are planned as shown in Table 3.5, where VLAN100 and VLAN101 are the network segment VLAN connecting the switch CORE with AGG1 and AGG2, VLAN2 is the VLAN subject to the test host PC1, and VLAN20 is the VLAN subject to the test host PC2.

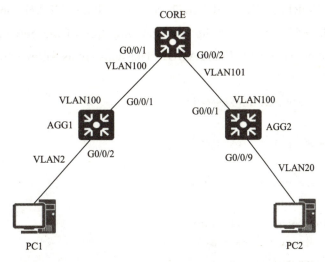

Fig. 3.42 Network topology for OSPF planning and configuration based on three-layer switches

Please configure single domain OSPF routing on CORE, AGG1 and AGG2 so that network interconnection can be achieved for routing view and testing.

Table 3.5 Planning of IP addresses

Device	Interface	IP address and mask	Gateway
CORE	Vlanif100	172.16.254.1/30	
	Vlanif101	172.16.254.5/30	
AGG1	Vlanif100	172.16.254.2/30	
	Vlanif2	172.16.2.1/24	
AGG2	Vlanif100	172.16.254.6/30	
	Vlanif20	172.16.20.1/24	
PC1	NIC	172.16.2.11/24	172.16.2.1
PC2	NIC	172.16.20.12/24	172.16.20.1

(1) Planning of OSPF

① Use single domain OSPF with the area number 0.

② Set CORE ROUTER-ID as 1.1.1.1, AGG1 ROUTER-ID as 2.2.2.2 and AGG3 ROUTER-ID as 3.3.3.3.

③ Configure the VLAN2 port of AGG1 and the VLAN20 port of AGG2 as silent ports, with no routing update forwarded.

④ Let the 172.16.254.0/30 and 172.16.254.4/30 network segment interfaces on CORE participate in the OSPF process; let the 172.16.254.0/30 and 172.16.2.0/24 network segment interfaces on AGG1 participate in the OSPF process; let the 172.16.254.4/30 and 172.16.20.0/24 network segment interfaces on AGG2 participate in the OSPF process.

(2) Configuration of IP and VLAN

On the switches CORE, AGG1 and AGG2, complete IP and VLAN configurations according to Fig. 3.42 and Table 3.5, and disable the port spanning tree of the network segments between CORE and AGG1, and between CORE and AGG2. The steps are as follows for configurations:

```
[CORE]vlan 100
[CORE-vlan100]description linktoAGG1
[CORE-vlan100]vlan 101
[CORE-vlan101]description linktoAGG2
[CORE-vlan101]quit
//Create a VLAN and configure its descriptions
[CORE]interface g0/0/1
[CORE-GigabitEthernet0/0/1]port link-type access
[CORE-GigabitEthernet0/0/1]port default vlan 100
[CORE-GigabitEthernet0/0/1]stp disable
```

```
[CORE-GigabitEthernet0/0/1]quit
//Configure the G0/0/1 interface as access interface, subject to VLAN100, and disable
//the spanning tree function on the interface
[CORE]interface g0/0/2
[CORE-GigabitEthernet0/0/2]port link-type access
[CORE-GigabitEthernet0/0/2]port default vlan 101
[CORE-GigabitEthernet0/0/2]stp disable
[CORE-GigabitEthernet0/0/2]quit
//Configure the G0/0/2 interface as access interface, subject to VLAN101, and disable the spanning
//tree function on the interface
[CORE]interface Vlanif 100
[CORE-Vlanif100]ip address 172.16.254.1 30
[CORE-Vlanif100]quit
[CORE]interface Vlanif 101
[CORE-Vlanif101]ip address 172.16.254.5 30
[CORE-Vlanif101]quit
//Configure IP addresses and masks for VLAN interfaces
[AGG1]vlan 2
[AGG1-vlan2]description linktopc1
[AGG1-vlan2]vlan 100
[AGG1-vlan100]description linktoCORE
[AGG1-vlan100]quit
//Create a VLAN and configure its descriptions
[AGG1]interface g0/0/1
[AGG1-GigabitEthernet0/0/1]port link-type access
[AGG1-GigabitEthernet0/0/1]port default vlan 100
[AGG1-GigabitEthernet0/0/1]stp disable
[AGG1-GigabitEthernet0/0/1]quit
//Configure the G0/0/1 interface as access interface, subject to VLAN100, and disable the spanning
//tree function on the interface
[AGG1]interface g0/0/2
[AGG1-GigabitEthernet0/0/2]port link-type access
[AGG1-GigabitEthernet0/0/2]port default vlan 2
[AGG1-GigabitEthernet0/0/2]quit
//Configure the G0/0/2 interface as access interface, subject to VLAN2
[AGG1]interface Vlanif 100
[AGG1-Vlanif100]ip address 172.16.254.2 30
[AGG1-Vlanif100]quit
[AGG1]interface Vlanif 2
[AGG1-Vlanif2]ip address 172.16.2.1 24
```

```
[AGG1-Vlanif2]quit
//Configure IP addresses and masks for VLAN interfaces
[AGG2]vlan 20
[AGG2-vlan20]description linktopc2
[AGG2-vlan20]vlan 100
[AGG2-vlan100]description linktoCORE
[AGG2-vlan100]quit
//Create a VLAN and configure its descriptions
[AGG2]interface g0/0/1
[AGG2-GigabitEthernet0/0/1]port link-type access
[AGG2-GigabitEthernet0/0/1]port default vlan 100
[AGG2-GigabitEthernet0/0/1]stp disable
[AGG2-GigabitEthernet0/0/1]quit
//Configure the G0/0/1 interface as access interface, subject to VLAN100, and disable
the spanning tree function on the interface
[AGG2]interface g0/0/9
[AGG2-GigabitEthernet0/0/9]port link-type access
[AGG2-GigabitEthernet0/0/9]port default vlan 20
[AGG2-GigabitEthernet0/0/9]quit
//Configure the G0/0/9 interface as access interface, subject to VLAN20
[AGG2]interface Vlanif 100
[AGG2-Vlanif100]ip address 172.16.254.6 30
[AGG2-Vlanif100]quit
[AGG2]interface Vlanif 20
[AGG2-Vlanif20]ip address 172.16.20.1 24
[AGG2-Vlanif20]quit
//Configure IP addresses and masks for VLAN interfaces
```

(3) Configuration of single domain OSPF

According to the OSPF plan, configure single domain OSPF on the switches CORE, AGG1 and AGG2, as follows:

```
[CORE]ospf 1 router-id 1.1.1.1
[CORE-ospf-1]area 0
[CORE-ospf-1-area-0.0.0.0]network 172.16.254.0 0.0.0.3
[CORE-ospf-1-area-0.0.0.0]network 172.16.254.4 0.0.0.3
[CORE-ospf-1-area-0.0.0.0]quit
[CORE-ospf-1]quit
//Create and run the OSPF process 1 with ROUTER-ID of 1.1.1.1, create and enter the
//OSPF area 0, let the network 172.16.254.0/30 and 172.16.254.4/30 interfaces partic
//ipate in the OSPF process and run in the area 0
```

```
[AGG1]ospf 1 router-id 2.2.2.2
[AGG1-ospf-1]area 0
[AGG1-ospf-1-area-0.0.0.0]network 172.16.2.0 0.0.0.255
[AGG1-ospf-1-area-0.0.0.0]network 172.16.254.0 0.0.0.3
[AGG1-ospf-1-area-0.0.0.0]quit
[AGG1-ospf-1]silent-interface Vlanif 2
[AGG1-ospf-1]quit
//Create and run the OSPF process 1 with ROUTER-ID of 2.2.2.2, create and enter the
//OSPF area 0, let the network 172.16.0.0/24 and 172.16.254.0/30 interfaces participate
//in the OSPF process and run in the area 0
[AGG2]ospf 1 router-id 3.3.3.3
[AGG2-ospf-1]area 0
[AGG2-ospf-1-area-0.0.0.0]network 172.16.20.0 0.0.0.255
[AGG2-ospf-1-area-0.0.0.0]network 172.16.254.4 0.0.0.3
[AGG2-ospf-1-area-0.0.0.0]quit
[AGG2-ospf-1]silent-interface Vlanif 20
[AGG2-ospf-1]quit
//Create and run the OSPF process 1 with ROUTER-ID of 3.3.3.3, create and enter the
//OSPF area 0, let the network 172.16.20.0/24 and 172.16.254.4/30 interfaces participate
//in the OSPF process and run in the area 0
```

(4) Verification and testing

① Execute the "display ospf peer brief" command on the switches CORE, AGG1 and AGG2 to view the OSPF neighbor table, as shown in Fig. 3.43 to Fig. 3.45. Pay attention to the neighbor ID value of each router and check if the status is FULL.

```
[CORE]display ospf peer brief

          OSPF Process 1 with Router ID 1.1.1.1
                 Peer Statistic Information
-----------------------------------------------------------
Area Id    Interface         Neighbor id    State
0.0.0.0    Vlanif100         2.2.2.2        Full
0.0.0.0    Vlanif101         3.3.3.3        Full
-----------------------------------------------------------
```

Fig. 3.43 Display results after executing the "display ospf peer brief" command on CORE

```
[AGG1]display ospf peer brief

          OSPF Process 1 with Router ID 2.2.2.2
                 Peer Statistic Information
-----------------------------------------------------------
Area Id    Interface         Neighbor id    State
0.0.0.0    Vlanif100         1.1.1.1        Full
-----------------------------------------------------------
```

Fig. 3.44 Display results after executing the "display ospf peer brief" command on AGG1

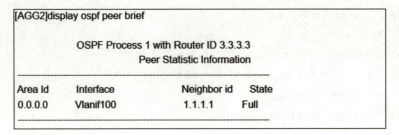

Fig. 3.45　Display results after executing the "display ospf peer brief" command on AGG2

② Execute the "display ip routing-table" command on the switches CORE, AGG1 and AGG2 to view the routing table, as shown in Fig. 3.46 to Fig. 3.48.

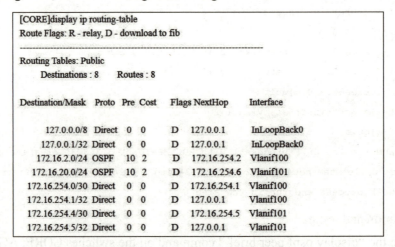

Fig. 3.46　CORE routing table

Fig. 3.47　AGG1 routing table

Project 3 Medium-sized Enterprise Internetworking Project   289

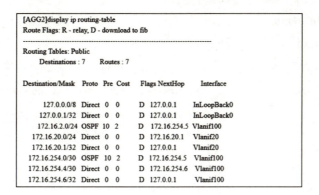

Fig. 3.48 AGG2 routing table

Task Completion

1. Task planning

According to the task description, network topology shown in Fig. 3.1, IP planning in Table 3.1, and network routing requirements of Project 3, the specific requirements are as follows for implementation of routing in this task:

① Configure the default route to the Internet on the outbound router EDGE-AR.

② Use single domain OSPF to achieve interconnection and interworking among all network segments of the corporate intranet.

③ Introduce the default route to the Internet into the OSPF routing protocol on EDGE-AR.

④ Achieve routing optimization.

See Table 3.6 for OSPF planning. The OSPF type of the link between CORE and router EDGE-AR, aggregation layer switch XZL-AGG, aggregation layer switch JXL-AGG and aggregation layer switch SXL-AGG has been changed to P2P.

Table 3.6 OSPF planning

Routing device	ROUTER-ID	Process number	Area number	Network segment participating in OSPF process	Silent interface
EDGE-AR	1.1.1.1	1	0.0.0.0	10.254.254.12/30	None
CORE	2.2.2.2	1	0.0.0.0	10.254.254.0/30, 10.254.254.4/30, 10.254.254.8/30, 10.254.254.12/30, 10.254.253.0/24	Vlanif 202
XZL-AGG	3.3.3.3	1	0.0.0.0	10.0.2.0/24, 10.0.3.0/24, 10.0.4.0/24, 10.0.5.0/24, 10.0.90.0/24, 10.0.91.0/24, 10.254.254.0/30	Vlanif 2, Vlanif 3, Vlanif 4, Vlanif 5, Vlanif 90 and Vlanif 91
JXL-AGG	4.4.4.4	1	0.0.0.0	10.1.12.0/24, 10.1.13.0/24, 10.1.92.0/24, 10.1.93.0/24 and 10.254.254.4/30	Vlanif 12, Vlanif 13, Vlanif 92 and Vlanif 93
SXL-AGG	5.5.5.5	1	0.0.0.0	10.2.22.0/24, 10.2.23.0/24, 10.2.94.0/24, 10.2.95.0/24 and 10.254.254.8/30	Vlanif 22, Vlanif 23, Vlanif 94 and Vlanif 95

2. Task implementation

(1) OSPF configuration on edge router EDGE-AR1

```
[EDGE-AR]ip route-static 0.0.0.0 0.0.0.0 125.71.28.1
//Configure the default route to the Internet. Set the next hop address as
//125.71.28.1
[EDGE-AR]interface g0/0/1
[EDGE-AR-GigabitEthernet0/0/1]ospf network-type p2p
[EDGE-AR-GigabitEthernet0/0/1]quit
//Configure the OSPF network type for interface G0/0/1 as P2P
[EDGE-AR]ospf 1 router-id 1.1.1.1
[EDGE-AR-ospf-1]area 0.0.0.0
[EDGE-AR-ospf-1-area-0.0.0.0]network 10.254.254.12 0.0.0.3
[EDGE-AR-ospf-1-area-0.0.0.0]quit
//Enable the OSPF process 1 with ROUTER-ID of 1.1.1.1 and let the network segment
10.254.254.12/30 participate in the OSPF process and run in the area 0
[EDGE-AR-ospf-1]default-route-advertise always
//Introduce the default route into the OSPF routing protocol. The parameter "always"
//indicates that the default route, only if configured on EDGE-AR, will be introduced
//into the OSPF routing protocol
```

(2) Configuration of OSPF on core layer and aggregation layer switches

① Complete OSPF configuration on the core layer switch CORE.

```
[CORE]interface Vlanif 199
[CORE-Vlanif199]ospf network-type p2p
[CORE-Vlanif199]quit
[CORE]interface Vlanif 200
[CORE-Vlanif200]ospf network-type p2p
[CORE-Vlanif200]quit
[CORE]interface Vlanif 201
[CORE-Vlanif201]ospf network-type p2p
[CORE-Vlanif201]quit
[CORE]interface Vlanif 203
[CORE-Vlanif203]ospf network-type p2p
[CORE-Vlanif203]quit
//Configure the OSPF network type for Vlanif199, Vlanif200, Vlanif201 and Vlanif203
//interfaces as P2P
[CORE]ospf 1 router-id 2.2.2.2
[CORE-ospf-1]area 0.0.0.0
[CORE-ospf-1-area-0.0.0.0]network 10.254.254.0 0.0.0.3
[CORE-ospf-1-area-0.0.0.0]network 10.254.254.4 0.0.0.3
```

```
[CORE-ospf-1-area-0.0.0.0]network 10.254.254.8 0.0.0.3
[CORE-ospf-1-area-0.0.0.0]network 10.254.254.12 0.0.0.3
[CORE-ospf-1-area-0.0.0.0]network 10.254.253.0 0.0.0.255
[CORE-ospf-1-area-0.0.0.0]quit
//Enable the OSPF process 1 with ROUTER-ID of 2.2.2.2 and let the network segments
//10.254.254.0/30, 10.254.254.3/30, 10.254.254.8/30, 10.254.254.12/30 and
//10.254.253.0/24 participate in the OSPF process and run in the area 0
[CORE-ospf-1]silent-interface Vlanif 202
[CORE-ospf-1]quit
//Configure the Vlanif202 interface as a silent interface
```

② Complete OSPF configuration on the aggregation layer switch XZL-AGG.

```
[XZL-AGG]interface Vlanif 199
[XZL-AGG-Vlanif199]ospf network-type p2p
[XZL-AGG-Vlanif199]quit
//Configure the OSPF network type of the Vlanif199 interface as P2P
[XZL-AGG]ospf 1 router-id 3.3.3.3
[XZL-AGG-ospf-1]area 0.0.0.0
[XZL-AGG-ospf-1-area-0.0.0.0]network 10.0.2.0 0.0.0.255
[XZL-AGG-ospf-1-area-0.0.0.0]network 10.0.3.0 0.0.0.255
[XZL-AGG-ospf-1-area-0.0.0.0]network 10.0.4.0 0.0.0.255
[XZL-AGG-ospf-1-area-0.0.0.0]network 10.0.5.0 0.0.0.255
[XZL-AGG-ospf-1-area-0.0.0.0]network 10.0.90.0 0.0.0.255
[XZL-AGG-ospf-1-area-0.0.0.0]network 10.0.91.0 0.0.0.255
[XZL-AGG-ospf-1-area-0.0.0.0]network 10.254.254.0 0.0.0.3
[XZL-AGG-ospf-1-area-0.0.0.0]quit
[XZL-AGG-ospf-1]
//Enable the OSPF process 1 with ROUTER-ID of 3.3.3.3 and let the network segments
//10.0.2.0/24, 10.0.3.0/24, 10.0.4.0/24, 10.0.5.0/24, 10.0.90.0/24, 10.0.91.0/24 and
//10.254.254.0/30 participate in the OSPF process and run in the area 0
[XZL-AGG-ospf-1]silent-interface Vlanif 2
[XZL-AGG-ospf-1]silent-interface Vlanif 3
[XZL-AGG-ospf-1]silent-interface Vlanif 4
[XZL-AGG-ospf-1]silent-interface Vlanif 5
[XZL-AGG-ospf-1]silent-interface Vlanif 90
[XZL-AGG-ospf-1]silent-interface Vlanif 91
[XZL-AGG-ospf-1]quit
//Configure the interfaces Vlanif2, Vlanif3, Vlanif4, Vlanif5, Vlanif90 and Vlanif91
//as silent interfaces
```

③ Complete OSPF configuration on the aggregation layer switch JXL-AGG.

```
[JXL-AGG]interface Vlanif 199
[JXL-AGG-Vlanif199]ospf network-type p2p
[JXL-AGG-Vlanif199]quit
//Configure the OSPF network type of the Vlanif199 interface as P2P
[JXL-AGG]ospf 1 router-id 4.4.4.4
[JXL-AGG-ospf-1-area-0.0.0.0]network 10.1.12.0 0.0.0.255
[JXL-AGG-ospf-1-area-0.0.0.0]network 10.1.13.0 0.0.0.255
[JXL-AGG-ospf-1-area-0.0.0.0]network 10.1.92.0 0.0.0.255
[JXL-AGG-ospf-1-area-0.0.0.0]network 10.1.93.0 0.0.0.255
[JXL-AGG-ospf-1-area-0.0.0.0]network 10.254.254.4 0.0.0.3
[JXL-AGG-ospf-1-area-0.0.0.0]quit
//Enable the OSPF process 1 with ROUTER-ID of 4.4.4.4 and let the network segments
//10.1.12.0/24, 10.1.13.0/24, 10.1.92.0/24, 10.1.93.0/24 and 10.254.254.4/30 participate
//in the OSPF process and run in the area 0
[JXL-AGG-ospf-1]silent-interface Vlanif 12
[JXL-AGG-ospf-1]silent-interface Vlanif 13
[JXL-AGG-ospf-1]silent-interface Vlanif 92
[JXL-AGG-ospf-1]silent-interface Vlanif 93
[JXL-AGG-ospf-1]quit
//Configure the interfaces Vlanif12, Vlanif13, Vlanif92 and Vlanif93 as silent interfaces
```

④ Complete OSPF configuration on the aggregation layer switch SXL-AGG.

```
[SXL-AGG]interface Vlanif 199
[SXL-AGG-Vlanif199]ospf network-type p2p
[SXL-AGG-Vlanif199]quit
//Configure the OSPF network type of the Vlanif199 interface as P2P
[SXL-AGG]ospf 1 router-id 5.5.5.5
[SXL-AGG-ospf-1]area 0.0.0.0
[SXL-AGG-ospf-1-area-0.0.0.0]network 10.2.22.0 0.0.0.255
[SXL-AGG-ospf-1-area-0.0.0.0]network 10.2.23.0 0.0.0.255
[SXL-AGG-ospf-1-area-0.0.0.0]network 10.2.94.0 0.0.0.255
[SXL-AGG-ospf-1-area-0.0.0.0]network 10.2.95.0 0.0.0.255
[SXL-AGG-ospf-1-area-0.0.0.0]network 10.254.254.8 0.0.0.3
[SXL-AGG-ospf-1-area-0.0.0.0]quit
//Enable the OSPF process 1 with ROUTER-ID of 5.5.5.5 and let the network segments
//10.2.22.0/24, 10.2.23.0/24, 10.2.94.0/24, 10.2.95.0/24 and 10.254.254.8/30 participate
//in the OSPF process and run in the area 0
[SXL-AGG-ospf-1]silent-interface Vlanif 22
[SXL-AGG-ospf-1]silent-interface Vlanif 23
[SXL-AGG-ospf-1]silent-interface Vlanif 94
```

```
[SXL-AGG-ospf-1]silent-interface Vlanif 95
[SXL-AGG-ospf-1]quit
[SXL-AGG]
//Configure the interfaces Vlanif22, Vlanif23, Vlanif94 and Vlanif95 as silent interfaces
```

(3) Verification and testing

① Execute the "display ospf peer brief" command on the router EDGE-AR, core switch CORE, aggregation layer switch XZL-AGG, aggregation layer switch JXL-AGG, and aggregation layer switch SXL-AGG respectively to check the OSPF neighbor table and see if the status of the OSPF neighbor router is FULL.

② Execute the "display ip routing-table" command on the router EDGE-AR, core switch CORE, aggregation layer switch XZL-AGG, aggregation layer switch JXL-AGG and aggregation layer switch SXL-AGG respectively to check the routing table of the router, with the results as shown in Fig. 3.49 to Fig. 3.52.

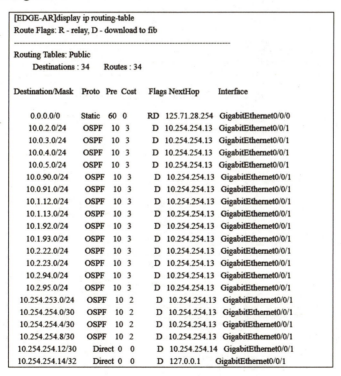

Fig. 3.49 Routing table for EDGE-AR

294　Enterprise Internetworking Technology (Bilingual)

```
[XZL-AGG]display ip routing-table
Route Flags: R - relay, D - download to fib
--------------------------------------------------------------

Routing Tables: Public
         Destinations : 29     Routes : 29
Destination/Mask    Proto  Pre  Cost    Flags NextHop      Interface
       0.0.0.0/0    O_ASE  150  1         D  10.254.254.1  Vlanif199
      10.0.2.0/24   Direct 0    0         D  10.0.2.254    Vlanif2
    10.0.2.254/32   Direct 0    0         D  127.0.0.1     Vlanif2
      10.0.3.0/24   Direct 0    0         D  10.0.3.254    Vlanif3
    10.0.3.254/32   Direct 0    0         D  127.0.0.1     Vlanif3
      10.0.4.0/24   Direct 0    0         D  10.0.4.254    Vlanif4
    10.0.4.254/32   Direct 0    0         D  127.0.0.1     Vlanif4
      10.0.5.0/24   Direct 0    0         D  10.0.5.254    Vlanif5
    10.0.5.254/32   Direct 0    0         D  127.0.0.1     Vlanif5
     10.0.90.0/24   Direct 0    0         D  10.0.90.254   Vlanif90
   10.0.90.254/32   Direct 0    0         D  127.0.0.1     Vlanif90
     10.0.91.0/24   Direct 0    0         D  10.0.91.254   Vlanif91
   10.0.91.254/32   Direct 0    0         D  127.0.0.1     Vlanif91
     10.1.12.0/24   OSPF   10   3         D  10.254.254.1  Vlanif199
     10.1.13.0/24   OSPF   10   3         D  10.254.254.1  Vlanif199
     10.1.92.0/24   OSPF   10   3         D  10.254.254.1  Vlanif199
     10.1.93.0/24   OSPF   10   3         D  10.254.254.1  Vlanif199
     10.2.22.0/24   OSPF   10   3         D  10.254.254.1  Vlanif199
     10.2.23.0/24   OSPF   10   3         D  10.254.254.1  Vlanif199
     10.2.94.0/24   OSPF   10   3         D  10.254.254.1  Vlanif199
     10.2.95.0/24   OSPF   10   3         D  10.254.254.1  Vlanif199
   10.254.253.0/24  OSPF   10   2         D  10.254.254.1  Vlanif199
   10.254.254.0/30  Direct 0    0         D  10.254.254.2  Vlanif199
   10.254.254.2/32  Direct 0    0         D  127.0.0.1     Vlanif199
   10.254.254.4/30  OSPF   10   2         D  10.254.254.1  Vlanif199
   10.254.254.8/30  OSPF   10   2         D  10.254.254.1  Vlanif199
   10.254.254.12/30 OSPF   10   2         D  10.254.254.1  Vlanif199
      127.0.0.0/8   Direct 0    0         D  127.0.0.1     InLoopBack0
      127.0.0.1/32  Direct 0    0         D  127.0.0.1     InLoopBack0
```

Fig. 3.50　Routing table for XZL-AGG

```
[JXL-AGG]display ip routing-table
Route Flags: R - relay, D - download to fib
--------------------------------------------------------------

Routing Tables: Public
         Destinations : 27     Routes : 27

Destination/Mask    Proto  Pre  Cost    Flags NextHop      Interface
       0.0.0.0/0    O_ASE  150  1         D  10.254.254.5  Vlanif199
      10.0.2.0/24   OSPF   10   3         D  10.254.254.5  Vlanif199
      10.0.3.0/24   OSPF   10   3         D  10.254.254.5  Vlanif199
      10.0.4.0/24   OSPF   10   3         D  10.254.254.5  Vlanif199
      10.0.5.0/24   OSPF   10   3         D  10.254.254.5  Vlanif199
     10.0.90.0/24   OSPF   10   3         D  10.254.254.5  Vlanif199
     10.0.91.0/24   OSPF   10   3         D  10.254.254.5  Vlanif199
     10.1.12.0/24   Direct 0    0         D  10.1.12.254   Vlanif12
   10.1.12.254/32   Direct 0    0         D  127.0.0.1     Vlanif12
     10.1.13.0/24   Direct 0    0         D  10.1.13.254   Vlanif13
   10.1.13.254/32   Direct 0    0         D  127.0.0.1     Vlanif13
     10.1.92.0/24   Direct 0    0         D  10.1.92.254   Vlanif92
   10.1.92.254/32   Direct 0    0         D  127.0.0.1     Vlanif92
     10.1.93.0/24   Direct 0    0         D  10.1.93.254   Vlanif93
   10.1.93.254/32   Direct 0    0         D  127.0.0.1     Vlanif93
     10.2.22.0/24   OSPF   10   3         D  10.254.254.5  Vlanif199
     10.2.23.0/24   OSPF   10   3         D  10.254.254.5  Vlanif199
     10.2.94.0/24   OSPF   10   3         D  10.254.254.5  Vlanif199
     10.2.95.0/24   OSPF   10   3         D  10.254.254.5  Vlanif199
   10.254.253.0/24  OSPF   10   2         D  10.254.254.5  Vlanif199
   10.254.254.0/30  OSPF   10   2         D  10.254.254.5  Vlanif199
   10.254.254.4/30  Direct 0    0         D  10.254.254.6  Vlanif199
   10.254.254.6/32  Direct 0    0         D  127.0.0.1     Vlanif199
   10.254.254.8/30  OSPF   10   2         D  10.254.254.5  Vlanif199
   10.254.254.12/30 OSPF   10   2         D  10.254.254.5  Vlanif199
      127.0.0.0/8   Direct 0    0         D  127.0.0.1     InLoopBack0
      127.0.0.1/32  Direct 0    0         D  127.0.0.1     InLoopBack0
```

Fig. 3.51　Routing Table for JXL-AGG

```
[SXL-AGG]display ip routing-table
Route Flags: R - relay, D - download to fib
------------------------------------------------------------
Routing Tables: Public
        Destinations : 27    Routes : 27

Destination/Mask    Proto   Pre  Cost    Flags  NextHop        Interface

        0.0.0.0/0   O_ASE   150  1         D    10.254.254.9   Vlanif199
       10.0.2.0/24  OSPF    10   3         D    10.254.254.9   Vlanif199
       10.0.3.0/24  OSPF    10   3         D    10.254.254.9   Vlanif199
       10.0.4.0/24  OSPF    10   3         D    10.254.254.9   Vlanif199
       10.0.5.0/24  OSPF    10   3         D    10.254.254.9   Vlanif199
      10.0.90.0/24  OSPF    10   3         D    10.254.254.9   Vlanif199
      10.0.91.0/24  OSPF    10   3         D    10.254.254.9   Vlanif199
      10.1.12.0/24  OSPF    10   3         D    10.254.254.9   Vlanif199
      10.1.13.0/24  OSPF    10   3         D    10.254.254.9   Vlanif199
      10.1.92.0/24  OSPF    10   3         D    10.254.254.9   Vlanif199
      10.1.93.0/24  OSPF    10   3         D    10.254.254.9   Vlanif199
     10.2.22.0/24   Direct  0    0         D    10.2.22.254    Vlanif22
    10.2.22.254/32  Direct  0    0         D    127.0.0.1      Vlanif22
     10.2.23.0/24   Direct  0    0         D    10.2.23.254    Vlanif23
    10.2.23.254/32  Direct  0    0         D    127.0.0.1      Vlanif23
     10.2.94.0/24   Direct  0    0         D    10.2.94.254    Vlanif94
    10.2.94.254/32  Direct  0    0         D    127.0.0.1      Vlanif94
     10.2.95.0/24   Direct  0    0         D    10.2.95.254    Vlanif95
    10.2.95.254/32  Direct  0    0         D    127.0.0.1      Vlanif95
   10.254.253.0/24  OSPF    10   2         D    10.254.254.9   Vlanif199
   10.254.254.0/30  OSPF    10   2         D    10.254.254.9   Vlanif199
   10.254.254.4/30  OSPF    10   2         D    10.254.254.9   Vlanif199
   10.254.254.8/30  Direct  0    0         D    10.254.254.10  Vlanif199
  10.254.254.10/32  Direct  0    0         D    127.0.0.1      Vlanif199
  10.254.254.12/30  OSPF    10   2         D    10.254.254.9   Vlanif199
       127.0.0.0/8  Direct  0    0         D    127.0.0.1      InLoopBack0
      127.0.0.1/32  Direct  0    0         D    127.0.0.1      InLoopBack0
```

Fig. 3.52 Routing table for SXL-AGG

③ After this task is completed, hosts on the corporate intranet can access the Internet and hosts on the Internet can access the Web and DNS services provided externally by the enterprise. Fig. 3.53 shows that the ping command is used on test host PC2 to test the IP connectivity to the test host PC1 on the Internet, and normal connection is confirmed.

```
PC>ping 125.71.28.1

Ping 125.71.28.1: 32 data bytes, Press Ctrl_C to break
From 125.71.28.1: bytes=32 seq=1 ttl=252 time=94 ms
From 125.71.28.1: bytes=32 seq=2 ttl=252 time=93 ms
From 125.71.28.1: bytes=32 seq=3 ttl=252 time=79 ms
From 125.71.28.1: bytes=32 seq=4 ttl=252 time=109 ms
From 125.71.28.1: bytes=32 seq=5 ttl=252 time=78 ms
```

Fig. 3.53 Intranet test host PC2 can successfully ping the external network

Exercises

Multiple choice

(1) In Huawei devices, the following (   ) is not the method for selecting ROUTER-ID via OSPF.

A. If a loopback interface is configured, select the largest IP address from the IP addresses

of the loopback interface as ROUTER-ID

B. If no loopback interface is configured, select the largest IP address from the IP addresses of other interfaces as ROUTER-ID

C. Use commands to define an arbitrary legal ROUTER-ID manually

D. Use the default 127. 0 0.1

(2) The following (   ) command is used to view the OSPF neighbor table.

    A. display ip routing-table　　　　　　B. display current-configuration

    C. display ospf peer　　　　　　　　　　D. display ip interface brief

(3) The metric for OSPF is (   ).

    A. cost　　　　　B. hops　　　　　C. load　　　　　D. delay

计算机网络互联技术（双语）

主编 阎 国 宋 敏 廖天津 李 宁
副主编 张春秋

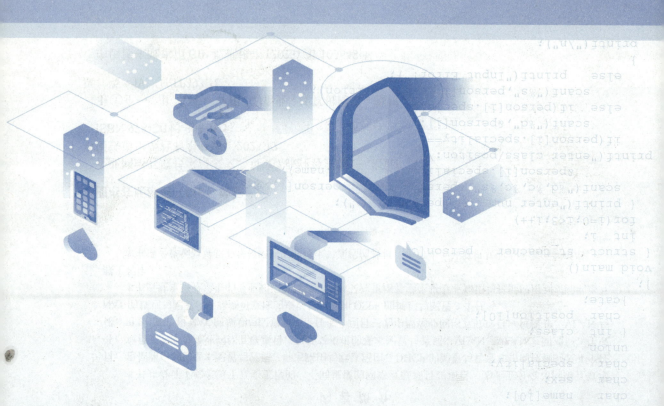

内 容 简 介

本书共设计了三个源于真实工程的、不同规模的网络互联项目。项目一为企业分公司网络互联项目，包括路由器的基本配置与管理、静态路由与默认路由、DHCP 的配置与管理和访问控制列表 4 个任务；项目二为小型企业网络互联项目，包括交换机的基本配置与管理，VLAN、VLAN 之间通信、生成树、PPP、RIP 和 NAT 的规划与配置等 7 个任务；项目三为中型企业网络互联项目，包括 IP、VLAN 与 NAT 的规划与配置，链路高可靠性的规划与配置和 OSPF 的规划与配置 3 个任务。

本书既注重培养学生的专业知识与技能，同时又注重培养学生将专业知识与技能应用到实际网络互联工程中。

本书适合作为高等职业学校网络互联技术等课程的教材，也可作为计算机从业人员等的参考书。

图书在版编目（CIP）数据

企业网络互联技术：汉文、英文 / 张纯容主编 .—北京：
中国铁道出版社有限公司，2023.12
ISBN 978-7-113-30310-5

Ⅰ.①企… Ⅱ.①张… Ⅲ.①企业 - 计算机网络 - 教材 -
汉、英 Ⅳ.① TP393.18

中国国家版本馆 CIP 数据核字 (2023) 第 103648 号

书 名：**企业网络互联技术（双语）**
QIYE WANGLUO HULIAN JISHU (SHUANGYU)
作 者：张纯容

策 划：潘晨曦 祁 云		编辑部电话：(010) 63549458
责任编辑：祁 云 绳 超		
封面设计：刘 颖		
责任校对：安海燕		
责任印制：樊启鹏		

出版发行：中国铁道出版社有限公司（100054，北京市西城区右安门西街 8 号）
网 址：http://www.tdpress.com/51eds/
印 刷：北京联兴盛业印刷股份有限公司
版 次：2023 年 12 月第 1 版 2023 年 12 月第 1 次印刷
开 本：850 mm×1 168 mm 1/16 印张：36 字数：781 千
书 号：ISBN 978-7-113-30310-5
定 价：118.00 元

版权所有 侵权必究

凡购买铁道版图书，如有印制质量问题，请与本社教材图书营销部联系调换。电话：(010) 63550836
打击盗版举报电话：(010) 63549461

前 言

本教材是双高计划专业群"软件技术专业群"中的三层次集群式项目课程体系建设与"融入职业认证与专业竞赛的网络工程课程群建设的研究与实践"教改项目的成果。本教材选取三个不同规模的真实网络项目,并将项目进行提取、精炼,结合华为1+X网络系统管理与运维等职业认证、网络专业竞赛的知识与技能,最终形成了本教材的3个项目,共14个任务。

本教材具有以下特色:

(1)人才培养目标定位准确。本教材以网络工程师、网络安全运维工程师、网络管理员等网络应用型专业人员为人才培养目标,通过深入的人才培养需求分析与技术调研,形成了对上述人才培养目标与培养规格的准确定位,明确了此类人才在网络互联技术知识、技能及工程素质方面的要求,并以此为基础确立了本教材的教学框架。

(2)教材开发模式创新。采取校内教师与业界资深工程师共同开发的模式。业界资深工程师提供真实的网络互联工程项目,校内教师与企业工程师共同分析并提炼教材知识点、技能点,由校内教师负责教材编写。这种形式有效保障了教材既能很好体现职业院校课程教学的特点、特色与规律,又能充分与网络主流技术和工程实际接轨。

(3)教材编写方式创新。教材设计了三个规模由小到大、来源于不同行业的项目。每个项目根据涉及的知识点与技能点,将其分解成对应的任务,每个任务由任务描述、任务目标、相关知识、相关技能、完成任务和习题组成,完成项目中的所有任务后即可完成整

企业网络互联技术（双语）

个项目的规划与部署。这种编写方式，既可培养学生掌握网络互联的相关知识与技能，更可培养学生掌握网络互联专业知识和技术在真实项目中的应用。

本教材由张纯容任主编，安宁、彭天炜、宋牧任副主编。其中，项目 1 的任务 1、任务 2 由安宁编写；项目 2 的任务 1 到任务 3 由彭天炜编写；项目 3 的任务 1、任务 2 由宋牧、张纯容编写；其余的项目任务均由张纯容编写。全书由张纯容修改、定稿。本教材的编写得到了华为技术有限公司韩江工程师的帮助，在此谨表由衷的谢意。

本教材立足于应用型网络技术人才的培养，在涉及教学内容的选择、编排及教学方法设计等方面做了一些改革创新尝试，我们非常欢迎并希望广大读者对本教材提出指正和建议。编者 E-mail：289302109@qq.com。

编　者

2023 年 5 月

目 录

项目1　企业分公司网络互联项目 1

任务1　路由器的基本配置与管理 2

任务2　静态路由与默认路由 30

任务3　DHCP的配置与管理 54

任务4　访问控制列表 67

项目2　小型企业网络互联项目 **85**

任务1　交换机的基本配置与管理 86

任务2　VLAN的规划与配置 101

任务3　VLAN之间通信的规划与配置 131

任务4　生成树的规划与配置 153

任务5　PPP的规划与配置 173

任务6　RIP的规划与配置 182

任务7　NAT的规划与配置 195

项目3 **中型企业网络互联项目** **202**

任务1　IP、VLAN与NAT的规划与配置 203

任务2　链路高可靠性的规划与配置 229

任务3　OSPF的规划与配置 249

项目 1

企业分公司网络互联项目

学习目标

知识目标

◎ 理解路由器的内部组成与物理接口，掌握路由器的基本配置与管理。

◎ 理解路由表的组成、路由表的插入规则，掌握路由器上静态路由、默认路由、浮动路由与汇总静态路由的规划与配置。

◎ 理解 DHCP 的工作过程，掌握路由器上 DHCP 的配置与管理。

◎ 理解 ACL 的工作过程，掌握 ACL 的规划与配置。

能力目标

◎ 具有基于路由器的分公司网络互联项目部署与实施能力，具体为：路由器的基本配置与管理，DHCP 服务的配置，静态路由、默认路由、浮动路由与汇总静态路由配置和 ACL 的配置。

◎ 具有一定的故障排除能力。

素质目标

◎ 具有团队协作能力。

◎ 具有分析问题与解决问题的能力。

项目描述

某企业分公司拟组建网络拓扑结构，如图 1.1 所示。分公司的预算部门、管理部门、服务器通过接入层交换机连接到路由器的三个不同端口实现分公司内部互联。分公司的路由器使用两条链路与总部的路由器相连，一条链路为千兆以太网链路，另一条链路为串行链路。分配给

2 企业网络互联技术（双语）

分公司的 IP 地址为"172.16.128.0/24"与"172.16.129.0/24"两个 IP 网段。分公司出口路由器与总部路由器的两个连接网段 IP 地址为"172.16.254.0/30"与"172.16.254.4/30"。建网需求如下：

◎分公司的预算部门有 140 台主机，管理部门有 100 台主机，分公司有 8 台服务器业务，预算部门、管理部门与服务器之间的数据需要广播隔离，它们不能直接通信，需要通过路由器实现通信。

◎分公司内网以及分公司与总部互联互通，分公司通过总部网络接入 Internet。

◎分公司预算部门与管理部门的主机 IP 地址采用 DHCP 自动获得，由分公司路由器提供 DHCP 服务。

◎连接到总部的分公司路由器出口与入口引入 ACL，禁用 TCP 与 UDP 协议的 135 ～ 139、445 与 3389 高危端口。

◎只有分公司管理部门的 PC3 和总部的 PC1 可使用 SSH 远程登录到"fenbu"路由器上对其进行远程管理。

◎完成配置后，将分公司与总部路由器的配置进行保存，并备份到总部的 FTP 服务器上。

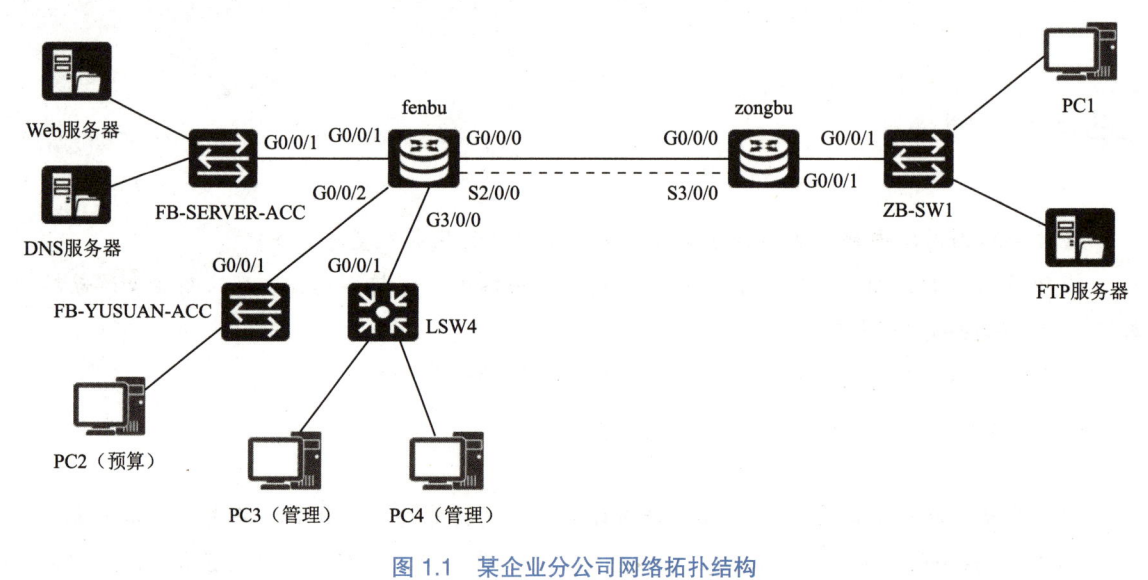

图 1.1 某企业分公司网络拓扑结构

任务1 路由器的基本配置与管理

任务描述

认知路由器的物理接口。在华为 ENSP 仿真环境中或者使用华为真实路由器、交换机搭建如图 1.1 所示的网络拓扑。完成该网络 IP 地址的规划，测试主机 PC 和服务器的 IP 配置。使用 Console 端口对网络中的路由器完成主机名、接口、Console 登录认证与 SSH 配置等基本配置及测试。对路由器配置文件进行备份与还原、密码恢复等管理工作。

项目 1　企业分公司网络互联项目　3

任务目标

● 掌握路由器的物理接口类型与功能。
● 掌握路由器初始化配置的连接方式。
● 掌握路由器命令行的用户接口（command line interface, CLI）的使用。
● 掌握路由器命令使用的在线帮助，熟悉路由器的快捷键使用，理解路由器的查看命令。
● 掌握路由器接口、控制端口认证、Telnet 和 SSH 的配置。
● 掌握路由器配置文件的管理与 Console 密码的清除。

相关知识

1. 路由器简介

路由器是工作在 OSI 模型第三层的网络互联设备。路由器的主要功能是最佳路径选择，即通常所说的路由。路由器通过网络层的协议地址，为第三层的数据包传输提供最佳路径的选择。

路由器本质上是一种特殊功能的计算机，因此有着与普通计算机相同的组成部件。路由器主要由主板、中央处理器（CPU）、闪存（flash memory）、随机存储器（random access memory, RAM）、只读存储器（read only memory, ROM）、操作系统（operating system, OS）、电源、底板、金属机壳和网络接口等组成。

2. 路由器的物理接口

路由器能实现异构网络的互联。路由器提供了不同类型的物理端口，以支持各种局域网技术、城域网技术和广域网技术。常见的局域网端口有快速以太网端口、GE（千兆以太网）端口、10GE 端口和 40GE 端口等。广域网接口有高速 A/S（异步/同步）串行接口等。路由器的接口可以是固定的（如图 1.2 中的 G0/0/1 接口），也可以根据用户的需求提供模块化（如图 1.3 中的 S0/1/1 接口）的选择。通常低端路由器采用固定方式，高端路由器则提供模块选择。图 1.2 与图 1.3 分别为华为 AR1220 与 NetEngine AR6000 Series 路由器的物理接口。

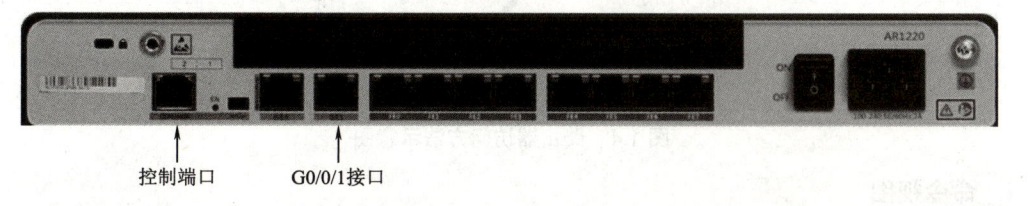

图 1.2　华为 AR1220 路由器的物理接口

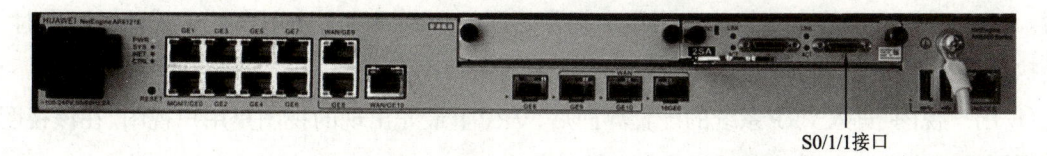

图 1.3　华为 NetEngine AR6000 Series 路由器的物理接口

为了区别路由器上的物理接口，引入了路由器接口的命名规则，为路由器上的每个物理接口赋予一个唯一的标识，以便于对路由器接口进行识别。

路由器接口的命名格式为"接口类型 插槽编号 / 模块编号 / 接口编号"。其中插槽编号、模块编号和接口编号一般从"0"开始，第一个编号为"0"，第二个编号为"1"，以此类推。

对于固定接口的路由器，其接口名称中的数字只包括接口编号，例如"Ethernet0（可缩写为 E0）"表示第一个以太网接口，"Serial1（可缩写为 S1）"表示第二个串口。

对于支持"在线插拔和删除"或具有动态更改物理接口配置功能的路由器，其接口名称的数字中至少包含插槽编号与接口编号两个数字。例如，在华为 AR2220 路由器中，"GigabitEthernet 0/1（可缩写为 G0/1）"代表位于"0"号插槽上的"1"号千兆位以太网端口。

对于在显卡上插有模块的路由器，其接口名称的数字中要包含插槽编号、模块编号和接口编号。例如，在华为 AR2220 路由器中，"Serial 2/0/0（可缩写为 S2/0/0）"是指"2"号插槽上"0"号模块的"0"号串口。

3. 路由器的访问方法

对路由器进行配置的方法主要有两种：一种是通过路由器的管理端口（如 Console 端口）对路由器进行配置；另一种是通过网络远程登录到路由器对其进行配置，如使用远程登录（telnet）、安全外壳（secure shell, SSH）协议或简单网络管理协议（simple network management protocol, SNMP）访问路由器，但通过网络远程访问路由器的方法需要路由器与 Telnet 客户端、SSH 客户端或 SNMP 网管工作站具有 IP 连通性。路由器访问方法示意图如图 1.4 所示。

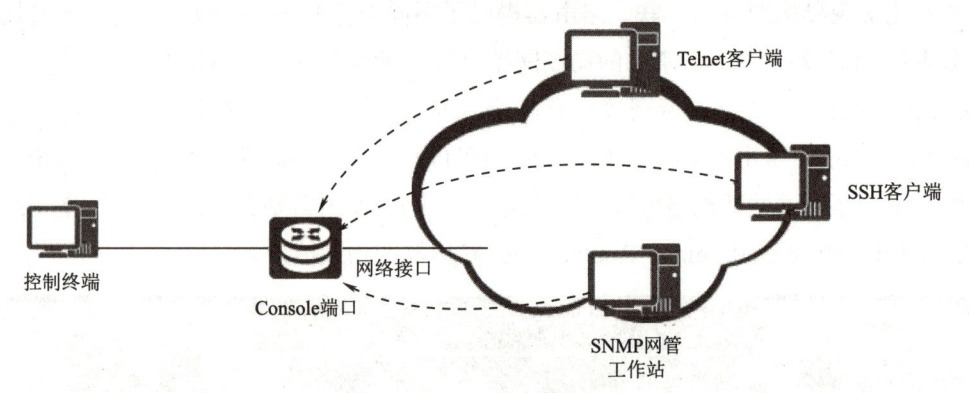

图 1.4　路由器访问方法示意图

4. 命令视图

通用路由平台 VRP（versatile routing platform）是华为公司数据通信产品的通用操作系统平台。VRP 采用分层的命令结构，为用户提供不同的命令视图。当用户处于某种视图中时，就只能执行该视图所允许的特定命令和操作，常见的命令视图有：

① 用户视图：进入 VRP 系统的配置界面后，VRP 上最先出现的视图是用户视图。在该视图下，用户视图可以查看路由器基本运行状态和统计信息。

② 系统视图：配置系统全局通用参数的视图。如果要修改系统参数，用户必须进入系统视图。用户还可以通过系统视图进入其他的功能配置视图，如接口视图、用户界面视图和协议视图等。

③ 接口视图：配置接口参数的视图称为接口视图。

④ 用户界面视图：用来管理工作在流方式下的异步接口。用户界面视图分为 Console 用户界面视图、AUX 用户界面视图、TTY 用户界面视图和 VTY 界面视图。

⑤ 路由协议视图：路由协议配置的大部分参数是在路由协议视图下进行的，后续的项目中会介绍路由协议，如 OSPF 协议视图、RIP 协议视图等。

路由器的每一条命令只能在特定的视图中执行。可以通过路由器的提示符判断当前所处的视图是什么。例如："< >"表示用户视图，"[ ]"表示除用户视图以外的其他视图。表 1.1 给出了华为路由器常见的命令视图及提示符。

表 1.1　华为路由器常见的命令视图及提示符

命 令 视 图	提示符示例	视图切换示例
用户视图	<Huawei>	路由器启动后默认视图
系统视图	[Huawei]	<Huawei>system-view
接口视图	[Huawei-GigabitEthernet0/0/0]	[Huawei]interface G0/0/0
用户界面视图	[Huawei-ui-console0]	[Huawei]user-interface console 0
路由协议视图	[Huawei-ospf-1]	[Huawei]ospf 1

5. 路由器命令使用的在线帮助

路由器为实现多种不同的功能，不仅提供了一个庞大的命令集，而且一个单一命令还可能提供了不同的使用参数，完全依靠用户的记忆进行命令的使用几乎是不可能的。为此，路由器提供了路由器命令使用的在线帮助功能，帮助用户完成相关的配置命令。在线帮助的基本使用方法如下：

① ？　//可显示当前视图下所能提供的全部命令集及简单描述。例如，在用户视图下输入"？"，
　　　　//可列出用户视图下所有可用的命令集及简单描述

② 字符串＋？　//可显示当前提示符下以字符串打头的所有命令集及简单描述。例如，输入
　　　　　　　//"dis？"则可显示当前视图下以字符串"dis"开始的所有命令及简单描述

③ 命令名＋空格＋？　//可显示当前提示符下此命令的全部关键字或参数及简单描述，如
　　　　　　　　//"display？"显示当前视图下"display"命令的全部关键字及简单描述

6. 命令历史功能的使用

在配置路由器或查看路由器状态时，经常会遇到需要重复输入同一条命令或输入类似命令的情形，为减少用户使用路由器时输入命令的工作量，路由器一般都提供命令历史功能，其可以为用户存储其刚刚用过的若干条命令。命令历史所能存储的最大命令条数取决于系统为此设定的最大存储空间，在系统限定的最大条数之内用户可以根据需要对命令历史的命令存储条数

6　**企业网络互联技术**（双语）

进行配置。使用命令历史可采用系统提供的查看命令，通常包括上翻命令（向前查）和下翻命令（向后查）。查看历史命令分别采用【Ctrl+P】或【Ctrl+N】组合键，或者使用小键盘上的【↑】或【↓】键。

7. 路由器的快捷键

除了使用帮助命令外，为了方便路由器的配置、监控和排除故障，路由器操作系统还提供了相关的快捷键。表 1.2 列出了路由器的快捷键。

表 1.2　路由器的快捷键

快　捷　键	功　能　描　述
Tab	用以把部分输入的命令项补全
↑（或 Ctrl+P）	在前面用过的命令列表中向后翻
↓（或 Ctrl+N）	在前面用过的命令列表中向前翻
Ctrl+C	放弃当前命令并退出
Ctrl+Z	直接返回到用户视图

在使用快捷键【Tab】时，要求输入的缩写命令或缩写参数包含足够字母使之可以和当前可用的任何其他命令或参数区分开，才可用【Tab】快捷方式自动补充该缩写命令或缩写参数剩下的部分。

8. 路由器的查看命令

在对路由器进行配置或配置更改之前、在对路由器进行配置效果的检查验证以及对路由器进行故障排除时，都需要提供路由器工作状态的检查功能。为此，所有的路由器都提供了系列用于查看路由器状态的命令。表 1.3 给出了华为路由器常用的查看命令。

表 1.3　华为路由器常用的查看命令

查　看　命　令	功　能　描　述
display version	显示系统版本
display current-configuration	显示正在运行的配置文件内容
display saved-configuration	显示保存在 Flash 中的配置文件内容
display interface	显示接口信息
display ip interface brief	显示包括 IP 地址和接口状态在内的简要的接口配置信息
display ip routing-table	显示路由表信息
display this	显示当前视图的运行配置
dir flash:	显示 Flash 中的内容

9. 路由器的文件管理方式

所有关于路由器的配置信息以路由器配置文件的形式存在。路由器当前运行的配置文件（current-configuration）被保存在路由器的 RAM 中，RAM 中的配置信息在掉电或路由器重启时会丢失，需要把配置文件保存到路由器的 Flash 中，路由器每次开机或重新启动时都将 Flash 中

的配置文件调用到 RAM 中来运行。除了上述配置文件保存方式外，也可以使用 TFTP、FTP、SFTP、SCP 以及 FTPS 方式把配置文件保存到其他主机可靠的磁盘上。

路由器在文件管理的过程中，可以充当服务器的角色，也可以充当客户端的角色。

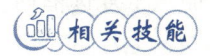

1. 路由器使用入门

完成路由器的初始化配置连接，熟悉路由器的命令视图，使用历史命令与帮助命令完成路由器的主机名配置。

（1）路由器的初始化配置连接

① 使用一条 USB 转 RJ-45 控制线作为控制终端的主机与路由器的 Console 端口相连。图 1.5 为笔记本计算机作为路由器控制台的连接示意图。USB 转 RJ-45 控制线的 USB 端插入笔记本计算机的 USB 口，USB 转 RJ-45 控制线的 RJ-45 端插入路由器的 Console 端口。

② 在主机上安装好 USB 转 RJ-45 控制线驱动程序后，右击桌面上的"此电脑"图标，在弹出的快捷菜单中选择"管理"命令，

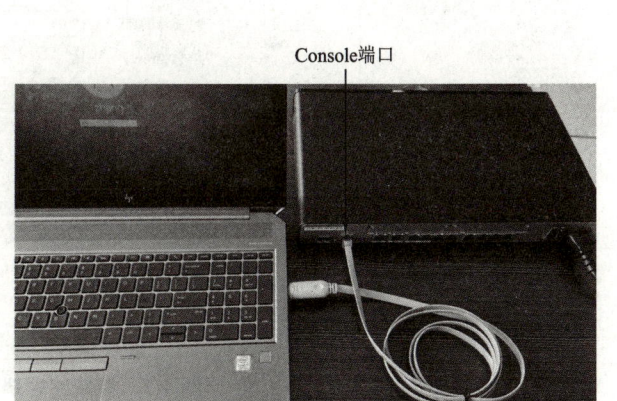

图 1.5　笔记本计算机作为路由器控制台的连接示意图

弹出"计算机管理"窗口，单击"设备管理器"。接着，单击右边列表中的"端口（COM 和 LPT）"，观察 USB-to-Serial Comm Port（COM3）括号中的 COM 端口数值，如图 1.6 所示。图中显示的 USB 转串口的 COM 端口为 COM3 端口。

图 1.6　查看控制终端的"USB-to-Serial Comm Port"的 COM 端口号

③ 在路由器控制台上运行 SecureCRT 软件，在弹出的"SecureCRT"窗口中，单击"文件"菜单，选择"快速连接"命令，在弹出的"快速连接"窗口中，协议选择"Serial"，端口选择"COM3"（**注**：COM 端口号要与设备管理器中查询的 USB 转串口的 COM 端口相同），波特率为 9600，数据位为 8 位，无奇偶校验，停止位为 1 位，没有流控，即流控下面三个复选框均取消勾选，如图 1.7 所示。

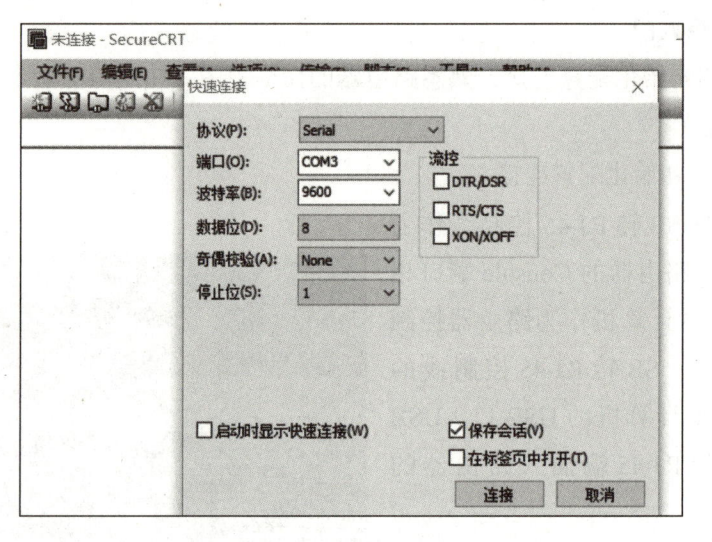

图 1.7　控制终端端口属性的设置

④ 单击"连接"按钮，再按【Enter】键输入默认的用户名与密码进入路由器的命令视图界面，如图 1.8 所示。

```
Press any key to get started

Login authentication

Username:admin
Password:
 Warning: There is a risk in the current configuration file. Please save configuration as soon as possible.
<Huawei>
 Warning: Auto-Config is working. Before configuring the device, stop Auto-Config. If you perform
configurations when Auto-Config is running, the DHCP, routing, DNS, and VTY configurations will be lost.
Do you want to stop Auto-Config? [y/n]:n
<Huawei>
```

图 1.8　路由器的初始化配置界面

（2）初步认识路由器的命令视图

通常路由器开机后默认提供给用户的是用户视图。华为路由器用户视图的提示符为"<Huawei>"，其中"<"与">"之间的"Huawei"是路由器的系统主机名，默认主机名为"Huawei"，路由器的主机名可以更改。在用户视图上输入"?"，可以查看该视图下所允许的命令及其功能，如图 1.9 所示。

项目 1　企业分公司网络互联项目　9

```
<Huawei>?  ◀────── 输入"?"显示当前视图下所有可用的命令与简单描述
User view commands:
 arp-ping          ARP-ping
 autosave           <Group> autosave command group
 backup            Backup  information
 cd               Change current directory
 clear             <Group> clear command group
 clock            Specify the system clock
 cls              Clear screen
 compare            Compare configuration file
 copy             Copy from one file to another
 debugging           <Group> debugging command group
 delete            Delete a file
 dialer            Dialer
 dir              List files on a filesystem
 display            Display information
 factory-configuration  Factory configuration
 fixdisk            Try to restory disk
 format            Format file system
 free              Release a user terminal interface
 ftp              Establish an FTP connection
 help             Description of the interactive help system
 hwtacacs-user        HWTACACS user
 license            <Group> license command group
 lldp             Link Layer Discovery Protocol
---- More ----  ◀────── "more"表示命令未显示完
```

图 1.9　路由器的用户视图

图 1.9 中，屏幕最下方的"----More----"表示屏幕命令还未显示完，此时可按【Enter】键或者空格键显示余下的命令。按【Enter】键，屏幕向下显示一行：按空格键，屏幕向下显示一屏；按键盘上的字母【Q】键，可直接退出，不再显示未显示的命令。按空格键显示下一屏的命令，再按【Enter】键找到进入系统视图的"system-view"，查看该命令的简单描述，按字母【Q】键退出帮助，再通过帮助命令"system-view?"查看其关键字，关键字中有"<cr>"，表示回车，即可执行该命令进入系统视图。系统视图的提示符为"[Huawei]"，如图 1.10 所示。

```
<Huawei>system-view ?
 <cr>  Please press ENTER to execute command
<Huawei>system-view
Enter system view, return user view with Ctrl+Z.
[Huawei]
```

图 1.10　使用帮助查看命令"system-view"的关键字并进入系统视图

在系统视图下，使用帮助命令"?"查看进入接口视图的命令，然后再通过帮助命令"interface ?"查看"interface"命令的关键字，进入 G0/0/0 接口的接口视图步骤如下。注意观察与记录 G0/0/0 接口的接口视图提示符。

```
[Huawei]?                              //列出当前视图下所有可用的命令
[Huawei]interface ?                    //查看"interface"的关键字或参数
[Huawei]interface GigabitEthernet 0/0/0    //进入 G0/0/0 接口的接口视图
```

10 企业网络互联技术（双语）

```
[Huawei-GigabitEthernet0/0/0]              //G0/0/0 接口的接口视图提示符
[Huawei-GigabitEthernet0/0/0]q?            // 列出以字母"q"开始的所有命令
[Huawei-GigabitEthernet0/0/0]quit          // 退出 G0/0/0 接口的接口视图，返回到系统视图
[Huawei]quit                               // 退出系统视图，返回到用户视图
<Huawei>                                   // 用户视图提示符
```

（3）使用历史命令与帮助命令完成路由器主机名的配置

在任何视图下均可使用小键盘上【↑】键调出刚使用的命令，请参考以下的命令练习使用历史命令与帮助命令完成路由器主机名的配置：

```
<Huawei> ↑
```

多次按小键盘上【↑】键调出已使用过的"system-view"命令，然后按【Enter】键执行。

```
[Huawei]?                                  // 列出系统视图下可用的命令
```

在命令显示后，可多次按空格键找到"sysname"命令。

```
[Huawei]sysname ?                          // 查看 sysname 的关键字或参数
[Huawei]sys ?                              // 查看字符串"sys"打头的所有命令
[Huawei]sys 再按 Tab 键                     // 将 sys 开头的命令补全
[Huawei]sysname  AR1                       // 配置路由器的主机名为"AR1"
[AR1]quit                                  // 退出系统视图，返回到用户视图
<AR1>                                      // 用户视图提示符
```

（4）路由器查看命令练习

在用户视图或在系统视图提示符下输入"display version"命令显示路由器的系统版本，如图 1.11 所示。

```
[AR1]display version
Huawei Versatile Routing Platform Software
VRP (R) software, Version 5.130 (AR2200 V200R003C00)
Copyright (C) 2011-2012 HUAWEI TECH CO., LTD
Huawei AR2220 Router uptime is 0 week, 0 day, 0 hour, 51 minutes
BKP 0 version information:
1. PCB      Version  : AR01BAK2A VER.NC
2. If Supporting PoE : No
3. Board    Type     : AR2220
4. MPU Slot Quantity : 1
5. LPU Slot Quantity : 6

MPU 0(Master) : uptime is 0 week, 0 day, 0 hour, 51 minutes
MPU version information :
1. PCB      Version  : AR01SRU2A VER.A
2. MAB      Version  : 0
3. Board    Type     : AR2220
4. BootROM  Version  : 0
```

图 1.11 "display version"命令显示路由器的系统版本

项目 1　企业分公司网络互联项目　11

2. 路由器基本配置

搭建如图 1.12 所示的网络拓扑，完成路由器接口的配置、路由器 Console 端口的配置、Telnet 服务的配置与 SSH 服务的配置，并进行验证与测试。

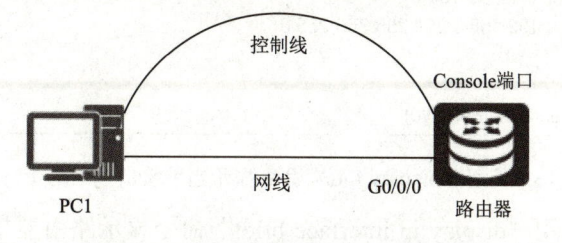

图 1.12　"路由器基本配置"网络拓扑

路由器接口 IP 与 PC 的 IP 规划见表 1.4。

表 1.4　路由器接口 IP 与 PC 的规划

设　备	接　口	IP 地址	网　关
AR1	G0/0/0	192.168.0.254/24	
PC1	NIC	192.168.0.10/24	192.168.0.254

根据图 1.12 所示的网络拓扑，使用一条 USB 转 RJ-45 控制线将主机 PC1 的 USB 端口与路由器 AR1 的 Console 端口相连，使用一条网线将主机 PC1 的网卡与路由器的 G0/0/0 接口相连。主机 PC1 既充当路由器的控制终端，也充当 Telnet 客户端和 SSH 客户端。

（1）路由器接口的配置与查看

路由器接口的配置任务通常包括：

①进入路由器需配置接口的接口视图下；

②在接口视图下完成接口的 IP 地址和子网掩码、接口描述、启用接口等配置内容。

注：华为路由器端口默认为启用状态。

路由器 G0/0/0 接口的 IP 地址与子网掩码、接口描述、启用接口的详细配置步骤如下：

```
[AR1]interface g0/0/0                               // 进入接口 G0/0/0 的端口视图
[AR1-GigabitEthernet0/0/0]ip address 192.168.0.254 24
// 配置 G0/0/0 接口的 IP 地址为 "192.168.0.254"，子网掩码长度为 "24"
```

注：子网掩码配置时也可以使用点分十进制。

```
[AR1-GigabitEthernet0/0/0]description linkto801room // 配置 G0/0/0 接口描述为 "linkto
                                                    //801room"
[AR1-GigabitEthernet0/0/0]undo shutdown             // 启用接口
```

完成配置后，使用以下查看命令查看接口的配置及详细信息：

①在 G0/0/0 接口的接口视图下，使用 "display this" 命令显示当前接口视图的运行配置，结果如图 1.13 所示。

12　企业网络互联技术（双语）

```
[AR1-GigabitEthernet0/0/0]display this
[V200R003C00]
#
interface GigabitEthernet0/0/0
 description linkto801room
 ip address 192.168.0.254 255.255.255.0
#
return
[AR1-GigabitEthernet0/0/0]
```

图 1.13　使用"display this"命令显示当前接口视图的运行配置

② 在任何视图下使用"display ip interface brief"命令显示路由器接口的简要配置信息，如图 1.14 所示。注意观察图中 G0/0/0 接口的 IP 地址是否正确配置，"Physical"与"Protocol"是否均为"up"。如果 G0/0/0 接口的"Physical"与"Protocol"均为"down"，通常是该接口网线损坏或者网线所连对端设备未加电。

```
[AR1]display ip interface brief
*down: administratively down
^down: standby
(l): loopback
(s): spoofing
The number of interface that is UP in Physical is 2
The number of interface that is DOWN in Physical is 2
The number of interface that is UP in Protocol is 2
The number of interface that is DOWN in Protocol is 2

Interface              IP Address/Mask    Physical   Protocol
GigabitEthernet0/0/0      192.168.0.254/24    up        up
GigabitEthernet0/0/1      unassigned          down      down
GigabitEthernet0/0/2      unassigned          down      down
NULL0                     unassigned          up        up(s)
[AR1]
```

图 1.14　使用"display ip interface brief"命令显示路由器接口的简要配置信息

③ 在任何视图下使用"display interface g0/0/0"命令显示路由器 G0/0/0 接口的详细信息，如图 1.15 所示。注意观察图 1.15 中加框的内容，G0/0/0 接口当前状态与协议状态是否为 UP，接口的 IP 地址与子网掩码是否配置正确。

（2）使用密码认证的控制端口配置与测试

用户通过控制端口访问路由器时，需要配置进行身份验证，该验证可以使用密码进行验证，也可以使用 AAA（authentication authorization accounting，认证、授权、审计）方式进行验证。

使用密码认证的控制端口配置任务为：

① 进入 Console0 的用户界面视图。

② 配置认证模式为密码认证。

③ 配置密码。

项目 1　企业分公司网络互联项目　　13

```
[AR1]display interface g0/0/0
GigabitEthernet0/0/0 current state : UP          ← 1. 接口状态为 "UP"
Line protocol current state : UP                 ← 2. 接口协议状态为 "UP"
Last line protocol up time : 2022-10-24 16:27:32 UTC-08:00
Description:linkto801room                         ← 3. 接口描述为 "linkto801room"
Route Port,The Maximum Transmit Unit is 1500
Internet Address is 192.168.0.254/24              ← 4. 接口IP地址与掩码为 "192.168.0.254/24"
IP Sending Frames' Format is PKTFMT_ETHNT_2, Hardware address is
00e0-fcab-115b
Last physical up time   : 2022-10-24 16:27:32 UTC-08:00
Last physical down time : 2022-10-24 14:40:33 UTC-08:00
Current system time: 2022-10-24 16:29:30-08:00
Port Mode: FORCE COPPER                           ← 5. 接口传输速率为 "1 000 Mbit/s"
Speed : 1000, Loopback: NONE
Duplex: FULL, Negotiation: ENABLE                 ← 6. 接口双工为全双工, 启用自动协商
Mdi  : AUTO
Last 300 seconds input rate 0 bits/sec, 0 packets/sec
Last 300 seconds output rate 0 bits/sec, 0 packets/sec
Input peak rate 0 bits/sec,Record time: -
Output peak rate 96 bits/sec,Record time: 2022-10-24 16:27:39

Input:  0 packets, 0 bytes
 Unicast:           0, Multicast:         0
 Broadcast:         0, Jumbo:             0
 Discard:           0, Total Error:       0

---- More ----
```

图 1.15　"display interface g0/0/0" 显示结果

请参考以下步骤完成使用密码认证的控制端口配置：

```
[AR1]user?                              // 查看以字符串 "user" 开始的所有命令
[AR1]user-interface ?                   // 查看 "user-interface" 的关键字或参数
[AR1]user-interface console ?           // 查看 "user-interface console" 的关键字或参数
[AR1]user-interface console 0           // 进入 console0 用户界面视图
[AR1-ui-console0]?                      // 列出 console0 用户界面视图下可用的命令
[AR1-ui-console0]authentication-mode?   // 查看 "authentication-mode" 的关键字或参数
[AR1-ui-console0]authentication-mode password  // 配置 console0 的认证模式为使用密码进行认证
```

在下面的交互信息中，输入需要配置的密码为 "Chengdu123"，具体如下所示：

```
Please configure the login password (maximum length 16):Chengdu123
[AR1-ui-console0]quit                   // 退出 console0 用户界面视图, 返回用户视图
[AR1]quit                               // 退出系统视图, 返回用户视图
<AR1>quit                               // 退出用户视图
```

完成 Console 的密码认证配置后，使用 Console 进入路由器的用户视图时，提示要进行 Console 登录认证，输入 Console 配置的认证密码 "Chengdu123" 后，才可进入用户视图，Console 密码登录认证如图 1.16 所示。

```
<AR1>quit
 Configuration console exit, please press any key to log on

Login authentication

Password: ←——— 在此输入配置的Console密码"Chengdu123", 注意
<AR1>              输入的密码不回显
```

图 1.16 Console 密码登录认证

（3）使用 AAA 认证的控制端口配置与测试

使用密码验证进行 Console 登录只能验证密码，不能验证用户的身份。可配置使用本地用户来进行 Console 登录的身份验证。

使用 AAA 认证的控制端口配置需要完成以下配置任务：

① 配置用于 Console 登录的合法 AAA 用户，具体包括配置合法的用户名与密码、授权用户的服务类型为"terminal"，用户的权限级别。

② 进入 Console 用户界面视图，配置认证模式为 AAA。

请参考以下步骤完成使用 AAA 认证的控制端口配置：

```
[AR1]aaa                                              // 进入 AAA 视图
[AR1-aaa]local-user zhangcr password cipher Chengdu123   // 创建本地用户，用户名为"zhangcr"，
                                                      // 密码为"Chengdu123"

[AR1-aaa]local-user zhangcr service-type terminal     // 授权用户"zhangcr"服务类型为
                                                      // "terminal"

[AR1-aaa]local-user zhangcr privilege level 15        // 配置用户"zhangcr"的权限级别为 15
[AR1-aaa]quit                                         // 返回到系统视图
[AR1]user-interface console 0                         // 进入 console0 用户界面视图
[AR1-ui-console0]authentication-mode aaa              // 配置 console0 的认证模式为使用
                                                      // AAA 进行认证

[AR1-ui-console0]quit                                 // 退出 console0 用户界面视图，返
                                                      // 回到用户视图

[AR1]quit                                             // 退出系统视图，返回到用户视图
<AR1>quit                                             // 退出用户视图
```

完成 AAA 认证的控制端口配置后，使用 Console 访问路由器时，提示要进行 Console 登录认证，输入合法的 Console 用户名与密码才能进入用户视图，如图 1.17 所示。

（4）Telnet 的配置与测试

远程主机可以使用 Telnet 远程登录到路由器上。远程主机为 Telnet 客户端，配置路由器为 Telnet 服务器端。路由器在进行 Telnet 服务配置之前，需要先完成路由器接口的配置，如果路由器与 Telnet 客户端中间还跨越了其他路由设备，还需要进行路由的配置，确保 Telnet 客户端与路由器具有 IP 的连通性。

```
<AR1>quit

 Configuration console exit, please press any key to log on


Login authentication


Username:zhangcr  ←——— 输入用户名"zhangcr"
Password:         ←——— 输入配置的终端用户"zhangcr"的密码"Chengdu123",
<AR1>                    注意输入的密码不回显
```

图 1.17　Console 本地用户认证

采用 AAA 认证的路由器 Telnet 服务配置包括以下配置任务：

① 启用 Telnet 服务。

② 配置 Telnet 登录的合法用户，具体包括进入 AAA 视图、创建用户名与密码、配置用户的服务类型为"Telnet"，配置用户的权限级别。

③ 进入到 VTY 用户界面视图，配置协议为 Telnet，认证模式为 AAA。

请参考以下步骤在路由器上完成 Telnet 的配置，其规划为：Telnet 会话为 VTY 0 到 VTY 4，共五个 Telnet 会话，认证采用 AAA 认证，Telnet 的用户名为"zhangcr"，密码为"Chengdu123"，权限级别为"15"。

```
[AR1]Telnet server enable                    // 启用 Telnet 服务
[AR1]aaa                                      // 进入 AAA 视图
[AR1-aaa]local-user zhangcr password cipher Chengdu123
// 创建本地用户，用户名为"zhangcr"，密码为"Chengdu123"
[AR1-aaa]local-user zhangcr service-type telnet  // 授权用户"zhangcr"服务类型为 Telnet 服务
[AR1-aaa]local-user zhangcr privilege level 15  // 配置用户"zhangcr"的权限级别为 15
[AR1-aaa]quit                                 // 返回到系统视图
[AR1]user-interface vty 0 4                   // 进入 VTY0 到 VTY4 的用户界面视图
[AR1-ui-vty0-4]protocol inbound telnet        // 配置 inbound 方向协议为 Telnet 协议
[AR1-ui-vty0-4]authentication-mode aaa        // 配置 VTY0 到 VTY4 五个虚拟终端会话的认
                                              // 证模式为使用 AAA 进行认证
[AR1-ui-vty0-4]quit                           // 退出 VTY0 到 VTY4 的用户界面视图，返
                                              // 回到用户视图
[AR1]quit                                     // 退出系统视图，返回到用户视图
<AR1>quit                                     // 退出用户视图
```

在 Telnet 客户端 PC1 上参考以下步骤完成 Telnet 测试：

① 打开"控制面板"，选择"网络和共享中心"中的"更改适配器设置"命令，在弹出的"网络连接"窗口中，右击"以太网"网卡，选择"属性"下的"网络和共享中心"命令，参照表 1.4 的参数完成主机 PC1 的 IP 地址、子网掩码、默认网关配置，如图 1.18 所示。

16 企业网络互联技术（双语）

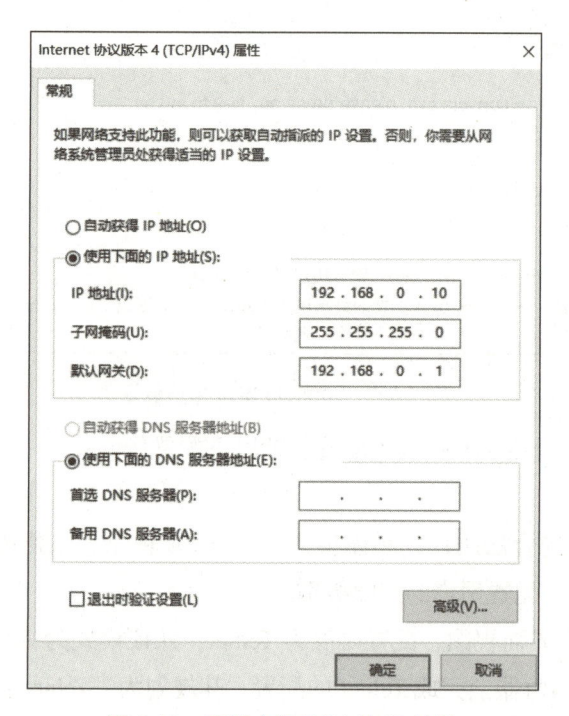

图 1.18 配置主机 PC1 的 IP 地址

②配置完成后，在 DOS 命令行下使用"ipconfig /all"命令查看配置的 IP 地址是否已生效，如图 1.19 所示。

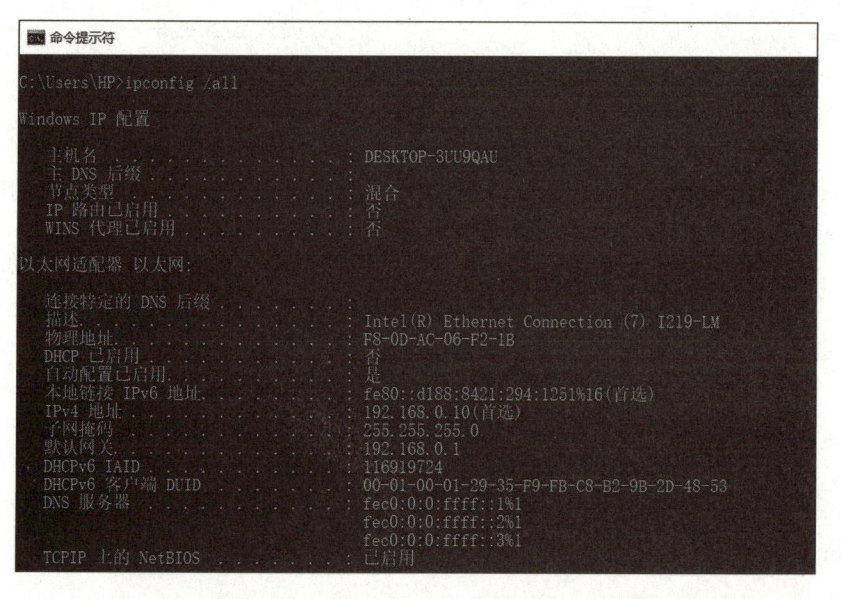

图 1.19 使用"ipconfig /all"命令查看主机 PC1 的 IP 信息

③ 在 PC1 的命令行下使用"ping 192.168.0.254"命令测试主机 PC1 与路由器 AR1 接口 G0/0/0 的 IP 连通性，如果收到目标主机"192.168.0.254"的回复，表示与路由器的 G0/0/0 接口具有 IP 连通性。图 1.20 表示已成功 ping 通"192.168.0.254"。

项目 1　企业分公司网络互联项目　17

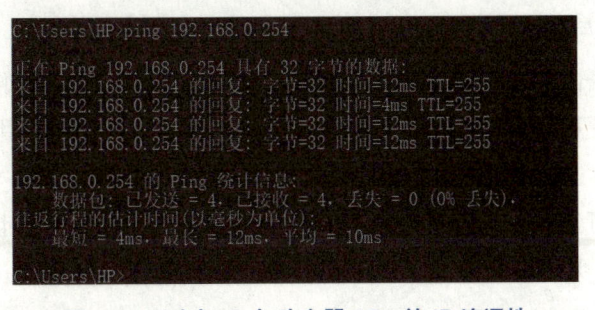

图 1.20　测试 PC 与路由器 AR1 的 IP 连通性

④在 PC1 上，运行 SecureCRT 虚拟终端软件。在 SecureCRT 软件窗口中，选择"文件"菜单下的"快速连接"命令，在弹出的"快速连接"窗口中，协议选择"Telnet"，主机名输入路由器 AR1 接口 G0/0/0 的 IP 地址"192.168.0.254"，端口为默认端口"23"，如图 1.21 所示。

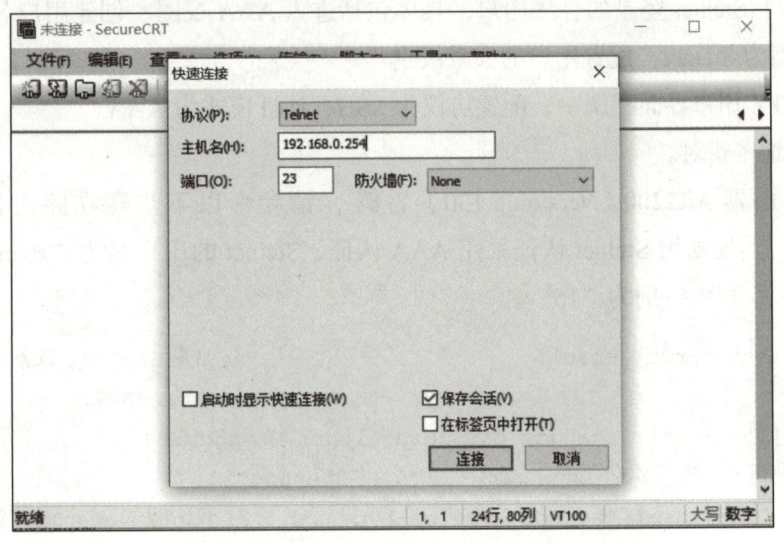

图 1.21　PC1 Telnet 到路由器 AR1

⑤单击图 1.21 中的"连接"按钮，然后根据提示输入用户名"zhangcr"，密码"Chengdu123"（注意，密码不回显）。按【Enter】键进入路由器的用户视图提示符，如图 1.22 所示。

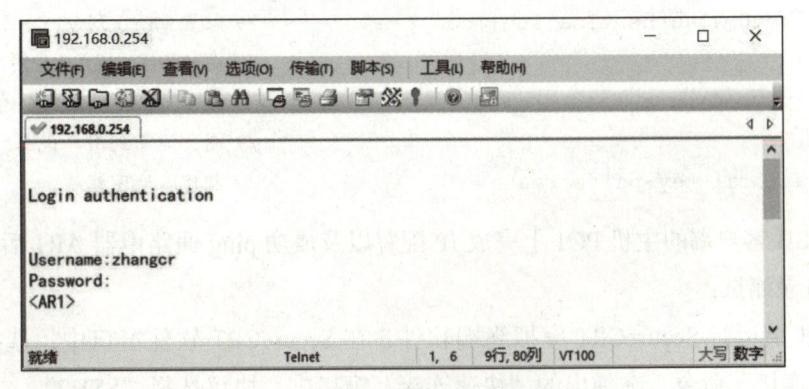

图 1.22　成功 Telnet 到路由器 AR1 上

18　企业网络互联技术（双语）

（5）SSH 的配置与测试

使用 Telnet 访问路由器存在许多的安全问题。其中，最为突出的问题是 Telnet 身份验证所需用户名与密码在线路上传输时是以明文的方式传输，网络入侵者可使用 Wireshark 等捕获包软件捕获到线路上传输的 Telnet 用户名与密码。

SSH 是目前较为可靠、专为远程登录和其他网络服务提供安全性的协议。在使用 SSH 访问路由器时，使用数字证书认证 SSH 客户端和路由器之间的连接，并加密传输身份认证密码。

SSH 协议有 SSH1 和 SSH2 两个版本，SSH2 在安全、功能和性能上都比 SSH1 有优势，目前广泛使用的是 SSH2。

华为路由器上 SSH 又称 Stelnet，其配置包括以下任务：

① 启用 Stelnet 服务。

② 配置用于 Stelnet 登录的合法用户，具体包括进入 AAA 视图、创建用户名与密码、配置用户的服务类型为 Stelnet，配置用户的权限级别。

③ 进入 VTY 用户界面视图中，配置协议为 SSH，认证模式为 AAA。

④ 生成本地密钥对。

以华为路由器 AR2200（Version 5.130）为例，请参考以下步骤在路由器 AR1 上完成 Stelnet 的配置，其规划为 Stelnet 认证采用 AAA 认证，Stelnet 的用户名为"zhangcr"，密码为"Chengdu123"，权限级别为"15"。

```
[AR1]stelnet server enable                        // 启用 Stelnet 服务
[AR1]aaa                                           // 进入 AAA 视图
[AR1-aaa]local-user zhangcr password cipher Chengdu123
// 创建本地用户，用户名为"zhangcr"，密码为"Chengdu123"
[AR1-aaa]local-user zhangcr service-type ssh       // 授权用户"zhangcr"服务类型为 SSH 服务
[AR1-aaa]local-user zhangcr privilege level 15     // 配置用户"zhangcr"的权限级别为 15
[AR1-aaa]quit                                      // 返回到系统视图
[AR1]user-interface vty 0 4                        // 进入 VTY0 到 VTY4 的用户界面视图
[AR1-ui-vty0-4]protocol inbound SSH                // 配置协议为 SSH 协议
[AR1-ui-vty0-4]authentication-mode aaa             // 配置 VTY0 到 VTY4 五个虚拟终端
                                                   // 会话的认证模式为使用 AAA 进行认证
[AR1-ui-vty0-4]quit                                // 退出 VTY0 到 VTY4 的用户界面视
                                                   // 图，返回到用户视图
[AR1]rsa local-key-pair create                     // 创建本地密钥对
```

在充当 SSH 客户端的主机 PC1 上完成 IP 配置以及成功 ping 通路由器 AR1 后，参考以下步骤完成 SSH 登录测试：

① 在 PC1 上运行 SecureCRT 虚拟终端软件，在 SecureCRT 软件窗口中，选择"文件"菜单下的"快速连接"命令，在弹出的"快速连接"窗口中，协议选择"SSH2"，主机名输入路

由器 AR1 接口 G0/0/0 的 IP 地址"192.168.0.254"，端口为默认端口"22"，用户名输入配置的
SSH 用户名"zhangcr"，如图 1.23 所示。

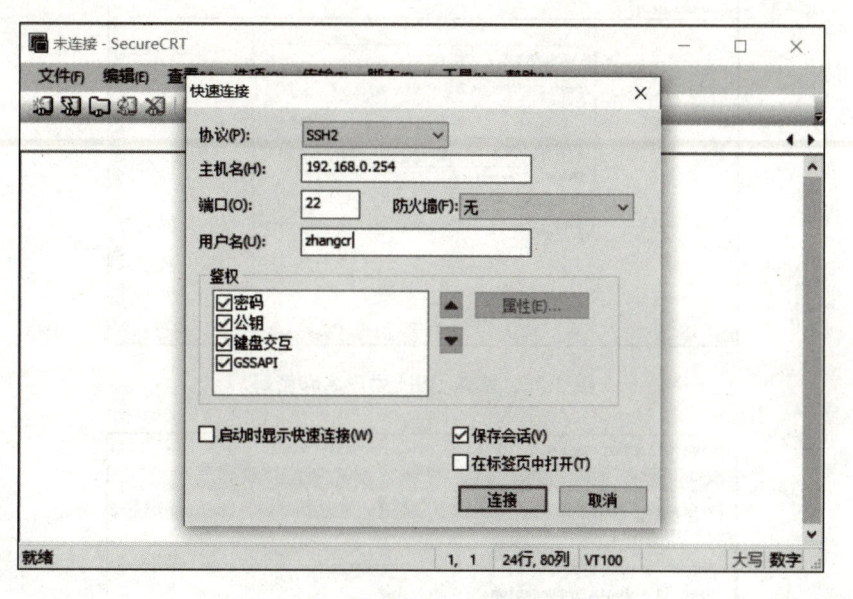

图 1.23　使用 SSH 连接到路由器

②单击图 1.23 中的"连接"按钮，在弹出的"新建主机密钥"窗口中，单击"接受并保存"
或者"只接受一次"按钮，如图 1.24 所示。

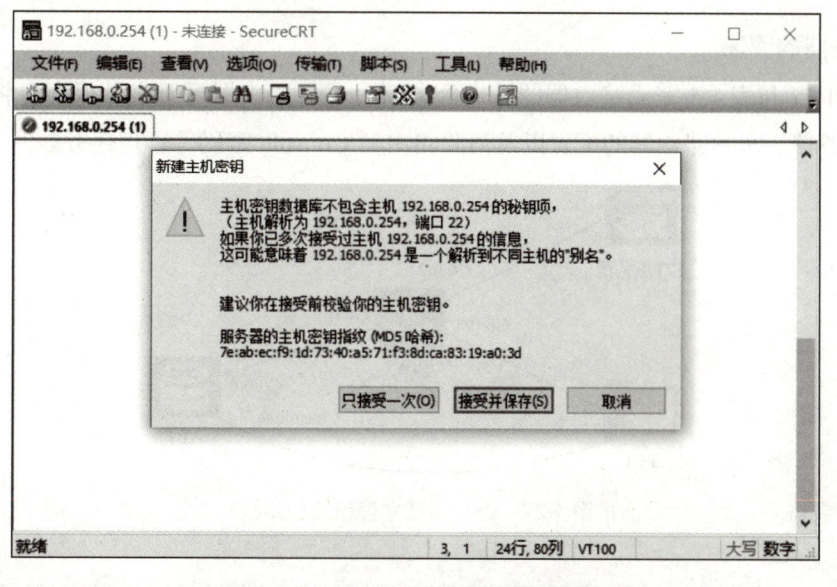

图 1.24　SSH "新建主机密钥"窗口

③根据提示，输入 SSH 用户名"zhangcr"，密码"Chengdu123"，如图 1.25 所示。单击"确
定"按钮登录到路由器 AR1 上，如图 1.26 所示。

20　企业网络互联技术（双语）

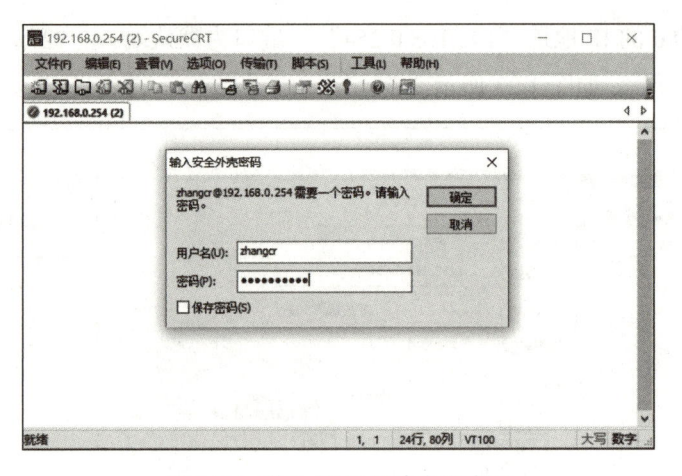

图 1.25　输入 SSH 用户名和密码

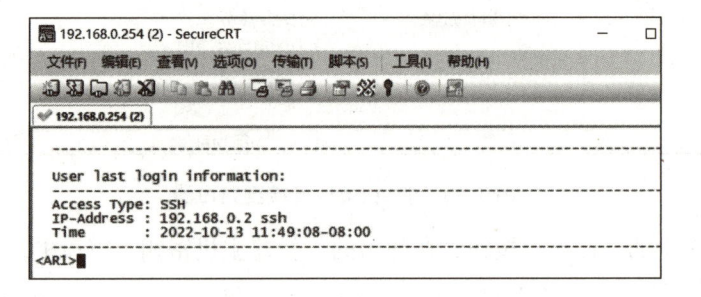

图 1.26　登录到路由器 AR1 上

3. 路由器的管理

根据图 1.27 与表 1.5 完成路由器的基本配置后，进行路由器配置文件保存、将配置文件备份到 FTP 服务器、清除路由器的配置以及清除路由器 Console 密码等管理任务。

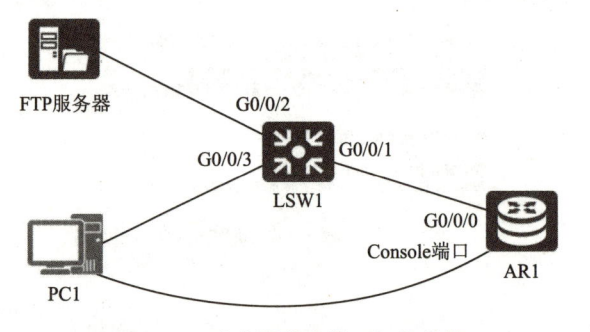

图 1.27　路由器基本管理拓扑结构

表 1.5　IP 规划

设　　备	接　　口	IP 地址	网　　关
AR1	G0/0/0	192.168.0.254/24	
PC1	NIC	192.168.0.10/24	192.168.0.254
FTP	NIC	192.168.0.1/24	192.168.0.254

项目 1　企业分公司网络互联项目　　21

（1）路由器配置文件管理

① 在路由器 AR1 上根据表 1.5 的 IP 规划完成 G0/0/0 接口的配置，配置步骤如下：

```
[AR1]interface g0/0/0
[AR1-GigabitEthernet0/0/0]ip address 192.168.0.254 24
[AR1-GigabitEthernet0/0/0]quit
```

② 在路由器 AR1 的用户视图下使用"dir"命令查看当前目录下的文件列表。

图 1.28 为在 AR1 上查看当前目录下文件列表示例。

```
<AR1>dir
Directory of flash:/

Idx  Attr   Size(Byte)  Date        Time(LMT)  FileName
  0  drw-           -   Mar 08 2023 09:41:22   dhcp
  1  -rw-     121,802   May 26 2014 09:20:58   portalpage.zip
  2  -rw-       2,263   Mar 08 2023 09:41:18   statemach.efs
  3  -rw-     828,482   May 26 2014 09:20:58   sslvpn.zip

1,090,732 KB total (784,464 KB free)
```

图 1.28　在 AR1 上查看当前目录下文件列表示例

③ 在 AR1 的命令视图下使用"save"命令保存配置，然后使用"dir flash:"命令查看 Flash 中已保存的配置文件，步骤与显示结果如图 1.29 所示。

```
<AR1>save
The current configuration will be written to the device.
Are you sure to continue? (y/n)[n]:y  ←——— 输入"y"确认保存配置并继续
It will take several minutes to save configuration file, please wait.......
Configuration file had been saved successfully
Note: The configuration file will take effect after being activated
<AR1>dir
Directory of flash:/

Idx  Attr   Size(Byte)  Date        Time(LMT)  FileName
  0  drw-           -   Mar 08 2023 09:41:22   dhcp
  1  -rw-     121,802   May 26 2014 09:20:58   portalpage.zip
  2  -rw-       2,263   Mar 08 2023 09:41:18   statemach.efs
  3  -rw-     828,482   May 26 2014 09:20:58   sslvpn.zip
  4  -rw-         249   Mar 08 2023 09:45:10   private-data.txt
  5  -rw-         562   Mar 08 2023 09:45:09   vrpcfg.zip

1,090,732 KB total (784,448 KB free)
```
保存的配置文件

图 1.29　保存配置并查看

④ 在 AR1 上使用"save"命令将当前的配置保存并命名为"AR1.cfg"，然后使用"dir flash:"命令查看 Flash 中的文件，如图 1.30 所示。

注：配置文件的文件名必须以".cfg"或".zip"作为扩展名。

```
<AR1>save AR1.cfg
Are you sure to save the configuration to AR1.cfg? (y/n)[n]:y      ← 输入 "y" 确认
It will take several minutes to save configuration file, please wait......   保存配置
Configuration file had been saved successfully
Note: The configuration file will take effect after being activated
<AR1>dir flash:
Directory of flash:/

Idx  Attr   Size(Byte)  Date        Time(LMT)  FileName
 0   -rw-        851    Mar 08 2023 10:11:28   ar1.cfg          ← 保存的配置文件
 1   drw-          -    Mar 08 2023 09:41:22   dhcp
 2   -rw-    121,802    May 26 2014 09:20:58   portalpage.zip
 3   -rw-      2,263    Mar 08 2023 09:41:18   statemach.efs
 4   -rw-    828,482    May 26 2014 09:20:58   sslvpn.zip
 5   -rw-        249    Mar 08 2023 09:45:10   private-data.txt
 6   -rw-        562    Mar 08 2023 09:45:09   vrpcfg.zip

1,090,732 KB total (784,452 KB free)
```

图 1.30　保存配置并命名为 "ar1.cfg"

⑤ 使用 "startup saved-configuration" 命令设置路由器下一次启动所使用的配置文件，然后验证设置。步骤如下：

```
<AR1>startup saved-configuration?          // 查看 "startup saved-configuration" 命令的
                                           // 关键字或参数
<AR1>startup saved-configuration ar1.cfg// 配置下一次启动所使用的配置文件为 "ar1.cfg"
<AR1>display startup                       // 显示设备本次及下次启动的操作系统、配置文件等信息
```

⑥ 根据表 1.5，完成 FTP 服务器的 IP 配置、FTP 服务的搭建与配置。在路由器 AR1 上输入 "ftp 192.168.0.1" 使路由器 AR1 作为 FTP 客户端连接到 FTP 服务器，把 Flash 中的 "ar1.cfg" 配置文件上传到 FTP 服务器上。步骤如图 1.31 所示。

```
<AR1>ftp 192.168.0.1
Trying 192.168.0.1 ...

Press CTRL+K to abort
Connected to 192.168.0.1.
220 FtpServerTry FtpD for free
User(192.168.0.1:(none)):anonymous    ← 输入匿名账号 "anonymous"
331 Password required for anonymous .
Enter password:    ← 输入任何的邮箱地址
230 User anonymous logged in , proceed
[AR1-ftp]put flash:/ar1.cfg
200 Port command okay.
150 Opening BINARY data connection for ar1.cfg
226 Transfer finished successfully. Data connection closed.
FTP: 0 byte(s) sent in 0.170 second(s) 0.00byte(s)/sec.
```

图 1.31　把 Flash 中的配置文件上传到 FTP 服务器

⑦ 输入 "reset saved-configuration" 命令，再输入 "reboot" 命令重启路由器，清空路由器的配置。步骤如图 1.32 所示。

项目 1　企业分公司网络互联项目　23

```
<AR1>reset saved-configuration
This will delete the configuration in the flash memory.

The device configuratio
ns will be erased to reconfigure.

Are you sure? (y/n)[n]:y  ←—— 输入 "y" 删除配置
 Clear the configuration in the device successfully.
<AR1>reboot
Info: The system is comparing the configuration, please wait.
Warning: All the configuration will be saved to the next startup configuration.
Continue ? [y/n]:n  ←—— 输入 "n" 不保存配置
System will reboot! Continue ? [y/n]:y  ←—— 输入 "y" 确认重启设备
```

图 1.32　清空路由器配置

（2）清除路由器 Console 密码

使用控制线把控制台与路由器的 Console 端口连接起来，确保连接正常。然后，先断开路由器的电源，再打开路由器的电源。观察路由器的启动过程，按以下步骤清除 Console 密码：

① 当出现如图 1.33 方框所示的 "Press Ctrl+B to break auto startup" 时，立即按键盘上的【Ctrl】加字母【B】键。

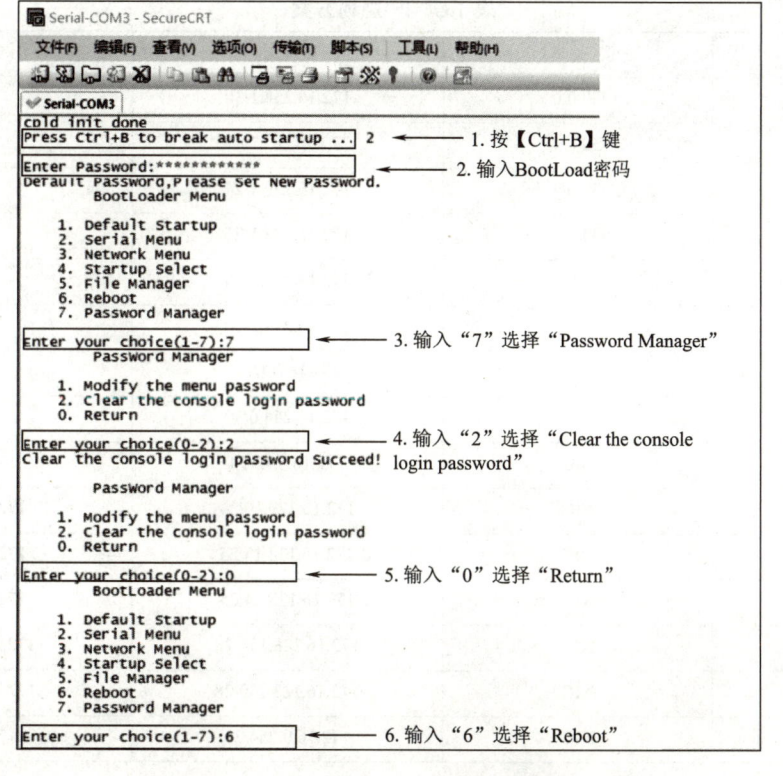

图 1.33　华为 AR1220 的 Console 密码清除步骤

② 在 "Enter password" 处输入 BootLoad 密码（默认情况下，华为 AR1220C 的默认密码为 "Admin@huawei"）。

24　企业网络互联技术（双语）

③在"BootLoader"菜单下输入选择"7"选择"Password Manager（密码管理）"。

④在"Password Manager"下输入"2"选择"Clear the console login password（清除 Console 登录密码）。

⑤在"Password Manager"下输入"0"选择"Return（返回）"，返回到"BootLoader"菜单。

⑥在"BootLoader"菜单下输入选择"6"选择"Reboot（重启）"，重新启动设备。步骤如图 1.33 所示。

注：不同的路由设备，其密码恢复步骤有细微差别。

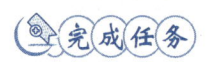

1. 任务规划

根据图 1.1 所示的项目 1 网络拓扑以及建网需求，分公司的预算部门有 140 台主机，管理部门有 100 台主机，分公司有 8 台服务器业务，预算部门、管理部门与服务器之间的数据需要广播隔离，它们不能直接通信，需要通过路由器实现通信。本任务使用表 1.6 所示的 IP 规划方案。

表 1.6　IP 规划方案

设 备	接 口	IP 地址	网 关
fenbu	G0/0/0	172.16.254.1/30	
	G0/0/1	172.16.128.129/25	
	G0/0/2	172.16.129.1/24	
	G3/0/0	172.16.128.1/25	
	S2/0/0	172.16.254.5/30	
zongbu	G0/0/0	172.16.254.2/30	
	G0/0/1	172.16.0.1/24	
	S3/0/0	172.16.254.6/30	
PC1	NIC	172.16.0.10/24	172.16.0.1
PC2	NIC	172.16.129.10/24	172.16.129.1
PC3	NIC	172.16.128.13/25	172.16.128.1
PC4	NIC	172.16.128.14/25	172.16.128.1
Web 服务器	NIC	172.16.128.131/28	172.16.128.129
DNS 服务器	NIC	172.16.128.130/28	172.16.128.129
FTP 服务器	NIC	172.16.0.254/24	172.16.0.1

①分部的路由器 Console 认证规划为：采用本地用户认证，用户名为 cdp，密码为 Chenghua123，用户命令行级别为 15。

② 分部的路由器 SSH 认证规划为：采用本地用户认证，用户名为 cdp，密码为 Chenghua123，用户命令行级别为 15。

③ 总部的路由器 Console 认证规划为：采用本地用户认证，用户名为 zhang，密码为 Chengdu123，用户命令行级别为 15。

④ 总部的路由器 SSH 认证规划为：采用本地用户认证，用户名为 zhang，密码为 Chengdu123，用户命令行级别为 15。

2. 任务实施

（1）路由器"fenbu"的配置

```
[Huawei]sysname fenbu
// 配置路由器的主机名为"fenbu"
[fenbu]interface g0/0/0
[fenbu-GigabitEthernet0/0/0]ip address 172.16.254.1 30
[fenbu-GigabitEthernet0/0/0]description linktozongbu
[fenbu-GigabitEthernet0/0/0]quit
// 配置路由器"fenbu"的 G0/0/0 接口
[fenbu]interface s2/0/0
[fenbu-Serial2/0/0]ip address 172.16.254.5 30
[fenbu-Serial2/0/0]description linktozongbu
[fenbu-Serial2/0/0]quit
// 配置路由器"fenbu"的 S2/0/0 接口
[fenbu]interface g0/0/1
[fenbu-GigabitEthernet0/0/1]ip address 172.16.128.129 28
[fenbu-GigabitEthernet0/0/1]description server
[fenbu-GigabitEthernet0/0/1]undo shutdown
[fenbu-GigabitEthernet0/0/1]quit
// 配置路由器"fenbu"的 G0/0/1 接口
[fenbu]interface g0/0/2
[fenbu-GigabitEthernet0/0/2]ip address 172.16.129.1 24
[fenbu-GigabitEthernet0/0/2]description yusuanbumen
[fenbu-GigabitEthernet0/0/2]quit
[fenbu]
// 配置路由器"fenbu"的 G0/0/2 接口
[fenbu]interface g3/0/0
[fenbu-GigabitEthernet3/0/0]description manager
[fenbu-GigabitEthernet3/0/0]ip address 172.16.128.1 25
[fenbu-GigabitEthernet3/0/0]quit
[fenbu]
// 配置路由器"fenbu"的 G3/0/0 接口
```

```
[fenbu]aaa
[fenbu-aaa]local-user cdp password cipher Chenghua123
[fenbu-aaa]local-user cdp service-type terminal ssh
// 配置路由器 "fenbu" 的本地用户 "cdp"，密码为 "Chenghua123"，授权服务类型为终端用户与 SSH 用户
[fenbu]user-interface console 0
[fenbu-ui-console0]authentication-mode aaa
[fenbu-ui-console0]user privilege level 15
[fenbu-ui-console0]quit
// 配置路由器 "fenbu" 的 Console 认证为 AAA 本地用户认证，用户命令行级别为 15 级
[fenbu]stelnet server enable
[fenbu]user-interface vty 0 4
[fenbu-ui-vty0-4]protocol inbound ssh
[fenbu-ui-vty0-4]authentication-mode aaa
[fenbu-ui-vty0-4]user privilege level 15
[fenbu-ui-vty0-4]quit
[fenbu]rsa local-key-pair create
// 配置路由器 "fenbu" 的 SSH，启用 Stelnet 服务，虚拟终端协议为 SSH，认证采用 AAA 本地用户认
// 证，用户命令行级别为 15 级，生成本地密钥对
[fenbu]quit
<fenbu>save // 保存配置
```

（2）路由器 "zongbu" 的配置

```
[Huawei]sysname zongbu
// 配置路由器的主机名为 "zongbu"
[zongbu]interface g0/0/0
[zongbu-GigabitEthernet0/0/0]ip address 172.16.254.2 30
[zongbu-GigabitEthernet0/0/0]quit
// 配置路由器 "zongbu" 的 G0/0/0 接口
[zongbu]interface s3/0/0
[zongbu-Serial3/0/0]ip address 172.16.254.6 30
[zongbu-Serial3/0/0]quit
// 配置路由器 "zongbu" 的 S3/0/0 接口
[zongbu]interface g0/0/1
[zongbu-GigabitEthernet0/0/1]ip address 172.16.0.1 24
[zongbu-GigabitEthernet0/0/1]quit
// 配置路由器 "zongbu" 的 G0/0/1 接口
[zongbu]aaa
[zongbu-aaa]local-user zhang password cipher Chengdu123
[zongbu-aaa]local-user zhang service-type terminal ssh
[zongbu-aaa]quit
```

项目 1　企业分公司网络互联项目　27

```
// 配置路由器"zongbu"的本地用户"zhang"，密码为"Chengdu123"，授权服务类型为终端用户与 SSH 用户
[zongbu]user-interface console 0
[zongbu-ui-console0]authentication-mode aaa
[zongbu-ui-console0]user privilege level 15
[zongbu-ui-console0]quit
// 配置路由器"zongbu"的 Console 认证为 AAA 本地用户认证，用户命令行级别为 15 级
[zongbu]stelnet server enable
[zongbu]user-interface vty 0 4
[zongbu-ui-vty0-4]protocol inbound ssh
[zongbu-ui-vty0-4]authentication-mode aaa
[zongbu-ui-vty0-4]user privilege level 15
[zongbu-ui-vty0-4]quit
[zongbu]rsa local-key-pair create
// 配置路由器"zongbu"的 SSH，启用 Stelnet 服务，虚拟终端协议为 SSH，认证采用 AAA 本地用户
// 认证，用户命令行级别为 15 级，生成本地密钥对
[zongbu]quit
<zongbu>save
 // 保存配置
```

（3）路由器查看与测试

① 完成分部路由器"fenbu"与总部路由器"zongbu"的主机名与 IP 地址配置后，在网络所有链路都正常的情况下，使用"display ip interface brief"查看路由器的接口 IP 是否已正确配置，物理（Physical）与协议（Protocol）是否已"UP"。路由器"fenbu"与路由器"zongbu"的接口简要信息分别如图 1.34、图 1.35 所示。

```
[fenbu]display ip interface brief
*down: administratively down
^down: standby
(l): loopback
(s): spoofing
The number of interface that is UP in Physical is 6
The number of interface that is DOWN in Physical is 2
The number of interface that is UP in Protocol is 6
The number of interface that is DOWN in Protocol is 2


Interface               IP Address/Mask      Physical   Protocol
GigabitEthernet0/0/0    172.16.254.1/30      up         up
GigabitEthernet0/0/1    172.16.128.129/28    up         up
GigabitEthernet0/0/2    172.16.129.1/24      up         up
GigabitEthernet3/0/0    172.16.128.1/25      up         up
GigabitEthernet4/0/0    unassigned           down       down
NULL0                   unassigned           up         up(s)
Serial2/0/0             172.16.254.5/30      up         up
Serial2/0/1             unassigned           down       down
```

图 1.34　链路正常情况下路由器"fenbu"的接口简要信息

28 企业网络互联技术（双语）

```
[zongbu]display ip interface brief
*down: administratively down
^down: standby
(l): loopback
(s): spoofing
The number of interface that is UP in Physical is 4
The number of interface that is DOWN in Physical is 3
The number of interface that is UP in Protocol is 4
The number of interface that is DOWN in Protocol is 3

Interface                    IP Address/Mask      Physical   Protocol
GigabitEthernet0/0/0         172.16.254.2/30      up         up
GigabitEthernet0/0/1         172.16.0.1/24        up         up
GigabitEthernet0/0/2         unassigned           down       down
GigabitEthernet4/0/0         unassigned           down       down
NULL0                        unassigned           up         up(s)
Serial3/0/0                  172.16.254.6/30      up         up
Serial3/0/1                  unassigned           down       down
```

图 1.35　链路正常情况下路由器"zongbu"的接口简要信息

② 分别在分部路由器"fenbu"与总部路由器"zongbu"上进行 Console 登录的测试，测试结果分别如图 1.36、图 1.37 所示。

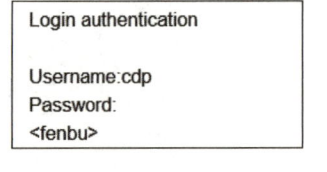

```
Login authentication

Username:cdp
Password:
<fenbu>
```

```
Login authentication

Username:zhang
Password:
 <zongbu>
```

图 1.36　路由器"fenbu"控制台登录认证 　　　图 1.37　路由器"zongbu"控制台登录认证

③ 在主机 PC1 上根据表 1.6 的规划完成 IP 配置，并在命令行使用 ping 测试到路由器"zongbu"G0/0/1 接口 IP 的 IP 连通性，确保连通性后，在 PC1 上运行 SecureCRT 虚拟终端软件，选择"文件"菜单下的"快速连接"命令，在弹出的"快速连接"窗口中，协议选择"SSH2"，主机名输入路由器 zongbu 接口 G0/0/1 的 IP 地址"172.16.0.1"，端口为默认端口"22"，用户名输入配置的 SSH 用户名"zhang"，单击"连接"按钮，在弹出的"新建主机密钥"窗口中，单击"接受并保存"按钮。在弹出的"输入安全外壳密码"窗口中，输入用户"zhang"的密码"Chengdu123"，如图 1.38 所示，使用 SSH 远程登录到路由器"zongbu"上。

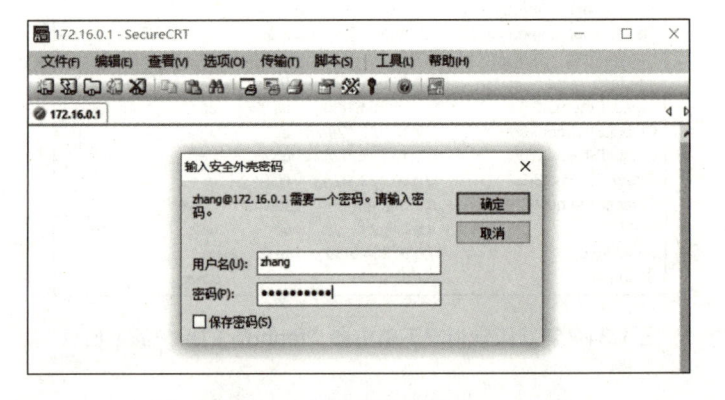

图 1.38　PC1 SSH 到路由器"zongbu"

项目 1　企业分公司网络互联项目　29

④ 在主机 PC2、主机 PC3 与主机 PC4 上根据表 1.6 的规划完成 IP 配置，并在命令行使用 ping 测试 PC2 到路由器"fenbu"G0/0/2 接口 IP（172.16.129.1）的 IP 连通性，PC3 与 PC4 到路由器"fenbu"G3/0/0 接口 IP（172.16.128.1）的 IP 连通性，确保连通性。然后在 PC2、PC3 与 PC4 上测试是否可以使用 SSH 登录到路由器"fenbu"上。图 1.39 为主机 PC2 使用路由器"fenbu"G3/0/0 的 IP（172.16.128.1）作为远程主机的目标地址，SSH 到路由器"fenbu"上。

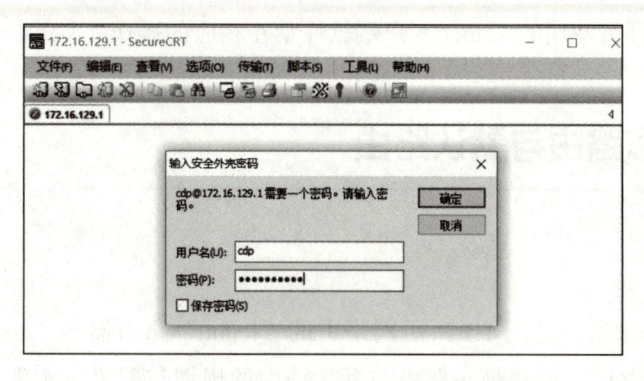

图 1.39　PC2 SSH 到路由器"fenbu"

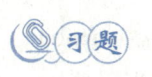 习题

选择题

（1）下面路由器存储设备（　　）的内容在断电或重启后会丢失。

　　A. RAM　　　　　　B. ROM　　　　　　C. NVRAM　　　　　D. Flash

（2）配置路由器 G0/0/0 接口的 IP 地址与掩码应该在提示符（　　）下进行。

　　A. [Huawei]　　　　　　　　　　B. <Huawei>

　　C. [Huawei-GigabitEthernet0/0/0]　　D. [Huawei-aaa]

（3）下面命令（　　）用于清空路由器的配置。

　　A. reset saved-configuration　　　　B. save

　　C. display version　　　　　　　　D. dir

（4）路由器命令"display ip interface brief"的作用是（　　）。

　　A. 显示路由器接口的简要配置信息

　　B. 配置接口

　　C. 检查是否建立连接

　　D. 进入接口配置模式

（5）路由器的初始化配置需要使用路由器的（　　）端口。

　　A. G0/0/0　　　　　B. S0/0/0　　　　　C. G0/0　　　　　D. Console

（6）下面命令（　　）可以修改华为路由器的设备名称。

　　A. hostname　　　　B. rename　　　　C. sysname　　　　D. domain

（7）网络管理员成功 Telnet 到路由器后，发现无法完成路由器接口 IP 地址的配置，可能的原因是（　　）。

    A. Telnet 用户级别配置错误

    B. Telnet 用户的认证方式配置错误

    C. 网络管理员主机不能 ping 通路由器

    D. 网络管理员使用的 Telnet 客户端软件禁止相应的操作

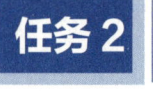

任务 2　静态路由与默认路由

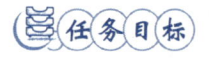

任务描述

完成本项目任务 1 后，在图 1.1 所示网络中的 "fenbu" 路由器与 "zongbu" 路由器上进行静态路由、浮动静态路由、汇总静态路由与默认路由的规划与配置。实现 "分公司内部与企业总部网络的互联互通，分公司通过总部接入 Internet" 的建网需求。在互联互通的基础上，路由器 "fenbu" 与路由器 "zongbu" 上的路由表条目尽可能少，不能有路由环路，路由器 "fenbu" 与路由器 "zongbu" 互联互通时，正常数据包走它们之间的千光以太网链路，只有当千兆以太网链路出现故障时，才走串行链路。完成配置后进行网络测试与故障排除。

任务目标

- 理解路由表的组成以及插入规则。
- 掌握静态路由的规划与配置。
- 掌握汇总静态路由的规划与配置。
- 掌握默认路由的配置。
- 掌握浮动路由的规划与配置。

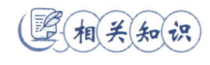

相关知识

1. 路由表

路由器可以通过管理员手工配置或者通过动态路由协议与其他路由器交换路由信息获得到达目标网络或节点的最佳路径，并将所获得的最佳路径信息以表的形式保存在路由器的 RAM 中，该表称为路由表。网络中的每个路由器根据其自身路由表中的选路信息对数据分组独立做出转发决定。图 1.40 所示为路由表示例。

① Destination/Mask：表示该路由条目的目标网络地址与子网掩码长度。

② Proto（protocol）：表示该路由条目的协议类型。表示路由器是通过什么路由协议获知该路由的。

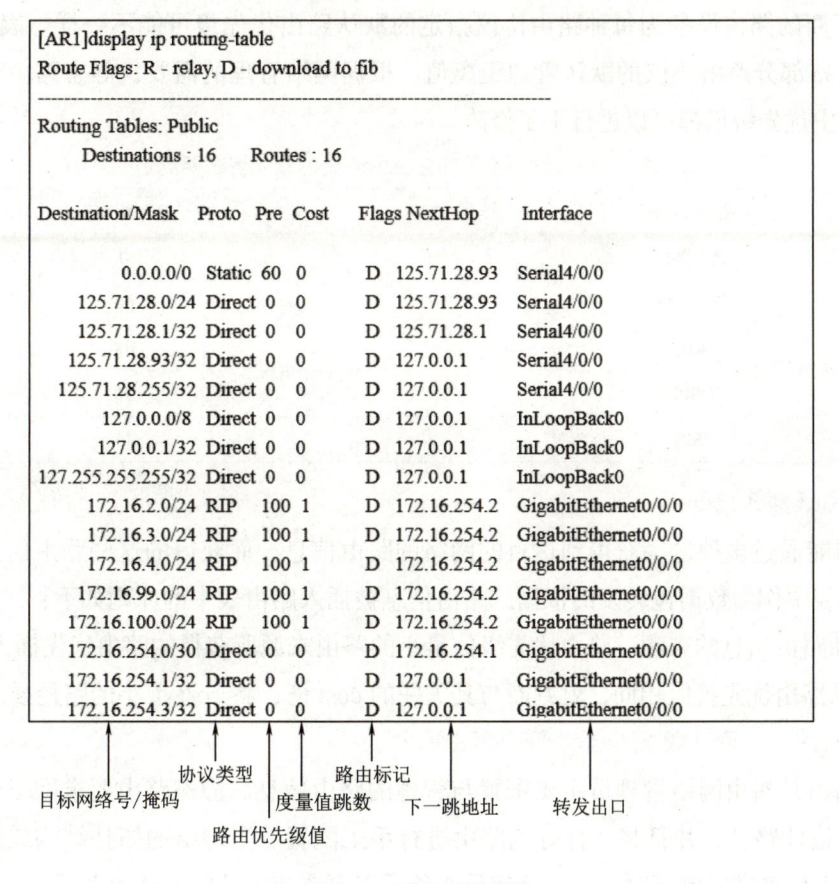

图 1.40　路由表示例

③ Pre（preference）：表示该路由条目的路由协议优先级。

④ Cost：路由开销。

⑤ Flags：路由标记。"D"表示该路由下发到 FIB。

⑥ NextHop：下一跳，其值为 IP 地址，值为到达目标网络所经过的本路由器直接相连的下一个路由器接口的 IP 地址。

⑦ Interface：表示该路由条目的转发出口。

2. 路由优先级

在构建路由表的过程中，当路由器可通过不同的来源（如各种动态路由协议、静态路由或直连路由）获取到关于同一目标网络的多条路由信息时，路由器使用路由优先级（preference）来标识不同路由源的可信度，并选择可信度最高的路由条目插入路由表中。

路由优先级被定义成"1 ～ 255"之间的整数值。路由优先级值越小，表示路由来源的可信度越高，被插入到路由表中的优先级也就越高。默认情况下，直连路由的路由优先级最小，通常被设置为"0"，直连路由的路由优先级不能手工更改，静态路由的路由优先级与动态路由协议的路由优先级均高于直连网络，它们被赋予不同的路由优先级值。

不同厂商的路由设备为每种路由协议指定的默认路由优先级可能不一样。表 1.7 中给出了华为路由设备部分路由协议的默认管理距离值。根据网络管理的需要，静态路由和各种动态路由协议的路由优先级值都可以进行手工修改。

表 1.7　华为路由设备部分路由协议的默认管理距离值

路 由 来 源	管理距离值
直连路由	0
静态路由	60
RIP	100
OSPF	10
ISIS	15

3. 路由表插入原则

路由器能通过多种途径获得到达目的网络的路由信息，而获得的这些路由信息并不一定全部插入路由表中作为数据包转发的依据。路由信息被插入路由表中的原则如下：

① 判断路由信息的来源，路由优先级值最小的路由来源所获得的路由优先插入路由表中。

② 如果路由优先级值相同，就判断与其关联的 cost 值，将 cost 值小的路径插入路由表中。

4. 静态路由、默认路由与浮动静态路由

静态路由是指由网络管理员手工配置与管理的路由信息。静态路由要求网络管理员根据网络拓扑选择最佳路径，并且将选择好的路由进行手工配置。当网络的拓扑结构或链路的状态发生变化，路由信息需要更新时，网络管理员必须手工更改相关的静态路由条目。

默认路由又称缺省路由，它是一种特殊的静态路由，默认路由给出了那些目的地址没有明确列在路由表中的数据包所对应的路由器转发接口或下一跳信息。通常引入默认路由可以有效地减少路由表的规模并降低路由表的维护开销。默认路由在路由表中的目标网络与子网掩码为 0.0.0.0/0。

浮动静态路由是作为备份路由存在的一种静态路由。在路由器上配置了一条或多条普通路由之后，浮动静态路由被配置成比某条或多条主路由具有更大的路由优先级值的静态路由，由于浮动静态路由的优先级值比主路由的优先级值大，这样当且仅当主路由失效后，浮动静态路由才会被插入路由表中。

5. 路由与数据转发的实现

数据包转发涉及两个过程，即"路由选择"与"交换"，其中"路由选择"是指为经过路由器的数据包选择一条到目的地的最佳路径；"交换"是指路由器从一个接口接收数据包并将其从另一个接口转发出去的过程。当路由器从一个接口收到一个需要到达另一个网络的数据包后，路由器路由与数据转发过程如下：

① 路由器根据数据帧的目的地址判断是否应该接收该数据帧，如果接收，路由器把它交给 IP 处理模块进行帧拆封，从中分离出相应的 IP 分组并交给路由模块。

② 路由模块通过目的地址与子网掩码的"与"运算从 IP 分组中提取出目标网络号，并将目标网络号与路由表的路由条目进行匹配，在匹配过程中采用最长匹配原则，即如果路由表中有多条路由条目都与目标网络号匹配，则选择路由表中与数据包的目的 IP 地址从最左侧开始存在最多匹配位数的路由作为首选路由。如果路由表中所有的路由条目与目标网络号都不匹配，路由器就将相应的 IP 分组丢弃；如果存在匹配，路由器便根据首选路由确定的出口将数据包封装成出口所需的数据帧格式并从该端口转发出去。

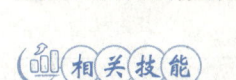 相关技能

1. 静态路由的规划与配置

搭建如图 1.41 所示的拓扑图，根据表 1.8 规划的 IP 地址，完成路由器接口与主机的 IP 配置；在路由器 AR1 与路由器 AR2 上完成静态路由的规划与配置，实现全网的互联互通，并进行网络连通性测试与故障排除。

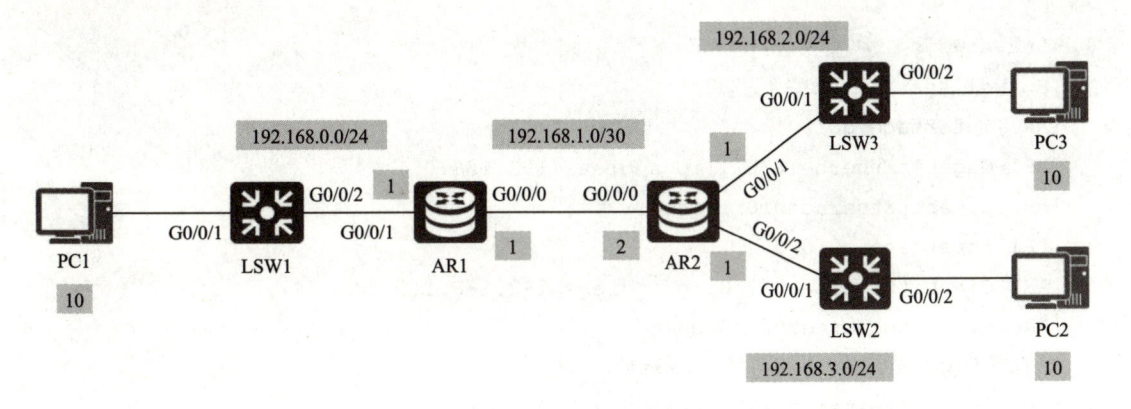

图 1.41 "静态路由的规划与配置"拓扑图

表 1.8 IP 的规划

设 备	接 口	IP 地址	网 关
AR1	G0/0/0	192.168.1.1/30	
	G0/0/1	192.168.0.1/24	
AR2	G0/0/0	192.168.1.2/30	
	G0/0/1	192.168.2.1/24	
	G0/0/2	192.168.3.1/24	
PC1	NIC	192.168.0.10/24	192.168.0.1
PC2	NIC	192.168.3.10/24	192.168.3.1
PC3	NIC	192.168.2.10/24	192.168.2.1

（1）路由器 AR1 静态路由的规划与配置

根据图 1.41 与表 1.8 可知，全网共有四个网段，对 AR1 来说，其有两个直连网段（192.168.0.0/24、192.168.1.0/30）和两个远程网段（192.168.2.0/24、192.168.3.0/24）。根据路由表插入原则，路由器连接直连网段的接口配置了 IP 地址与子网掩码后，其物理接口与链路协议状态均为"UP"时，其接口所在的网络会直接插入路由表中，不需要其他额外的配置。远程网段"192.168.2.0/24"和"192.168.3.0/24"则需要路由器使用静态路由或动态路由获得其选路信息。

因此，路由器 AR1 需要为两个远程网段"192.168.2.0/24"与"192.168.3.0/24"使用静态路由选路，规划见表 1.9。

表 1.9 路由器 AR1 的静态路由规划

目标网络	目标网络掩码 / 掩码长度	下一跳
192.168.2.0	24	192.168.1.2
192.168.3.0	24	192.168.1.2

完成静态路由的配置只需要根据规划在系统视图模式下使用"ip route-static"命令手工配置。参考以下步骤完成 AR1 静态路由的配置：

```
<Huawei>sys
[Huawei]sysname AR1
[AR1]interface g0/0/1
[AR1-GigabitEthernet0/0/1]ip address 192.168.0.1 24
[AR1-GigabitEthernet0/0/1]quit
[AR1]interface g0/0/0
[AR1-GigabitEthernet0/0/0]ip address 192.168.1.1 30
[AR1-GigabitEthernet0/0/0]quit
// 配置路由器 AR1 的主机名与接口 IP 地址
[AR1]ip route-static 192.168.2.0 24 192.168.1.2
// 配置到目标网络"192.168.2.0/24"的静态路由，所选路由使用下一跳表示，下一跳地址为
// "192.168.1.2"
[AR1]ip route-static 192.168.3.0 24 192.168.1.2
// 配置到目标网络"192.168.3.0/24"的静态路由，所选路由使用下一跳表示，下一跳地址为
// "192.168.1.2"
[AR1]quit
<AR1>save
// 保存配置
```

（2）路由器 AR2 静态路由的规划与配置

对 AR2 来说，其有三个直连网段（192.168.1.0/30、192.168.2.0/24、192.168.3.0/24）和一个远程网段（192.168.0.0/24）。根据路由表插入原则，路由器连接直连网段的接口配置了 IP 地址与子网掩码后，其物理接口与链路协议状态均为"UP"时，其接口所在的网络会直接插入路由表中，

项目 1 企业分公司网络互联项目 35

不需要其他额外的配置。远程网段则需要路由器使用静态路由或动态路由获得其选路信息。

因此，路由器 AR2 需要为远程网段"192.168.0.0/24"使用静态路由选路，规划见表 1.10。

表 1.10 路由器 AR2 的静态路由规划

目标网络	目标网络掩码 / 掩码长度	下 一 跳
192.168.0.0	24	192.168.1.1

完成静态路由的配置只需要根据规划在系统视图模式下使用"ip route-static"命令手工配置。
参考以下步骤完成 AR2 静态路由的配置：

```
<Huawei>sys
[Huawei]sysname AR2
[AR2]interface g0/0/0
[AR2-GigabitEthernet0/0/0]ip address 192.168.1.2 30
[AR2-GigabitEthernet0/0/0]quit
[AR2]interface g0/0/1
[AR2-GigabitEthernet0/0/1]ip address 192.168.2.1 24
[AR2-GigabitEthernet0/0/1]quit
[AR2]interface g0/0/2
[AR2-GigabitEthernet0/0/2]ip address 192.168.3.1 24
[AR2-GigabitEthernet0/0/2]quit
// 配置路由器 AR2 的主机名与接口 IP 地址
[AR2]ip route-static 192.168.0.0 24 192.168.1.1
// 配置到目标网络"192.168.0.0/24"的静态路由，所选路由使用下一跳表示，下一跳地址为
// "192.168.1.1"
[AR2]quit
<AR2>save
// 保存配置
```

（3）网络连通性测试与排障

参照表 1.8 的 IP 规划完成网络中所有 PC 的 IP 地址、子网掩码、默认网关配置，配置
完成后在 DOS 命令行下使用"ipconfig"查看其 IP 配置信息是否正确。图 1.42 所示为主机
PC1"ipconfig"命令显示的 IP 配置信息。

```
PC>ipconfig

Link local IPv6 address...............: fe80::5689:98ff:fe25:3e0f
IPv6 address..........................: :: / 128
IPv6 gateway..........................: ::
IPv4 address..........................: 192.168.0.10
Subnet mask...........................: 255.255.255.0
Gateway...............................: 192.168.0.1
Physical address......................: 54-89-98-25-3E-0F
DNS server............................:
```

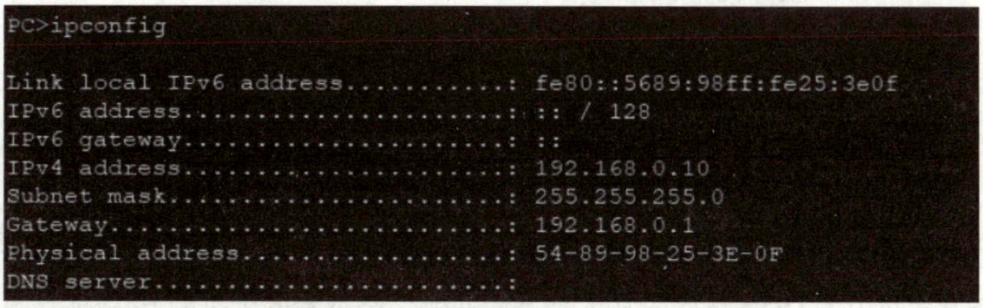

图 1.42 主机 PC1"ipconfig"命令显示的 IP 配置信息

在网络中的任何一台主机上，进入 DOS 命令行，使用 ping 实用程序测试到其他目标主机的 IP 连通性。图 1.43 所示为主机 PC1 可以 ping 通主机 PC2 与 PC3 的 IP 地址。

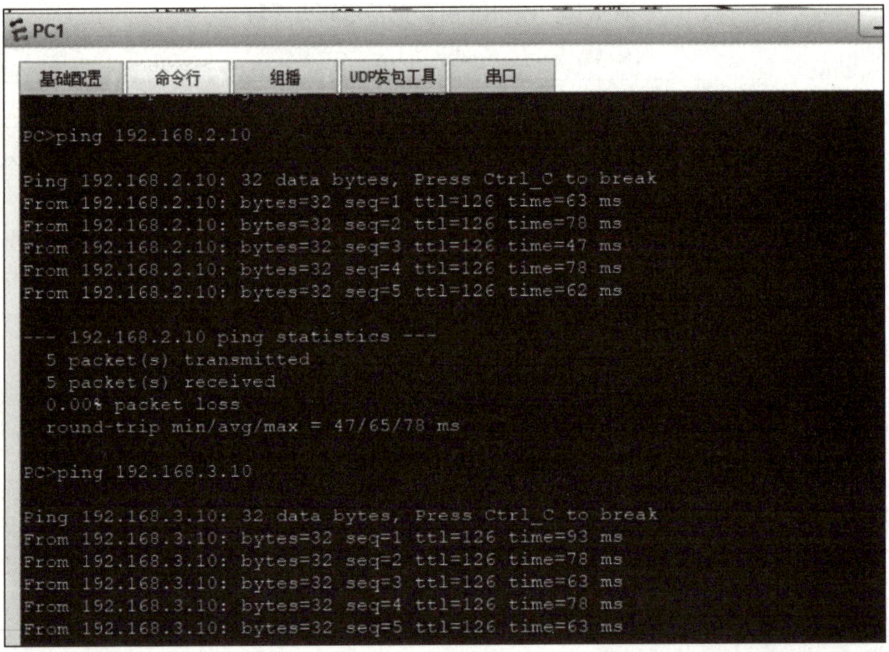

图 1.43　主机 PC1 可以 ping 通主机 PC2 与 PC3 的 IP 地址

还可以在路由器上使用扩展 ping 命令进行网络连通性测试，如图 1.44 所示。在 AR1 上使用"ping -a 192.168.0.1 192.168.3.1"显示结果。其中参数"-a 192.168.0.1"表示 ping 发送的 ICMP 包使用的源地址为路由器 AR1 的 G0/0/0 接口地址"192.168.0.1"，目的地址为路由器 AR2 的 G0/0/2 接口地址"192.168.3.1"。

```
<AR1>ping -a 192.168.0.1 192.168.3.1
 PING 192.168.3.1: 56  data bytes, press CTRL_C to break
  Reply from 192.168.3.1: bytes=56 Sequence=1 ttl=255 time=50 ms
  Reply from 192.168.3.1: bytes=56 Sequence=2 ttl=255 time=20 ms
  Reply from 192.168.3.1: bytes=56 Sequence=3 ttl=255 time=30 ms
  Reply from 192.168.3.1: bytes=56 Sequence=4 ttl=255 time=30 ms
  Reply from 192.168.3.1: bytes=56 Sequence=5 ttl=255 time=30 ms

 --- 192.168.3.1 ping statistics ---
 5 packet(s) transmitted
 5 packet(s) received
 0.00% packet loss
 round-trip min/avg/max = 20/32/50 ms
```

图 1.44　扩展 ping 指定路由器源地址"192.168.0.1" ping 通"192.168.3.1"

如果测试时网络不能连通，请使用以下步骤完成排障：

① 在所有主机上的 DOS 命令行下使用"ipconfig"命令查看主机的 IP 地址、子网掩码、默

认网关是否正确配置。

② 在路由器 AR1 与 AR2 上使用"display ip interface brief"命令查看路由器的接口 IP 是否正确配置,物理接口与链路协议状态是否已"UP"。

③ 在路由器 AR1 与 AR2 上使用"display current-configuration | include ip route-static"命令查看当前正在运行的配置文件中关于静态路由的配置是否正确,如图 1.45、图 1.46 所示。

```
[AR1]display current-configuration | include ip route-static
ip route-static 192.168.2.0 255.255.255.0 192.168.1.2
ip route-static 192.168.3.0 255.255.255.0 192.168.1.2
```

图 1.45 在 AR1 上查看当前正在运行的配置文件中关于静态路由的配置

```
[AR2]display current-configuration | include ip route-static
ip route-static 192.168.0.0 255.255.255.0 192.168.1.1
```

图 1.46 在 AR2 上查看当前正在运行的配置文件中关于静态路由的配置

注意观察正在运行的配置文件中关于静态路由的配置,不要有错误或多余的静态路由配置,如果有错误或多余的静态路由配置,则在原来的静态路由配置命令之前加上"undo"命令删除多余或错误的静态路由配置,再重新使用"ip route-static"命令配置正确的静态路由。删除错误的静态路由配置示例如下:

```
[AR1]undo ip route-static 192.168.3.0 24 192.168.0.2
```

④ 在路由器上使用"display ip routing-table"命令显示路由表内容,检查是否为网络中所有网段选路,所选的路径是否正确,如图 1.47、图 1.48 所示。

```
[AR1]display ip routing-table
Route Flags: R - relay, D - download to fib
------------------------------------------------------------
Routing Tables: Public
        Destinations : 12    Routes : 12

Destination/Mask    Proto   Pre Cost    Flags NextHop        Interface

      127.0.0.0/8       Direct 0   0        D  127.0.0.1      InLoopBack0
      127.0.0.1/32      Direct 0   0        D  127.0.0.1      InLoopBack0
127.255.255.255/32  Direct 0   0        D  127.0.0.1      InLoopBack0
   192.168.0.0/24    Direct 0   0        D  192.168.0.1    GigabitEthernet0/0/1
   192.168.0.1/32    Direct 0   0        D  127.0.0.1      GigabitEthernet0/0/1
 192.168.0.255/32   Direct 0   0        D  127.0.0.1      GigabitEthernet0/0/1
   192.168.1.0/30    Direct 0   0        D  192.168.1.1    GigabitEthernet0/0/0
   192.168.1.1/32    Direct 0   0        D  127.0.0.1      GigabitEthernet0/0/0
   192.168.1.3/32    Direct 0   0        D  127.0.0.1      GigabitEthernet0/0/0
   192.168.2.0/24    Static 60  0        RD 192.168.1.2    GigabitEthernet0/0/0
   192.168.3.0/24    Static 60  0        RD 192.168.1.2    GigabitEthernet0/0/0
255.255.255.255/32  Direct 0   0        D  127.0.0.1      InLoopBack0
```

图 1.47 AR1 路由表内容

38 企业网络互联技术（双语）

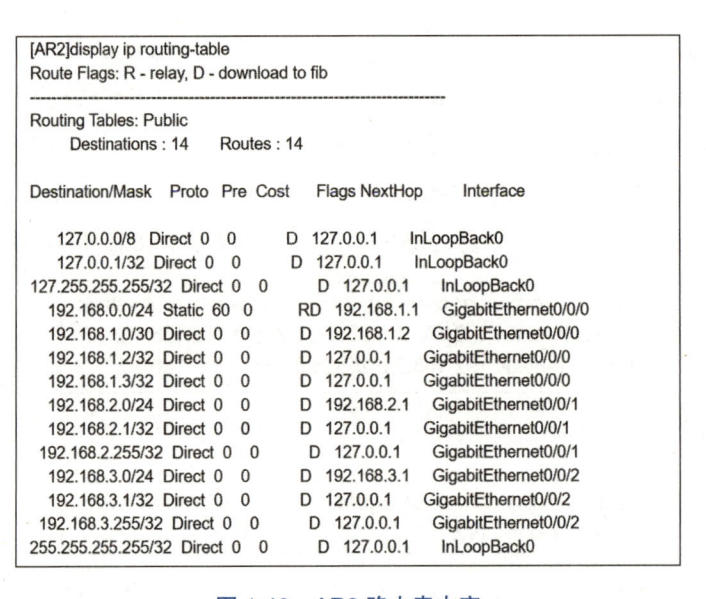

```
[AR2]display ip routing-table
Route Flags: R - relay, D - download to fib
------------------------------------------------------------------
Routing Tables: Public
         Destinations : 14      Routes : 14

Destination/Mask    Proto   Pre  Cost     Flags NextHop      Interface

      127.0.0.0/8   Direct  0    0          D   127.0.0.1    InLoopBack0
     127.0.0.1/32   Direct  0    0          D   127.0.0.1    InLoopBack0
 127.255.255.255/32 Direct  0    0          D   127.0.0.1    InLoopBack0
   192.168.0.0/24   Static  60   0          RD  192.168.1.1  GigabitEthernet0/0/0
   192.168.1.0/30   Direct  0    0          D   192.168.1.2  GigabitEthernet0/0/0
   192.168.1.2/32   Direct  0    0          D   127.0.0.1    GigabitEthernet0/0/0
   192.168.1.3/32   Direct  0    0          D   127.0.0.1    GigabitEthernet0/0/0
   192.168.2.0/24   Direct  0    0          D   192.168.2.1  GigabitEthernet0/0/1
   192.168.2.1/32   Direct  0    0          D   127.0.0.1    GigabitEthernet0/0/1
 192.168.2.255/32   Direct  0    0          D   127.0.0.1    GigabitEthernet0/0/1
   192.168.3.0/24   Direct  0    0          D   192.168.3.1  GigabitEthernet0/0/2
   192.168.3.1/32   Direct  0    0          D   127.0.0.1    GigabitEthernet0/0/2
 192.168.3.255/32   Direct  0    0          D   127.0.0.1    GigabitEthernet0/0/2
 255.255.255.255/32 Direct  0    0          D   127.0.0.1    InLoopBack0
```

图 1.48　AR2 路由表内容

2. 浮动静态路由的规划与配置

搭建如图 1.49 所示的拓扑图，根据表 1.11 规划的 IP 地址，完成路由器与主机的 IP 配置；在路由器 AR1 与路由器 AR2 上规划与配置到远程目标网络的浮动静态路由，数据包正常走 AR1 与 AR2 的千兆以太网链路，当 AR1 与 AR2 的千兆以太网链路出现故障时，数据包走 AR1 与 AR2 之间的串行链路，实现全网的互联互通，并进行网络连通性测试与故障排除。

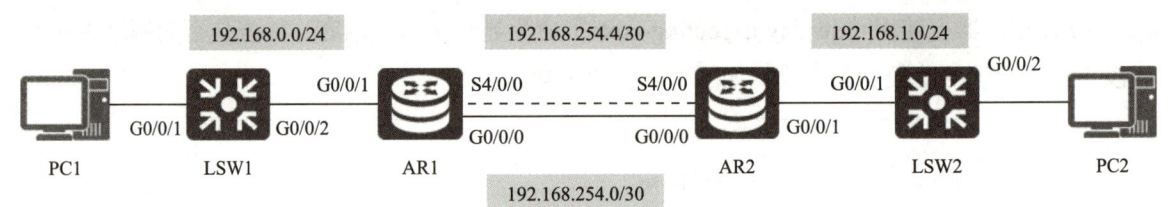

图 1.49　"浮动静态路由的规划与配置"拓扑图

表 1.11　IP 的规划

设　　备	接　　口	IP 地址	网　　关
AR1	G0/0/0	192.168.254.1/30	
	G0/0/1	192.168.0.1/24	
	S4/0/0	192.168.254.5/30	
AR2	G0/0/0	192.168.254.2/30	
	G0/0/1	192.168.1.1/24	
	S4/0/0	192.168.254.6/30	
PC1	NIC	192.168.0.10/24	192.168.0.1
PC2	NIC	192.168.1.10/24	192.168.1.1

项目 1　企业分公司网络互联项目　39

根据任务需求，AR1 与 AR2 的千兆以太网链路为 AR1 与 AR2 浮动静态路由的主路由，AR1
与 AR2 之间的串行链路为 AR1 与 AR2 浮动静态路由的备份路由。根据路由表插入规则，可以将
主路由的静态路由路由优先级值设置为比备份路由的静态路由路由优先级值小就可以实现。

AR1 与 AR2 的浮动静态路由规划见表 1.12。

表 1.12　AR1 与 AR2 的浮动静态路由规划

设　备	目标网络 / 掩码长度	下一跳或转发出口	路由优先级值	备　注
AR1	192.168.1.0/24	192.168.254.2	默认值 60	主路由
	192.168.1.0/24	S4/0/0	70	备用路由
AR2	192.168.0.0/24	192.168.254.1	默认值 60	主路由
	192.168.0.0/24	S4/0/0	70	备用路由

（1）路由器 AR1 的配置

参考以下步骤完成 AR1 的配置，实现网络的互联互通。

```
<Huawei>sys
[Huawei]sysname AR1
[AR1]interface g0/0/0
[AR1-GigabitEthernet0/0/0]ip address 192.168.254.1 30
[AR1-GigabitEthernet0/0/0]quit
[AR1]interface g0/0/1
[AR1-GigabitEthernet0/0/1]ip address 192.168.0.1 24
[AR1-GigabitEthernet0/0/1]quit
[AR1]interface s4/0/0
[AR1-Serial4/0/0]ip address 192.168.254.5 30
[AR1-Serial4/0/0]quit
// 配置路由器 AR1 的主机名与接口 IP 地址
[AR1]ip route-static 192.168.1.0 24 192.168.254.2
[AR1]ip route-static 192.168.1.0 24 s4/0/0 preference 70
// 配置到目标网络 "192.168.1.0/24" 的浮动静态路由，主路由路径下一跳地址为 "192.168.
//254.2"；备份路由的转发出口为 S4/0/0 接口
[AR1]quit
<AR1>save
// 保存配置
```

（2）路由器 AR2 的配置

```
<Huawei>sys
[Huawei]sysname AR2
[AR2]
[AR2]interface g0/0/0
```

```
[AR2-GigabitEthernet0/0/0]ip address 192.168.254.2 30
[AR2-GigabitEthernet0/0/0]quit
[AR2]interface g0/0/1
[AR2-GigabitEthernet0/0/1]ip address 192.168.1.1 24
[AR2-GigabitEthernet0/0/1]quit
[AR2]interface S4/0/0
[AR2-Serial4/0/0]ip address 192.168.254.6 30
[AR2-Serial4/0/0]quit
// 配置路由器 AR2 的主机名与接口 IP 地址
[AR2]ip route-static 192.168.0.0 24 192.168.254.1
[AR2]ip route-static 192.168.0.0 24 s4/0/0 preference 70
// 配置到目标网络 "192.168.0.0/24" 的浮动静态路由, 主路由路径下一跳地址为 "192.168.
//254.1"; 备份路由的转发出口为 S4/0/0 接口
[AR2]quit
<AR2>sa
<AR2>save
// 保存配置
```

（3）路由测试与故障排除

在路由器 AR1 与 AR2 上使用 "display ip routing-table" 命令显示路由表内容, 检查路由表中显示的静态路由是否为配置的主路由路径。图 1.50 显示 AR1 路由表中到目标网络 "192.168.1.0/24" 的路径下一跳地址为 "192.168.254.2", 出口为 "GigabitEthernet0/0/0" 接口。图 1.51 显示 AR2 路由表中到目标网络 "192.168.0.0/24" 的路径下一跳地址为 "192.168.254.1", 出口为 "GigabitEthernet0/0/0" 接口。

```
[AR1]display ip routing-table
Route Flags: R - relay, D - download to fib
------------------------------------------------------------------------
Routing Tables: Public
        Destinations : 15      Routes : 15

Destination/Mask    Proto  Pre Cost    Flags NextHop        Interface

        127.0.0.0/8   Direct 0   0       D  127.0.0.1       InLoopBack0
        127.0.0.1/32  Direct 0   0       D  127.0.0.1       InLoopBack0
127.255.255.255/32    Direct 0   0       D  127.0.0.1       InLoopBack0
      192.168.0.0/24  Direct 0   0       D  192.168.0.1     GigabitEthernet0/0/1
      192.168.0.1/32  Direct 0   0       D  127.0.0.1       GigabitEthernet0/0/1
    192.168.0.255/32  Direct 0   0       D  127.0.0.1       GigabitEthernet0/0/1
      192.168.1.0/24  Static 60  0      RD  192.168.254.2   GigabitEthernet0/0/0
    192.168.254.0/30  Direct 0   0       D  192.168.254.1   GigabitEthernet0/0/0
    192.168.254.1/32  Direct 0   0       D  127.0.0.1       GigabitEthernet0/0/0
    192.168.254.3/32  Direct 0   0       D  127.0.0.1       GigabitEthernet0/0/0
    192.168.254.4/30  Direct 0   0       D  192.168.254.5   Serial4/0/0
    192.168.254.5/32  Direct 0   0       D  127.0.0.1       Serial4/0/0
    192.168.254.6/32  Direct 0   0       D  192.168.254.6   Serial4/0/0
    192.168.254.7/32  Direct 0   0       D  127.0.0.1       Serial4/0/0
  255.255.255.255/32  Direct 0   0       D  127.0.0.1       InLoopBack0
```

图 1.50　AR1 路由表中静态路由路径条目

项目 1　企业分公司网络互联项目　41

```
[AR2]display ip routing-table
Route Flags: R - relay, D - download to fib
-------------------------------------------------------------------------
Routing Tables: Public
        Destinations : 15    Routes : 15

Destination/Mask   Proto  Pre  Cost    Flags NextHop       Interface

        127.0.0.0/8    Direct 0    0         D   127.0.0.1     InLoopBack0
        127.0.0.1/32   Direct 0    0         D   127.0.0.1     InLoopBack0
  127.255.255.255/32   Direct 0    0         D   127.0.0.1     InLoopBack0
      192.168.0.0/24   Static 60   0         RD  192.168.254.1 GigabitEthernet0/0/0
      192.168.1.0/24   Direct 0    0         D   192.168.1.1   GigabitEthernet0/0/1
      192.168.1.1/32   Direct 0    0         D   127.0.0.1     GigabitEthernet0/0/1
    192.168.1.255/32   Direct 0    0         D   127.0.0.1     GigabitEthernet0/0/1
    192.168.254.0/30   Direct 0    0         D   192.168.254.2 GigabitEthernet0/0/0
    192.168.254.2/32   Direct 0    0         D   127.0.0.1     GigabitEthernet0/0/0
    192.168.254.3/32   Direct 0    0         D   127.0.0.1     GigabitEthernet0/0/0
    192.168.254.4/30   Direct 0    0         D   192.168.254.6 Serial4/0/0
    192.168.254.5/32   Direct 0    0         D   192.168.254.5 Serial4/0/0
    192.168.254.6/32   Direct 0    0         D   127.0.0.1     Serial4/0/0
    192.168.254.7/32   Direct 0    0         D   127.0.0.1     Serial4/0/0
  255.255.255.255/32   Direct 0    0         D   127.0.0.1     InLoopBack0
```

图 1.51　AR2 路由表中静态路由路径条目

参照表 1.11 规划的参数完成测试主机 PC1 与测试主机 PC2 的 IP 地址、子网掩码、默认网关配置，配置完成后在 DOS 命令行下使用 "ipconfig" 命令查看其 IP 配置信息是否正确。

在主机 PC1 的 DOS 命令行下使用 "tracert PC2 的 IP 地址" 跟踪 PC2 访问 PC1 的 IP 数据包所经过的路径，如图 1.52 所示。可以看出，PC1 到 PC2 的数据包所走路径为 "192.168.0.1" → "192.168.254.2" → "192.168.1.10"，与主路由的路径相同。

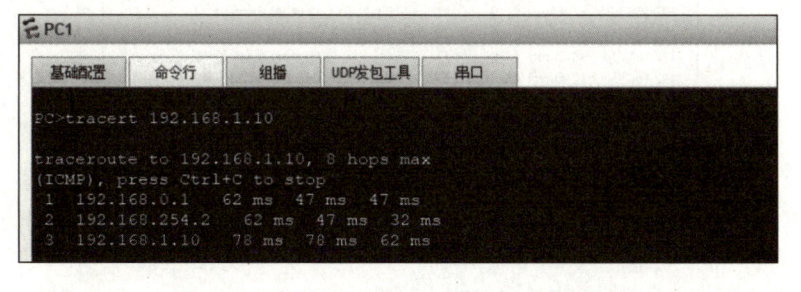

图 1.52　主机 PC2 "tracert" PC1 测试结果

把路由器 AR1 与 AR2 之间的千兆以太网链路删除，再使用 "display ip routing-table" 命令显示路由表内容，检查路由表中显示的静态路由是否已变为配置的备份路由路径，如图 1.53 所示，此时 AR1 路由表中到目标网络 "192.168.1.0/24" 的路径转发出口为 "Serial4/0/0" 接口。如图 1.54 所示，此时 AR2 路由表中到目标网络 "192.168.0.0/24" 的路径转发出口为 "Serial4/0/0" 接口。

重新在主机 PC1 的 DOS 命令行下使用 "tracert PC2 的 IP 地址" 跟踪 PC2 访问 PC1 的 IP 数据包所经过的路径，如图 1.55 所示，可以看出 PC1 到 PC2 的数据包所走路径为 "192.168.0.1" → "192.168.254.6" → "192.168.1.10"，与备份路由的路径相同。

```
[AR1]display ip routing-table
Route Flags: R - relay, D - download to fib
-------------------------------------------------------------------------
Routing Tables: Public
       Destinations : 15    Routes : 15

Destination/Mask    Proto  Pre Cost    Flags NextHop      Interface

          127.0.0.0/8  Direct 0  0       D   127.0.0.1    InLoopBack0
          127.0.0.1/32 Direct 0  0       D   127.0.0.1    InLoopBack0
 127.255.255.255/32 Direct 0  0          D   127.0.0.1    InLoopBack0
       192.168.0.0/24 Direct 0  0        D   192.168.0.1  GigabitEthernet0/0/1
       192.168.0.1/32 Direct 0  0        D   127.0.0.1    GigabitEthernet0/0/1
     192.168.0.255/32 Direct 0  0        D   127.0.0.1    GigabitEthernet0/0/1
       192.168.1.0/24 Static 70 0        D   192.168.254.5 Serial4/0/0
     192.168.254.0/30 Direct 0  0        D   192.168.254.1 GigabitEthernet0/0/0
     192.168.254.1/32 Direct 0  0        D   127.0.0.1    GigabitEthernet0/0/0
     192.168.254.3/32 Direct 0  0        D   127.0.0.1    GigabitEthernet0/0/0
     192.168.254.4/30 Direct 0  0        D   192.168.254.5 Serial4/0/0
     192.168.254.5/32 Direct 0  0        D   127.0.0.1    Serial4/0/0
     192.168.254.6/32 Direct 0  0        D   192.168.254.6 Serial4/0/0
     192.168.254.7/32 Direct 0  0        D   127.0.0.1    Serial4/0/0
 255.255.255.255/32 Direct 0  0          D   127.0.0.1    InLoopBack0
```

图 1.53　AR1 路由表

```
[AR2]display ip routing-table
Route Flags: R - relay, D - download to fib
-------------------------------------------------------------------------
Routing Tables: Public
       Destinations : 15    Routes : 15

Destination/Mask    Proto  Pre Cost    Flags NextHop      Interface

          127.0.0.0/8  Direct 0  0       D   127.0.0.1    InLoopBack0
          127.0.0.1/32 Direct 0  0       D   127.0.0.1    InLoopBack0
 127.255.255.255/32 Direct 0  0          D   127.0.0.1    InLoopBack0
       192.168.0.0/24 Static 70 0        D   192.168.254.6 Serial4/0/0
       192.168.1.0/24 Direct 0  0        D   192.168.1.1  GigabitEthernet0/0/1
       192.168.1.1/32 Direct 0  0        D   127.0.0.1    GigabitEthernet0/0/1
     192.168.1.255/32 Direct 0  0        D   127.0.0.1    GigabitEthernet0/0/1
     192.168.254.2/30 Direct 0  0        D   192.168.254.2 GigabitEthernet0/0/0
     192.168.254.2/32 Direct 0  0        D   127.0.0.1    GigabitEthernet0/0/0
     192.168.254.3/32 Direct 0  0        D   127.0.0.1    GigabitEthernet0/0/0
     192.168.254.4/30 Direct 0  0        D   192.168.254.6 Serial4/0/0
     192.168.254.5/32 Direct 0  0        D   192.168.254.5 Serial4/0/0
     192.168.254.6/32 Direct 0  0        D   127.0.0.1    Serial4/0/0
     192.168.254.7/32 Direct 0  0        D   127.0.0.1    Serial4/0/0
 255.255.255.255/32 Direct 0  0          D   127.0.0.1    InLoopBack0
```

图 1.54　AR2 路由表

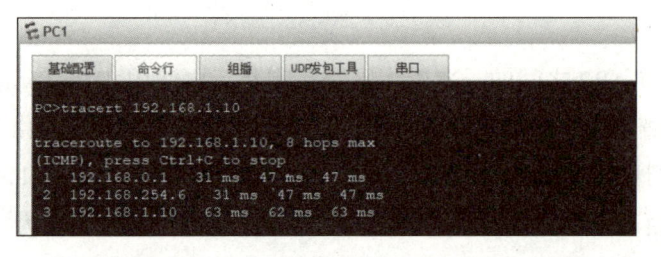

图 1.55　主机 PC2 "tracert" PC1 测试结果

使用一条网线将路由器 AR1 的 G0/0/0 接口与 AR2 的 G0/0/0 接口相连，再使用 "display ip routing-table" 命令显示路由表内容，检查路由表中显示的静态路由是否已恢复为配置的主路由路径。

项目 1　企业分公司网络互联项目　43

如果测试时网络不能连通，或者跟踪路径时，发现选路有故障，请使用以下步骤完成排障：

① 在所有主机上的 DOS 命令行下使用"ipconfig"命令查看主机的 IP 地址、子网掩码、默认网关是否正确配置。

② 在路由器 AR1 与 AR2 上使用"display ip interface brief"命令查看路由器的接口 IP 是否正确配置，物理接口与链路协议状态是否已"UP"。

③ 如果路由表中没有正确选路，请在路由器 AR1 与 AR2 上使用"display current-configuration | include ip route-static"命令查看当前正在运行的配置文件，AR1 与 AR2 中的浮动静态路由配置是否正确，如图 1.56、图 1.57 所示。

```
[AR1]display current-configuration | include ip route-static
ip route-static 192.168.1.0 255.255.255.0 192.168.254.2
ip route-static 192.168.1.0 255.255.255.0 Serial4/0/0 preference 70
```

图 1.56　AR1 正在运行的配置文件中浮动静态路由配置信息

```
[AR2]display current-configuration | include ip route-static
ip route-static 192.168.0.0 255.255.255.0 192.168.254.1
ip route-static 192.168.0.0 255.255.255.0 Serial4/0/0 preference 70
```

图 1.57　AR2 正在运行的配置文件中浮动静态路由配置信息

注意观察正在运行的配置文件中关于浮动静态路由的配置，不要有错误或多余的静态路由配置，如果有错误或多余的静态路由配置，使用"undo"命令删除多余或错误的静态路由配置，再重新使用"ip route-static"命令配置正确的浮动静态路由。

3. 汇总静态路由与默认路由的规划与配置

搭建如图 1.58 所示的拓扑图，根据表 1.13 规划的 IP 地址，完成路由器与主机的 IP 配置；在路由器 AR1 上规划与配置到 Internet 的默认路由，在 Internet 上的 ISP 路由器上规划与配置到企业内网的汇总静态路由，实现全网的互联互通。使得路由器 AR1 与 ISP 上的路由表条目尽可能减少，且不能有路由环路；并进行网络连通性测试与故障排除。

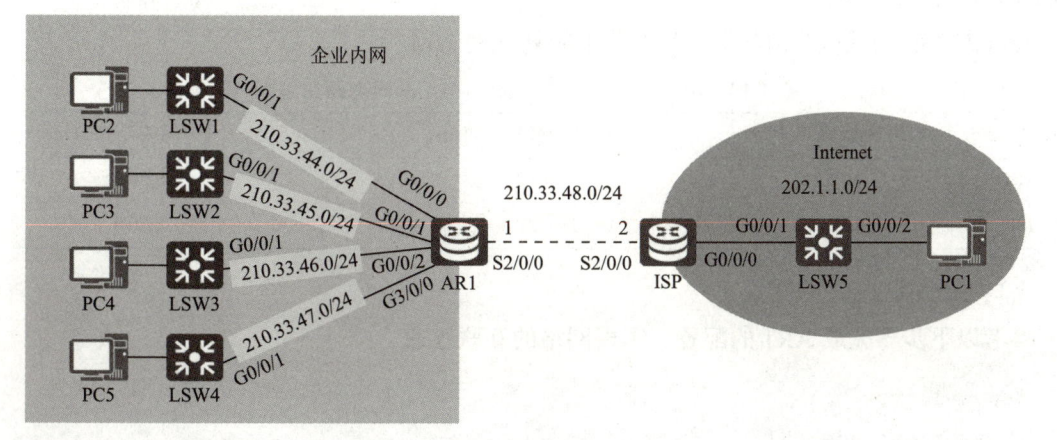

图 1.58　"汇总静态路由与默认路由的规划与配置"拓扑图

表 1.13　IP 的规划

设　备	接　口	IP 地址	网　关
AR1	G0/0/0	210.33.44.1/24	
	G0/0/1	210.33.45.1/24	
	G0/0/2	210.33.46.1/24	
	G3/0/0	210.33.47.1/24	
	S2/0/0	210.33.48.1/30	
ISP	S2/0/0	210.33.48.2/30	
	G0/0/0	202.1.1.1/24	
PC1	NIC	202.1.1.10/24	202.1.1.1
PC2	NIC	210.33.44.10/24	210.33.44.1
PC3	NIC	210.33.45.10/24	210.33.45.1
PC4	NIC	210.33.46.10/24	210.33.46.1
PC5	NIC	210.33.47.10/24	210.33.47.1

对 AR1 来说，"210.33.44.0/24"、"210.33.45.0/24"、"210.33.46.0/24"、"210.33.47.0/24"和"210.33.48.0/24"五个网络均为直连网络，它只需要规划与配置到 Internet 的默认路由（注：子任务拓扑图中只列出 Internet 上的一个网段，而实际的 Internet 上目标网络数量非常庞大，路由器中不可能为其逐个选路）。所选路径可以使用本地转发出口"S2/0/0"接口表示，也可以使用下一跳地址"210.33.48.2"。此处规划使用本地转发出口"S2/0/0"。

对 Internet 上的 ISP 路由器来说，远程网络有"210.33.44.0/24"、"210.33.45.0/24"、"210.33.46.0/24"和"210.33.47.0/24"四个网络，需要配置到这四个远程网络的静态路由。由于四个远程网络所选的路一样，转发出口均为串口"S2/0/0"接口，且四个目标网络号连续，可以将四个远程网络的网络号汇总成为"210.33.44.0/22"，从而减少路由器 ISP 的路由表条目，降低路由表的维护开销。

AR1 与 ISP 的默认路由与汇总静态路由规划见表 1.14。

表 1.14　AR1 与 ISP 的默认路由与汇总静态路由规划

设　备	目标网络	目标网络掩码 / 掩码长度	下一跳或转发出口
AR1	0.0.0.0	0.0.0.0	S2/0/0
ISP	210.33.44.0	22	S2/0/0

（1）路由器 AR1 的配置

参考以下步骤完成 AR1 的配置，实现网络的互联互通。

```
<Huawei>sys
[Huawei]sysname AR1
[AR1]interface G0/0/0
```

项目 1　企业分公司网络互联项目　45

```
[AR1-GigabitEthernet0/0/0]IP address 210.33.44.1 24
[AR1-GigabitEthernet0/0/0]quit
[AR1]interface g0/0/1
[AR1-GigabitEthernet0/0/1]IP address 210.33.45.1 24
[AR1-GigabitEthernet0/0/1]quit
[AR1]interface g0/0/2
[AR1-GigabitEthernet0/0/2]IP address 210.33.46.1 24
[AR1-GigabitEthernet0/0/2]quit
[AR1]interface g3/0/0
[AR1-GigabitEthernet3/0/0]IP address 210.33.47.1 24
[AR1-GigabitEthernet3/0/0]quit
[AR1]interface s2/0/0
[AR1-Serial2/0/0]IP address 210.33.48.1 30
[AR1-Serial2/0/0]description linktointernet
[AR1-Serial2/0/0]quit
// 配置路由器 AR1 的主机名与接口 IP 地址
[AR1]ip route-static 0.0.0.0 0.0.0.0 s2/0/0
// 配置到 Internet 的默认路由，所选路径使用本地转发出口表示，转发出口为 S2/0/0 接口
```

（2）路由器 ISP 的配置

```
<Huawei>sys
[Huawei]sysname ISP
[ISP]interface s2/0/0
[ISP-Serial2/0/0]ip address 210.33.48.2 30
[ISP-Serial2/0/0]quit
[ISP]interface g0/0/0
[ISP-GigabitEthernet0/0/0]ip address 202.1.1.1 24
[ISP-GigabitEthernet0/0/0]quit
// 配置路由器 AR2 的主机名与接口 IP 地址
[ISP]ip route-static 210.33.44.0 22 s2/0/0
// 配置到企业内网 "210.33.44.0/22" 的汇总静态路由，转发出口为 S2/0/0 接口
```

（3）网络测试与排障

　　参照表 1.13 规划的参数完成 Internet 上一台测试主机 PC1 与企业内网测试主机 PC2 的 IP 地址、子网掩码、默认网关配置。配置完成后，在 DOS 命令行下使用 "ipconfig" 命令查看其 IP 配置信息是否正确。

　　在主机 PC1 的 DOS 命令行下，使用 ping 实用程序测试到企业内网（路由器 AR1 的四个千兆以太网口 G0/0/0、G0/0/1、G0/0/2 与 G3/0/0 接口）的 IP 连通性。图 1.59 所示为主机 PC1 ping 通路由器 AR1 的 G0/0/0 与 G0/0/1 的测试结果。

46 企业网络互联技术（双语）

```
PC1
 基础配置   命令行   组播   UDP发包工具   串口

PC>ping 210.33.44.1

Ping 210.33.44.1: 32 data bytes, Press Ctrl_C to break
From 210.33.44.1: bytes=32 seq=1 ttl=254 time=62 ms
From 210.33.44.1: bytes=32 seq=2 ttl=254 time=31 ms
From 210.33.44.1: bytes=32 seq=3 ttl=254 time=16 ms
From 210.33.44.1: bytes=32 seq=4 ttl=254 time=15 ms
From 210.33.44.1: bytes=32 seq=5 ttl=254 time=31 ms

--- 210.33.44.1 ping statistics ---
  5 packet(s) transmitted
  5 packet(s) received
  0.00% packet loss
  round-trip min/avg/max = 15/31/62 ms

PC>ping 210.33.45.1

Ping 210.33.45.1: 32 data bytes, Press Ctrl_C to break
From 210.33.45.1: bytes=32 seq=1 ttl=254 time=32 ms
From 210.33.45.1: bytes=32 seq=2 ttl=254 time=31 ms
From 210.33.45.1: bytes=32 seq=3 ttl=254 time=62 ms
From 210.33.45.1: bytes=32 seq=4 ttl=254 time=16 ms
From 210.33.45.1: bytes=32 seq=5 ttl=254 time=31 ms
```

图 1.59　主机 PC1 ping 通路由器 AR1 的 G0/0/0 与 G0/0/1 的测试结果

在主机 PC2 的 DOS 命令行下，使用 ping 实用程序测试到 Internet 的 IP 连通性。图 1.60 所示为主机 PC2 可以 ping 通 Internet 上的测试主机 PC1。

```
PC2
 基础配置   命令行   组播   UDP发包工具   串口

PC>ping 202.1.1.10

Ping 202.1.1.10: 32 data bytes, Press Ctrl_C to break
From 202.1.1.10: bytes=32 seq=1 ttl=126 time=93 ms
From 202.1.1.10: bytes=32 seq=2 ttl=126 time=78 ms
From 202.1.1.10: bytes=32 seq=3 ttl=126 time=79 ms
From 202.1.1.10: bytes=32 seq=4 ttl=126 time=62 ms
From 202.1.1.10: bytes=32 seq=5 ttl=126 time=78 ms

--- 202.1.1.10 ping statistics ---
  5 packet(s) transmitted
  5 packet(s) received
  0.00% packet loss
  round-trip min/avg/max = 62/78/93 ms
```

图 1.60　主机 PC2 ping 通 Internet 上的测试主机 PC1

在主机 PC2 的 DOS 命令行下，使用 "tracert" 路由跟踪实用程序，确定 PC2 访问 PC1 的 IP 数据包经过的路径，如图 1.61 所示。

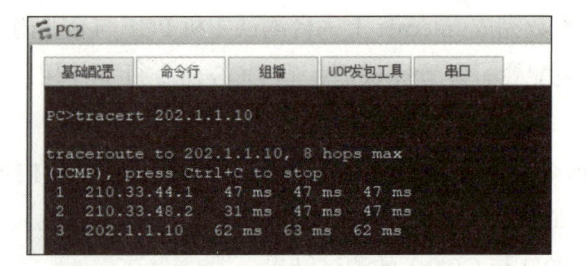

```
PC2
 基础配置   命令行   组播   UDP发包工具   串口

PC>tracert 202.1.1.10

traceroute to 202.1.1.10, 8 hops max
(ICMP), press Ctrl+C to stop
 1  210.33.44.1   47 ms   47 ms   47 ms
 2  210.33.48.2   31 ms   47 ms   47 ms
 3  202.1.1.10    62 ms   63 ms   62 ms
```

图 1.61　主机 PC2 "tracert" PC1 测试结果

项目 1　企业分公司网络互联项目　47

如果测试时网络不能连通，请使用以下步骤完成排障，使得网络具有 IP 连通性。

① 在所有主机上的 DOS 命令行下使用 "ipconfig" 命令查看主机的 IP 地址、子网掩码、默认网关是否正确配置。

② 在路由器 AR1 与 ISP 上使用 "display ip interface brief" 命令查看路由器的接口 IP 是否正确配置，物理接口与链路协议状态是否已 "UP"。

③ 在路由器上使用 "display ip routing-table" 命令显示路由表内容，检查是否为网络中所有网段选路，所选的路径是否正确，如图 1.62、图 1.63 所示。

```
[AR1]display ip routing-table
Route Flags: R - relay, D - download to fib
-----------------------------------------------------------------
Routing Tables: Public
        Destinations : 21      Routes : 21
Destination/Mask    Proto  Pre  Cost    Flags NextHop        Interface
        0.0.0.0/0   Static 60   0        D    210.33.48.1    Serial2/0/0
      127.0.0.0/8   Direct 0    0        D    127.0.0.1      InLoopBack0
      127.0.0.1/32  Direct 0    0        D    127.0.0.1      InLoopBack0
127.255.255.255/32  Direct 0    0        D    127.0.0.1      InLoopBack0
    210.33.44.0/24  Direct 0    0        D    210.33.44.1    GigabitEthernet0/0/0
    210.33.44.1/32  Direct 0    0        D    127.0.0.1      GigabitEthernet0/0/0
  210.33.44.255/32  Direct 0    0        D    127.0.0.1      GigabitEthernet0/0/0
    210.33.45.0/24  Direct 0    0        D    210.33.45.1    GigabitEthernet0/0/1
    210.33.45.1/32  Direct 0    0        D    127.0.0.1      GigabitEthernet0/0/1
  210.33.45.255/32  Direct 0    0        D    127.0.0.1      GigabitEthernet0/0/1
    210.33.46.0/24  Direct 0    0        D    210.33.46.1    GigabitEthernet0/0/2
    210.33.46.1/32  Direct 0    0        D    127.0.0.1      GigabitEthernet0/0/2
  210.33.46.255/32  Direct 0    0        D    127.0.0.1      GigabitEthernet0/0/2
    210.33.47.0/24  Direct 0    0        D    210.33.47.1    GigabitEthernet3/0/0
    210.33.47.1/32  Direct 0    0        D    127.0.0.1      GigabitEthernet3/0/0
  210.33.47.255/32  Direct 0    0        D    127.0.0.1      GigabitEthernet3/0/0
    210.33.48.0/30  Direct 0    0        D    210.33.48.1    Serial2/0/0
    210.33.48.1/32  Direct 0    0        D    127.0.0.1      Serial2/0/0
    210.33.48.2/32  Direct 0    0        D    210.33.48.2    Serial2/0/0
    210.33.48.3/32  Direct 0    0        D    127.0.0.1      Serial2/0/0
255.255.255.255/32  Direct 0    0        D    127.0.0.1      InLoopBack0
```

图 1.62　AR1 路由表内容

```
[ISP]display ip routing-table
Route Flags: R - relay, D - download to fib
-----------------------------------------------------------------
Routing Tables: Public
        Destinations : 12      Routes : 12

Destination/Mask    Proto  Pre  Cost    Flags NextHop        Interface

      127.0.0.0/8   Direct 0    0        D    127.0.0.1      InLoopBack0
      127.0.0.1/32  Direct 0    0        D    127.0.0.1      InLoopBack0
127.255.255.255/32  Direct 0    0        D    127.0.0.1      InLoopBack0
     202.1.1.0/24   Direct 0    0        D    202.1.1.1      GigabitEthernet0/0/0
     202.1.1.1/32   Direct 0    0        D    127.0.0.1      GigabitEthernet0/0/0
   202.1.1.255/32   Direct 0    0        D    127.0.0.1      GigabitEthernet0/0/0
    210.33.44.0/22  Static 60   0        D    210.33.48.2    Serial2/0/0
    210.33.48.0/30  Direct 0    0        D    210.33.48.2    Serial2/0/0
    210.33.48.1/32  Direct 0    0        D    210.33.48.1    Serial2/0/0
    210.33.48.2/32  Direct 0    0        D    127.0.0.1      Serial2/0/0
    210.33.48.3/32  Direct 0    0        D    127.0.0.1      Serial2/0/0
255.255.255.255/32  Direct 0    0        D    127.0.0.1      InLoopBack0
```

图 1.63　ISP 路由表内容

48 企业网络互联技术（双语）

如果路由表中没有正确选路，请在路由器 AR1 与 ISP 上使用"display current-configuration | include ip route-static"命令查看当前正在运行的配置文件，AR1 中的默认路由配置、ISP 的汇总静态路由的配置是否正确，如图 1.64、图 1.65 所示。

```
[AR1]display current-configuration | include ip route-static
ip route-static 0.0.0.0 0.0.0.0 Serial2/0/0
```

图 1.64　AR1 正在运行的配置文件中默认路由配置信息

```
[ISP]display current-configuration | include ip route-static
ip route-static 210.33.44.0 255.255.252.0 Serial2/0/0
```

图 1.65　ISP 正在运行的配置文件中汇总静态路由配置信息

注意观察正在运行的配置文件中关于默认路由与汇总静态路由的配置，不要有错误或多余的路由配置，如果有错误或多余的路由配置，使用"undo"命令删除多余或错误的静态路由配置，再重新使用"ip route-static"命令配置正确的默认路由与汇总静态路由。

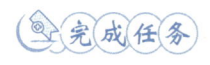 完成任务

1. 任务规划

根据图 1.1 的网络拓扑结构和项目 1 网络互联互通建网需求，对分部的路由器"fenbu"来说，远程网络只有总部的网络（但分部的路由器不知总部的网段数量，拓扑图中只列了总部的一个网络示例），且到总部的路径有两条，路由要求只有千兆以太网链路出现故障时，才走串行链路，因此需要规划到总部的浮动默认路由来实现。

对总部的路由器"zongbu"来说，有三个远程网络"172.16.128.0/25"、"172.16.128.129/28"与"172.16.129.0/24"，到这三个网络的路由选路相同，即千兆以太网链路为主链路，串行链路为备份链路，且三个目标网络号连续成块，可以使用 CIDR（classless inter-domain routing，无类别域间路由）进行路由汇总。

路由器"fenbu"与路由器"zongbu"的静态默认路由与汇总静态路由规划见表 1.15。

表 1.15　路由器"fenbu"与路由器"zongbu"的静态默认路由与汇总静态路由规划表

设　备	目标网络	目标网络掩码 / 掩码长度	下一跳或转发出口	路由优先级值
fenbu	0.0.0.0	0.0.0.0	172.16.254.2	默认值 60
	0.0.0.0	0.0.0.0	S2/0/0	70
zongbu	172.16.128.0	23	172.16.254.1	默认值 60
	172.16.128.0	23	S3/0/0	70

2. 任务实施

注：本任务是在本项目任务 1 "完成任务"基础上实施完成的。

（1）路由器"fenbu"的配置

请参考以下路由配置步骤，实现网络的互联互通。

```
<fenbu>sys
[fenbu]ip route-static 0.0.0.0 0.0.0.0 172.16.254.2
[fengbu]ip route-static 0.0.0.0 0.0.0.0 s2/0/0 preference 70
// 配置到总部网络的浮动默认路由，主路由路径下一跳地址为"172.16.254.2"，即路径为千兆以太
// 网链路；备份路由的转发出口为 S2/0/0 接口
```

（2）路由器"zongbu"的配置

```
<zongbu>sys
[zongbu]ip route-static 172.16.128.0 23 172.16.254.1
[zongbu]ip route-static 172.16.128.0 23 s3/0/0 preference 70
// 配置到分部网络的浮动汇总静态路由，主路由路径下一跳地址为"172.16.254.1"，即路径为千兆
// 以太网链路；备份路由的转发出口为 S3/0/0 接口
```

（3）网络测试与排障

① 在路由器"fenbu"与路由器"zongbu"上使用"display ip routing-table | include Static"命令显示路由表中静态路由条目，检查路由表中显示的静态路由信息是否与规划一致。图 1.66 所示为路由器"fenbu"路由表中的默认路由的路径为下一跳地址为"172.16.254.2"，出口为"GigabitEthernet0/0/0"接口。图 1.67 所示为路由器"zongbu"路由表中到目标网络"172.16.128.0/23"的路径为下一跳地址为"172.16.254.1"，出口为"GigabitEthernet0/0/0"接口。根据路由表中选路情况，从数据转发逻辑上分析是否存在路由环路。

```
[fenbu]display ip routing-table | include Static
Route Flags: R - relay, D - download to fib
------------------------------------------------------------------
Routing Tables: Public
        Destinations : 21     Routes : 21


Destination/Mask  Proto  Pre Cost    Flags NextHop      Interface

0.0.0.0/0  Static 60  0        RD  172.16.254.2  GigabitEthernet0/0/0
```

图 1.66　路由器"fenbu"路由表中静态路由条目

```
[zongbu]display ip routing-table | include Static
Route Flags: R - relay, D - download to fib
------------------------------------------------------------------
Routing Tables: Public
        Destinations : 15     Routes : 15


Destination/Mask  Proto  Pre Cost    Flags NextHop      Interface

172.16.128.0/23 Static 60  0        RD  172.16.254.1  GigabitEthernet0/0/0
```

图 1.67　路由器"zongbu"路由表中静态路由条目

② 参照表 1.6 规划的参数完成测试主机 PC1、PC2、Web 服务器、DNS 服务器与 FTP 服务器的 IP 地址、子网掩码、默认网关配置，配置完成后在 DOS 命令行下使用"ipconfig"命令查看其 IP 配置信息是否正确。

③ 在主机 PC1 的 DOS 命令行下，使用 ping 实用程序测试到网络中其他主机与服务器的 IP 连通性。图 1.68 所示分公司的主机 PC2 ping 通了总部的主机 PC1。

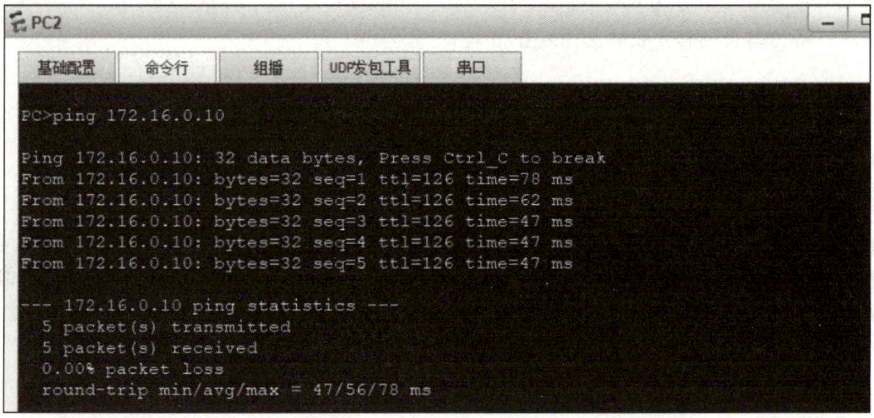

图 1.68　分公司的主机 PC2 ping 通了总部的主机 PC1

④ 在分公司主机 PC2 的 DOS 命令行下，使用"tracert PC1 的 IP 地址"跟踪分公司主机 PC2 访问总公司主机 PC1 的 IP 数据包所经过的路径，如图 1.69 所示，可以看出 PC2 到 PC1 的数据包所走路径为"172.16.129.1"→"172.16.254.2"→"172.16.0.10"，与规划的浮动路由中主路由的路径相同。

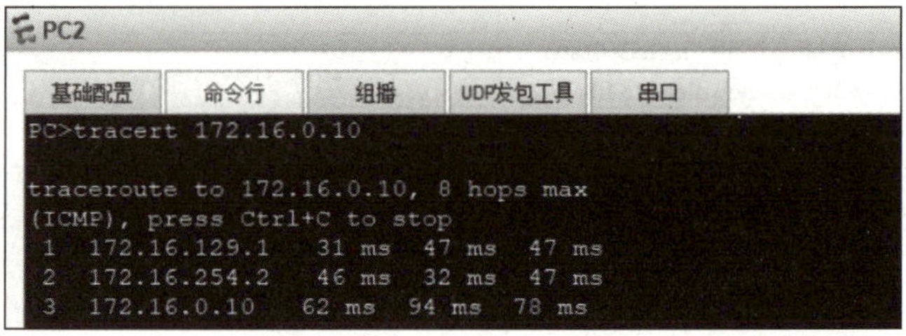

图 1.69　主机 PC2 "tracert" PC1 测试结果

⑤ 在分公司主机 PC2 的 DOS 命令行下，使用"tracert"命令跟踪分公司主机 PC2 到一个网络图中不存在的 IP 地址数据包所经过的路径，查看是否有路由环路存在（即数据包不停地在两个或两个以上的路由器之间来回转发，形成环路）。图 1.70 所示的跟踪结果显示不存在路由环路，而图 1.71 所示的跟踪结果显示存在路由环路（数据包在"172.16.254.1"与"172.16.254.2"两个路由设备之间来回转发）。

项目 1　企业分公司网络互联项目　51

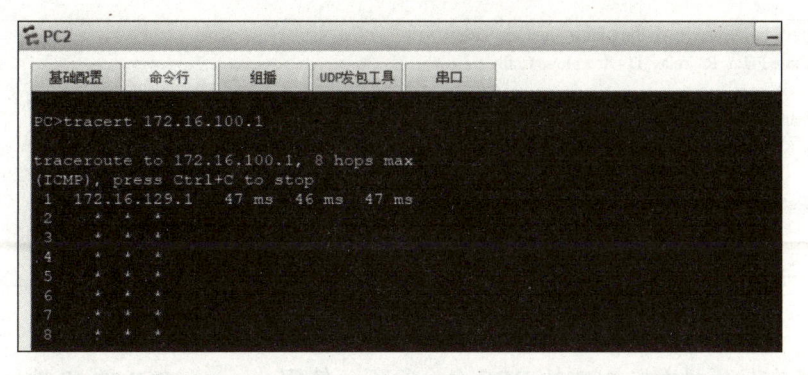

图 1.70　主机 PC2 "tracert" 不存在的 IP 地址结果显示不存在路由环路

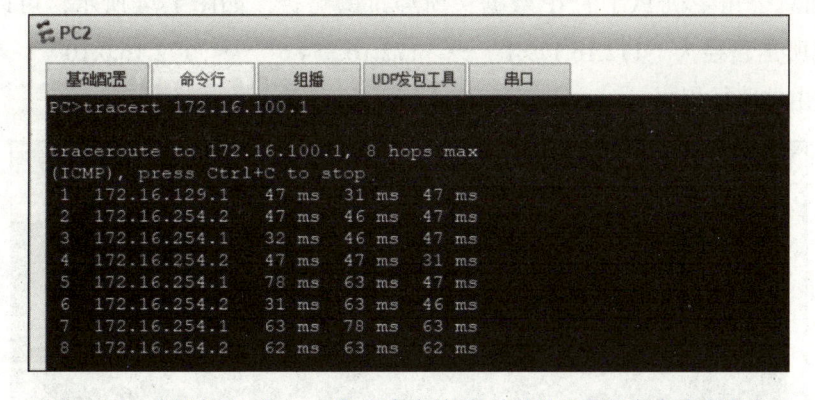

图 1.71　主机 PC2 "tracert" 不存在的 IP 地址结果显示存在路由环路

⑥ 把路由器 "fenbu" 与路由器 "zongbu" 之间的千兆以太网链路删除，再使用 "display ip routing-table | include Static" 命令显示路由表中静态路由条目，检查路由表中显示的静态路由是否已变为配置的备份路由路径。如图 1.72 所示，此时路由器 "fenbu" 路由表中默认路由的路径转发出口为 "Serial2/0/0" 接口。如图 1.73 所示，此时路由器 "zongbu" 路由表中到分公司 "172.16.128.0/23" 的路径转发出口为 "Serial3/0/0" 接口。

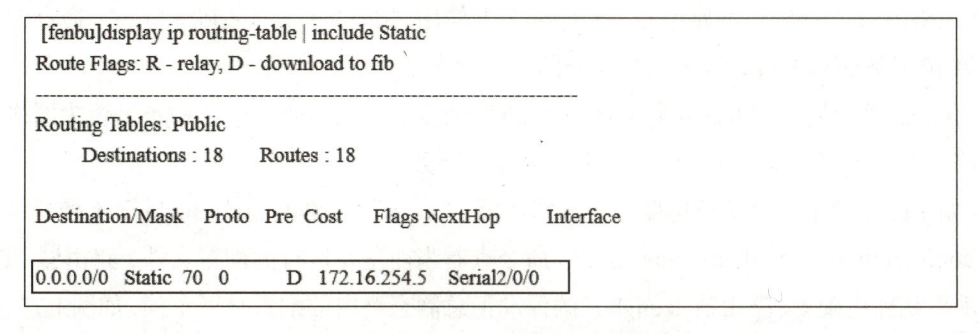

图 1.72　路由器 "fenbu" 路由表中静态路由条目

52   企业网络互联技术（双语）

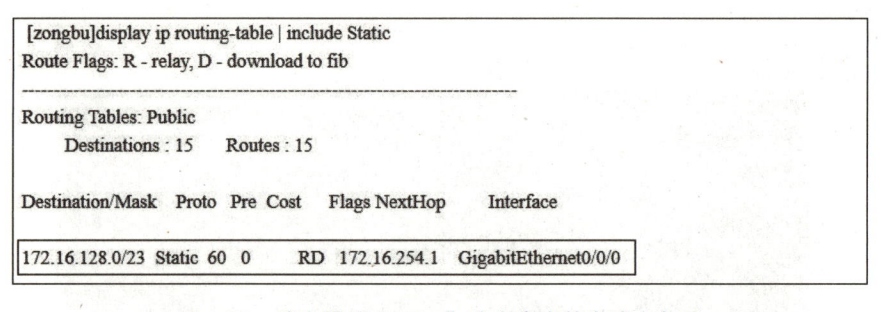

图 1.73   路由器"zongbu"路由表中静态路由条目

⑦ 再重新在分公司主机 PC2 的 DOS 命令行下，使用"tracert PC1 的 IP 地址"跟踪分公司主机 PC2 访问总公司主机 PC1 的 IP 数据包所经过的路径，如图 1.74 所示，可以看出 PC2 到 PC1 的数据包所走路径为"172.16.129.1"→"172.16.254.6"→"172.16.0.10"，与规划的浮动路由中备份路由的路径相同。

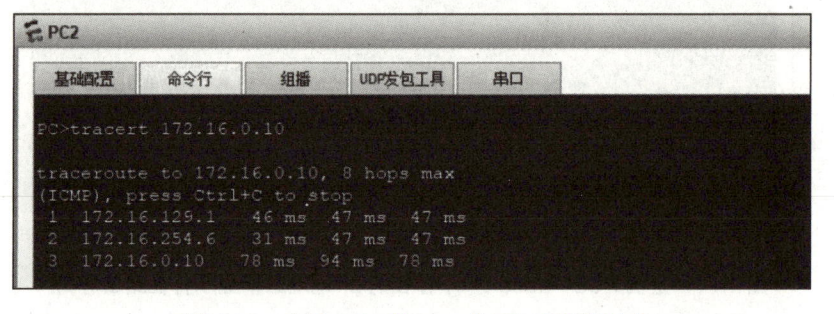

图 1.74   主机 PC2"tracert"PC1 测试结果

如果测试时网络不能连通，或者路由出现环路，请参考以下步骤完成排障，使得网络具有 IP 连通性且网络中无路由环路：

① 在所有主机上的 DOS 命令行下使用"ipconfig"命令查看主机的 IP 地址、子网掩码、默认网关是否正确配置。

② 在路由器"fenbu"与路由器"zongbu"上使用"display ip interface brief"命令查看路由器的接口 IP 是否正确配置，物理接口与链路协议状态是否已"UP"。

③ 在路由器上使用"display ip routing-table"命令显示路由表内容，检查是否为网络中所有网段选路，所选的路径是否正确，逻辑上是否存在环路。

④ 如果路由表中没有正确选路，请在路由器"fenbu"与路由器"zongbu"上使用"display current-configuration | include ip route-static"命令查看当前正在运行的配置文件，路由器"fenbu"中的浮动默认路由配置、路由器"zongbu"的浮动汇总静态路由的配置是否正确，如图 1.75、图 1.76 所示。

注意观察正在运行的配置文件中关于浮动默认路由与浮动汇总静态路由的配置，不要有错误或多余的路由配置，如果有错误或多余的路由配置，使用"undo"命令删除多余或错误

的静态路由配置，再重新使用"ip route-static"命令配置正确的浮动默认静态与浮动汇总静态路由。

```
[fenbu]display current-configuration | include ip route-static
ip route-static 0.0.0.0 0.0.0.0 172.16.254.2
ip route-static 0.0.0.0 0.0.0.0 Serial2/0/0 preference 70
```

图 1.75　路由器"fenbu"正在运行的配置文件中默认浮动路由配置信息

```
[zongbu]display current-configuration | include ip route-static
ip route-static 172.16.128.0 255.255.254.0 172.16.254.1
ip route-static 172.16.128.0 255.255.254.0 Serial3/0/0 preference 70
```

图 1.76　路由器"zongbu"正在运行的配置文件中浮动汇总静态路由配置信息

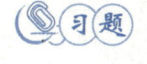 习题

选择题

（1）华为路由器上静态路由默认的路由优先级值为（　　　）。

A. 60 　　　　　　B. 70 　　　　　　C. 110 　　　　　　D. 10

（2）华为路由器使用下面（　　）命令来配置静态路由。

A. ip routing 　　　　　　　　B. ip route-static

C. AAA 　　　　　　　　　　D. route

（3）下面（　　　）命令用于显示华为路由器的路由表内容。

A. show ip route 　　　　　　B. display ip routing-table

C. display interface 　　　　　D. display ip interface brief

（4）配置默认路由时，目标网络号/掩码为（　　　）。

A. 0. 0. 0. 0/32 　　　　　　　B. 0. 0. 0. 0/0

C. 255. 255. 255. 255/0 　　　　D. 255. 255. 255. 255/32

（5）关于命令"ip route-static 192.168.0.0 22 192.168.1.1 preference 70"，下列说法不正确的是（　　　）。

A. 该路由的目标网络号为 192. 168. 0. 0/24

B. 该路由的目标网络号为 192. 168. 0. 0/22

C. 该路由的下一跳地址为 192. 168. 1. 1

D. 该路由优先级值为 70

（6）在 VRP 平台上，静态路由的默认协议优先级值为（　　　）。

A. 1 　　　　　　　B. 10 　　　　　　C. 60 　　　　　　D. 70

54　企业网络互联技术（双语）

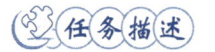 **任务 3** **DHCP 的配置与管理**

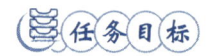 **任务描述**

完成本项目任务 1 与任务 2 后，在图 1.1 所示网络中的"fenbu"路由器上配置 DHCP 服务，为分公司的预算部门与管理部门的主机自动分配 IP 地址、默认网关、DNS 服务器地址，为管理部门主机 PC3 绑定静态的 IP 为"172.16.128.13"，实现项目 1 的"DHCP 服务"建网需求，并进行 DHCP 的测试与排障。

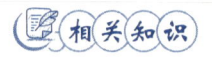 **任务目标**

- 理解 DHCP 的工作过程。
- 掌握路由器上 DHCP 服务的规划与配置。
- 掌握 DHCP 中继的规划与配置。
- 具有 DHCP 排障的能力。

相关知识

1. DHCP 的工作过程

动态主机配置协议（dynamic host configuration protocol, DHCP）是一种用于集中对用户 IP 地址进行动态管理和配置的技术。DHCP 采用客户端 / 服务器模式配置为自动获取 IP 地址的主机为 DHCP 客户端，提供 DHCP 服务的为服务器。通常，路由器、三层交换机、硬件防火墙、无线控制器、无线路由器等设备均可配置成 DHCP 服务器。DHCP 工作过程如图 1.77 所示。

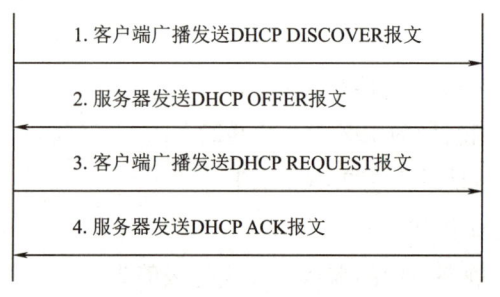

图 1.77　DHCP 工作过程

2. DHCP 中继

由于 DHCP 客户端在 IP 地址动态获取过程中采用广播方式发送 DHCP DISCOVER 报文，

项目 1　企业分公司网络互联项目　55

当 DHCP 客户端和 DHCP 服务器处于不同物理网段时，由于路由器不转发广播报文，因此需要使用 DHCP 中继功能。DHCP 客户端可以通过 DHCP 中继与其他网段的 DHCP 服务器通信，最终获取 IP 地址。

DHCP 中继的工作过程如图 1.78 所示。

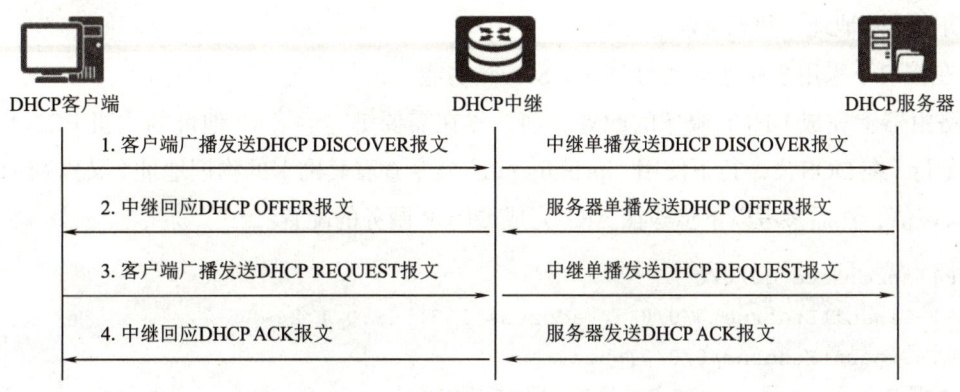

图 1.78　DHCP 中继的工作过程

相关技能

1. DHCP 服务的规划与配置

搭建如图 1.79 所示的拓扑图，路由器 AR1 G0/0/0 接口的 IP 地址为 192.168.0.1/24，请在 AR1 上部署 DHCP 服务，为 PC1 与 PC2 所在网段中的主机自动分配 IP 地址，并进行测试与故障排除。

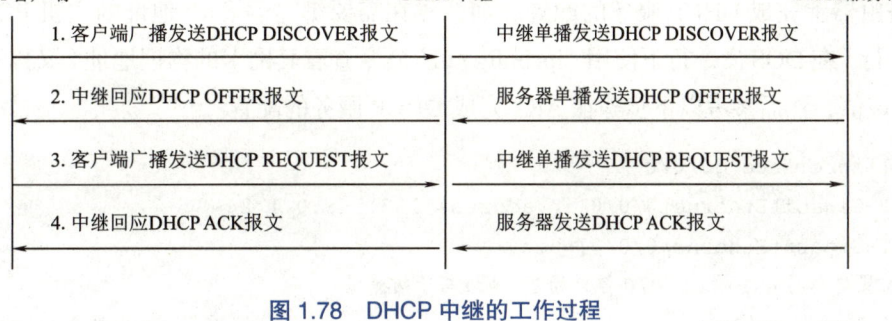

图 1.79　"DHCP 服务的规划与配置"拓扑图

DHCP 服务的规划如下：

① DHCP 服务器提供的 IP 地址池名称为 student。

② 地址池"student"分配的 IP 地址网络号为 192.168.0.0/24；网关地址为 192.168.0.1/24；主 DNS 与备用 DNS 地址为 172.16.5.25 与 172.16.5.26。

③ 地址池"student"中排除地址 192.168.0.240 至 192.168.0.254，这些地址不参与自动分配。

④ 地址租期为 4 h。

⑤ 为 PC2 绑定静态的 IP 为 192.168.0.2。

56　企业网络互联技术（双语）

（1）路由器上 DHCP 服务的配置

路由器上 DHCP 服务的配置任务为

① 启用 DHCP 服务。

② 创建规划的 IP 地址池，并配置地址池相关参数（地址范围、默认网关地址、DNS 地址、地址租期、排除地址、静态绑定地址等）。

③ 在接口下采用全局地址池的 DHCP Server 功能。

在路由器上完成 DHCP 服务的配置之前，先在需要绑定静态 IP 地址的主机 PC2 上，进入 DOS 命令行，在 DOS 命令行下使用"ipconfig /all"命令查看其网卡的物理地址（又称 MAC 地址）值，记录该值，然后参考以下步骤在 AR1 完成 DHCP 服务的配置。

```
[AR1]interface g0/0/0
[AR1-GigabitEthernet0/0/0]ip address 192.168.0.1 24
[AR1-GigabitEthernet0/0/0]quit
// 配置路由器 AR1 的 G0/0/0 接口的 IP 地址与子网掩码
[AR1]dhcp enable                          // 启用 DHCP 服务
[AR1]ip pool student                      // 创建 IP 地址池，地址池名为"student"
[AR1-ip-pool-student]network 192.168.0.0 mask 24 // 配置地址池范围
[AR1-ip-pool-student]gateway-list 192.168.0.1     // 配置默认网关地址
[AR1-ip-pool-student]dns-list 172.16.5.25 172.16.5.26 // 配置 DNS 服务器地址
[AR1-ip-pool-student]excluded-ip-address 192.168.0.240 192.168.0.254
// 配置排除地址范围为 192.168.0.240 至 192.168.0.254
[AR1-ip-pool-student]lease day 0 hour 4           // 配置地址租用期限为 4 h
[AR1-ip-pool-student]static-bind ip-address 192.168.0.2 mac-address
5489-98a5-2eab  // 配置为主机 PC2 绑定的 IP 地址为 192.168.0.2（注："5489-98a5-2eab"
                // 为查看的 PC2 的 MAC 地址）
[AR1]interface g0/0/0
[AR1-GigabitEthernet0/0/0]dhcp select global      // 开启 G0/0/0 接口采用全局地址
                                                   // 池的 DHCP Server 功能
[AR1-GigabitEthernet0/0/0]quit
[AR1]quit
<AR1>save
```

（2）DHCP 测试与排障

在主机 PC1 与 PC2 上，配置其 IP 地址与 DNS 地址为自动获取，具体步骤为：选择"控制面板"→"网络与共享中心"→"更改适配器设置"命令，右击"以太网网络连接"，选择"属性"→"Internet 协议版本 4（TCP/IPv4）"命令，在"属性"中选中"自动获得 IP 地址"与"自动获得 DNS 地址"，单击"确定"按钮，关闭"网络连接属性"窗口。

在主机 PC1 上使用"ipconfig /all"命令查看 IP 信息，查看是否获得了正确的 IP 地址、子网掩

项目 1　企业分公司网络互联项目　57

码、默认网关、DNS 服务器地址以及租期时间。

注：在华为仿真软件 ENSP 中测试主机没有"ipconfig /all"命令，只能使用"ipconfig"命令查看 IP 信息。图 1.80 为在 ENSP 中的测试主机查看自动获得的 IP 信息。

```
PC1
  基础配置    命令行    组播    UDP发包工具    串口

PC>ipconfig

Link local IPv6 address...........: fe80::5689:98ff:fe2f:64e1
IPv6 address.....................: :: / 128
IPv6 gateway.....................: ::
IPv4 address.....................: 192.168.0.239
Subnet mask......................: 255.255.255.0
Gateway..........................: 192.168.0.1
Physical address.................: 54-89-98-2F-64-E1
DNS server.......................: 172.16.5.25
                                   172.16.5.26
```

图 1.80　主机 PC1 自动获得的 IP 信息

在主机 PC2 上使用"ipconfig /all"命令查看 IP 信息，查看获得的 IP 地址是否为绑定的 IP 地址"192.168.0.2"。

如图 1.81 所示 IP 信息中，PC2 获得的 IP 地址为静态绑定的"192.168.0.2"。

```
PC2
  基础配置    命令行    组播    UDP发包工具    串口

PC>ipconfig

Link local IPv6 address...........: fe80::5689:98ff:fea5:2eab
IPv6 address.....................: :: / 128
IPv6 gateway.....................: ::
IPv4 address.....................: 192.168.0.2
Subnet mask......................: 255.255.255.0
Gateway..........................: 192.168.0.1
Physical address.................: 54-89-98-A5-2E-AB
DNS server.......................: 172.16.5.25
                                   172.16.5.26
```

图 1.81　主机 PC2 自动获得的 IP 信息

还可以在路由器上使用"display ip pool name student used"查看地址池"student"中已使用的地址情况，如图 1.82 所示。图中显示"192.168.0.2"静态绑定给 MAC 地址为"5489-98a5-2eab"的主机，另外"192.168.0.239"已动态分配给主机。

如果测试时主机不能获得 IP 地址，参考以下步骤完成排障：

① 在所有主机上查看 IP 地址与 DNS 地址是否已配置为自动获得。

② 在路由器 AR1 上使用"display ip interface brief"命令查看路由器的接口 IP 是否已正确配置，物理接口与链路协议状态是否已"UP"。

③ 在路由器 AR1 上使用"<AR1>display current-configuration | begin ip pool"命令查看当前正在运行的配置文件中关于 IP 地址池"student"的配置是否正确，如图 1.83 所示。

```
[AR1]display ip pool name student used
  Pool-name     : student
  Pool-No       : 0
  Lease         : 0 Days 4 Hours 0 Minutes
  Domain-name   : -
  DNS-server0   : 172.16.5.25
  DNS-server1   : 172.16.5.26
  NBNS-server0  : -
  Netbios-type  : -
  Position      : Local       Status      : Unlocked
  Gateway-0     : 192.168.0.1
  Mask          : 255.255.255.0
  VPN instance  : --


    Start       End    Total Used Idle(Expired) Conflict Disable
  ----------------------------------------------------------------
    192.168.0.1 192.168.0.254 253  2    236(0)       0      15

  ----------------------------------------------------------------

  Network section :
  ----------------------------------------------------------------
  Index    IP        MAC       Lease  Status
  ----------------------------------------------------------------
     1   192.168.0.2   5489-98a5-2eab    -   Static-bind used
   238   192.168.0.239 5489-982f-64e1  2653  Used
```

图 1.82 查看地址池 "student" 中已使用的地址情况

```
<AR1>display current-configuration | begin ip pool
ip pool student
gateway-list 192.168.0.1
network 192.168.0.0 mask 255.255.255.0
static-bind ip-address 192.168.0.2 mac-address 5489-98a5-2eab
excluded-ip-address 192.168.0.240 192.168.0.254
lease day 0 hour 4 minute 0
dns-list 172.16.5.25 172.16.5.26
```

图 1.83 在路由器上查看当前配置文件中关于 IP 地址池 "student" 的配置

④ 在路由器 AR1 上，进入 G0/0/0 接口的接口视图下，使用 "display this" 命令显示当前接口视图下的配置，检查该接口是否配置了采用全局地址池的 DHCP Server 功能，如图 1.84 所示。

```
[AR1-GigabitEthernet0/0/0]display this
[V200R003C00]
#
interface GigabitEthernet0/0/0
 ip address 192.168.0.1 255.255.255.0
 dhcp select global
#
return
```

图 1.84 使用 "display this" 命令显示接口视图下的配置

2. DHCP 中继的规划与配置

搭建如图 1.85 所示的拓扑图，主机 PC1 与 PC2 由路由器 AR2 提供 DHCP 服务，为其动态

分配 IP 地址，请完成 DHCP 服务及中继的配置，并进行测试与故障排除。

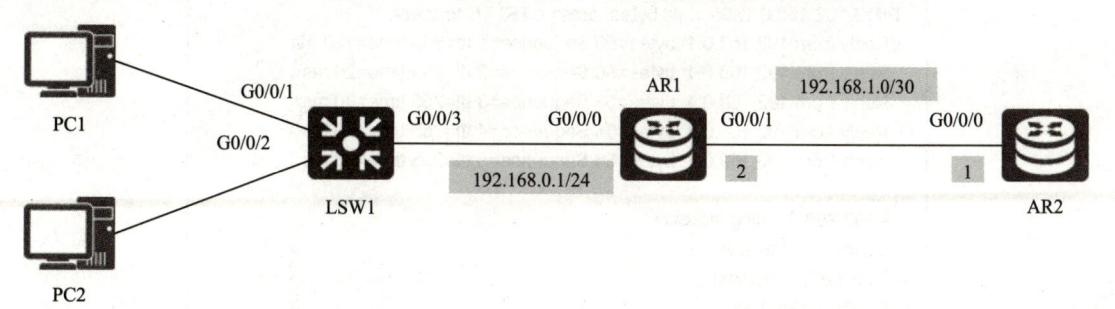

图 1.85 "DHCP 中继的规划与配置"拓扑图

DHCP 服务的规划如下：

① DHCP 服务器提供的 IP 地址池名称为 student。

② 地址池"student"分配的 IP 地址网络号为 192.168.0.0/24；网关地址为 192.168.0.1/24；主 DNS 与备用 DNS 地址为 172.16.5.25 与 172.16.5.26。

③ 地址池"student"中排除地址"192.168.0.240"至"192.168.0.254"，这些地址不参与自动分配。

④ 地址租期为 4 h。

⑤ 为 PC2 绑定静态的 IP 为"192.168.0.2"。

DHCP 中继的规划如下：

① 路由器 AR1 的 G0/0/0 接口配置 DHCP 中继功能。

② DHCP 服务器地址为 192.168.1.1。

（1）路由器接口与路由的配置

参考以下步骤，在路由器 AR1 与路由器 AR2 上配置接口的 IP 地址以及在路由器 AR2 上配置到目标网络"192.168.0.0/24"的静态路由，实现网络中所有网段的互联互通。

```
[AR1]interface g0/0/0
[AR1-GigabitEthernet0/0/0]ip address 192.168.0.1 24
[AR1-GigabitEthernet0/0/0]quit
[AR1]interface g0/0/1
[AR1-GigabitEthernet0/0/1]ip address 192.168.1.2 30
[AR1-GigabitEthernet0/0/1]quit
[AR2]interface g0/0/0
[AR2-GigabitEthernet0/0/0]ip address 192.168.1.1 30
[AR2-GigabitEthernet0/0/0]quit
[AR2]ip route-static 192.168.0.0 24 192.168.1.2
```

完成路由器接口 IP 与路由的配置后，在路由器 AR2 上 ping 路由器 AR1 的 G0/0/0 接口的 IP 地址（192.168.0.1），测试路由器 AR2 到远程网络的连通性，如图 1.86 所示。

60 企业网络互联技术（双语）

```
[AR2]PING 192.168.0.1
  PING 192.168.0.1: 56  data bytes, press CTRL_C to break
  Reply from 192.168.0.1: bytes=56 Sequence=1 ttl=255 time=120 ms
  Reply from 192.168.0.1: bytes=56 Sequence=2 ttl=255 time=20 ms
  Reply from 192.168.0.1: bytes=56 Sequence=3 ttl=255 time=30 ms
  Reply from 192.168.0.1: bytes=56 Sequence=4 ttl=255 time=30 ms
  Reply from 192.168.0.1: bytes=56 Sequence=5 ttl=255 time=30 ms

  --- 192.168.0.1 ping statistics ---
  5 packet(s) transmitted
  5 packet(s) received
  0.00% packet loss
  round-trip min/avg/max = 20/46/120 ms
```

图 1.86　AR2 ping AR1 的 G0/0/0 接口 IP

（2）在路由器 AR2 上完成 DHCP 服务的配置

请参考以下步骤在 AR2 上完成 DHCP 服务的配置。

```
[AR2]dhcp enable                               // 启用 DHCP 服务
[AR2]ip pool student                           // 创建 IP 地址池，地址池名为 "student"
[AR2-ip-pool-student]network 192.168.0.0 mask 24// 配置地址池范围
[AR2-ip-pool-student]gateway-list 192.168.0.1
// 配置为 DHCP 客户端分配的默认网关地址
[AR2-ip-pool-student]dns-list 172.16.5.25 172.16.5.26
// 配置为 DHCP 客户端分配的 DNS 服务器地址
[AR2-ip-pool-student]excluded-ip-address 192.168.0.240 192.168.0.254
// 配置排除地址范围为 192.168.0.240 至 192.168.0.254
[AR2-ip-pool-student]lease day 0 hour 4
// 配置为 DHCP 客户端分配的 IP 地址租用期限为 4 h
[AR2-ip-pool-student]static-bind ip-address 192.168.0.2 mac-address 5489-985e-6cc6
// 为 MAC 地址为 "5489-985e-6cc6" 的主机 PC2 绑定静态的 IP 地址 "192.168.0.2"
[AR2]interface g0/0/0
[AR2-GigabitEthernet0/0/0]dhcp select global
// 开启 G0/0/0 接口采用全局地址池的 DHCP Server 功能
[AR2-GigabitEthernet0/0/0]quit
[AR2]quit
<AR2>save
```

（3）配置路由器 AR1 作为 DHCP 中继

配置路由器 AR1 作为 DHCP 中继，需完成的配置任务为

① 启用 DHCP 服务。

② 进入可接收到 DHCP 客户端发送的 DHCP DISCOVERY 广播包的三层端口（配置有 IP 地址的端口），启用接口的 DHCP 中继功能，并配置 DHCP 服务器的 IP 地址。

项目 1 企业分公司网络互联项目 61

参考以下步骤配置路由器 AR1 作为 DHCP 中继，实现 PC1 与 PC2 所在网段的主机从路由器 AR2 自动获得 IP 地址。

```
[AR1]dhcp enable                                        // 启用 DHCP 服务
[AR1]interface g0/0/0
[AR1-GigabitEthernet0/0/0]dhcp select relay             // 启用接口的 DHCP 中继功能
[AR1-GigabitEthernet0/0/0]dhcp relay server-ip 192.168.1.1
// 配置 DHCP 服务器的地址为 "192.168.1.1"，该地址为路由器 AR2 的 G0/0/0 接口的 IP 地址
[AR1-GigabitEthernet0/0/0]quit
[AR1]quit
<AR1>
```

（4）DHCP 测试与排障

参考前述 DHCP 测试与排障方法进行 DHCP 中继的测试与排障。

如果测试时主机不能获得 IP 地址，还需要在作为 DHCP 中继的路由器 AR1 上，参考以下步骤完成排障：

① 在路由器 AR1 上与 AR2 上使用 "display ip interface brief" 查看路由器的接口 IP 是否已正确配置，物理接口与链路协议状态是否已 "UP"。

② 在路由器 AR2 上使用 "display ip routing-table" 命令显示路由表内容，检查是否为远程网段选路，所选的路径是否正确，如图 1.87 所示。

```
[AR2]display ip routing-table
Route Flags: R - relay, D - download to fib
------------------------------------------------------------
Routing Tables: Public
        Destinations : 8      Routes : 8

Destination/Mask      Proto  Pre  Cost   Flags NextHop      Interface

        127.0.0.0/8      Direct 0    0      D    127.0.0.1    InLoopBack0
        127.0.0.1/32     Direct 0    0      D    127.0.0.1    InLoopBack0
127.255.255.255/32    Direct 0    0      D    127.0.0.1    InLoopBack0
    192.168.0.0/24     Static 60   0      RD   192.168.1.2  GigabitEthernet0/0/0
    192.168.1.0/30     Direct 0    0      D    192.168.1.1  GigabitEthernet0/0/0
    192.168.1.1/32     Direct 0    0      D    127.0.0.1    GigabitEthernet0/0/0
    192.168.1.3/32     Direct 0    0      D    127.0.0.1    GigabitEthernet0/0/0
255.255.255.255/32    Direct 0    0      D    127.0.0.1    InLoopBack0
```

图 1.87 在路由器 AR2 上查看路由表

③ 在路由器 AR1 上使用 "display dhcp relay interface g0/0/0" 命令查看 G0/0/0 接口的 DHCP 中继配置信息是否正确，如图 1.88 所示。

```
[AR1]display dhcp relay interface g0/0/0
DHCP relay agent running information of interface GigabitEthernet0/0/0 :
Server IP address [01] : 192.168.1.1
Gateway address in use : 192.168.0.1
```

图 1.88 在路由器 AR1 上查看 G0/0/0 接口的 DHCP 中继配置信息

62　企业网络互联技术（双语）

④ 在路由器 AR1 上使用"display dhcp statistics"命令查看 DHCP 收发报文统计信息，如图 1.89 所示。

```
[AR1]display dhcp statistics
Input: total 12 packets, discarded 0 packets
  Bootp request       :       0,  Bootp reply    :       0
  Discover            :       6,  Offer          :       2
  Request             :       2,  Ack            :       2
  Release             :       0,  Nak            :       0
  Decline             :       0,  Inform         :       0

Output: total 8 packets, discarded 0 packets
```

图 1.89　使用"display dhcp statistics"命令查看 DHCP 收发报文统计信息

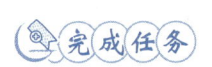

1. 任务规划

根据图 1.1 所示的网络拓扑结构和项目 1 的 DHCP 服务需求，在"fenbu"路由器上配置 DHCP 服务，需提供两个 IP 地址池。DHCP 服务的规划见表 1.16。

表 1.16　DHCP 服务的规划

地址池名	IP 地址网络号	网关地址	DNS 地址	地址租期	绑定静态的 IP
yusuan	172.16.129.0/24	172.16.129.1	172.16.128.253	8 h	
manage	172.16.128.0/25	172.16.128.1	172.16.128.253	8 h	主机 PC2 绑定静态 IP "172.16.128.13"

2. 任务实施

（1）查看并记录主机 PC3 的 MAC 地址

在主机 PC3 上，进入 DOS 命令行，使用"ipconfig /all"命令，查看主机 PC3 的 MAC 地址并记录，如图 1.90 所示的显示结果中，主机 PC3 的 MAC 地址为 54-89-98-5E-6B-B6。

图 1.90　查看并记录主机 PC3 的 MAC 地址

（2）"fenbu"路由器上 DHCP 服务的配置

参考以下步骤完成"fenbu"路由器上的 DHCP 服务的配置。

```
[fenbu]dhcp enable
// 启用 DHCP 服务
[fenbu]ip pool yusuan
[fenbu-ip-pool-yusuan]network 172.16.129.0 mask 24
[fenbu-ip-pool-yusuan]gateway-list 172.16.129.1
[fenbu-ip-pool-yusuan]dns-list 172.16.128.253
[fenbu-ip-pool-yusuan]lease day 0 hour 8
[fenbu-ip-pool-yusuan]quit
// 配置 DHCP 地址池"yusuan", 网络号为 172.16.129.0/24; 网关为 172.16.129.1; DNS 为
//172.16.128.253, 地址租期为 8 h
[fenbu]ip pool manage
[fenbu-ip-pool-manage]network 172.16.128.0 mask 25
[fenbu-ip-pool-manage]gateway-list 172.16.128.1
[fenbu-ip-pool-manage]dns-list 172.16.128.253
[fenbu-ip-pool-manage]lease day 0 hour 8
[fenbu-ip-pool-manage]static-bind ip-address 172.16.128.13 mac-address 5489-985e-6bb6
[fenbu-ip-pool-manage]quit
// 配置 DHCP 地址池"manage", 网络号为 172.16.128.0/25; 网关为 172.16.128.1; DNS 为
//172.16.128.253, 地址租期为 8 h, 为 MAC 地址为"5489-985e-6bb6"的主机 PC3 绑定静态的
//IP 地址"172.16.128.13"
[fenbu]interface g0/0/2
[fenbu-GigabitEthernet0/0/2]dhcp select global
[fenbu-GigabitEthernet0/0/2]quit
[fenbu]interface g3/0/0
[fenbu-GigabitEthernet3/0/0]dhcp select global
[fenbu-GigabitEthernet3/0/0]quit
// 开启 G0/0/2 与 G3/0/0 接口采用全局地址池的 DHCP Server 功能
[fenbu]quit
<fenbu>save
// 保存配置
```

（3）DHCP 测试与排障

① 在预算部门的主机 PC2 上、管理部门的主机 PC3 与 PC4 上，配置其 IP 地址与 DNS 地址为自动获取。

② 在主机 PC2、PC3 与 PC4 上使用"ipconfig /all"命令查看 IP 信息，查看是否获得了正确的 IP 地址、子网掩码、默认网关、DNS 服务器地址以及租期时间，注意观察管理部门的主机 PC3 获得的 IP 地址是否为静态绑定的地址"172.16.128.13"，如图 1.91 所示。

64 企业网络互联技术（双语）

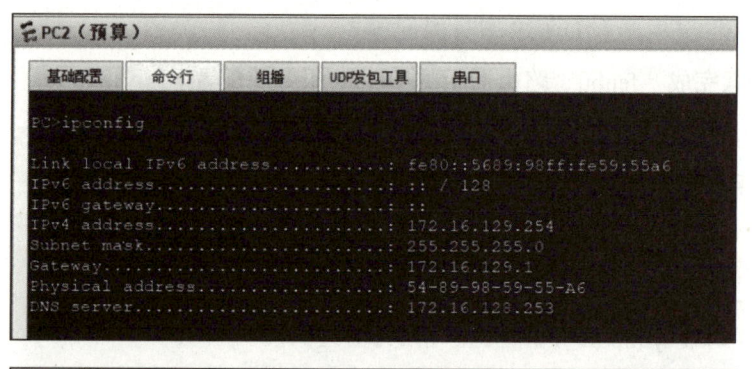

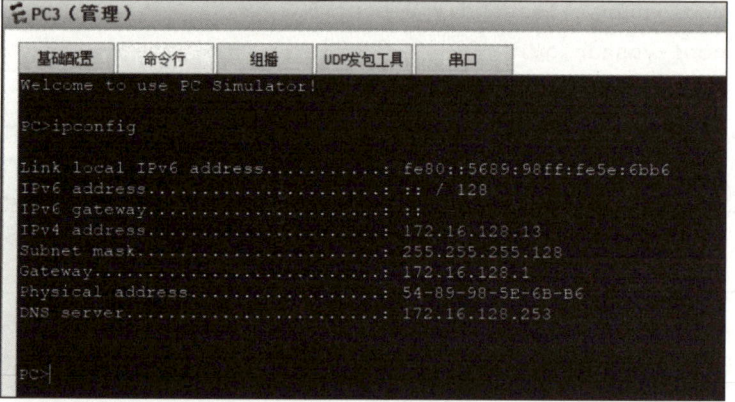

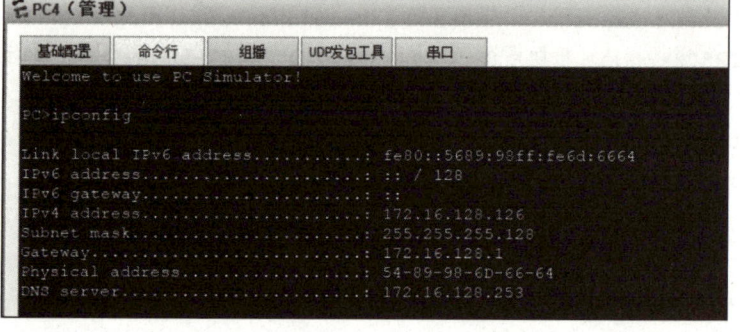

图 1.91　主机 PC2、PC3 与 PC4 上自动获得的 IP 地址信息

③ 在"fenbu"路由器上使用"display ip pool"命令查看地址池及地址统计信息，如图 1.92 所示。

④ 在"fenbu"路由器上使用"display ip pool name yusuan used"命令查看地址池"yusuan"及已使用地址统计信息，如图 1.93 所示。

⑤ 在"fenbu"路由器上使用"display ip pool name manage used"命令查看地址池"manage"及已使用地址统计信息，如图 1.94 所示。

⑥ 最后在主机 PC2、PC3 与 PC4 的 DOS 命令行下使用"tracert PC1 的 IP 地址"跟踪访问 PC1 的 IP 数据包所经过的路径。图 1.95 所示为在主机 PC3 上 tracert PC1 的结果，可以看出 PC3 到 PC1 的数据包所走路径为"172.16.128.1"→"172.16.254.2"→"172.16.0.10"。

项目 1　企业分公司网络互联项目　　65

如果测试时主机 PC2、PC3 与 PC4 不能获得 IP 地址，参考以下步骤完成排障：

① 在所有主机上查看 IP 地址与 DNS 地址是否已配置为自动获得。

② 在"fenbu"路由器上使用"display ip interface brief"命令查看路由器的接口（观察 G0/0/0 与 G3/0/0 接口）IP 是否已正确配置，物理接口与链路协议状态是否已"UP"。

③ 在"fenbu"路由器上使用"display current-configuration | begin ip pool"命令查看当前正在运行的配置文件中关于 IP 地址池的配置是否正确，如图 1.96 所示。

```
[fenbu]display ip pool
---------------------------------------------------------------

Pool-name    : yusuan
Pool-No      : 0
Position     : Local       Status      : Unlocked
Gateway-0    : 172.16.129.1
Mask         : 255.255.255.0
VPN instance : --


---------------------------------------------------------------

Pool-name    : manage
Pool-No      : 1
Position     : Local       Status      : Unlocked
Gateway-0    : 172.16.128.1
Mask         : 255.255.255.128
VPN instance : --


IP address Statistic
  Total    :378
  Used     :3        Idle     :375
  Expired  :0        Conflict :0        Disable  :0
[fenbu]
```

图 1.92　查看地址池及地址统计信息

```
[fenbu]display ip pool name yusuan used
  Pool-name    : yusuan
  Pool-No      : 0
  Lease        : 0 Days 8 Hours 0 Minutes
  Domain-name  : --
  DNS-server0  : 172.16.128.253
  NBNS-server0 : --
  Netbios-type : --
  Position     : Local       Status      : Unlocked
  Gateway-0    : 172.16.129.1
  Mask         : 255.255.255.0
  VPN instance : --
  ------------------------------------------------------------
    Start       End     Total Used Idle(Expired) Conflict Disable
  ------------------------------------------------------------
    172.16.129.1 172.16.129.254  253   1      252(0)       0        0
  ------------------------------------------------------------

  Network section :

  ------------------------------------------------------------
  Index     IP        MAC     Lease  Status
  ------------------------------------------------------------
    253  172.16.129.254  5489-9859-55a6    3423  Used
```

图 1.93　查看地址池"yusuan"已使用的地址情况

```
[fenbu]display ip pool name manage used
  Pool-name     : manage
  Pool-No       : 1
  Lease         : 0 Days 8 Hours 0 Minutes
  Domain-name   : -
  DNS-server0   : 172.16.128.253
  NBNS-server0  : -
  Netbios-type  : -
  Position      : Local        Status        : Unlocked
  Gateway-0     : 172.16.128.1
  Mask          : 255.255.255.128
  VPN instance  : --

  ----------------------------------------------------------------------
    Start      End     Total Used Idle(Expired) Conflict Disable
  ----------------------------------------------------------------------
   172.16.128.1 172.16.128.126  125    2    123(0)       0        0
  ----------------------------------------------------------------------

  Network section :
  ----------------------------------------------------------------------
  Index      IP          MAC      Lease  Status
  ----------------------------------------------------------------------
    12  172.16.128.13   5489-985e-6bb6    -   Static-bind used
   125  172.16.128.126  5489-986d-6664   945  Used
```

图 1.94 查看地址池 "manage" 已使用的地址情况

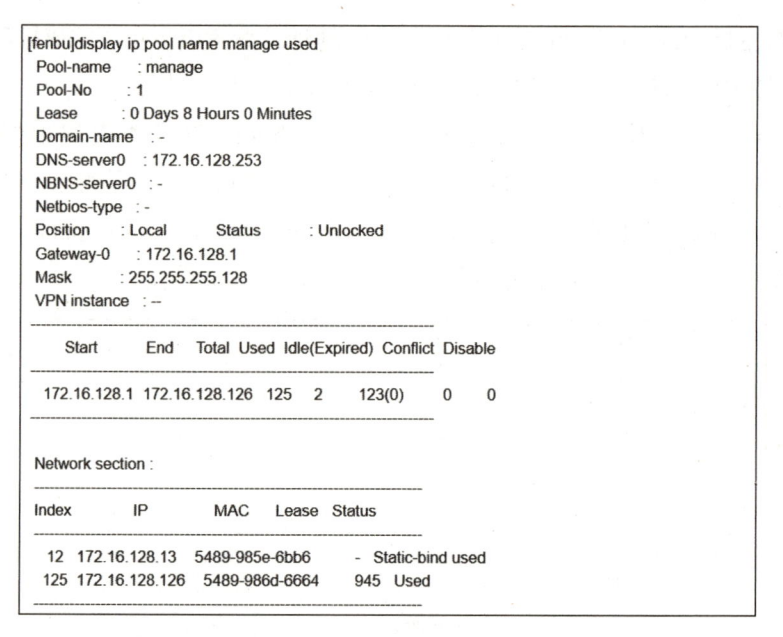

图 1.95 主机 PC3 "tracert" PC1 测试结果

```
[fenbu]display current-configuration | begin ip pool
ip pool yusuan
 gateway-list 172.16.129.1
 network 172.16.129.0 mask 255.255.255.0
 lease day 0 hour 8 minute 0
 dns-list 172.16.128.253
#
ip pool manage
 gateway-list 172.16.128.1
 network 172.16.128.0 mask 255.255.255.128
 static-bind ip-address 172.16.128.13 mac-address 5489-985e-6bb6
 lease day 0 hour 8 minute 0
 dns-list 172.16.128.253
```

图 1.96 查看当前配置文件中关于 IP 地址池的配置

④ 在 "fenbu" 路由器上，进入 G0/0/2 接口与 G3/0/0 的接口视图下，使用 "display ip this" 命令显示当前接口视图下的配置，检查该接口是否配置了采用全局地址池的 DHCP Server 功能。

项目 1　企业分公司网络互联项目　67

习题

选择题

（1）DHCP DISCOVER 报文的目的 IP 地址为（　　）。

    A. 0. 0. 0. 0　　　　　B. 224. 0. 0. 9　　　　C. 127. 0. 0. 1　　　　D. 255. 255. 255. 255

（2）华为路由器中，下面（　　）命令可显示地址池中已分配的 IP 地址和对应的 MAC 地址。

    A. display ip pool name yusuan used　　　　B. display ip pool

    C. display ip pool name yusuan　　　　　　D. display dhcp

（3）下面（　　）命令用于显示华为路由器的 IP 地址池。

    A. display ip pool　　　　　　　　　　　B. display dhcp pool

    C. display interface　　　　　　　　　　D. display ip interface brief

（4）某台路由器 DHCP 地址池配置信息如下：

```
#
ip pool jwc
  network 192.168.10.0 mask 255.255.255.0
  gateway-list 192.168.10.1
  dns-list 172.16.5.25
#
```

下面说法不正确的是（　　）。

    A. 该地址池网关地址为 192.168.10.1　　B. 该地址池 DNS 地址为 192.168.10.1

    C. 该地址池 DNS 地址为 172.16.5.25　　D. 该地址池网络地址为 192.168.10.0/24

任务 4　访问控制列表

任务描述

完成本项目 1 的任务 1 到任务 3 后，在图 1.1 所示的网络拓扑结构中的分公司 "fenbu" 路由器上完成 ACL（访问控制列表）规划与配置任务，并将分公司路由器的配置备份到总部的 FTP 服务器上，实现项目 1 的 "连接到总部的分公司路由器出口与入口引入 ACL，禁用 TCP 与 UDP 协议的 135~139、445 与 3389 高危端口"、"只有分公司管理部门的 PC3 和总部的 PC1 可使用 SSH 远程登录到 "fenbu" 路由器上对其进行远程管理" 和 "完成配置后，将分公司与总部路由器的配置进行保存，并备份到总部的 FTP 服务器上" 的建网需求。

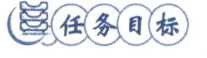

● 理解 ACL 的工作原理。

● 掌握基本 ACL 的规划与配置。

● 掌握高级 ACL 的规划与配置。

● 具有测试与故障排除的能力。

1. ACL 的工作原理

访问控制列表（access control list, ACL）是作用于路由器等设备接口的指令列表，它根据数据包报头中的条件来决定是允许还是拒绝数据包通过，又称包过滤（packet filtering）。ACL 既是控制网络通信流量的手段，也是网络安全策略的一个组成部分，是实现网络边界安全的重要机制之一。ACL 既可以在路由器上实现，也可以在防火墙等设备上实现。

ACL 通常被应用于路由器等设备接口。根据 ACL 所在接口的位置，它既可实现对进入路由器接口的数据包进行过滤，也可对从路由器接口出去的数据包进行过滤，前者称为"入（inbound）"方向的包过滤，后者称为"出（outbound）"方向的包过滤。图 1.97 所示为 ACL 的工作原理。

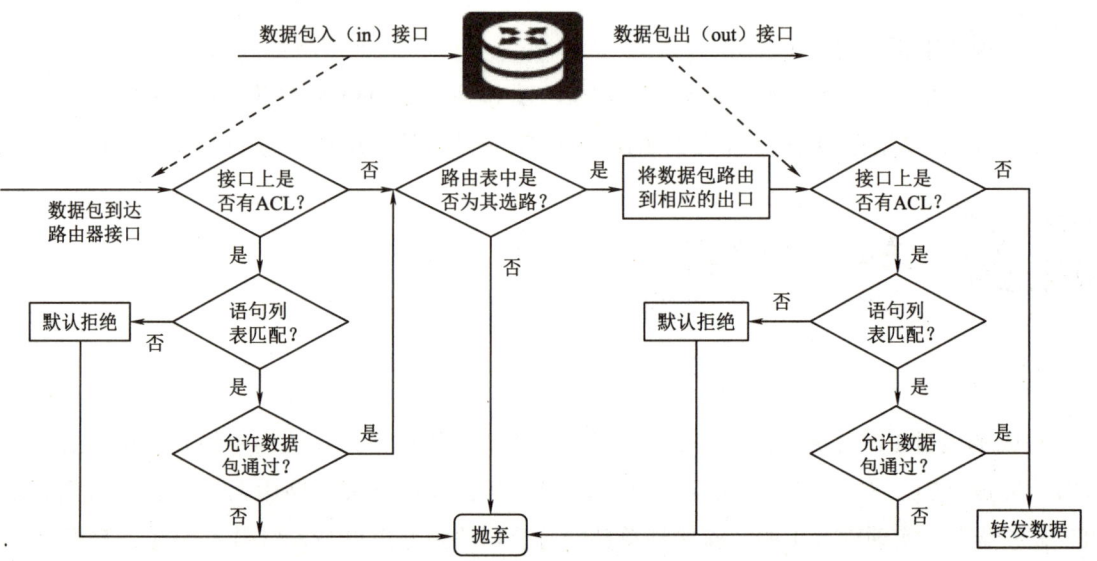

图 1.97 ACL 的工作原理

2. ACL 类别与标识

根据访问控制功能的不同，ACL 被分成基本 ACL、高级 ACL 和二层 ACL 等。基本 ACL 根据源 IP 地址、分片信息和生效时间段等信息来定义规则，对 IPv4 报文进行过滤。如果只需要根据源 IP 地址对报文进行过滤，可以配置基本 ACL，但基本 ACL 无法对来自同一网络或主机

项目 1　企业分公司网络互联项目　　**69**

的不同类型流量进行区分，也就是说，它不能根据应用层协议或应用服务的不同进行有选择的过滤。高级 ACL 可以根据检查数据包中的协议类型、源地址、目的地址和端口号等进行接收或拒绝数据包。

　　用户在创建 ACL 时必须为其指定编号，不同的编号对应不同类型的 ACL，见表 1.17。同时，为了便于记忆和识别，用户在创建 ACL 时还可选择是否为其设置名称。ACL 一旦创建，便不允许用户再为其设置名称、修改或删除其原有名称。

表 1.17　华为 ACL 分类

ACL 类型	编号范围	规则制订依据
基本 ACL	2000～2999	报文的源地址
高级 ACL	3000～3999	报文的源地址、目的地址、协议类型、端口号等
二层 ACL	4000～4999	报文的源 MAC 地址、目的 MAC 地址、802.1p 优先级等

3. 通配符掩码

　　通配符掩码（wildcard-mask）与 IP 地址成对使用，用来说明 IP 地址中的相应位是否需要被检查与匹配。通配符掩码中 "1" 表示所对应的 IP 地址相应二进制位不需要被匹配，"0" 表示所对应的 IP 地址相应二进制位需要匹配。与子网掩码类似，通配符掩码的长度为 32 位（二进制），用点分十进制表示。例如，使用通配符掩码 "0.255.255.255" 表示对应的 IP 地址中的前 8 位需要检查，后 24 位可以忽略。表 1.18 给出了更多通配符掩码的示例。

表 1.18　通配符掩码的示例

测试条件	IP 地址	通配符掩码
10.0.0.0/8	10.0.0.0	0.255.255.255
172.31.0.0/16	172.31.0.0	0.0.255.255
202.33.44.0/24	202.33.44.0	0.0.0.255
192.168.1.64/26	192.168.1.64	0.0.0.63
210.33.44.254/32	210.33.44.254	0.0.0.0

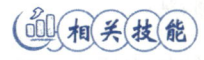

1. 基本 ACL 的规划与配置

　　搭建如图 1.98 所示的拓扑图，在完成全网的互联互通前提下，实施基本 ACL，实现：

　　① 除了 "192.168.2.0/24" 网段的主机 PC2 不能访问服务器网段，网络中所有主机网段均可访问服务器网段。

　　② 只有 PC1 可 SSH 到 AR1 上，并进行测试与故障排除。

70 企业网络互联技术（双语）

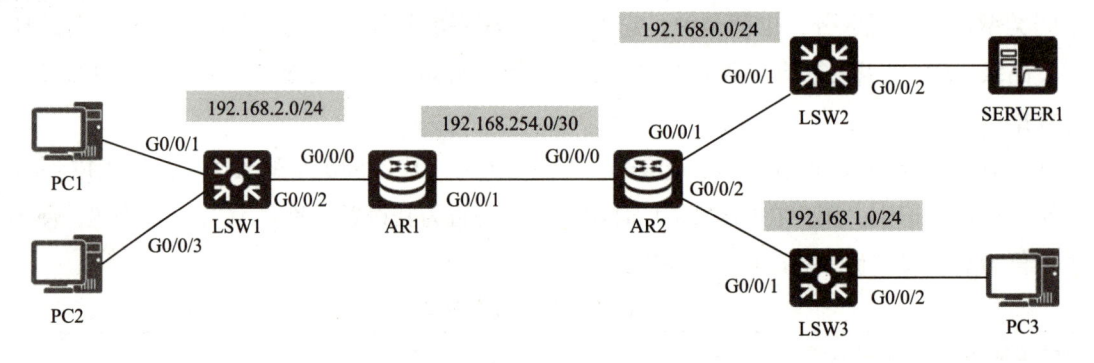

图 1.98 "基本 ACL 的规划与配置"拓扑图

路由器接口 IP 与 PC 的 IP 规划见表 1.19。

表 1.19 IP 规划

设 备	接 口	IP 地址	网 关
AR1	G0/0/0	192.168.2.1/24	
	G0/0/1	192.168.254.1/30	
AR2	G0/0/0	192.168.254.2/30	
	G0/0/1	192.168.0.1/24	
	G0/0/2	192.168.1.1/24	
PC1	NIC	192.168.2.11	192.168.2.1
PC2	NIC	192.168.2.12	192.168.2.1
PC3	NIC	192.168.1.13	192.168.1.1
SERVER1	NIC	192.168.0.10	192.168.0.1

使用静态路由实现全网的互联互通。静态路由的规划见表 1.20。

表 1.20 静态路由的规划

设 备	目标网络	目标网络掩码 / 掩码长度	下一跳或转发出口
AR1	192.168.0.0	23	192.168.254.2
AR2	192.168.2.0	24	192.168.254.1

根据需求：除了"192.168.2.0/24"网段的主机 PC2 不能访问服务器网段，网络中所有主机网段均可访问服务器网段。由于基本 ACL 只能检查数据包中的源 IP 地址，因此要实现该需求，需要在靠近目的端（服务器网段）的路由器 AR2 上创建一个基本 ACL（该列表号为 2000），并应用到路由器 AR2 的 G0/0/1 接口的出口方向上。

根据需求：只有 PC1 可 SSH 到 AR1 上。需要在路由器 AR1 上创建一个基本 ACL，应用到路由器 AR1 的 VTY 0 4 的用户接口视图。

ACL 的规划见表 1.21。

项目 1　企业分公司网络互联项目　71

表 1.21　ACL 的规划

设　　备	列 表 号	ACL 规则	应用接口与方向
AR1	2000	允许源地址为 192.168.2.11 的主机通过；禁止所有通过	VTY 0 4 的 inbound 方向
AR2	2001	禁止源地址为 192.168.2.12 的主机通过；允许网络 192.168.2.0/24 通过；允许网络 192.168.1.0/24 通过	AR2 的 G0/0/1 接口出口

（1）路由器基本配置与静态路由配置

参考以下步骤，在路由器 AR1 与路由器 AR2 上完成接口配置与静态路由的配置，实现网络中所有网段的互联互通。

```
[AR1]interface g0/0/0
[AR1-GigabitEthernet0/0/0]ip address 192.168.2.1 24
[AR1-GigabitEthernet0/0/0]quit
[AR1]interface g0/0/1
[AR1-GigabitEthernet0/0/1]ip address 192.168.254.1 30
[AR1-GigabitEthernet0/0/1]quit
[AR1]ip route-static 192.168.0.0 23 192.168.254.2
[AR1]quit
<AR1>save
// 配置路由器 AR1 的接口与到远程网络的汇总静态路由，并保存配置
[AR2]interface g0/0/0
[AR2-GigabitEthernet0/0/0]ip address 192.168.254.2 30
[AR2-GigabitEthernet0/0/0]quit
[AR2]interface g0/0/1
[AR2-GigabitEthernet0/0/1]ip address 192.168.0.1 24
[AR2-GigabitEthernet0/0/1]quit
[AR2]interface g0/0/2
[AR2-GigabitEthernet0/0/2]ip address 192.168.1.1 24
[AR2-GigabitEthernet0/0/2]quit
[AR2]ip route-static 192.168.2.0 24 192.168.254.1
[AR2]quit
<AR2>save
// 配置路由器 AR2 的接口与到远程网络的静态路由，并保存配置
```

（2）路由器 AR1 上 SSH 的配置

参考以下步骤在路由器 AR1 上完成 Stelnet 的配置，其规划为：Stelnet 认证采用 AAA 认证，Stelnet 的用户名为"zhangcr"，密码为"Chengdu123"，权限级别为"15"。

```
[AR1]stelnet server enable                              // 启用 Stelnet 服务
[AR1]aaa                                                // 进入 AAA 视图
[AR1-aaa]local-user zhangcr password cipher Chengdu123  // 创建本地用户，用户名为"zhangcr"，
```

72　企业网络互联技术（双语）

```
                                                  // 密码为"Chengdu123"
[AR1-aaa]local-user zhangcr service-type ssh       // 配置用户"zhangcr"授权 SSH 服务
[AR1-aaa]local-user zhangcr privilege level 15     // 配置用户"zhangcr"的权限级别为 15
[AR1-aaa]quit                                      // 返回到系统视图
[AR1]user-interface vty 0 4                        // 进入 VTY0 到 VTY4 的用户界面视图
[AR1-ui-vty0-4]protocol inbound SSH                // 配置协议为 SSH 协议
[AR1-ui-vty0-4]authentication-mode aaa             // 配置 VTY0 到 VTY4 五个虚拟终端
                                                   // 会话的认证模式为使用 AAA 进行认证
[AR1-ui-vty0-4]quit    // 退出 VTY0 到 VTY4 的用户界面视图，返回到用户视图
[AR1]rsa local-key-pair create                     // 使用 RSA 算法创建本地密钥对
```

（3）网络连通性与 SSH 测试

参照表 1.19 规划的 IP 参数完成网络中所有 PC 与服务器的 IP 地址、子网掩码、默认网关配置，配置完成后在 DOS 命令行下使用"ipconfig"命令查看其 IP 配置信息是否正确，并使用"ping"命令测试全网的连通性。（**注**：此时网络所有主机之间均具有 IP 连通性）图 1.99 所示为主机 PC1 成功 ping 通服务器与 PC3。

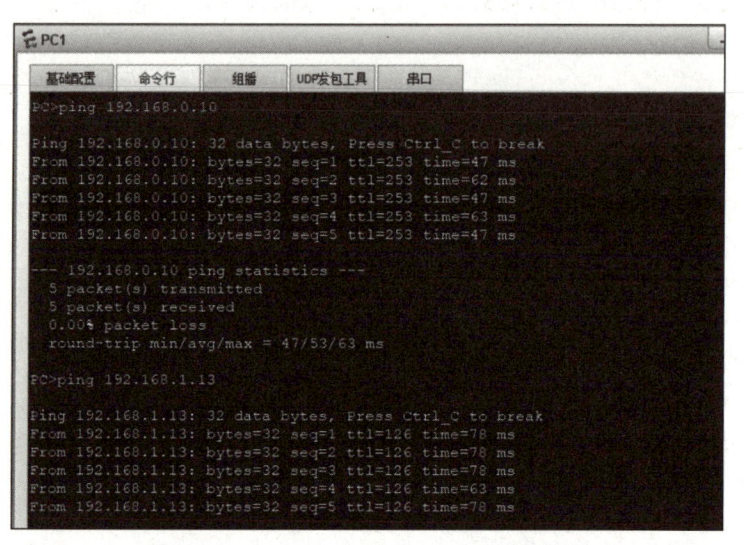

图 1.99　主机 PC1 成功 ping 通服务器与 PC3

在 PC1、PC2，运行 SecureCRT 虚拟终端软件，使用 SSH 访问路由器 AR1 G0/0/0 接口的 IP 地址"192.168.2.1"。

在 PC3 上，运行 SecureCRT 虚拟终端软件，使用 SSH 访问路由器 AR1 G0/0/1 接口的 IP 地址"192.168.254.1"。此时，PC1、PC2 与 PC3 等所有主机均可成功使用 SSH 登录到路由器 AR1。

如果出现主机与服务器之间相互不通，主机不能使用 SSH 登录到路由器 AR1 上，请排除故障。

（4）基本 ACL 的配置

根据表 1.21，参考以下步骤在路由器 AR2 上完成基本 ACL 的配置，实现除了"192.168.2.0/24"

网段的主机 PC2 不能访问服务器网段，网络中所有主机网段均可访问服务器网段。

```
[AR2]acl 2001    // 创建编号为 2001 的基本 ACL
[AR2-acl-basic-2001]rule deny source 192.168.2.12 0.0.0.0
// 配置拒绝主机 "192.168.2.12" 通过的规则
[AR2-acl-basic-2001]rule permit source 192.168.2.0 0.0.0.255
// 配置允许网络 "192.168.2.0/24" 通过的规则
[AR2-acl-basic-2001]rule permit source 192.168.1.0 0.0.0.255
// 配置允许网络 "192.168.1.0/24" 通过的规则
[AR2-acl-basic-2001]quit
[AR2]interface g0/0/1
[AR2-GigabitEthernet0/0/1]traffic-filter outbound acl 2001
// 将 ACL 2001 应用到 G0/0/1 接口的出口方向
[AR2-GigabitEthernet0/0/1]quit
```

根据表 1.21，参考以下步骤在路由器 AR1 上完成基本 ACL 的配置，实现只有 PC1 可以 SSH 到 AR1 上。

```
[AR1]acl 2000    // 创建编号为 2000 的基本 ACL
[AR1-acl-basic-2000]rule permit source 192.168.2.11 0.0.0.0
// 配置允许主机 "192.168.2.11" 通过的规则
[AR1-acl-basic-2000]rule deny source any    // 配置拒绝所有流量通过的规则
[AR1]user-interface vty 0 4
[AR1-ui-vty0-4]acl 2000 inbound             // 将 ACL 2000 应用到 VTY 0 到 VTY 4 流量
                                            // 进入方向
[AR1-ui-vty0-4]quit
[AR1]
```

（5）ACL 测试与排障

分别在 PC1、PC2 与 PC3 的 DOS 命令行下，使用 ping 实用程序测试到服务器的 IP 连通性。图 1.100 所示为主机 PC1 与 PC3 可以 ping 通服务器，但主机 PC2 不能 ping 通服务器。

分别在主机 PC1、PC2 与 PC3 上，运行 SecureCRT 虚拟终端软件，测试是否可以 SSH 访问路由器 AR1。实现了 ACL 后，只有主机 PC1 可以成功使用 SSH 访问路由器 AR1，其他主机均不能使用 SSH 访问路由器 AR1。

如果测试结果不是除了 "192.168.2.0/24" 网段的主机 PC2 不能访问服务器网段，网络中所有主机网段均可访问服务器网段。请使用以下步骤完成排障：

在路由器 AR2 上使用 "display acl 2001" 命令显示基本 ACL 2001 的配置和运行情况，如图 1.101 所示。

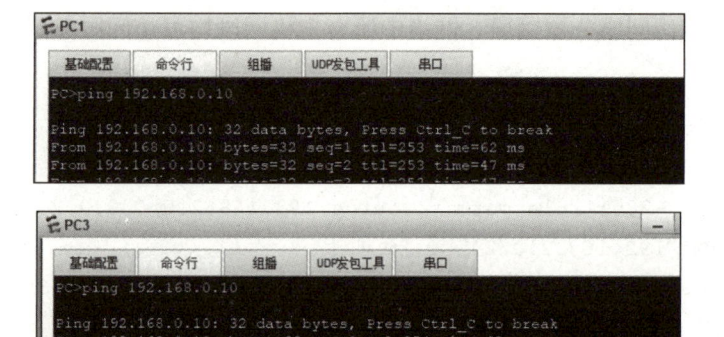

图 1.100　主机 PC1 与 PC3 可以 ping 通服务器

```
<AR2>display acl 2001
Basic ACL 2001, 3 rules
Acl's step is 5
 rule 5 deny source 192.168.2.12 0 (13 matches)
 rule 10 permit source 192.168.2.0 0.0.0.255
 rule 15 permit source 192.168.1.0 0.0.0.255
```

图 1.101　“display acl 2001”命令的显示结果

在路由器 AR2 上使用“display current-configuration | begin acl”命令查看 ACL 的配置，如图 1.102 所示。

```
<AR2>display current-configuration | begin acl
acl number 2001
 rule 5 deny source 192.168.2.12 0
 rule 10 permit source 192.168.2.0 0.0.0.255
 rule 15 permit source 192.168.1.0 0.0.0.255
```

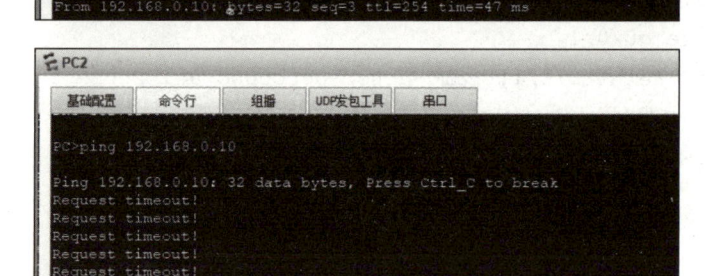

图 1.102　“display current-configuration | begin acl”命令的显示结果

如果配置的 ACL 出现错误，请参考以下命令删除 ACL 2001，并重新配置正确的 ACL。

```
[AR2]undo acl 2001     //删除 ACL 2001
```

在路由器 AR2 上使用“display traffic-filter applied-record”命令查看 ACL 的应用信息，如图 1.103 所示。

项目 1　企业分公司网络互联项目　75

```
[AR2]display traffic-filter applied-record
----------------------------------------------------
Interface              Direction  AppliedRecord
----------------------------------------------------
GigabitEthernet0/0/1     outbound   acl 2001
```

图 1.103　"display traffic-filter applied-record" 命令查看 ACL 的应用信息

如果显示的不是 G0/0/1 的出口（outbound）方向应用了 ACL 2001，则先进入配置错误的接口视图，使用 "undo traffic-filter" 命令删除应用的 ACL，再进入 G0/0/1 接口，将 ACL 2001 应用到 G0/0/1 的出口方向。下面命令示例了删除 G0/0/1 接口出口方向应用的 ACL。

```
[AR2-GigabitEthernet0/0/1]undo traffic-filter outbound// 删除 G0/0/1 接口出口方向应用的 ACL
```

如果测试结果不是只有 PC1 可 SSH 到 AR1 上。可使用以下步骤完成排障：

在路由器 AR1 上使用 "display acl 2000" 命令显示基本 ACL 2000 的配置和运行情况，如图 1.104 所示。

```
<AR1>display acl 2000
Basic ACL 2000, 2 rules
Acl's step is 5
 rule 5 permit source 192.168.2.11 0 (4 matches)
 rule 10 deny (8 matches)
```

图 1.104　"display acl 2000" 命令的显示结果

在路由器 AR2 上使用 "display current-configuration | begin acl" 命令查看 ACL 的配置，如图 1.105 所示。

```
<AR2>display current-configuration | begin acl
acl number 2001
 rule 5 deny source 192.168.2.12 0
 rule 10 permit source 192.168.2.0 0.0.0.24
 rule 15 permit source 192.168.1.0 0.0.0.24
```

图 1.105　"display current-configuration | begin acl" 命令的显示结果

如果配置的 ACL 出现错误，请删除错误的 ACL，并重新配置正确的 ACL。

在路由器 AR1 VTY 0 4 用户界面视图下，使用 "display this" 命令查看 VTY 0 4 用户界面视图下的配置信息，查看是否配置了将 ACL 2000 应用到 VTY 0 4 进入方向上，如图 1.106 所示。

```
[AR1-ui-vty0-4]display this
[V200R003C00]
#
user-interface con 0
 authentication-mode password
user-interface vty 0 4
 acl 2000 inbound
 authentication-mode aaa
 protocol inbound ssh
```

图 1.106　"display this" 显示 VTY 0 4 用户界面视图下的配置

如果应用的 ACL 方向错误，使用"undo acl inbound"命令删除 VTY 0 4 下应用的 ACL，再应用正确的 ACL 到进入方向（inbound）。下面的命令为删除 VTY 0 4 接口应用的错误 ACL。

```
[AR1-ui-vty0-4]undo acl inbound      // 删除应用进入方向的 ACL
[AR1-ui-vty0-4]undo acl outbound     // 删除应用出去方向的 ACL
```

2. 高级 ACL 的规划与配置

搭建如图 1.107 所示的拓扑网，在完成全网的互联互通的前提下，实施高级 ACL，实现："192.168.2.0/24"网段中的所有主机只能访问 Web 服务器的 WWW 服务，不能 ping Web 服务器，但访问其他无限制。

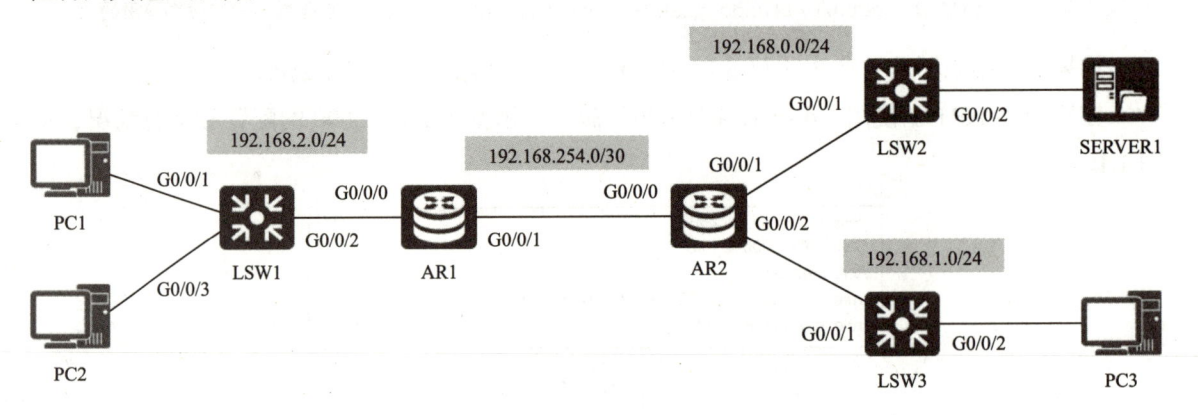

图 1.107　"高级 ACL 的规划与配置"拓扑图

路由器接口 IP 与 PC 的 IP 规划见表 1.22。

表 1.22　IP 规划

设　　备	接　　口	IP 地址	网　　关
AR1	G0/0/0	192.168.2.1/24	
	G0/0/1	192.168.254.1/30	
AR2	G0/0/0	192.168.254.2/30	
	G0/0/1	192.168.0.1/24	
	G0/0/2	192.168.1.1/24	
PC1	NIC	192.168.2.11/24	192.168.2.1
PC2	NIC	192.168.2.12/24	192.168.2.1
PC3	NIC	192.168.1.13/24	192.168.1.1
SERVER1	NIC	192.168.0.10/24	192.168.0.1

使用静态路由实现全网的互联互通，静态路由的规划见表 1.23。

表 1.23　静态路由的规划

设　　备	目标网络	目标网络掩码 / 掩码长度	下一跳或转发出口
AR1	192.168.0.0	23	192.168.254.2
AR2	192.168.2.0	24	192.168.254.1

项目 1　企业分公司网络互联项目　77

根据需求："192.168.2.0/24"网段中的所有主机只能访问 Web 服务器的 WWW 服务，不能 ping Web 服务器，但访问其他无限制。实现该需求，需要在靠近源端的路由器 AR1 上创建一个高级 ACL（该列表号为 3000），并应用到路由器 AR1 的 G0/0/0 接口的入口方向上。ACL 的规划见表 1.24。

表 1.24　ACL 的规划

设　　备	列 表 号	ACL 规则	应用接口与方向
AR1	3000	允许源网络是"192.168.2.0/24"的主机到主机"192.168.0.10"的 TCP 协议、端口号为 www（80）的流量通过； 禁止源网络是"192.168.2.0/24"的主机到主机"192.168.0.10"的 ICMP 协议（类型为 echo）的流量通过； 允许 IP 协议任何源到任何目的通过	路由器 AR1 的 G0/0/0 的 inbound 方向

（1）路由器基本配置与静态路由配置

参考以下步骤，在路由器 AR1 与路由器 AR2 上完成接口配置与静态路由的配置，实现网络中所有网段的互联互通。

```
[AR1]interface g0/0/0
[AR1-GigabitEthernet0/0/0]ip address 192.168.2.1 24
[AR1-GigabitEthernet0/0/0]quit
[AR1]interface g0/0/1
[AR1-GigabitEthernet0/0/1]ip address 192.168.254.1 30
[AR1-GigabitEthernet0/0/1]quit
[AR1]ip route-static 192.168.0.0 23 192.168.254.2
[AR1]quit
<AR1>save
// 配置路由器 AR1 的接口与到远程网络的汇总静态路由，并保存配置
[AR2]interface g0/0/0
[AR2-GigabitEthernet0/0/0]ip address 192.168.254.2 30
[AR2-GigabitEthernet0/0/0]quit
[AR2]interface g0/0/1
[AR2-GigabitEthernet0/0/1]ip address 192.168.0.1 24
[AR2-GigabitEthernet0/0/1]quit
[AR2]interface g0/0/2
[AR2-GigabitEthernet0/0/2]ip address 192.168.1.1 24
[AR2-GigabitEthernet0/0/2]quit
[AR2]ip route-static 192.168.2.0 24 192.168.254.1
[AR2]quit
<AR2>save
// 配置路由器 AR2 的接口与到远程网络的静态路由，并保存配置
```

（2）网络连通性

参照表1.20规划的IP参数完成网络中所有PC与服务器的IP地址、子网掩码、默认网关配置，配置完成后在DOS命令行下使用"ipconfig"命令查看其IP配置信息是否正确，并使用"ping"命令测试全网的连通性。（注：此时网络所有主机之间均具有IP连通性）图1.108所示为主机PC1成功ping通服务器与PC3。

如果出现主机与服务器之间相互不通，请排除故障。

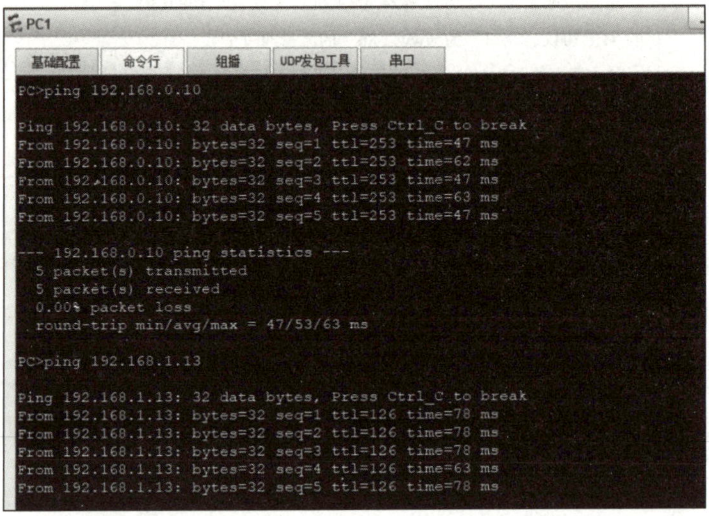

图 1.108　主机 PC1 成功 ping 通服务器与 PC3

（3）高级 ACL 的配置

根据表1.24，参考以下步骤在路由器AR1上完成高级ACL的配置。

```
[AR1]acl 3000    // 创建编号为 3000 的高级 ACL
[AR1-acl-adv-3000]rule permit tcp source 192.168.2.0 0.0.0.255 destination
192.168.0.10 0.0.0.0 destination-port eq www
    // 允许源网络 "192.168.2.0/24" 到主机 "192.168.0.10" 目标端口号为 www(80) 的 TCP 协议
    // 流量通过
    [AR1-acl-adv-3000]rule deny icmp source 192.168.2.0 0.0.0.255 destination
192.168.0.10 0.0.0.0 icmp-type echo
    // 禁止源网络 "192.168.2.0/24" 到主机 "192.168.0.10" 类型为 echo 的 ICMP 协议流量通过
    [AR1-acl-adv-3000]rule permit ip source any destination any
    // 允许 IP 协议任何源到任何目的流量通过
    [AR1-acl-adv-3000]quit
    [AR1]interface g0/0/0
    [AR1-GigabitEthernet0/0/0]traffic-filter inbound acl 3000
    // 将 ACL 3000 应用到入口方向
    [AR1-GigabitEthernet0/0/0]quit
    [AR1]
```

项目 1　企业分公司网络互联项目　79

（4）ACL 测试与排障

在网络"192.168.2.0/24"的任何一台主机上，打开浏览器，访问 Web 服务器提供的 Web 服务，然后在 DOS 命令行下，使用 ping 实用程序测试到 Web 服务器的 IP 连通性。图 1.109、图 1.110 所示为网段"192.168.2.0/24"中主机 PC1 测试结果，它访问 Web 服务器提供的 Web 服务，但不能 ping 通 Web 服务器。

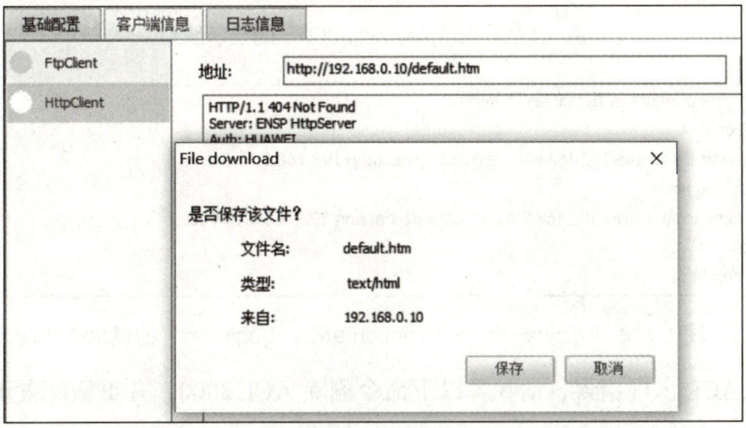

图 1.109　主机 PC1 可以访问 Web 服务器提供的 Web 服务

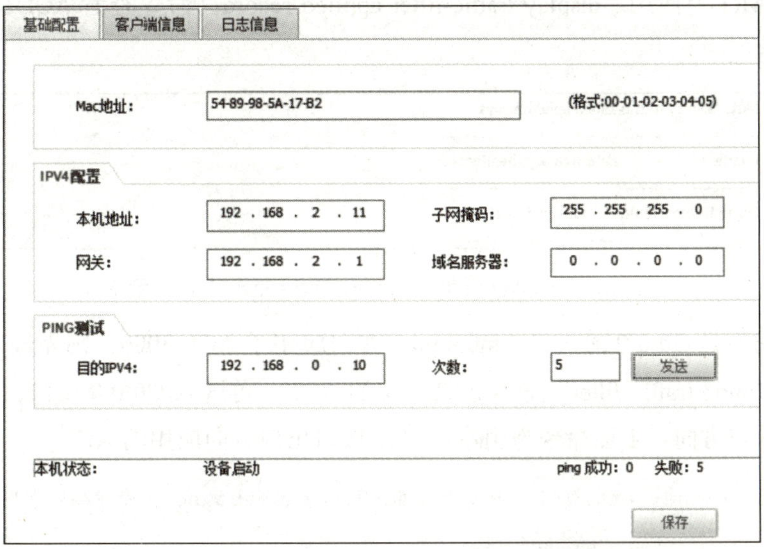

图 1.110　主机 PC1 不能 ping 通 Web 服务器

如果测试结果不满足子任务需求。请使用以下步骤完成排障：

在路由器 AR1 上使用"display acl 3000"命令显示高级 ACL 3000 的配置和运行情况，如图 1.111 所示。

在路由器 AR1 上使用"display current-configuration | begin acl"命令查看 ACL 的配置，如图 1.112 所示。

80　企业网络互联技术（双语）

```
[AR1]display acl 3000
Advanced ACL 3000, 3 rules
Acl's step is 5
 rule 5 permit tcp source 192.168.2.0 0.0.0.255 destination 192.168.0.10 0 desti
nation-port eq www
 rule 10 deny icmp source 192.168.2.0 0.0.0.255 destination 192.168.0.10 0 icmp-
type echo
 rule 15 permit ip (32 matches)
```

图 1.111　"display acl 3000"的显示结果

```
[AR1]display current-configuration | begin acl
acl number 3000
 rule 5 permit tcp source 192.168.2.0 0.0.0.255 destination 192.168.0.10 0 desti
nation-port eq www
 rule 10 deny icmp source 192.168.2.0 0.0.0.255 destination 192.168.0.10 0 icmp-
type echo
 rule 15 permit ip
```

图 1.112　"display current-configuration | begin acl"的显示结果

如果配置的 ACL 出现错误，请参考以下命令删除 ACL 3000，并重新配置正确的 ACL。

```
[AR1]undo acl 3000      // 删除 ACL 3000
```

在路由器 AR1 上使用"display traffic-filter applied-record"命令查看 ACL 的应用信息，如图 1.113 所示。

```
[AR1]display traffic-filter applied-record
-------------------------------------------------
Interface          Direction  AppliedRecord
-------------------------------------------------
GigabitEthernet0/0/0    inbound   acl 3000
```

图 1.113　"display traffic-filter applied-record"的显示结果

如果显示的不是 G0/0/0 的入口（inbound）方向应用了 ACL 3000，则先进入配置错误的接口视图，使用"undo traffic-filter"命令删除应用的 ACL，再进入 G0/0/0 接口，将 ACL 3000 应用到 G0/0/0 的入口方向。下面命令为删除 G0/0/1 接口出口方向应用的 ACL。

```
[AR1-GigabitEthernet0/0/0]undo traffic-filter outbound// 删除接口出口方向应用的 ACL
```

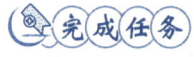 完成任务

1. 任务规划

根据任务描述以及项目 1 的建网安全需求，结合图 1.1 所示的网络拓扑，在"fenbu"路由器上规划表 1.25 所示的一个基本 ACL 和一个高级 ACL 实现项目的安全需求。

项目 1　企业分公司网络互联项目　81

表 1.25　ACL 的规划

设 备 名	列 表 号	ACL 规则	应用接口与方向
fenbu	2000	允许源地址为 172.16.128.13 的主机通过。 禁止所有流量通过	VTY 0 4 的 inbound 方向
fenbu	3000	禁用任何源到任何目的 TCP 协议，端口号为 135～139、445 和 3389 端口的流量通过。 禁用任何源到任何目的 UDP 协议，端口号为 135～139、445 和 3389 端口的流量通过。 允许任何源到任何目的 IP 协议流量通过	"g0/0/0" 与 "s2/0/0" 接口的 inbound 与 outbound 方向

2. 任务实施

（1）基本 ACL 的配置

根据表 1.25，参考以下步骤在 "fenbu" 路由器上完成基本 ACL 的配置，实现只有管理部门的 PC3 可 SSH 到 "fenbu" 路由器上进行远程管理。

```
[fenbu]acl 2000
[fenbu-acl-basic-2000]rule permit source 172.16.128.13 0.0.0.0
[fenbu-acl-basic-2000]rule deny source any
// 创建基本 ACL，只允许源地址为 172.16.128.13 的主机通过
[fenbu]user-interface vty 0 4
[fenbu-ui-vty0-4]acl 2000 inbound
[fenbu-ui-vty0-4]quit
// 将 ACL 2000 应用到 VTY 0 至 VTY 4 流量进入方向
```

（2）高级 ACL 的配置

根据表 1.25，参考以下步骤在 "fenbu" 路由器上完成高级 ACL 的配置，实现连接到总部的分公司路由器出口与入口均引入 ACL，禁用 TCP 与 UDP 协议的 135～139、445、3389 端口。

```
[fenbu]acl 3000
[fenbu-acl-adv-3000]rule deny tcp source any destination any destination-port eq 135
[fenbu-acl-adv-3000]rule deny tcp source any destination any destination-port eq 136
[fenbu-acl-adv-3000]rule deny tcp source any destination any destination-port eq 137
[fenbu-acl-adv-3000]rule deny tcp source any destination any destination-port eq 138
[fenbu-acl-adv-3000]rule deny tcp source any destination any destination-port eq 139
[fenbu-acl-adv-3000]rule deny tcp source any destination any destination-port eq 445
```

```
    [fenbu-acl-adv-3000]rule deny tcp source any destination any destination-
port eq 3389
    [fenbu-acl-adv-3000]rule deny udp source any destination any destination-
port eq 135
    [fenbu-acl-adv-3000]rule deny udp source any destination any destination-
port eq 136
    [fenbu-acl-adv-3000]rule deny udp source any destination any destination-
port eq 137
    [fenbu-acl-adv-3000]rule deny udp source any destination any destination-
port eq 138
    [fenbu-acl-adv-3000]rule deny udp source any destination any destination-
port eq 139
    [fenbu-acl-adv-3000]rule deny udp source any destination any destination-
port eq 445
    [fenbu-acl-adv-3000]rule deny udp source any destination any destination-
port eq 3389
    [fenbu-acl-adv-3000]rule permit ip
    [fenbu-acl-adv-3000]quit
// 配置高级 ACL，禁用 TCP 与 UDP 协议的 135~139、445、3389 端口
    [fenbu]interface g0/0/0
    [fenbu-GigabitEthernet0/0/0]traffic-filter inbound acl 3000
    [fenbu-GigabitEthernet0/0/0]traffic-filter outbound acl 3000
    [fenbu-GigabitEthernet0/0/0]quit
    [fenbu]interface s2/0/0
    [fenbu-Serial2/0/0]traffic-filter inbound acl 3000
    [fenbu-Serial2/0/0]traffic-filter outbound acl 3000
    [fenbu-Serial2/0/0]quit
    [fenbu]
// 将 ACL 3000 应用到"fenbu"路由器的入口与出口方向
```

（3）ACL 测试与查看

① 在预算部门的主机 PC2、管理部门 PC3 与 PC4 上，运行 SecureCRT 虚拟终端软件，测试是否只有管理部门 PC3 可以 SSH 访问"fenbu"路由器。预算部门的主机 PC2 和管理部门 PC4 以及其他主机均不能使用 SSH 访问"fenbu"路由器。

② 在"fenbu"路由器上使用"display acl 2000"命令查看基本 ACL 2000 的配置和运行情况，如图 1.114 所示。

③ 在"fenbu"路由器上使用"display acl 3000"命令查看高级 ACL 3000 的配置和运行情况，如图 1.115 所示。

```
[fenbu]display acl 2000
Basic ACL 2000, 2 rules
Acl's step is 5
 rule 5 permit source 172.16.128.13 0
 rule 10 deny
[fenbu]
```

图 1.114　使用"display acl 2000"命令查看基本 ACL 2000 的配置和运行情况

```
[fenbu]display acl 3000
Advanced ACL 3000, 14 rules
Acl's step is 5
 rule 5 deny tcp destination-port eq 135
 rule 10 deny tcp destination-port eq 136
 rule 15 deny tcp destination-port eq 137
 rule 20 deny tcp destination-port eq 138
 rule 25 deny tcp destination-port eq 139
 rule 30 deny tcp destination-port eq 445
 rule 35 deny tcp destination-port eq 3389
 rule 40 deny udp destination-port eq 135
 rule 45 deny udp destination-port eq 136
 rule 50 deny udp destination-port eq netbios-ns
 rule 55 deny udp destination-port eq netbios-dgm
 rule 60 deny udp destination-port eq netbios-ssn
 rule 65 deny udp destination-port eq 445
 rule 70 deny udp destination-port eq 3389
[fenbu]
```

图 1.115　使用"display acl 3000"命令查看高级 ACL 3000 的配置和运行情况

④ 在"fenbu"路由器上使用"display current-configuration | begin acl"命令查看 ACL 的配置，如图 1.116 所示。

```
[fenbu]display current-configuration | begin acl
acl number 2000
 rule 5 permit source 172.16.128.13 0
 rule 10 deny
#
acl number 3000
 rule 5 deny tcp destination-port eq 135
 rule 10 deny tcp destination-port eq 136
 rule 15 deny tcp destination-port eq 137
 rule 20 deny tcp destination-port eq 138
 rule 25 deny tcp destination-port eq 139
 rule 30 deny tcp destination-port eq 445
 rule 35 deny tcp destination-port eq 3389
 rule 40 deny udp destination-port eq 135
 rule 45 deny udp destination-port eq 136
 rule 50 deny udp destination-port eq netbios-ns
 rule 55 deny udp destination-port eq netbios-dgm
 rule 60 deny udp destination-port eq netbios-ssn
 rule 65 deny udp destination-port eq 445
 rule 70 deny udp destination-port eq 3389
```

图 1.116　"display current-configuration | begin acl"命令的显示结果

如果配置的 ACL 出现错误，请删除错误的 ACL，并重新配置正确的 ACL。

⑤ 在"fenbu"路由器上使用"display traffic-filter applied-record"命令查看 ACL 的应用信息，

84 企业网络互联技术（双语）

如图 1.117 所示。

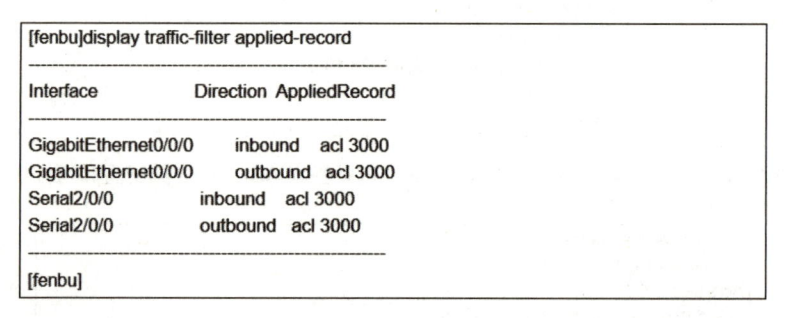

图 1.117 "display traffic-filter applied-record" 命令的显示结果

（4）路由器配置的备份

```
<fenbu>save fenbu.cfg
// 将当前的配置保存并命名为 "fenbu.cfg"
<fenbu>ftp 192.168.0.1
[fenbu-ftp]put fenbu.cfg
// 路由器作为 FTP 客户端连接到 FTP 服务器，并把 Flash 中的 "fenbu.cfg" 配置文件上传到 FTP
// 服务器上
```

习题

选择题

（1）192.168.0.0/25 的通配符掩码为（　　）。

　　A. 255. 255. 255. 128　　　　　　　B. 0. 0. 0. 127

　　C. 0. 0. 0. 1　　　　　　　　　　　D. 255. 255. 255. 0

（2）（　　）可以过滤 ICMP 的流量。

　　A. 基本 ACL　　　　　　　　　　　B. 扩展 ACL

（3）基本 ACL 用于流量过滤时，尽量应用于网络中的（　　）。

　　A. 目的端　　　　　　　　　　　　B. 源端

（4）高级 ACL 用于流量过滤时，尽量应用于网络中的（　　）。

　　A. 目的端　　　　　　　　　　　　B. 源端

（5）高级 ACL 的编号范围是（　　）。

　　A. 2000 ～ 2999　　B. 3000 ～ 2999　　C. 4000 ～ 4999　　D. 1000 ～ 1999

（6）基本 ACL 的编号范围是（　　）。

　　A. 2000 ～ 2999　　B. 3000 ～ 2999　　C. 4000 ～ 4999　　D. 1000 ～ 1999

项目 2

小型企业网络互联项目

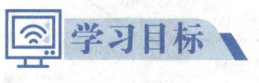 **学习目标**

知识目标

◎ 理解小型企业网络互联项目的网络架构。

◎ 理解 VLAN 的作用与工作过程，理解使用单臂路由和使用三层交换机实现 VLAN 之间通信过程。

◎ 理解生成树的工作过程。

◎ 理解 PPP 工作过程，PAP 和 CHAP 认证。

◎ 理解 RIP 工作过程。

◎ 理解 NAT 作用、工作过程和 NAT 分类。

能力目标

◎ 具有独立部署、实施一个小型企业网络互联项目的能力。包括：交换机的配置与管理、VLAN 的配置、生成树的配置、VLAN 之间通信的配置、DHCP 的配置、RIP 的配置、PPP 的配置以及 NAT 的配置。

素质目标

◎ 具有团队协作能力。

◎ 具有分析问题与解决问题的能力。

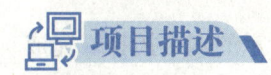 **项目描述**

某小型企业（位于一幢四层的楼宇内）拟组建如图 2.1 所示的网络，网络采用接入层加核

心层的两层网络架构，网络拓扑图中共有四台楼层交换机，其中一层、二层与三层的接入交换机连接到核心交换机上，核心交换机与企业的出口路由器 AR 相连，四层的楼层交换机直接连接到出口路由器 AR，企业的门户网站等服务器直接连接到核心交换机上。出口路由器使用串行链路连接到 Internet 服务提供端的 ISP 路由器上。

企业共有研发、销售、售后、人力资源四个部门。销售部门位于一层和二层，售后部门位于二层与三层，研发部门位于三层，企业领导与人力资源部门位于四层。每层均安装数字摄像头实现监控。

企业网络部署需求如下：

◎ 企业不同部门间业务数据需完全隔离。

◎ 企业内网需互联互通。

◎ 企业只申请到一个公有地址 125.71.28.93，企业内网采用私有地址寻址，在出口路由器上部署 NAT，实现企业内网所有主机均可访问 Internet，Internet 上的主机可以访问企业对外的门户网站服务器。

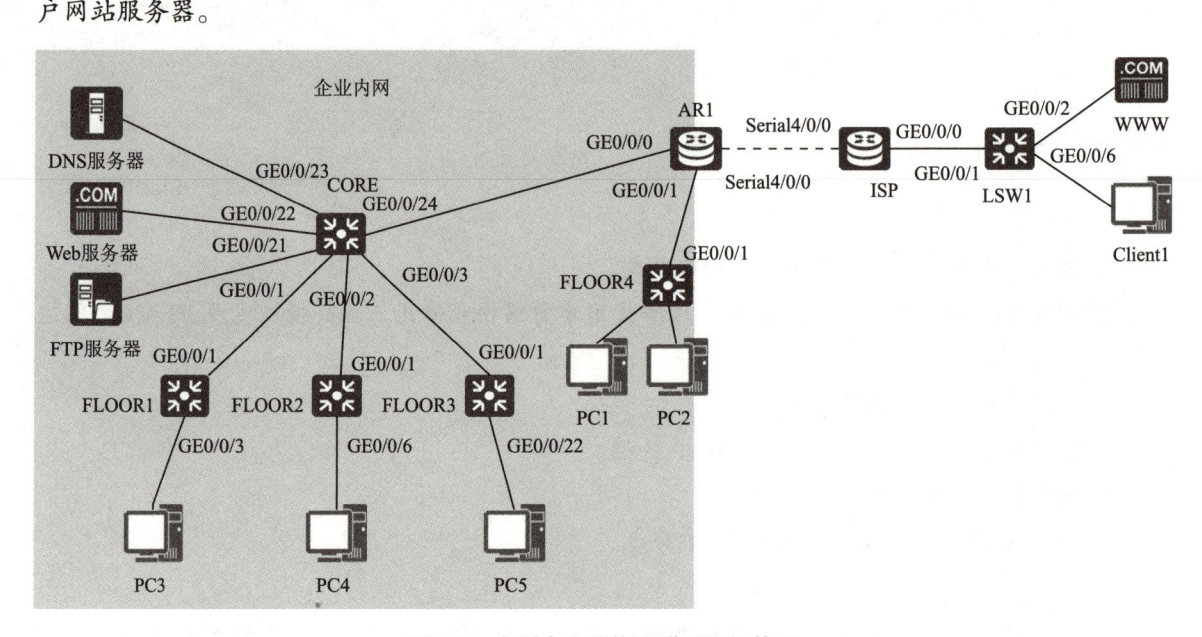

图 2.1　小型企业网络互联项目拓扑图

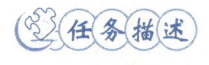

任务1 交换机的基本配置与管理

任务描述

使用华为 ENSP 仿真环境或者使用华为的路由器交换机搭建如图 2.1 所示的网络拓扑。完成该网络 IP 地址的规划，测试主机 PC 与服务器的 IP 配置、交换机主机名配置、Console 端口配置、SSH 的配置，并进行网络验证与测试。

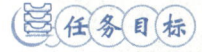

- 理解交换机的工作原理、交换机的转发方式。
- 具有交换机初始化配置的能力。
- 具有交换机配置文件备份的能力。
- 具有清除交换机 Console 密码的能力。

1. 交换机的工作原理

以太网交换机是以太网最主要的联网设备。在以太网交换机中需要维持一张 MAC 地址表（address table），简称 MAC 表。该表给出了关于交换机不同端口所连主机的 MAC 地址信息。MAC 表内容可由管理员手工添加，也可以通过学习交换机收到的数据帧中的源 MAC 地址建立。图 2.2 所示为华为交换机 MAC 表示例。

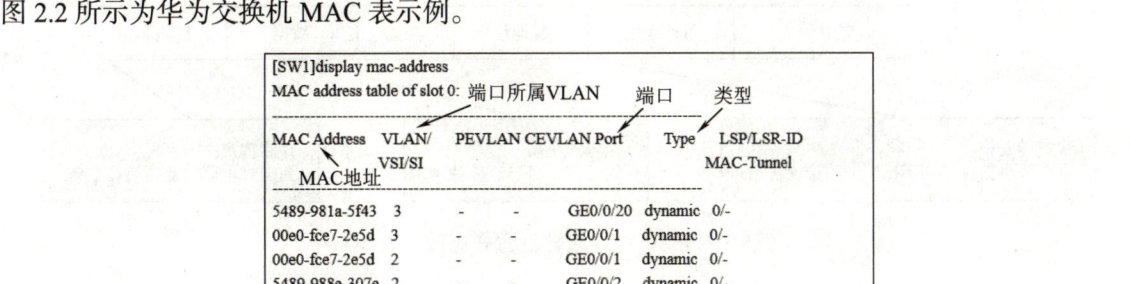

图 2.2　华为交换机 MAC 表示例

交换机的 MAC 表通常包括 MAC 地址（MAC Address）、VLAN、端口（Port）和类型（Type）等字段。其中，类型字段表示交换机获取 MAC 地址与对应端口条目的方式。当 MAC 地址条目是交换机动态学习到的时，其类型值为"dynamic"；当 MAC 地址是管理员手工静态指定的时，其类型值为"static"。交换机动态学习的 MAC 条目默认老化时间（aging time）是 300 s，如果某 MAC 条目在老化时间到期之前一直没有刷新，则该 MAC 条目将从 MAC 表中删除。管理员静态配置的 MAC 表条目不受地址老化时间的影响。

当交换机刚启动时，交换机的 MAC 表是空的。交换机通过学习收到的数据帧中源 MAC 地址来建立 MAC 表，并根据数据帧中的目的 MAC 地址做出转发决定。

当交换机从某一端口收到数据帧时，交换机检查数据帧的源 MAC 地址，如果源 MAC 地址在 MAC 表中不存在，交换机将其添加到 MAC 表中。其中，VLAN 为对应端口所属 VLAN，端口为收到该数据帧的交换机端口。如果源 MAC 地址在 MAC 表中存在，则刷新其老化时间。交换机根据所接收数据帧中的目的 MAC 地址，查找 MAC 表，并根据以下规则做出转发决定：

① 如果数据帧的目的 MAC 地址为组播地址或者广播地址，则洪泛（flooding）该数据帧，

即向除了接收到该数据帧的源端口之外的其他所有交换机端口转发该帧。

② 如果数据帧的目的 MAC 地址为单播地址，但目的 MAC 地址在 MAC 表中不存在，也洪泛该数据帧。

③ 如果数据帧的目的 MAC 地址为单播地址，且目的 MAC 地址与源 MAC 地址对应于交换机相同的端口，则不转发该帧。

④ 如果数据帧的目的 MAC 地址为单播地址，且目的 MAC 地址与源 MAC 地址对应于交换机不同的端口，则从目的 MAC 地址所对应的交换机端口转发该帧。

2. 交换机的转发方式

交换机转发数据帧有存储转发（store-and-forward）与直通交换（cut-through）两种交换转发方式。其中，直通交换方式又进一步分成快速转发（fast-forward）与无碎片（fragment-free）交换两种方式，如图 2.3 所示。

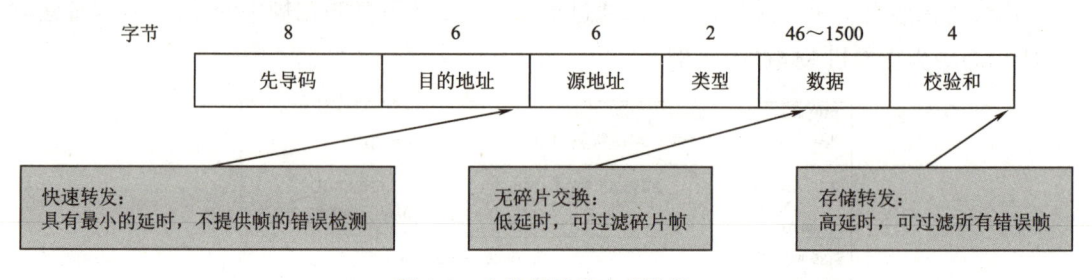

图 2.3 交换机转发方式比较

（1）存储转发

存储转发是指交换机收完一个完整的数据帧之后，进行 CRC（cyclic redundancy check，循环冗余校验码）校验，确认数据帧无差错后再根据数据帧头中的目的 MAC 地址做出转发决定，进行转发。

（2）直通交换

直通交换是指交换机只在收到数据帧的目的 MAC 地址时（此时没有收到完整的数据帧），根据数据帧中的目的 MAC 地址做出转发决定，进行转发。直通交换进一步分为快速转发和无碎片交换两种方式。其中，快速转发方式是指交换机只要检测到数据帧中的目的 MAC 地址，就立即查找 MAC 表做出转发决定，进行转发；而无碎片交换方式在要求已收到的数据帧必须大于最小帧长（64 字节），再做转发，任何长度小于 64 字节的数据帧都会被立即丢弃。

3. 交换机的分类

交换机有多种不同的分类标准。交换机常见的分类有：

（1）固定端口交换机与模块化交换机

固定端口交换机的端口是固定的，不能扩充，固定端口交换机的端口数量通常为 8 端口、16 端口、24 端口和 48 端口等。模块化交换机配置了额外的开放性插槽，可以通过插入模块来扩充交换机的端口数量，用户可以配置不同数量、不同速率和不同接口类型的模块来适应不同

网络的需求。模块化交换机一般都有较强的容错能力，支持冗余的交换模块，支持可热插拔的双电源。模块化交换机拥有更大的灵活性和可扩充性，但它的价格比固定端口交换机贵很多，一般用于大型网络中核心层与汇聚层。图 2.4 给出了华为固定端口交换机与模块化交换机的外形。

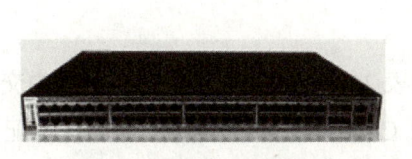　　　　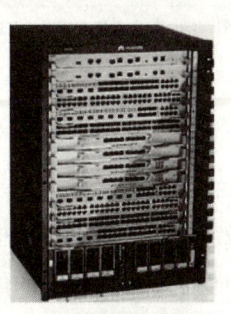

（a）固定端口交换机　　　　　　　　　　　　（b）模块化交换机

图 2.4　华为固定端口交换机与模块化交换机的外形

（2）可堆叠交换机与不可堆叠交换机

堆叠技术主要是为了增加交换机的端口密度，在单个交换机的端口数不能满足组网需求时，可以考虑采用堆叠交换机。堆叠在一起的多个交换机从逻辑上变成一台交换设备，作为一个整体参与数据转发，简化了网络组网，提高了网络的可靠性。

4．二层交换机与三层交换机

根据交换机完成的功能对应于 OSI 参考模型，可将交换机分为二层交换机、三层交换机、四层交换机等。二层交换机工作在 OSI 参考模型的第二层（即数据链路层），根据数据帧的目的 MAC 地址对数据进行转发和过滤；三层交换机工作在 OSI 参考模型的第三层（即网络层），它不仅可以根据数据帧的目的 MAC 地址信息来做出转发决策，还可以根据分组中的第三层地址（如 IP 地址）做出转发决策。三层交换机具有路由功能，在企业网络内部通常会选择使用三层交换机实现不同网段之间的路由。

5．交换机的访问方法与命令视图

可以通过 Console 端口、Telnet、SSH、Web 等方法访问交换机。使用 Console 端口访问交换机通常用于交换机的初始化配置和管理工作，以及对交换机的状态进行监控和一些灾难性恢复工作。其物理连接及访问方法与 Console 端口访问路由器类似。

与路由器类似，交换机提供了多种不同的命令视图，以满足不同权限用户执行不同的访问功能。表 2.1 给出了交换机的主要命令视图及提示符。

表 2.1　交换机的主要命令视图及提示符

命 令 视 图	提示符示例	视图切换示例
用户视图	<Huawei>	路由器启动后默认视图
系统视图	[Huawei]	<Huawei>system-view
接口视图	[Huawei-GigabitEthernet0/0/0]	[Huawei]interface G0/0/0

90 企业网络互联技术（双语）

续表

命令视图	提示符示例	视图切换示例
VLAN 视图	[Huawei-vlan1]	[Huawei]vlan 1
用户界面视图	[Huawei-ui-console0]	[Huawei]user-interface console 0

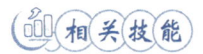 相关技能

1. 交换机的初始化配置

搭建如图 2.5 所示的拓扑图。根据拓扑图与表 2.2 所示的 IP 规划完成网络中交换机主机名、交换机管理 IP 地址与默认网关的配置、交换机 Console 端口的配置、Telnet 的配置、SSH 的配置，并进行网络测试，以具有交换机的基本配置能力。

```
                G0/0/2           G0/0/0          G0/0/6
  [PC1]━━━━━━━[LSW1]━━━━━━━━━[AR1]━━━━━━━━━[LSW2]━━━━━━━━[FTP服务器]
              G0/0/6           G0/0/1          G0/0/5
  PC1         LSW1            AR1            LSW2        FTP服务器
```

图 2.5 "交换机基本配置"拓扑图

表 2.2 IP 规划

设　　备	接　　口	IP 地址 / 子网掩码	网　　关
LSW1	Vlanif1	192.168.0.1/24	192.168.0.254
LSW2	Vlanif1	192.168.1.1/24	192.168.1.254
AR1	G0/0/0	192.168.0.254/24	
	G0/0/1	192.168.1.254/24	
FTP 服务器	NIC	192.168.1.2/24	192.168.1.254
PC1	NIC	192.168.0.2/24	192.168.0.254

（1）交换机的基本配置

① 使用一条 USB 转串口线将主机 PC1 的 USB 端口与交换机的 Console 端口相连，使用 UTP 直连线将主机 PC1 的网卡与交换机的 G0/0/2 接口相连、FTP 服务器的网卡与交换机的 G0/0/5 接口相连。

② 在主机 PC1 上，右击桌面上的"此电脑"图标，在弹出的快捷菜单中，依次选择"管理"→"设备管理器"→"端口（COM 和 LPT）"，观察并记录"USB-to-Serial Comm Port"的端口号数值。

③ 在主机 PC1 上，运行 SecureCRT 软件，选择"文件"→"快速连接"→"Serial"协议，选择记录的 USB 转串口的端口号，波特率值选择"9600"，数据位为 8 位，无奇偶校验，停止位为 1 位，取消勾选流控下面所有三个复选框。单击"连接"按钮，进入交换机的命令视图。

④ 在交换机命令行中，参考以下步骤完成交换机 LSW1 的主机名、交换机管理 IP 地址与

默认网关的配置：

```
<Huawei>sys                              // 进入系统视图
[Huawei]sysname LSW1                      // 配置交换机的主机名为"LSW1"
[LSW1]interface Vlanif 1                   // 创建并进入 VLAN1 接口视图
[LSW1-Vlanif1]ip address 192.168.0.1 24   // 配置VLAN1接口的IP地址为"192.168.0.1"，
                                          // 子网掩码长度为"24"
[LSW1-Vlanif1]undo shutdown               // 启用接口
[LSW1-Vlanif1]quit                        // 返回到系统视图
[LSW1]ip route-static 0.0.0.0 0.0.0.0 192.168.0.254    // 配置默认网关地址为192.168.0.254
```

⑤ 在交换机命令行中，参考以下步骤完成交换机 LSW2 的主机名、交换机管理 IP 地址与默认网关的配置：

```
<Huawei>sys                              // 进入系统视图
[Huawei]sysname LSW2                      // 配置交换机的主机名为"LSW2"
[LSW2]interface Vlanif 1                   // 创建并进入 VLAN1 接口视图
[LSW2-Vlanif1]ip address 192.168.1.1 24   // 配置 VLAN1 接口的 IP 地址为"192.168.
                                          // 1.1"，子网掩码长度为"24"
[LSW2-Vlanif1]undo shutdown               // 启用接口
[LSW2-Vlanif1]quit                        // 返回到系统视图
[LSW2]ip route-static 0.0.0.0 0.0.0.0 192.168.1.254    // 配置默认网关地址为192.168.1.254
```

⑥ 分别在交换机 LSW1 与 LSW2 上，使用"display ip interface brief"命令查看接口 IP 的简要信息，观察接口的 IP 地址与掩码是否正确，接口的物理与协议状态是否为"UP"。图 2.6 为在交换机 LSW1 上使用"display ip interface brief"命令显示的接口 IP 简要信息示例。

```
[LSW1]display ip interface brief
*down: administratively down
^down: standby
(l): loopback
(s): spoofing
The number of interface that is UP in Physical is 2
The number of interface that is DOWN in Physical is 1
The number of interface that is UP in Protocol is 2
The number of interface that is DOWN in Protocol is 1

Interface          IP Address/Mask    Physical  Protocol
MEth0/0/1          unassigned         down      down
NULL0              unassigned         up        up(s)
Vlanif1            192.168.0.1/24     up        up
```

图 2.6　交换机 LSW1 上显示的接口 IP 简要信息

⑦ 分别在交换机 LSW1 与 LSW2 上，使用"display ip routing-table"命令查看路由表信息，图 2.7 为交换机 LSW1 上显示的路由表信息。

注：默认路由下一跳地址为"192.168.0.254"。

```
[LSW1]display ip routing-table
Route Flags: R - relay, D - download to fib
--------------------------------------------------------------------------------
Routing Tables: Public
         Destinations : 5       Routes : 5

Destination/Mask    Proto   Pre  Cost    Flags NextHop       Interface

        0.0.0.0/0   Static  60   0        RD   192.168.0.254  Vlanif1
      127.0.0.0/8   Direct  0    0         D   127.0.0.1      InLoopBack0
     127.0.0.1/32   Direct  0    0         D   127.0.0.1      InLoopBack0
   192.168.0.0/24   Direct  0    0         D   192.168.0.1    Vlanif1
   192.168.0.1/32   Direct  0    0         D   127.0.0.1      Vlanif1
```

图 2.7　交换机 LSW1 上显示的路由表信息

⑧ 在路由器 AR1 上根据表 2.2 的 IP 规划，参考以下步骤完成路由器主机名、接口 IP 地址与子网掩码的配置：

```
<Huawei>sys
[Huawei]sysname AR1
// 配置路由器的主机名
[AR1]interface g0/0/0
[AR1-GigabitEthernet0/0/0]ip address 192.168.0.254 24
[AR1-GigabitEthernet0/0/0]quit
[AR1]
// 配置路由器 AR1 的 G0/0/0 接口的 IP 地址与子网掩码
[AR1]interface g0/0/1
[AR1-GigabitEthernet0/0/1]ip address 192.168.1.254 24
[AR1-GigabitEthernet0/0/1]quit
[AR1]
// 配置路由器 AR1 的 G0/0/1 接口的 IP 地址与子网掩码
```

⑨ 完成主机 PC1 与 FTP 服务器的 IP 地址与子网掩码配置，使用 ping 实用程序测试主机 PC1、交换机 LSW1、交换机 LSW2 和 FTP 服务器之间的 IP 连通性。此时，它们之间可以相互 ping 通。图 2.8 所示为交换机 LSW1 成功 ping 通交换机 LSW2 的 IP 地址。

```
[LSW1]ping 192.168.1.1
 PING 192.168.1.1: 56  data bytes, press CTRL_C to break
  Reply from 192.168.1.1: bytes=56 Sequence=1 ttl=254 time=50 ms
  Reply from 192.168.1.1: bytes=56 Sequence=2 ttl=254 time=50 ms
  Reply from 192.168.1.1: bytes=56 Sequence=3 ttl=254 time=50 ms
  Reply from 192.168.1.1: bytes=56 Sequence=4 ttl=254 time=40 ms
  Reply from 192.168.1.1: bytes=56 Sequence=5 ttl=254 time=70 ms

 --- 192.168.1.1 ping statistics ---
  5 packet(s) transmitted
  5 packet(s) received
  0.00% packet loss
  round-trip min/avg/max = 40/52/70 ms
```

图 2.8　交换机 LSW1 成功 ping 通交换机 LSW2 的 IP 地址

（2）Console 的配置

交换机 Console 的配置与路由器 Console 的配置步骤相似，使用 AAA 认证进行控制端口的配置任务为：

① 配置用于 Console 登录的合法 AAA 用户，具体包括配置合法的用户名与密码、授权用户的服务类型为 terminal，用户的权限级别。

② 进入 Console 用户界面视图，配置认证模式为 AAA。

在交换机 LSW1 上完成 Console 的配置步骤如下：

```
[LSW1]aaa
[LSW1-aaa]local-user cdp password cipher cdp123456
[LSW1-aaa]local-user cdp service-type terminal
[LSW1-aaa]local-user cdp privilege level 15
[LSW1-aaa]quit
// 创建用于 Console 登录的用户，配置用户的用户名为"cdp"、密码为"cdp123456"、用户的服
务类型为"terminal"、权限级别为"15"
[LSW1]user-interface console0
[LSW1-ui-console0]authentication-mode aaa
[LSW1-ui-console0]quit
[LSW1]quit
<LSW1>
// 配置 Console0 的认证模式为使用 AAA 进行认证
```

在交换机 LSW1 上完成 Console 配置后，使用"quit"命令退出用户视图，出现"Please Press ENTER"提示，按回车键，出现登录认证提示，输入合法的 Console 用户名"cdp"与对应的密码"cdp123456"，登录进入用户视图，如图 2.9 所示。

```
Please Press ENTER.

Login authentication

Username:cdp  ◀——  输入 Console 用户名"cdp"
Password:     ◀——  输入用户的密码"cdp123456"，注意：密码不回显
<LSW1>
```

图 2.9　交换机 LSW1 Console 登录认证测试

（3）Telnet 的配置与测试

交换机 Console 的配置与路由器 Console 的配置步骤相似。交换机上 Telnet 配置包括以下任务：

① 启用 Telnet。

② 配置用于 Telnet 登录的合法用户，具体包括进入 AAA 视图、创建用户名与密码、配置用户的服务类型为 Telnet，配置用户的权限级别。

94　企业网络互联技术（双语）

③ 进入 VTY 用户界面视图，配置协议为 Telnet，认证模式为 AAA。

参考以下步骤在交换机 LSW1 上完成 Telnet 的配置。

```
[LSW1]aaa
[LSW1-aaa]local-user cdptelnet password cipher cdp123456
[LSW1-aaa]local-user cdptelnet service-type telnet
[LSW1-aaa]local-user cdptelnet privilege level 15
[LSW1-aaa]quit
// 配置用于 Telnet 登录的用户名，用户名为 "cdptelnet"，密码为 "cdp123456"，用户服务类
// 型为 "telnet"，权限级别为 "15"
[LSW1]telnet server enable
// 启用 Telnet 服务
[LSW1]user-interface vty 0 4
[LSW1-ui-vty0-4]protocol inbound telnet
[LSW1-ui-vty0-4]authentication-mode aaa
[LSW1-ui-vty0-4]quit
// 配置 VTY 0 4 的协议为 Telnet，认证模式为使用 AAA 进行认证
```

在交换机上完成 Telnet 配置后，在主机 PC1 上使用 PUTTY 或者 SecureCRT 等 Telnet 客户端软件，使用 Telnet 协议远程登录访问交换机 LSW1（其管理地址为 "192.168.0.1"）。也可以在路由器 AR1 上与交换机 LSW2 上使用 Telnet 命令远程登录访问交换机 LSW1。图 2.10 所示为在交换机 LSW2 上使用 Telnet 远程登录到交换机 LSW1。

```
<LSW2>telnet 192.168.0.1
Trying 192.168.0.1 ...
Press CTRL+K to abort
Connected to 192.168.0.1 ...

Login authentication

Username:cdptelnet ◀——— 输入telnet的用户名 "cdptelnet"
Password:         ◀——— 输入用户的密码 "cdp123456"，注意：密码不回显
Info: The max number of VTY users is 5, and the number
      of current VTY users on line is 1.
      The current login time is 2022-11-06 10:33:32.
<LSW1>
```

图 2.10　在交换机 LSW2 上使用 Telnet 远程登录到交换机 LSW1 上

（4）SSH 的配置与测试

不是所有的交换机都支持 SSH。交换机必须有加密特征和操作软件版本支持才可以配置 SSH，并且不同型号交换机支持的生成本地密钥对的算法有差异。通常，在华为交换机上配置使用本地用户名与密码登录 SSH 需要完成：

① 启用 Stelnet 服务。

② 配置用于 Stelnet 登录的合法用户，具体包括进入 AAA 视图、创建用户名与密码、配置

用户的服务类型为 Stelnet，配置用户的权限级别等步骤。

③ 配置 SSH 的用户与服务类型、SSH 认证类型。

④ 进入 VTY 用户界面视图中，配置协议为 SSH，认证模式为 AAA。

⑤ 生成本地密钥对。

对应上述步骤，下面给出的是以华为 S5700-28C-HI 交换机为例的 SSH 配置与验证命令：

```
[LSW1]stelnet server enable
// 启用 Stelnet 服务
[LSW1]aaa
[LSW1-aaa]local-user cdpssh password cipher cdp123456
[LSW1-aaa]local-user cdpssh service-type ssh
[LSW1-aaa]local-user cdpssh privilege level 15
[LSW1-aaa]quit
// 配置用于 SSH 登录的用户名，用户名为 "cdpssh"，密码为 "cdp123456"，用户服务类型为
// "ssh"，权限级别为 "15"
[LSW1]ssh user cdpssh
[LSW1]ssh user cdpssh service-type stelnet
[LSW1]ssh user cdpssh authentication-type password
// 配置 SSH 的用户为 cdpssh，服务类型为 Stelnet，使用密码认证
[LSW1]user-interface vty 0 4
[LSW1-ui-vty0-4]protocol inbound ssh
[LSW1-ui-vty0-4]authentication-mode aaa
[LSW1-ui-vty0-4]quit
// 配置 VTY 0 4 的协议为 SSH，认证模式为使用 AAA 进行认证
```

最后，使用 "rsa local-key-pair create" 命令生成本地密钥对，如图 2.11 所示。

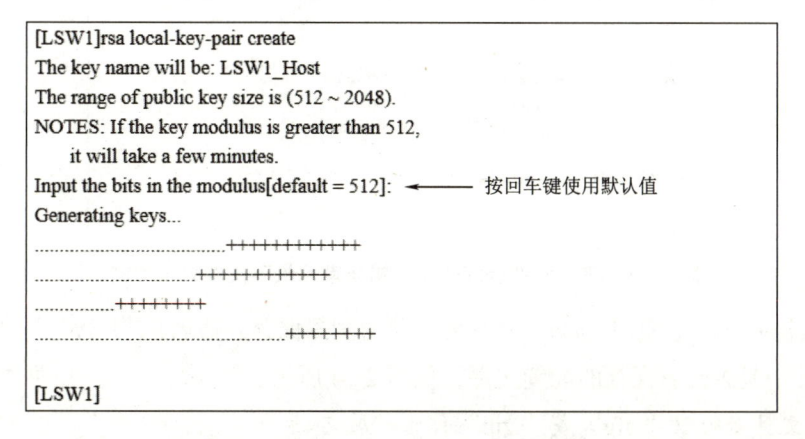

图 2.11 使用 RSA 算法创建本地密钥对

在交换机上完成 SSH 配置后，可以在主机 PC1 上使用 PUTTY 或者 SecureCRT 等 SSH 客户端软件使用 SSH 协议远程登录访问交换机 LSW1（其管理地址为 "192.168.0.1"）。也可以在

路由器 AR1 上或者交换机 LSW2 上使用"stelnet"命令远程登录到交换机 LSW1。

注： 在路由器或交换机上使用"stelnet"命令之前，需要先使用"ssh client first-time enable"使能 SSH 客户端首次认证。

图 2.12 所示为在交换机 LSW2 上首次使用 stelnet 远程访问交换机 LSW1。

```
[LSW2]ssh client first-time enable        ←—— 使能SSH客户端首次认证
[LSW2]stelnet 192.168.0.1
Please input the username:cdpssh          ←—— 输入SSH用户名"cdpssh"
Trying 192.168.0.1 ...
Press CTRL+K to abort
Connected to 192.168.0.1 ...
The server is not authenticated. Continue to access it? [Y/N] :y  ←—— 输入"Y"
Save the server's public key? [Y/N] :y    ←—— 输入"Y"
The server's public key will be saved with the name 192.168.0.1. Please wait...
Enter password:   ←—— 输入对应的密码"cdp123456"
Info: The max number of VTY users is 5, and the number
      of current VTY users on line is 1.
      The current login time is 2022-11-06 11:56:30.
<LSW1>
```

图 2.12　在交换机 LSW2 上首次使用 stelnet 远程访问交换机 LSW1

2. 交换机的管理

在完成"交换机的基本配置"基础上，将图 2.5 所示网络中交换机 LSW1 与 LSW2 的配置文件备份到 Flash 和 FTP 服务器上，并熟悉清除交换机 Console 密码的操作。

（1）交换机文件系统管理

① 在交换机 LSW1 与 LSW2 上使用"dir"命令查看当前目录下的文件列表。图 2.13 所示为在交换机 LSW1 上查看当前目录下文件列表示例。

```
<LSW1>dir
Directory of flash:/

Idx  Attr   Size(Byte) Date      Time     FileName
 0   drw-        -     Aug 06 2015 21:26:42  src
 1   drw-        -     Feb 15 2023 21:46:14  compatible

32,004 KB total (31,972 KB free)

<LSW1>
```

图 2.13　在交换机 LSW1 上查看当前目录下文件列表示例

② 在交换机 LSW1 上使用"save"命令将当前的配置保存并命名为"LSW1"，然后使用"dir"命令查看 Flash 中是否已有保存的配置文件，如图 2.14 所示。

注： 配置文件必须以".cfg"或".zip"作为扩展名。

③ 使用"startup saved-configuration"命令设置交换机下一次启动所使用的配置文件，步骤如下：

```
<LSW1>startup saved-configuration?    //查看"startup saved-configuration"
                                      //命令的关键字或参数
<LSW1>startup saved-configuration LSW1cfg.zip      //配置下一次启动所使用的配置文件
                                                   //为"LSW1cfg.zip"
<LSW1>display startup               //显示设备本次及下次启动的操作系统、配置文件等信息
```

```
<LSW1>save LSW1cfg.zip
Are you sure to save the configuration to flash:/LSW1cfg.zip?[Y/N]:Y ←—— 输入"Y"
Now saving the current configuration to the slot 0.
Save the configuration successfully.
<LSW1>dir
Directory of flash:/

  Idx  Attr   Size(Byte) Date       Time      FileName
   0   drw-          -  Aug 06 2015 21:26:42  src
   1   drw-          -  Feb 15 2023 21:46:14  compatible
   2   -rw-        452  Feb 15 2023 21:59:07  LSW1cfg.zip

32,004 KB total (31,968 KB free)

<LSW1>
```

图 2.14 保存并命名配置文件示例

④ 根据表 2.2 所示的 IP 规划，完成 FTP 服务器的 IP 配置、FTP 服务的搭建与配置。在交换机 LSW1 上配置输入"ftp 192.168.1.2"，使交换机作为 FTP 客户端连接到 FTP 服务器，把 Flash 中的"LSW1cfg.zip"配置文件上传到 FTP 服务器上，步骤如图 2.15 所示。

```
<LSW1>ftp 192.168.1.2
Trying 192.168.1.2 ...
Press CTRL+K to abort
Connected to 192.168.1.2.
220 FtpServerTry FtpD for free
User(192.168.1.2:(none)):anonymous ←—— 输入匿名账号"anonymous"
331 Password required for anonymous .
Enter password: ←—— 输入任何的邮箱地址作为密码
230 User anonymous logged in , proceed

[ftp]put LSW1cfg.zip
200 Port command okay.
150 Opening BINARY data connection for LSW1cfg.zip

100%
226 Transfer finished successfully. Data connection closed.
FTP: 681 byte(s) sent in 0.290 second(s) 2.34Kbyte(s)/sec.
```

图 2.15 把交换机 Flash 中的配置文件上传到 FTP 服务器上

⑤ 完成文件的备份后，输入"reset saved-configuration"命令，清空交换机的配置，然后输入"reboot"命令重启交换机，步骤如图 2.16 所示。

```
<LSW1>reset saved-configuration
Warning: The action will delete the saved configuration in the device.
The configuration will be erased to reconfigure. Continue? [Y/N]:Y  ← 输入 "Y"
Warning: Now clearing the configuration in the device.
Feb 15 2023 22:17:55-08:00 LSW1 %%01CFM/4/RST_CFG(l)[0]:The user chose Y
when de
ciding whether to reset the saved configuration.
Info: Succeeded in clearing the configuration in the device.
<LSW1>reboot
Info: The system is now comparing the configuration, please wait.
Warning: All the configuration will be saved to the configuration file for the n
ext startup:, Continue?[Y/N]:N  ← 输入 "N"
Info: If want to reboot with saving diagnostic information, input 'N' and then e
xecute 'reboot save diagnostic-information'.
System will reboot! Continue?[Y/N]:Y  ← 输入 "Y"
```

图 2.16　清空交换机配置并重启交换机

（2）清除交换机 Console 密码

以华为交换机 FutureMatrix S5736-S24T4XC 为例，清除交换机 Console 密码的步骤如下：

① 使用控制线把控制台与交换机的 Console 端口连接起来，确保连接正常。然后，断开交换机的电源，同时重新打开交换机的电源。观察交换机的启动过程，当出现如图 2.17 所示的 "Press Ctrl+B or Ctrl+E to enter BootLoad menu" 时，及时（3 s 内）按下快捷键【Ctrl+B】或【Ctrl+E】，进入 BootLoad 菜单。

```
Last reset type: Watchdog
Press Ctrl+B or Ctrl+E to enter BootLoad menu: 1
Info: The password is empty. For security purposes, change the password.
New password:  ← 1. 设置BootLoad密码
Verify:  ← 2. 确认BootLoad密码
Modify password ok.
    BootLoad Menu

  1. Boot with default mode
  2. Enter startup submenu
  3. Enter ethernet submenu
  4. Enter filesystem submenu
  5. Enter password submenu
  6. Clear password for console user
  7. Reboot
  (Press Ctrl+E to enter diag menu)

Enter your choice(1-7):6  ← 3. 输入 "6" 选择 "clear password for console user"
Note: Clear password for console user? Yes or No(Y/N): Y  ← 4. 输入 "Y"，确认清除
Clear password for console user successfully.                    Console用户密码
Note: Choose "1. Boot with default mode" to boot, then set a new password.

Note: If the device is restarted during startup, you need to perform this operation again.

    BootLoad Menu

  1. Boot with default mode
  2. Enter startup submenu
  3. Enter ethernet submenu
  4. Enter filesystem submenu
  5. Enter password submenu
  6. Clear password for console user
  7. Reboot
  (Press Ctrl+E to enter diag menu)

Enter your choice(1-7): 1  ← 5. 输入 "1" 直接从当前阶段继续启动
```

图 2.17　交换机清除 Console 用户密码的步骤

② 在"New Password"处输入新设置的 BootLoad 密码。在"Verify"处再次输入设置的 BootLoad 密码。

注：华为交换机 FutureMatrix S5736-S24T4XC 首次进入 BootLoad 时，需新设置 BootLoad 密码。

③ 在"BootLoader"菜单下输入"6"选择"Clear password for console user"，清除 Console 用户密码。

④ 在"Note: Clear password for console user? Yes or No(Y/N)"处输入"Y"，确认清除 Console 用户密码。

⑤ 清除 Console 用户密码成功后，在接下来的"BootLoader"菜单中输入"1"选择"1. Boot with default mode"直接从当前阶段启动。启动之后直接设置 Console 新密码即可进入用户系统模式。

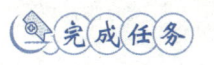 完成任务

1. 任务规划

根据任务描述和项目 2 的建网需求，图 2.1 所示网络拓扑中企业内网所有接入层与核心层交换机 Console 端口与 SSH 的规划如下：

① Console 端口采用 AAA 认证，用户名为"cdp"，密码为"cdp123456"，用户命令行级别为"15"。

② SSH 认证规划为：采用本地用户认证，用户名为"cdp"，密码为"cdp123456"，用户命令行级别为"15"。

2. 任务实施

（1）交换机的基本配置

① 在接入层交换机"FLOOR1"上完成主机名、Console 登录认证与 SSH 认证的配置，配置步骤如下：

```
[Huawei]sysname FLOOR1
[FLOOR1]stelnet server enable
// 启用 Stelnet 服务
[FLOOR1]aaa
[FLOOR1-aaa]local-user cdp password cipher cdp123456
[FLOOR1-aaa]local-user cdp service-type terminal ssh
[FLOOR1-aaa]local-user cdp privilege level 15
[FLOOR1-aaa]quit
// 创建用于 Console 与 SSH 登录的用户，配置用户的用户名为"cdp"、密码为"cdp123456"、用户的服务类型为"terminal"与"SSH"、权限级别为"15"
[FLOOR1]user-interface console 0
[FLOOR1-ui-console0]authentication-mode aaa
```

100 企业网络互联技术（双语）

```
[FLOOR1-ui-console0]quit
// 配置 Console0 的认证模式为使用 AAA 进行认证
[FLOOR1]ssh user cdp
[FLOOR1]ssh user cdp service-type stelnet
[FLOOR1]ssh user cdp authentication-type password
// 配置 SSH 的用户名为 "cdp"，服务类型为 "stelnet"，使用密码认证
[FLOOR1]user-interface vty 0 4
[FLOOR1-ui-vty0-4]quit
[FLOOR1-ui-vty0-4]authentication-mode aaa
[FLOOR1-ui-vty0-4]protocol inbound ssh
// 配置 VTY 0 4 的协议为 SSH，认证模式为 AAA
[FLOOR1]rsa local-key-pair create
// 生成本地密钥对
[FLOOR1]quit
<FLOOR1>save
// 保存配置
```

② 参考以上步骤完成图 2.1 中企业内网中所有交换机（CORE、FLOOR2、FLOOR3、FLOOR4）的主机名、Console 登录认证与 SSH 认证的配置。

（2）Console 配置验证

在企业内网中所有交换机上退出用户视图，按回车键重新通过 Console 连接交换机，验证是否需要登录认证提示，输入合法的 Console 用户名 "cdp" 与对应的密码 "cdp123456" 后是否能进入交换机的用户视图。

注： 由于企业内网所有交换机此时还未完成管理 IP 地址的配置，因此需等到完成本项目任务 3 的 "完成任务" 后才能验证交换机配置的 SSH 功能。

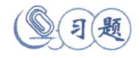 **习题**

选择题

（1）二层交换机根据（　　　）做出转发决定。

    A. 源 MAC 地址　　　　　　　　　　B. 目的 MAC 地址

    C. 源 IP 地址　　　　　　　　　　　　D. 目的 IP 地址

（2）下面（　　　）命令用于显示设备本次及下次启动的操作系统、配置文件等信息。

    A. display startup　　　　　　　　　　B. display current

    C. display interface brief　　　　　　D. display version

（3）当交换机收到的数据帧的目的 MAC 地址是单播地址，但该 MAC 地址在交换机的 MAC 表中不存在时，交换机通过（　　　）转发该数据帧。

A. 丢弃该数据帧   B. 洪泛

C. 从交换机所有端口发送出去   D. 从交换机的一个端口发送出去

（4）下面交换机转发方式具有最大延迟的是（　　）。

A. 无碎片交换   B. 快速转发   C. 存储转发   D. 直通交换

任务2　VLAN 的规划与配置

任务描述

在完成本项目任务 1 的基础上，根据图 2.1 所示的网络拓扑图，在企业内网交换机上完成 VLAN 的规划与配置，实现企业内网中不同部门间业务数据流的完全隔离的网络部署需求。

任务目标

- 理解 VLAN 的工作过程。
- 具有 VLAN 的配置与管理能力。
- 具有 VLAN 的故障排除能力。

相关知识

1. VLAN 概述

以太网经历了从共享式以太网到交换式以太网的发展。当前主流的园区网络为交换式以太网。由于交换机可以分割冲突域，因此，交换式以太网网络性能大大提高。但是，根据交换机的工作原理，交换机不能分割广播域，交换网络中任何一台主机发出的组播或广播帧，局域网中所有的主机都能接收到，有可能导致网络中存在大量广播流量，产生广播风暴现象。如图 2.18 所示的网络中，主机 PC1 发出一个 DHCP DISCOVER 广播帧，网络中所有主机均能接收到该广播帧。

为了解决广播域带来的问题，引入了虚拟局域网（virtual local area network, VLAN）技术。VLAN 是在局域网交换机上通过交换机软件实现根据功能、部门、应用等将设备或用户组成虚拟工作组或逻辑网段的技术。

一个 VLAN 就是一个广播域，属于同一 VLAN 的节点之间可以直接相互通信，不同 VLAN 的节点需要通过三层设备进行通信；VLAN 可以不受网络用户的物理位置限制而根据用户需求进行网络逻辑划分，从而可显著减少用户增加、删除或移动时的网络管理开销。采用 VLAN 划分逻辑网段如图 2.19 所示。

102 企业网络互联技术（双语）

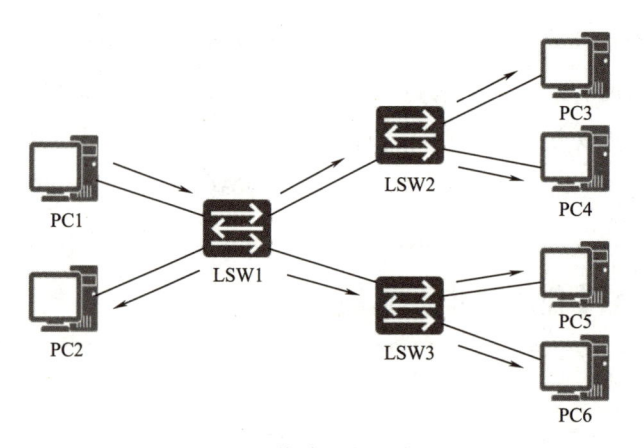

图 2.18　传统以太网的问题

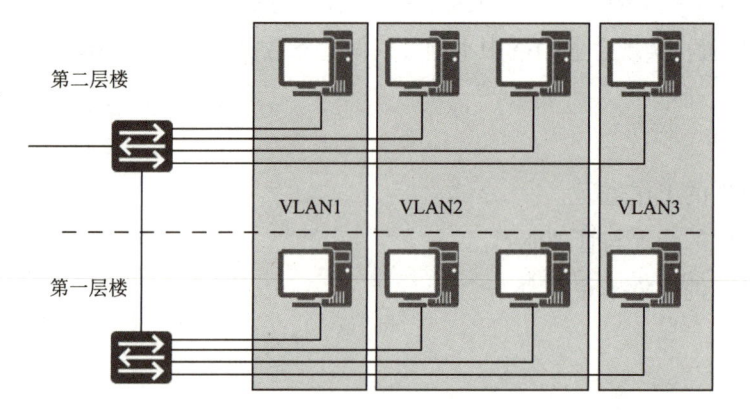

图 2.19　采用 VLAN 划分逻辑网段

2. VLAN 的类型

（1）数据 VLAN

数据 VLAN 是指用于传送各类用户数据流量的 VLAN，有时也被称为"用户 VLAN"。

（2）语音 VLAN

IP 电话、IP 软交换等 IP 语音终端所产生的通信流量称为 VoIP 流量。VoIP 流量对传输的实时性要求较高，通常在网络 QoS 策略中需要将其设置为高优先级，以便相关的网络设备能够优先传输语音通信流量。因此，在进行 IP 语音部署时，通常会将 VoIP 通信流量放在一个独立的 VLAN 中，这个独立的 VLAN 就被称为语音 VLAN。

（3）默认 VLAN

默认 VLAN 又称 PVID（port default VLAN ID，端口默认 VLAN ID）。当交换机接口接收到一个不带标签（untagged）帧，交换机根据 PVID 给此数据帧添加等于 PVID 的标签（tag），然后再交给交换机内部处理。交换机接口发送数据帧时，如果发现此数据帧的 Tag 的 VID 值与 PVID 相同，则交换机会将 Tag 去掉，再从此接口发送出去。每个接口都有一个默认 VLAN。默认情况下，所有接口的默认 VLAN 均为 VLAN1，但用户可以根据需要进行配置。VLAN1 不能被删除。

项目 2　小型企业网络互联项目　103

（4）管理 VLAN

管理 VLAN 是配置用于访问交换机管理功能的 VLAN。通过为管理 VLAN 分配 IP 地址、子网掩码和网关地址，用户可基于 HTTP、Telnet、SSH 或 SNMP 等协议通过 IP 网络实现对交换机的远程管理。默认情况下，VLAN1 被指定为管理 VLAN。但为了安全起见，建议配置除 VLAN1 以外的其他 VLAN 作为管理 VLAN。

3. 中继协议

如果在有多个交换机的网络中划分了多个 VLAN，交换机与交换机之间的链路可能会承载来自多个不同 VLAN 的数据，则需要将承载多个源于不同 VLAN 数据的物理线路配置为中继链路，并允许需要的 VLAN 数据可以通过。如图 2.20 所示的交换网络中，交换机 LSW1 与 LSW2 之间的链路为中继链路，允许 VLAN2 和 VLAN3 的数据帧在该链路上传输。

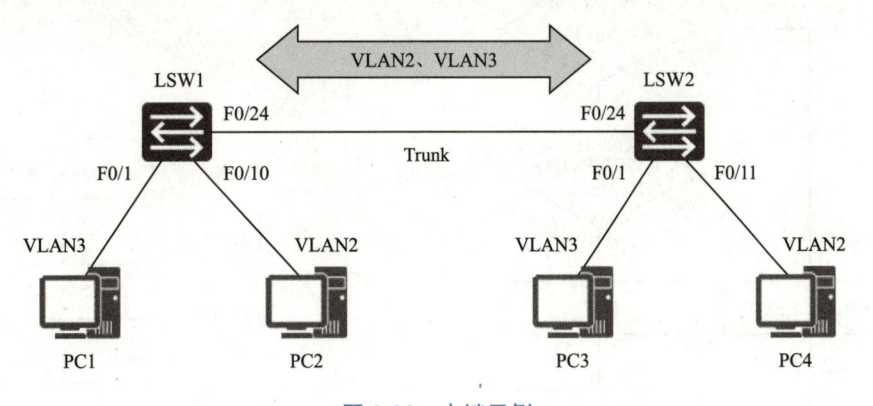

图 2.20　中继示例

由于在中继链路上承载了来自多个 VLAN 的不同数据，在交换机上需要有一种机制使交换机能够识别中继链路上的数据帧来自哪个 VLAN，以进行正确的转发。通常采用在以太网帧的帧头中插入一个标签的方法来识别中继链路中的数据帧属于哪个 VLAN，最常用的 VLAN 中继协议是 IEEE 802.1Q，它是一个开放的封装协议，所有厂商都支持。IEEE 802.1Q 的帧格式如图 2.21 所示，它相当于在标准的以太网帧头添加了 4 字节，成为带有 VLAN 标签的帧。在添加的 4 字节中，2 字节为标记协议标识符（tag protocol indentifier, TPID），2 字节为标签控制信息段（tag control information，TCI）。在 IEEE 802.1Q 帧结构中，VLAN 标签字段的含义如下：

8字节	6字节	6字节	2字节	2字节	2字节	46～1 500字节	4字节
前导码	目的MAC地址	源MAC地址	标记协议标志	标签控制信息	类型	数据	帧校验序列

图 2.21　IEEE 802.1Q 的帧格式

① 标记协议标识符（TPID）：长度为 2 字节，值为 "0x8100" 时，表示为 IEEE 802.1Q 帧。

② 标签控制信息字段：长度为 2 字节，包含的是帧的控制信息。它包括 3 bit 优先级（PRI）、1 bit 标准格式指示符（CFI）和 12 bit VLAN 标识符（VLAN ID）。其中，PRI 用以标识帧的优先级，

主要用于 QoS。在以太网环境中，CFI 字段的值为 0；VLAN ID 用以标识该帧所属的 VLAN。

4．交换机的链路类型

（1）接入（access）接口

接入接口连接像用户主机和服务器等不能识别 Tag 的用户终端设备。接入接口必须配置 PVID（交换机默认的 PVID 为 VLAN1），数据帧进入接入接口的 VLAN 标签处理过程如图 2.22 所示。数据帧从接入接口发送时，如果数据帧中的 VLAN ID 与接入接口配置的 PVID 相同，则剥离标签，将不带标签的以太网帧从链路上发送出去，如图 2.23 所示。

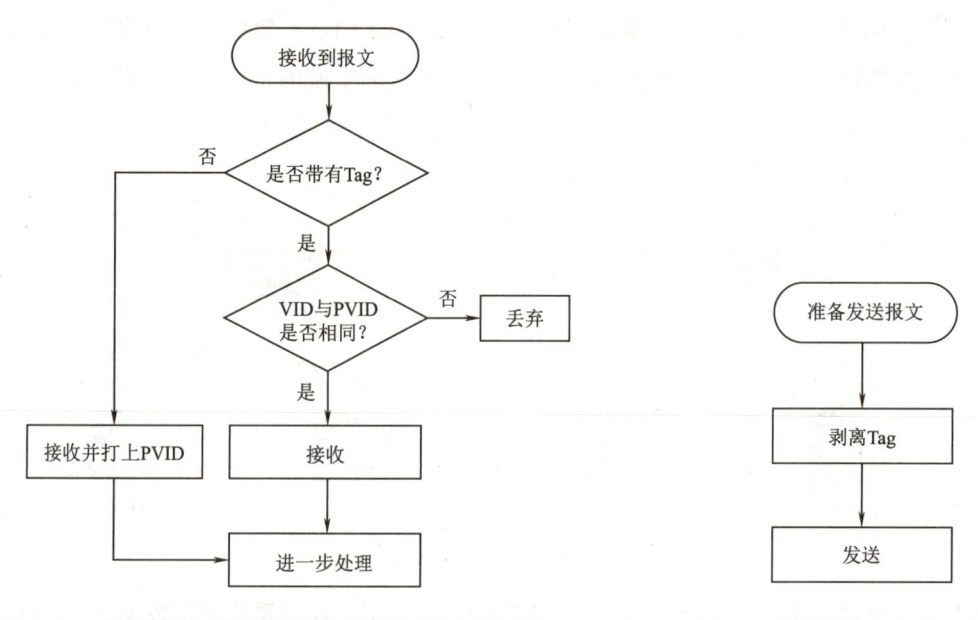

图 2.22　数据帧进入接入接口的 VLAN 标签处理过程

图 2.23　数据帧从接入接口发送时的 VLAN 标签处理过程

（2）中继（trunk）接口

中继接口又称干道（trunk）接口。配置为中继的接口允许多个 VLAN 的数据帧通过，这些数据帧通过 802.1Q 的标签进行区分，但只允许一个 VLAN 的帧从该类接口上发出时不带 Tag（即剥离 Tag）。中继接口通常用于交换机之间、交换机与路由器之间等。数据帧进入中继接口的 VLAN 标签处理过程如图 2.24 所示。数据帧从中继接口发送时的 VLAN 标签处理过程如图 2.25 所示。

注：中继接口 PVID 的 VLAN 数据在中继链路上不携带标签。

（3）混合（hybrid）接口

混合接口与中继接口类似，也允许多个 VLAN 的数据帧通过，这些数据帧通过 802.1Q 的标签实现区分。用户可以灵活指定混合接口在发送某个（或某些）VLAN 的数据帧时是否携带标签。混合接口可传输两种帧：带 VLAN 信息的帧和不带 VLAN 信息的帧。数据帧进入混合接

口的 VLAN 标签处理过程如图 2.26 所示。数据帧从混合接口发送时的 VLAN 标签处理过程如图 2.27 所示。

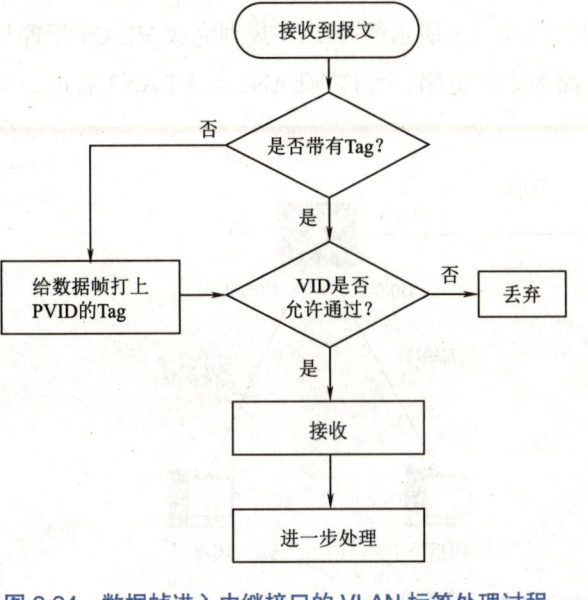

图 2.24　数据帧进入中继接口的 VLAN 标签处理过程

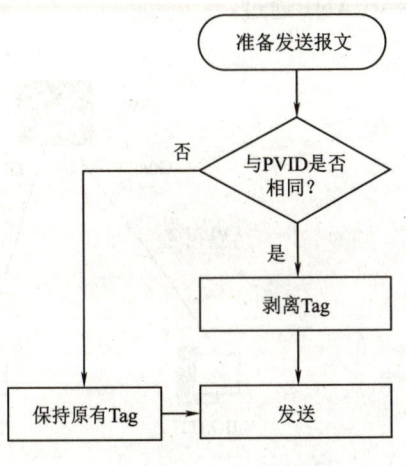

图 2.25　数据帧从中继接口发送时的 VLAN 标签处理过程

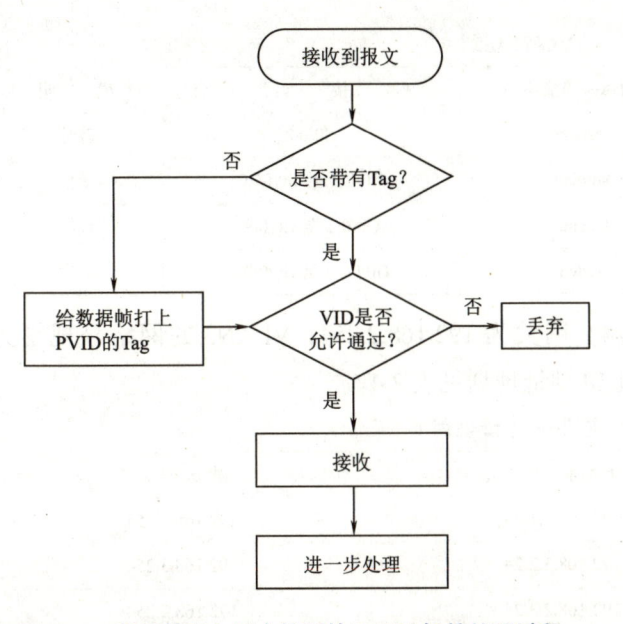

图 2.26　数据帧进入混合接口的 VLAN 标签处理过程

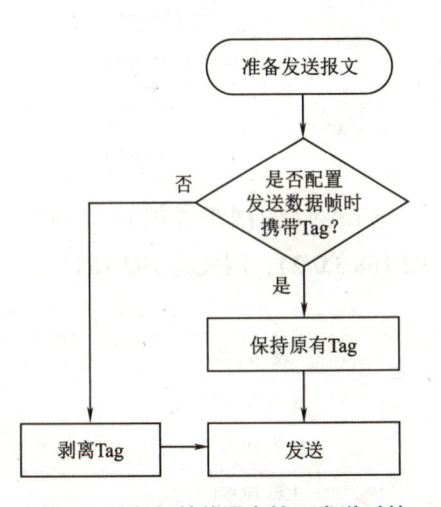

图 2.27　数据帧从混合接口发送时的 VLAN 标签处理过程

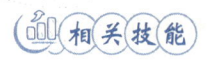 相关技能

1. VLAN 的规划与配置

搭建如图 2.28 所示的拓扑图，根据拓扑图与表 2.3 所示的 VLAN 规划完成 VLAN 配置与验证，其中，交换机 LSW1 与 LSW2 之间的链路为中继链路，允许 VLAN2 与 VLAN3 通过，不允许 VLAN1 通过。

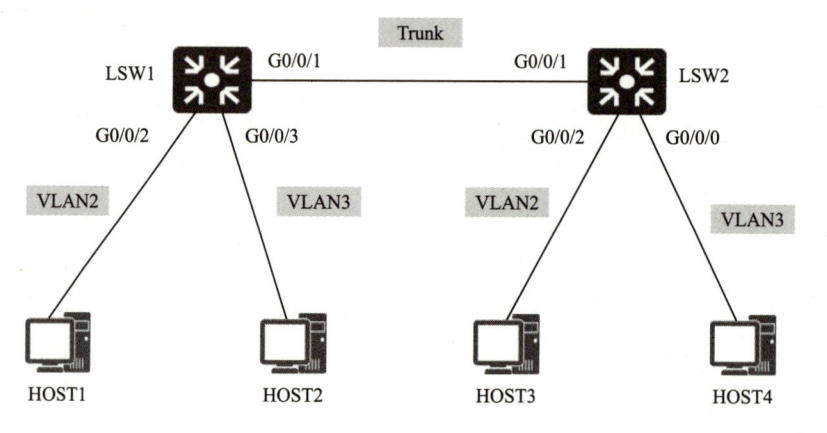

图 2.28 VLAN 的规划与配置拓扑图

交换网络中 VLAN 规划见表 2.3。

表 2.3 VLAN 规划

设 备	接口或 VLAN	VLAN 的描述	接 口	说 明
LSW1	VLAN2	teacher	G0/0/2	教师
	VLAN3	student	G0/0/3 至 G0/0/20	学生
LSW2	VLAN2	teacher	G0/0/2 至 G0/0/9	教师
	VLAN3	student	G0/0/10 至 G0/0/20	学生

VLAN2 的网络号规划为 192.168.2.0/24，网关为 192.168.2.254；VLAN3 的网络号规划为 192.168.3.0/24，网关为 192.168.3.254。主机 IP 地址规划见表 2.4。

表 2.4 主机 IP 地址规划

设 备	IP 地址与掩码	默 认 网 关
主机 HOST1	192.168.2.1/24	192.168.2.254
主机 HOST2	192.168.3.2/24	192.168.3.254
主机 HOST3	192.168.2.3/24	192.168.2.254
主机 HOST4	192.168.3.4/24	192.168.3.254

VLAN 的配置一般需要三步：

在局域网所有的交换机上根据规划的 VLAN 创建 VLAN 并配置 VLAN 的描述或者名称；

在局域网所有的交换机上配置接入接口，并指定接入接口所属 VLAN；

项目 2　小型企业网络互联项目　107

在局域网所有的交换机上配置中继接口，并配置允许哪些 VLAN 通过以及配置中继接口的 PVID。

（1）交换机 LSW1 的 VLAN 配置

① 创建 VLAN。根据表 2.3 的 VLAN 规划，在交换机 LSW1 上需创建两个 VLAN，分别为 VLAN2 与 VLAN3，以华为 S5700-28C-HI 交换机为例，在 LSW1 交换机上创建 VLAN，并配置 VLAN 的描述步骤如下：

```
[LSW1]vlan 2                        // 创建 VLAN2, 进入 VLAN2 视图模式
[LSW1-vlan2]description teacher      // 配置 VLAN2 的描述为 "teacher"
[LSW1-vlan2]vlan 3                   // 创建 VLAN3, 进入 VLAN3 视图模式
[LSW1-vlan3]description student      // 配置 VLAN3 的描述为 "student"
[LSW1-vlan3]quit                     // 返回到系统视图
```

② 配置接入接口。接入接口的配置通常需完成两个任务：一是进入接入接口的接口视图，配置接口为接入接口；二是配置接口属于哪个 VLAN。

以华为 S5700-28C-HI 交换机为例，将交换机 LSW1 的 G0/0/2 接口配置为接入接口，并属于 VLAN2 的配置步骤如下：

```
[LSW1]interface g0/0/2                                // 进入 G0/0/2 接口的接口视图
[LSW1-GigabitEthernet0/0/2]port link-type access     // 配置 G0/0/2 接口为接入接口
[LSW1-GigabitEthernet0/0/2]port default vlan 2        // 配置 G0/0/2 接口属于 VLAN2
[LSW1-GigabitEthernet0/0/2]quit                       // 返回到系统视图
```

根据表 2.3 的 VLAN 规划，交换机 LSW1 的 G0/0/3 至 G0/0/20 接口属于 VLAN3，如果每个接入接口均单独进行配置，则配置命令重复较多且配置命令数量大。由于 G0/0/3 至 G0/0/20 接口的所有接口配置命令均相同，可以将这些接口加入一个端口组进行配置，以减少配置工作量。

以华为 S5700-28C-HI 交换机为例，把交换机 LSW1 的 G0/0/3 至 G0/0/20 接口加入端口组 1，配置为接入接口，并属于 VLAN3 的配置步骤如下：

```
[LSW1]port-group 1                                // 创建并进入端口组 1 的接口视图
[LSW1-port-group-1]group-member g0/0/3 to g0/0/20 // 把 G0/0/3 至 G0/0/20 接口添加为端口组
                                                  // 1 的成员
[LSW1-port-group-1]port link-type access          // 配置端口组 1 中的所有接口为接入接口
[LSW1-port-group-1]port default vlan 3            // 配置端口组 1 中的所有接口属于 VLAN3
[LSW1-port-group-1]quit                           // 返回到系统视图
```

③ 配置中继接口。中继接口的配置通常需完成三个任务：一是进入中继接口的接口视图，配置其为中继接口。二是配置中继接口允许哪些 VLAN 的数据帧通过。最后指定中继接口的 PVID。

根据表 2.3 的 VLAN 规划，交换机 LSW1 的 G0/0/1 接口为中继接口，允许 VLAN2、

VLAN3 通过，不允许 VLAN1 通过。由于华为交换机中继接口默认允许 VLAN1 通过，因此需要禁用 VLAN1 通过。具体配置步骤如下：

```
[LSW1]interface g0/0/1                                    // 进入 G0/0/1 接口的接口视图
[LSW1-GigabitEthernet0/0/1]port link-type trunk   // 配置 G0/0/1 接口为中继接口
[LSW1-GigabitEthernet0/0/1]port trunk allow-pass vlan 2 3
// 配置 G0/0/1 接口允许 VLAN2 与 VLAN3 通过
[LSW1-GigabitEthernet0/0/1]undo port trunk allow-pass vlan 1
// 配置 G0/0/1 接口不允许 VLAN1 通过
[LSW1-GigabitEthernet0/0/1]quit                           // 返回到系统视图
```

注： 华为交换机中继接口默认的 PVID 为 VLAN1。

（2）交换机 LSW2 的 VLAN 配置

① 创建 VLAN。根据表 2.3 的 VLAN 规划，以华为 S5700-28C-HI 交换机为例，在 LSW2 交换机上创建 VLAN2 与 VLAN3，并配置 VLAN 的描述步骤如下：

```
[LSW2]vlan 2                             // 创建 VLAN2，进入 VLAN2 视图模式
[LSW2-vlan2]description teacher          // 配置 VLAN2 的描述为 "teacher"
[LSW2-vlan2]vlan 3                       // 创建 VLAN3，进入 VLAN3 视图模式
[LSW2-vlan3]description student          // 配置 VLAN3 的描述为 "student"
[LSW2-vlan3]quit                         // 返回到系统视图
```

② 配置接入接口。根据表 2.3 的 VLAN 规划，交换机 LSW2 的 G0/0/2 至 G0/0/20 接口为接入接口，其中 G0/0/2 至 G0/0/9 接口属于 VLAN2，G0/0/10 至 G0/0/20 接口属于 VLAN3，由于 G0/0/2 至 G0/0/9 接口配置命令相同，将其加入一个端口组 1 进行配置，G0/0/10 至 G0/0/20 接口配置命令相同，将其加入端口组 2 进行配置。

以华为 S5700-28C-HI 交换机为例，使用端口组进行交换机 LSW2 接入接口的配置步骤如下：

```
[LSW2]port-group 1                                  // 创建并进入端口组 1 的接口视图
[LSW2-port-group-1]group-member g0/0/2 to g0/0/9    // 把 G0/0/2 至 G0/0/9 接口添加为端口组 1
                                                    // 的成员
[LSW2-port-group-1]port link-type access            // 配置端口组 1 中的所有接口为接入接口
[LSW2-port-group-1]port default vlan 2              // 配置端口组 1 中的所有接口属于 VLAN2
[LSW2-port-group-1]quit                             // 返回到系统视图
[LSW2]port-group 2                                  // 创建并进入端口组 2 的接口视图
[LSW2-port-group-2]group-member g0/0/10 to g0/0/20  // 把 G0/0/10 至 G0/0/20 接口添加为端口组
                                                    // 2 的成员
[LSW2-port-group-2]port link-type access            // 配置端口组 2 中的所有接口为接入接口
[LSW2-port-group-2]port default vlan 3              // 配置端口组 2 中的所有接口属于 VLAN3
```

③ 配置中继接口。根据表 2.3 的 VLAN 规划，交换机 LSW2 的 G0/0/1 接口为中继接口，允许 VLAN2、VLAN3 通过，不允许 VLAN1 通过。具体配置步骤如下：

```
[LSW2]interface g0/0/1                              // 进入 G0/0/1 接口的接口视图
[LSW2-GigabitEthernet0/0/1]port link-type trunk  // 配置 G0/0/1 接口为中继接口
[LSW2-GigabitEthernet0/0/1]port trunk allow-pass vlan 2 3
// 配置 G0/0/1 接口允许 VLAN2 与 VLAN3 通过
[LSW2-GigabitEthernet0/0/1]undo port trunk allow-pass vlan 1
// 配置 G0/0/1 接口不允许 VLAN1 通过
[LSW2-GigabitEthernet0/0/1]quit                     // 返回到系统视图
```

（3）VLAN 的验证与测试

在交换机上完成 VLAN 的创建、接入接口的配置与中继接口的配置后，可使用 "display" 命令查看验证 VLAN 是否配置正确。

① 执行 "display vlan" 命令，观察已创建的 VLAN 信息是否存在、正确，接口是否已加入正确的 VLAN 中。图 2.29 所示为交换机 LSW1 上的 VLAN 信息，图 2.30 所示为交换机 LSW2 上的 VLAN 信息。

图 2.29　交换机 LSW1 上的 VLAN 信息

② 执行 "display port vlan" 命令可查看交换机接口 VLAN 信息。图 2.31 中，交换机 LSW1 的 G0/0/1 接口为 trunk（中继），只允许 VLAN2 与 VLAN3 通过；G0/0/2 接口为 access（接入），PVID 为 "2"，即属于 VLAN2；G0/0/3 至 G0/0/20 接口为 access（接入），PVID 为 "3"，即属于 VLAN3；G0/0/21 至 G0/0/24 接口为默认值，即链路类型为 hybrid，PVID 为 "1"，即默认属于 VLAN1，与规划配置一致。

```
[LSW2]display vlan
The total number of vlans is : 3
--------------------------------------------------------------------
U: Up;          D: Down;         TG: Tagged;         UT: Untagged;
MP: Vlan-mapping;                ST: Vlan-stacking;
#: ProtocolTransparent-vlan;     *: Management-vlan;
--------------------------------------------------------------------

VID  Type   Ports
--------------------------------------------------------------------
1    common  UT:GE0/0/21(D)    GE0/0/22(D)     GE0/0/23(D)     GE0/0/24(D)

2    common  UT:GE0/0/2(U)     GE0/0/3(D)      GE0/0/4(D)      GE0/0/5(D)

                GE0/0/6(D)      GE0/0/7(D)      GE0/0/8(D)      GE0/0/9(D)
              TG:GE0/0/1(U)

3    common  UT:GE0/0/10(U)    GE0/0/11(D)     GE0/0/12(D)     GE0/0/13(D)

                GE0/0/14(D)     GE0/0/15(D)     GE0/0/16(D)     GE0/0/17(D)
                GE0/0/18(D)     GE0/0/19(D)     GE0/0/20(D)
              TG:GE0/0/1(U)


VID  Status  Property     MAC-LRN Statistics Description
--------------------------------------------------------------------

1    enable  default      enable  disable    VLAN 0001
2    enable  default      enable  disable    teacher
3    enable  default      enable  disable    student
```

图 2.30　交换机 LSW2 上的 VLAN 信息

```
[LSW1]display port vlan
Port                    Link Type    PVID   Trunk VLAN List
GigabitEthernet0/0/1    trunk        1      2-3
GigabitEthernet0/0/2    access       2      -
GigabitEthernet0/0/3    access       3      -
GigabitEthernet0/0/4    access       3      -
GigabitEthernet0/0/5    access       3      -
GigabitEthernet0/0/6    access       3      -
GigabitEthernet0/0/7    access       3      -
GigabitEthernet0/0/8    access       3      -
GigabitEthernet0/0/9    access       3      -
GigabitEthernet0/0/10   access       3      -
GigabitEthernet0/0/11   access       3      -
GigabitEthernet0/0/12   access       3      -
GigabitEthernet0/0/13   access       3      -
GigabitEthernet0/0/14   access       3      -
GigabitEthernet0/0/15   access       3      -
GigabitEthernet0/0/16   access       3      -
GigabitEthernet0/0/17   access       3      -
GigabitEthernet0/0/18   access       3      -
GigabitEthernet0/0/19   access       3      -
GigabitEthernet0/0/20   access       3      -
GigabitEthernet0/0/21   hybrid       1      -
GigabitEthernet0/0/22   hybrid       1      -
GigabitEthernet0/0/23   hybrid       1      -
GigabitEthernet0/0/24   hybrid       1      -
```

图 2.31　交换机 LSW1 的接口 VLAN 信息

　　图 2.32 中，交换机 LSW2 的 G0/0/1 接口为 trunk（中继），只允许 VLAN2 与 VLAN3 通过；G0/0/2 至 G0/0/9 接口为 access（接入），PVID 为"2"，即属于 VLAN2；G0/0/10 至 G0/0/20 接口为 access（接入），PVID 为"3"，即属于 VLAN3。与规划配置一致。

```
[LSW2]display port vlan
Port                    Link Type   PVID  Trunk VLAN List
------------------------------------------------------------
GigabitEthernet0/0/1    trunk        1    2-3
GigabitEthernet0/0/2    access       2    -
GigabitEthernet0/0/3    access       2    -
GigabitEthernet0/0/4    access       2    -
GigabitEthernet0/0/5    access       2    -
GigabitEthernet0/0/6    access       2    -
GigabitEthernet0/0/7    access       2    -
GigabitEthernet0/0/8    access       2    -
GigabitEthernet0/0/9    access       2    -
GigabitEthernet0/0/10   access       3    -
GigabitEthernet0/0/11   access       3    -
GigabitEthernet0/0/12   access       3    -
GigabitEthernet0/0/13   access       3    -
GigabitEthernet0/0/14   access       3    -
GigabitEthernet0/0/15   access       3    -
GigabitEthernet0/0/16   access       3    -
GigabitEthernet0/0/17   access       3    -
GigabitEthernet0/0/18   access       3    -
GigabitEthernet0/0/19   access       3    -
GigabitEthernet0/0/20   access       3    -
GigabitEthernet0/0/21   hybrid       1    -
GigabitEthernet0/0/22   hybrid       1    -
GigabitEthernet0/0/23   hybrid       1    -
GigabitEthernet0/0/24   hybrid       1    -
```

图 2.32　交换机 LSW2 的端口 VLAN 信息

③ 在交换机的接口视图下，执行 "display this" 命令查看交换机接口的配置信息。交换机 LSW1 的 G0/0/1 接口的配置信息如图 2.33 所示。

```
[LSW1-GigabitEthernet0/0/1]display this
#
interface GigabitEthernet0/0/1
 port link-type trunk
 undo port trunk allow-pass vlan 1
 port trunk allow-pass vlan 2 to 3
```

图 2.33　交换机 LSW1 的 G0/0/1 接口的配置信息

④ 参考表 2.4 规划的参数完成测试主机 HOST1、HOST2、HOST3 与 HOST4 的 IP 地址、子网掩码、默认网关配置。配置完成后，在 DOS 命令行下使用 "ipconfig" 命令查看其 IP 配置信息是否正确。（注：由于本任务中的网络拓扑无路由设备，网关是否配置不影响下面的测试。）

⑤ 根据 IP 与 VLAN 的规划，同一 VLAN 的主机可以不经过路由设备直接通信，不同 VLAN 间的主机没有路由设备不能直接通信。由于 HOST1 与 HOST3 同属于 VLAN2，HOST2 与 HOST4 同属于 VLAN3，因此，本任务中，理论上主机 HOST1 可以 ping 通 HOST3，但 HOST1 不能 ping 通 HOST2 与 HOST4。图 2.34 所示为主机 HOST1 可以 ping 通 HOST3，但不能 ping 通 HOST2 的连通性测试示例。

2. 混合接口的应用与配置

某公司混合接口应用拓扑图如图 2.35 所示。网络中没有路由设备。公司有销售与行政两个部门。请使用混合接口实现：公司的销售部门、行政部门与服务器之间相互需要广播隔离，销售部门与行政部门之间相互不能通信，但销售部门与行政部门可以访问公司的服务器。

112　企业网络互联技术（双语）

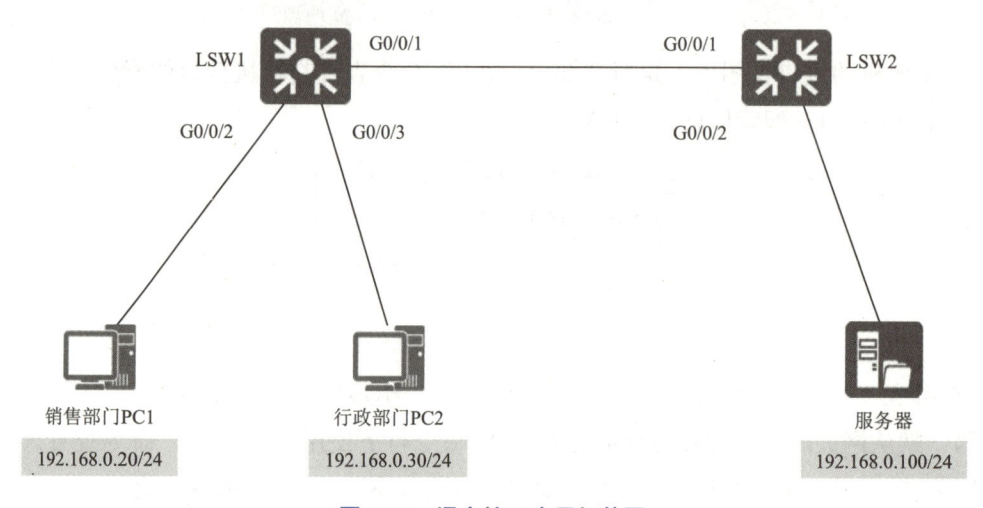

图 2.34　主机 HOST1 可以 ping 通 HOST3，但不能 ping 通 HOST2 的连通性测试

　　根据任务需求，将销售部门、行政部门与服务器规划属于不同的 VLAN，实现广播隔离的需求。其中，销售部门属于 VLAN20，行政部门属于 VLAN30，服务器属于 VLAN100，所有部门 IP 地址的网络号均为 "192.168.0.0/24"，测试主机 PC1、PC2 与服务器的 IP 地址见表 2.5。

表 2.5　主机与服务器的 IP 地址

设　　备	IP 地址与掩码
主机 PC1	192.168.0.20/24
主机 PC2	192.168.0.30/24
服务器	192.168.0.100/24

项目 2　小型企业网络互联项目　113

将交换机 LSW1 与 LSW2 的端口规划为混合接口，控制通过的打标签与不打标签 VLAN 可实现任务需求。LSW1 与 LSW2 的混合接口规划见表 2.6。

表 2.6　LSW1 与 LSW2 的混合接口规划

设　备	接　口	链路类型	PVID	允许通过的 VLAN 列表
LSW1	G0/0/1	hybrid	VLAN1	打标签（tagged）的 VLAN：VLAN20、VLAN30、VLAN100
	G0/0/2	hybrid	VLAN20	不打标签（untagged）的 VLAN：VLAN20、VLAN100
	G0/0/3	hybrid	VLAN30	不打标签（untagged）的 VLAN：VLAN30、VLAN100
LSW2	G0/0/1	hybrid	VLAN1	打标签（tagged）的 VLAN：VLAN20、VLAN30、VLAN100
	G0/0/2	hybrid	VLAN100	不打标签（untagged）的 VLAN：VLAN20、VLAN30、VLAN100

（1）交换机 LSW1 的 VLAN 配置

① 创建 VLAN。交换机 LSW1 的 G0/0/1 接口需要通过的打标签（tagged）VLAN 有 VLAN20、VLAN30 和 VLAN100，因此在交换机 LSW1 上需要创建 VLAN20、VLAN30 和 VLAN100，可以使用本任务相关技能中"1.VLAN 的规划与配置"中创建 VLAN 的方法一个一个创建这三个 VLAN，也可以批量创建 VLAN20、VLAN30 与 VLAN100。批量创建 VLAN20、VLAN30 与 VLAN100 的步骤如下：

```
[LSW1]vlan batch 20 30 100        // 批量创建 VLAN20、VLAN30 和 VLAN100
```

② 混合接口配置。根据表 2.6 所示的混合接口规划，交换机 LSW1 的混合接口配置步骤如下：

```
[LSW1]interface g0/0/1                        // 进入 G0/0/1 接口的接口视图
[LSW1-GigabitEthernet0/0/1]port link-type hybrid // 配置 G0/0/2 接口为 (hybrid) 接口
[LSW1-GigabitEthernet0/0/1]port hybrid tagged vlan 20 30 100
   // 配置 G0/0/1 混合接口加入的打标签 VLAN 为 VLAN20、VLAN30 与 VLAN100
[LSW1-GigabitEthernet0/0/1]quit                // 返回到系统视图
[LSW1]interface g0/0/2                        // 进入 G0/0/2 接口的接口视图
[LSW1-GigabitEthernet0/0/2]port link-type hybrid // 配置 G0/0/2 接口为混合接口
[LSW1-GigabitEthernet0/0/2]port hybrid pvid vlan 20 // 配置 G0/0/2 端口的 PVID 为 VLAN20
[LSW1-GigabitEthernet0/0/2]port hybrid untagged vlan 20 100
   // 配置 G0/0/1 混合接口加入的不打标签 VLAN 为 VLAN20 与 VLAN100
[LSW1-GigabitEthernet0/0/2]quit                // 返回到系统视图
[LSW1]interface g0/0/3                        // 进入 G0/0/3 接口的接口视图
[LSW1-GigabitEthernet0/0/3]port link-type hybrid // 配置 G0/0/3 接口为混合接口
```

```
[LSW1-GigabitEthernet0/0/3]port hybrid pvid vlan 30      // 配置 G0/0/3 接口的 PVID 为 VLAN30
[LSW1-GigabitEthernet0/0/3]port hybrid untagged vlan 30 100
    // 配置 G0/0/3 混合接口加入的不打标签 VLAN 为 VLAN30 与 VLAN100
[LSW1-GigabitEthernet0/0/3]quit                          // 返回到系统视图
```

（2）交换机 LSW2 的 VLAN 配置

① 创建 VLAN。交换机 LSW2 的 G0/0/1 接口需要通过的打标签（tagged）的 VLAN 有 VLAN20、VLAN30 和 VLAN100，因此在交换机 LSW2 上需要创建 VLAN20、VLAN30 和 VLAN100，批量创建 VLAN20、VLAN30 与 VLAN100 的步骤如下：

```
[LSW2]vlan batch 20 30 100                    // 批量创建 VLAN20、VLAN30 和 VLAN100
```

② 交换机 LSW2 上混合接口配置。根据表 2.6 所示的混合接口规划，交换机 LSW2 的混合接口配置步骤如下：

```
[LSW2]interface g0/0/1                                    // 进入 G0/0/1 接口的接口视图
[LSW2-GigabitEthernet0/0/1]port link-type hybrid // 配置 G0/0/2 接口为混合接口
[LSW2-GigabitEthernet0/0/1]port hybrid tagged vlan 20 30 100
    // 配置 G0/0/1 混合接口加入的打标签 VLAN 为 VLAN20、VLAN30 与 VLAN100
[LSW2-GigabitEthernet0/0/1]quit                          // 返回到系统视图
[LSW2]interface g0/0/2                                    // 进入 G0/0/2 接口的接口视图
[LSW2-GigabitEthernet0/0/2]port link-type hybrid        // 配置 G0/0/2 接口为混合接口
[LSW2-GigabitEthernet0/0/2]port hybrid pvid vlan 100    // 配置 G0/0/2 接口的 PVID 为 VLAN100
[LSW2-GigabitEthernet0/0/2]port hybrid untagged vlan 20 30 100
    // 配置 G0/0/1 混合接口加入的不打标签 VLAN 为 VLAN20、VLAN30 与 VLAN100
[LSW2-GigabitEthernet0/0/2]quit                          // 返回到系统视图
```

（3）VLAN 的验证与测试

完成 VLAN 的创建与混合接口的配置后，可使用以下 display 命令验证 VLAN 是否配置正确。

① 在交换机上执行"display vlan"命令查看 VLAN 信息，观察创建的 VLAN 信息是否正确。图 2.36 所示为在交换机 LSW1 上使用"display vlan"命令查看 VLAN 信息的显示结果：VLAN20 是 G0/0/2 接口加入的不打标签 VLAN，是 G0/0/1 接口加入的打标签 VLAN；VLAN30 是 G0/0/3 接口加入的不打标签 VLAN，是 G0/0/1 接口加入的打标签 VLAN；VLAN100 是 G0/0/2 与 G0/0/3 接口加入的不打标签 VLAN，是 G0/0/1 接口加入的打标签 VLAN。

图 2.37 所示为在交换机 LSW2 上使用"display vlan"命令查看 VLAN 信息的显示结果：VLAN20 是 G0/0/2 接口加入的不打标签 VLAN，是 G0/0/1 接口加入的打标签 VLAN；VLAN30 是 G0/0/2 接口加入的不打标签 VLAN，是 G0/0/1 接口加入的打标签 VLAN；VLAN100 是 G0/0/2 接口加入的不打标签 VLAN，是 G0/0/1 接口加入的打标签 VLAN。

② 使用 "display port vlan" 命令，可查看交换机混合接口 PVID 和打标签的 VLAN 信息，如图 2.38 所示。交换机 LSW1 上 G0/0/1 接口为混合接口，允许打标签通过的 VLAN 为 VLAN20、VLAN30 与 VLAN100。

```
[LSW1]display vlan
The total number of vlans is : 4
--------------------------------------------------------------------------------
U: Up;          D: Down;          TG: Tagged;          UT: Untagged;
MP: Vlan-mapping;            ST: Vlan-stacking;
#: ProtocolTransparent-vlan;    *: Management-vlan;
--------------------------------------------------------------------------------
VID Type   Ports
--------------------------------------------------------------------------------
1    common UT:GE0/0/1(U)      GE0/0/2(U)      GE0/0/3(U)      GE0/0/4(D)
                GE0/0/5(D)      GE0/0/6(D)      GE0/0/7(D)      GE0/0/8(D)
                GE0/0/9(D)      GE0/0/10(D)     GE0/0/11(D)     GE0/0/12(D)
                GE0/0/13(D)     GE0/0/14(D)     GE0/0/15(D)     GE0/0/16(D)
                GE0/0/17(D)     GE0/0/18(D)     GE0/0/19(D)     GE0/0/20(D)
                GE0/0/21(D)     GE0/0/22(D)     GE0/0/23(D)     GE0/0/24(D)
20   common UT:GE0/0/2(U)
                TG:GE0/0/1(U)
30   common UT:GE0/0/3(U)
                TG:GE0/0/1(U)
100  common UT:GE0/0/2(U)     GE0/0/3(U)
                TG:GE0/0/1(U)
```

图 2.36　在交换机 LSW1 上使用 "display vlan" 命令显示结果

```
[LSW2]display vlan
The total number of vlans is : 4
--------------------------------------------------------------------------------
U: Up;          D: Down;          TG: Tagged;          UT: Untagged;
MP: Vlan-mapping;            ST: Vlan-stacking;
#: ProtocolTransparent-vlan;    *: Management-vlan;
--------------------------------------------------------------------------------
VID Type   Ports
--------------------------------------------------------------------------------
1    common UT:GE0/0/1(U)      GE0/0/2(U)      GE0/0/3(D)      GE0/0/4(D)
                GE0/0/5(D)      GE0/0/6(D)      GE0/0/7(D)      GE0/0/8(D)
                GE0/0/9(D)      GE0/0/10(D)     GE0/0/11(D)     GE0/0/12(D)
                GE0/0/13(D)     GE0/0/14(D)     GE0/0/15(D)     GE0/0/16(D)
                GE0/0/17(D)     GE0/0/18(D)     GE0/0/19(D)     GE0/0/20(D)
                GE0/0/21(D)     GE0/0/22(D)     GE0/0/23(D)     GE0/0/24(D)
20   common UT:GE0/0/2(U)
                TG:GE0/0/1(U)
30   common UT:GE0/0/2(U)
                TG:GE0/0/1(U)
100  common UT:GE0/0/2(U)
                TG:GE0/0/1(U)
```

图 2.37　在交换机 LSW2 上使用 "display vlan" 命令显示结果

```
[LSW1]display port vlan
Port          Link Type  PVID  Trunk VLAN List
-------------------------------------------------------------
GigabitEthernet0/0/1   hybrid     1    20 30 100
GigabitEthernet0/0/2   hybrid    20    -
GigabitEthernet0/0/3   hybrid    30    -
GigabitEthernet0/0/4   hybrid     1    -
GigabitEthernet0/0/5   hybrid     1    -
GigabitEthernet0/0/6   hybrid     1    -
GigabitEthernet0/0/7   hybrid     1    -
GigabitEthernet0/0/8   hybrid     1    -
GigabitEthernet0/0/9   hybrid     1
```

图 2.38　在交换机 LSW1 上使用"display port vlan"命令显示结果

如图 2.39 所示，交换机 LSW2 上 G0/0/1 接口为 Hybrid 接口，允许打标签通过的 VLAN 为 VLAN20、VLAN30 与 VLAN100。

```
[LSW2]display port vlan
Port          Link Type  PVID  Trunk VLAN List
-------------------------------------------------------------
GigabitEthernet0/0/1   hybrid     1    20 30 100
GigabitEthernet0/0/2   hybrid   100    -
GigabitEthernet0/0/3   hybrid     1    -
GigabitEthernet0/0/4   hybrid     1    -
GigabitEthernet0/0/5   hybrid     1    -
GigabitEthernet0/0/6   hybrid     1    -
GigabitEthernet0/0/7   hybrid     1    -
GigabitEthernet0/0/8   hybrid     1    -
GigabitEthernet0/0/9   hybrid     1    -
```

图 2.39　在交换机 LSW2 上使用"display port vlan"命令显示结果

③ 在交换机不打标签的 Hybrid 接口的接口视图下，执行"display this"命令查看该接口的配置信息。图 2.40 所示为在交换机 LSW1 的 G0/0/2 接口的接口视图下使用"display this"命令查看的该接口的配置。

图 2.41 所示为在交换机 LSW2 的 G0/0/2 接口的接口视图下使用"display this"命令查看的该接口的配置。

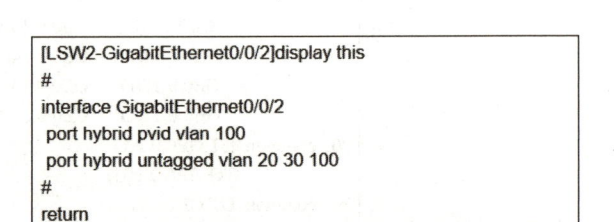

图 2.40　交换机 LSW1 G0/0/2 接口的配置　　　图 2.41　交换机 LSW2 G0/0/2 接口的配置

④ 参考表 2.6 规划的参数完成测试主机 PC1、PC2 和服务器的 IP 地址与子网掩码配置，配置完成后在 DOS 命令行下使用"ipconfig"命令查看其 IP 配置信息是否正确。（注：本任务中，

测试主机 PC1、PC2 与服务器的 IP 地址网络必须相同，不需要配置默认网关地址。）

图 2.42 所示为在测试主机 PC1 上使用"ipconfig"命令查看的 IP 配置信息，注意观察：IP 地址是否为规划的 IP 地址；子网掩码是否为规划的子网掩码。

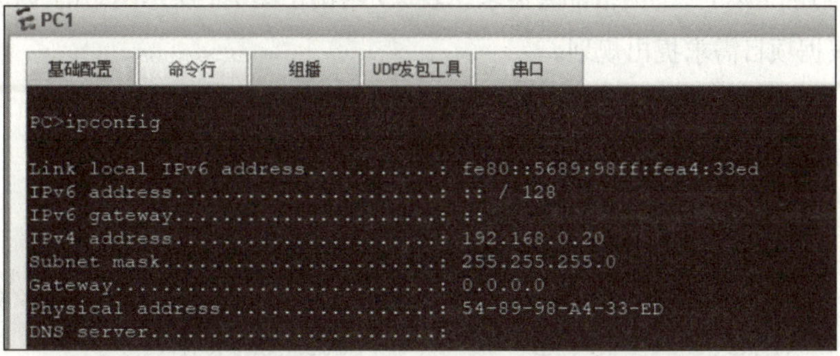

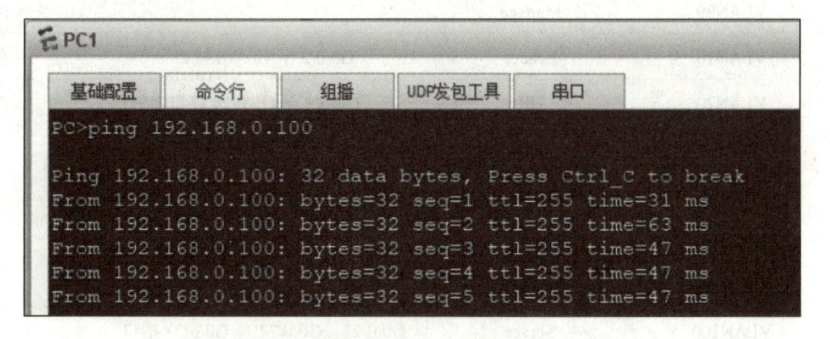

图 2.42　使用"ipconfig"命令查看 IP 配置信息

⑤ 根据规划，测试主机 PC1 与 PC2 相互不能通信，但 PC1 与 PC2 均可访问服务器。图 2.43、图 2.44 所示为测试主机 PC1 与 PC2、服务器之间连通性的测试结果。（注：使用 ping 进行测试时，如果目标主机开启了个人防火墙，则需要在防火墙上放行 ICMP 回显请求的流量才能 ping 通。）

图 2.43　测试主机 PC1 可以 ping 通服务器

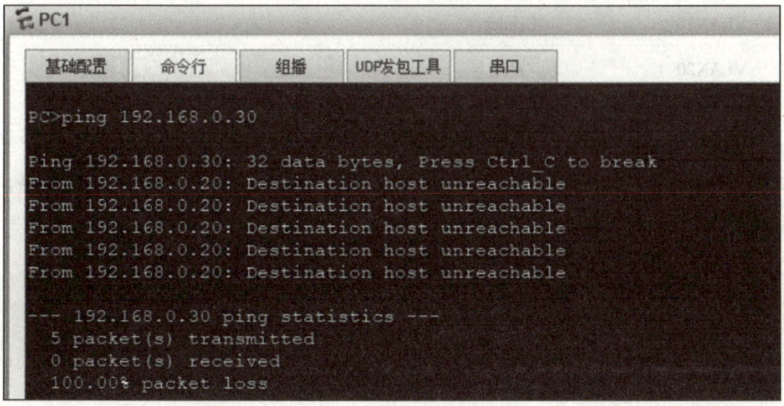

图 2.44　测试主机 PC1 不能 ping 通测试主机 PC2

118　**企业网络互联技术**（双语）

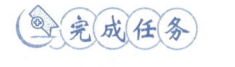 完成任务

1. 任务规划

根据任务描述和项目 2 的网络部署需求，表 2.7 给出了图 2.1 所示网络的 VLAN 参考规划，读者也可自己根据项目需求提出规划。

表 2.7　VLAN 的一种参考规划

设　备	接口或 VLAN	VLAN 的描述	接　　口	说　　明
FLOOR1	VLAN3	Sales	G0/0/2 至 G0/0/14 接口	销售
	VLAN4	Service	G0/0/15 至 G0/0/24 接口	售后
	VLAN99	Manage		交换机管理 VLAN
FLOOR2	VLAN2	RD	G0/0/2 至 G0/0/10 接口	研发
	VLAN3	Sales	G0/0/11 至 G0/0/20 接口	销售
	VLAN4	Service	G0/0/21 至 G0/0/24 接口	售后
	VLAN99	Manage		交换机管理 VLAN
FLOOR3	VLAN2	RD	G0/0/2 至 G0/0/16 接口	研发
	VLAN4	Service	G0/0/17 至 G0/0/24 接口	售后
	VLAN99	Manage		交换机管理 VLAN
FLOOR4	VLAN10	Leader	G0/0/2 至 G0/0/8 接口	企业领导
	VLAN20	HR	G0/0/9 至 G0/0/24 接口	人力资源
CORE	VLAN2	RD		
	VLAN3	Sales		
	VLAN4	Service		
	VLAN99	Manage		
	VLAN100	Server	G0/0/21、G0/0/22 与 G0/0/23 接口	服务器 VLAN
	VLAN101	linktorouter	G0/0/24 接口	到路由器的连接网段
FLOOR4	VLAN10	Leader		领导
	VLAN20	HR		人力资源

结合图 2.1 与表 2.7，交换机 CORE 与交换机 FLOOR1、交换机 FLOOR2 与交换机 FLOOR3 之间的链路以及路由器 AR1 与交换机 FLOOR4 之间的链路需要配置为中继链路。VLAN 中继接口的规划见表 2.8。

表 2.8　VLAN 中继接口的规划

交换机名	中继端口	允许通过的 VLAN	PVID
CORE	G0/0/1	VLAN1、VLAN3、VLAN4 与 VLAN99	VLAN1
	G0/0/2	VLAN1、VLAN2、VLAN3、VLAN4 与 VLAN99	VLAN1
	G0/0/3	VLAN1、VLAN2、VLAN4 与 VLAN99	VLAN1

项目 2　小型企业网络互联项目　119

续表

交换机名	中继端口	允许通过的 VLAN	PVID
FLOOR1	G0/0/1	VLAN1、VLAN3、VLAN4 与 VLAN99	VLAN1
FLOOR2	G0/0/1	VLAN1、VLAN2、VLAN3、VLAN4 与 VLAN99	VLAN1
FLOOR3	G0/0/1	VLAN1、VLAN2、VLAN4 与 VLAN99	VLAN1
FLOOR4	G0/0/1	VLAN1、VLAN10 与 VLAN20	VLAN1

注：中继接口 PVID 的 VLAN 数据在中继链路传输时不携带标签。

2. 任务实施

（1）交换机 CORE 的 VLAN 配置

① 创建 VLAN。以华为 S5700-28C-HI 交换机为例，根据表 2.7 完成交换机 CORE 的 VLAN 配置，步骤如下：

```
[CORE]vlan 2                            // 创建 VLAN2
[CORE-vlan2]description RD              // 配置 VLAN2 的描述为 "RD"
[CORE-vlan2]vlan 3                      // 创建 VLAN3
[CORE-vlan3]description Sales           // 配置 VLAN3 的描述为 "Sales"
[CORE-vlan3]vlan 4                      // 创建 VLAN4
[CORE-vlan4]description Service         // 配置 VLAN4 的描述为 "Service"
[CORE-vlan4]vlan 99                     // 创建 VLAN99
[CORE-vlan99]description Manage         // 配置 VLAN99 的描述为 "Manage"
[CORE-vlan99]vlan 100                   // 创建 VLAN100
[CORE-vlan100]description Server        // 配置 VLAN100 的描述为 "Server"
[CORE-vlan99]vlan 101                   // 创建 VLAN101
[CORE-vlan101]description linktorouter  // 配置 VLAN100 的描述为 "linktorouter"
[CORE-vlan101]quit
```

② 接入接口的配置。以华为 S5700-28C-HI 交换机为例，将交换机 CORE 的 G0/0/21、G0/0/22 与 G0/0/23 接口指派给 VLAN 100，G0/0/24 接口指派给 VLAN100 的配置步骤如下：

```
[CORE]interface g0/0/21                                // 进入 G0/0/21 接口的接口视图
[CORE-GigabitEthernet0/0/21]port link-type access      // 配置 G0/0/21 接口链路模式为接入
[CORE-GigabitEthernet0/0/21]port default vlan 100      // 配置 G0/0/21 接口的 PVID 为 VLAN100
[CORE-GigabitEthernet0/0/21]quit
[CORE]interface g0/0/22                                // 进入 G0/0/22 接口的接口视图
[CORE-GigabitEthernet0/0/22]port link-type access      // 配置 G0/0/22 接口链路模式为接入
[CORE-GigabitEthernet0/0/22]port default vlan 100      // 配置 G0/0/22 接口的 PVID 为 VLAN100
[CORE-GigabitEthernet0/0/22]quit
[CORE]interface g0/0/23                                // 进入 G0/0/23 接口的接口视图
```

```
[CORE-GigabitEthernet0/0/23]port link-type access        // 配置 G0/0/23 接口链路模式为接入
[CORE-GigabitEthernet0/0/23]port default vlan 100        // 配置 G0/0/23 接口的 PVID 为 VLAN100
[CORE-GigabitEthernet0/0/23]quit
[CORE]interface g0/0/24                                  // 进入 G0/0/24 接口的接口视图
[CORE-GigabitEthernet0/0/24]port link-type access        // 配置 G0/0/24 接口链路模式为接入
[CORE-GigabitEthernet0/0/24]port default vlan 101        // 配置 G0/0/24 接口的 PVID 为 VLAN101
```

③ 中继接口的配置。根据规划，交换机 CORE 的 G0/0/1 接口、G0/0/2 接口与 G0/0/3 接口需配置为中继接口，其中 G0/0/1 接口允许 VLAN1、VLAN3、VLAN4 与 VLAN99 通过；G0/0/2 接口允许 VLAN1、VLAN2、VLAN3、VLAN4 与 VLAN99 通过；G0/0/3 接口允许 VLAN1、VLAN2、VLAN4 与 VLAN99 通过。具体配置步骤如下：

```
[CORE]interface g0/0/1                                   // 进入 G0/0/1 接口的接口视图
[CORE-GigabitEthernet0/0/1]port link-type trunk          // 配置 G0/0/1 接口链路模式为中继
[CORE-GigabitEthernet0/0/1]port trunk allow-pass vlan 3 4 99
// 配置 G0/0/1 接口允许 VLAN3、VLAN4 与 VLAN99 通过
[CORE-GigabitEthernet0/0/1]quit                          // 返回到系统视图
[CORE]interface g0/0/2                                   // 进入 G0/0/2 接口的接口视图
[CORE-GigabitEthernet0/0/2]port link-type trunk          // 配置 G0/0/2 接口链路模式为中继
[CORE-GigabitEthernet0/0/2]port trunk allow-pass vlan 2 3 4 99
// 配置 G0/0/2 接口允许 VLAN2、VLAN3、VLAN4 与 VLAN99 通过
[CORE-GigabitEthernet0/0/2]quit                          // 返回到系统视图
[CORE]interface g0/0/3                                   // 进入 G0/0/3 接口的接口视图
[CORE-GigabitEthernet0/0/3]port link-type trunk          // 配置 G0/0/3 接口链路模式为中继
[CORE-GigabitEthernet0/0/3]port trunk allow-pass vlan 2 4 99
// 配置 G0/0/3 接口允许 VLAN2、VLAN4 与 VLAN99 通过
[CORE-GigabitEthernet0/0/3]quit                          // 返回到系统视图
```

注：华为交换机中继接口 VLAN1 默认允许通过。

（2）交换机 FLOOR1 的 VLAN 配置

① 创建 VLAN。以华为 S5700-28C-HI 交换机为例，根据表 2.7，在交换机 FLOOR1 上创建 VLAN 步骤如下：

```
[FLOOR1]vlan 3                          // 创建 VLAN3
[FLOOR1-vlan2]description Sales         // 配置 VLAN3 的描述为 "Sales"
[FLOOR1-vlan3]vlan 4                    // 创建 VLAN4
[FLOOR1-vlan4]description  Service      // 配置 VLAN4 的描述为 "Service"
[FLOOR1-vlan4]vlan 99                   // 创建 VLAN99
[FLOOR1-vlan99]description  Manage      // 配置 VLAN99 的描述为 "Manage"
```

② 接入接口的配置。根据表 2.7，交换机 FLOOR1 的 G0/0/2 至 G0/0/14 接口属于 VLAN3，配置相同；G0/0/15 至 G0/0/24 接口属于 VLAN4，配置相同；因此可以采用把相同配置的接口加入同一个端口组来完成配置，将交换机 FLOOR1 的 G0/0/2 至 G0/0/14 接口添加到端口组 1 中，将交换机 FLOOR1 的 G0/0/15 至 G0/0/24 接口添加到端口组 2 中。以华为 S5700-28C-HI 交换机为例，使用端口组将交换机 FLOOR1 的 G0/0/2 至 G0/0/14 接口配置为属于 VLAN2，G0/0/15 至 G0/0/24 接口配置为属于 VLAN4 的配置步骤如下：

```
[FLOOR1]port-group 1                         // 创建并进入端口组 1 的接口视图
[FLOOR1-port-group-1]group-member g0/0/2 to g0/0/14 // 把 G0/0/2 至 G0/0/14 接口加入
                                             // 端口组 1 中
[FLOOR1-port-group-1]port link-type access   // 配置端口组 1 中的所有接口链路模式为接入
[FLOOR1-port-group-1]port default vlan 3     // 配置端口组 1 中的所有接口的 PVID 为 VLAN3
[FLOOR1-port-group-1]quit                    // 返回到系统视图
[FLOOR1]port-group 2                          // 创建并进入端口组 2 的接口视图
[FLOOR1-port-group-2]group-member g0/0/15 to g0/0/24 // 把 G0/0/15 至 G0/0/24 接口加入
                                             // 端口组 2 中
[FLOOR1-port-group-2]port link-type access   // 配置端口组 1 中的所有接口链路模式为接入
[FLOOR1-port-group-2]port default vlan 4     // 配置端口组 1 中的所有接口的 PVID 为 VLAN4
```

③ 中继接口的配置。根据规划，交换机 FLOOR1 的 G0/0/1 接口需要允许 VLAN3、VLAN4 与 VLAN99 通过，因此需要将其配置为中继接口，具体配置步骤如下：

```
[FLOOR1]interface g0/0/1                              // 进入 G0/0/1 接口的接口视图
[FLOOR1-GigabitEthernet0/0/1]port link-type trunk // 配置 G0/0/1 接口链路模式为中继
[FLOOR1-GigabitEthernet0/0/1]port trunk allow-pass vlan 3 4 99
// 配置 G0/0/1 接口允许 VLAN3、VLAN4 与 VLAN99 通过
[CORE-GigabitEthernet0/0/1]quit                      // 返回到系统视图
```

（3）交换机 FLOOR2 的 VLAN 配置

① 创建 VLAN。以华为 S5700-28C-HI 交换机为例，根据表 2.7，在交换机 FLOOR2 上创建 VLAN 步骤如下：

```
[FLOOR2]vlan 2                    // 创建 VLAN2
[FLOOR2-vlan2]description RD      // 配置 VLAN2 的描述为 "RD"
[FLOOR2]vlan 3                    // 创建 VLAN3
[FLOOR2-vlan2]description Sales   // 配置 VLAN3 的描述为 "Sales"
[FLOOR2-vlan3]vlan 4             // 创建 VLAN4
[FLOOR2-vlan4]description Service // 配置 VLAN4 的描述为 "Service"
[FLOOR2-vlan4]vlan 99           // 创建 VLAN99
[FLOOR2-vlan99]description  Manage // 配置 VLAN99 的描述为 "Manage"
```

② 接入接口的配置。根据表 2.7，交换机 FLOOR2 的 G0/0/2 至 G0/0/10 接口属于 VLAN2；G0/0/11 至 G0/0/20 接口属于 VLAN3；G0/0/21 至 G0/0/24 接口属于 VLAN4；配置时采用将交换机 FLOOR2 的 G0/0/2 至 G0/0/10 接口添加到端口组 1 中，G0/0/11 至 G0/0/20 接口添加到端口组 2 中，G0/0/21 至 G0/0/24 接口添加到端口组 3 中，进行接入接口的配置步骤如下：

```
[FLOOR2]port-group 1                              // 创建并进入端口组 1 的接口视图
[FLOOR2-port-group-1]group-member g0/0/2 to g0/0/10 // 把 G0/0/2 至 G0/0/10 接口加入
                                                  // 端口组 1 中
[FLOOR2-port-group-1]port link-type access        // 配置端口组 1 中的所有接口链路模式为接入
[FLOOR2-port-group-1]port default vlan 2          // 配置端口组 1 中的所有接口的 PVID 为 VLAN2
[FLOOR2-port-group-1]quit                          // 返回到系统视图
[FLOOR2]port-group 2                              // 创建并进入端口组 2 的接口视图
[FLOOR2-port-group-2]group-member g0/0/11 to g0/0/20// 把 G0/0/11 至 G0/0/20 接口加入
                                                  // 端口组 2 中
[FLOOR2-port-group-2]port link-type access        // 配置端口组 2 中的所有接口链路模式为接入
[FLOOR2-port-group-2]port default vlan 3          // 配置端口组 2 中的所有接口的 PVID 为 VLAN3
[FLOOR2-port-group-2]quit                          // 返回到系统视图
[FLOOR2]port-group 3                              // 创建并进入端口组 3 的接口视图
[FLOOR2-port-group-3]group-member g0/0/21 to g0/0/24// 把 G0/0/21 至 G0/0/24 接口加入
                                                  // 端口组 3 中
[FLOOR2-port-group-3]port link-type access        // 配置端口组 3 中的所有接口链路模式为接入
[FLOOR2-port-group-3]port default vlan 4          // 配置端口组 3 中的所有接口的 PVID 为 VLAN4
```

③ 中继接口的配置。根据规划，交换机 FLOOR2 的 G0/0/1 接口需要允许 VLAN2、VLAN3、VLAN4 与 VLAN99 通过，因此需要将其配置为中继接口，具体配置步骤如下：

```
[FLOOR2]interface g0/0/1                              // 进入 G0/0/1 接口的接口视图
[FLOOR2-GigabitEthernet0/0/1]port link-type trunk     // 配置 G0/0/1 接口链路模式为中继
[FLOOR2-GigabitEthernet0/0/1]port trunk allow-pass vlan 2 3 4 99
// 配置 G0/0/1 接口允许 VLAN2、VLAN3、VLAN4 与 VLAN99 通过
[FLOOR2-GigabitEthernet0/0/1]quit                     // 返回到系统视图
```

（4）交换机 FLOOR3 的 VLAN 配置

① 创建 VLAN。根据表 2.7，以华为 S5700-28C-HI 交换机为例，在交换机 FLOOR3 上创建 VLAN 步骤如下：

```
[FLOOR3]vlan batch 2 4 99            // 批量创建 VLAN2、VLAN4 与 VLAN99
[FLOOR3]vlan 2                       // 进入 VLAN2 视图
[FLOOR3-vlan2]description RD         // 配置 VLAN2 的描述为 "RD"
[FLOOR3-vlan2]vlan 4                 // 进入 VLAN4 视图
```

```
[FLOOR3-vlan4]description  Service    // 配置 VLAN4 的描述为 "Service"
[FLOOR3-vlan4]vlan 99                  // 进入 VLAN99 视图
[FLOOR3-vlan99]description  Manage    // 配置 VLAN99 的描述为 "Manage"
```

② 接入接口的配置。根据表 2.7，交换机 FLOOR3 的 G0/0/2 至 G0/0/16 接口属于 VLAN2；G0/0/17 至 G0/0/24 接口属于 VLAN4；配置时采用将交换机 FLOOR3 的 G0/0/2 至 G0/0/16 接口添加到端口组 1 中，G0/0/17 至 G0/0/24 接口添加到端口组 2 中，进行接入接口的配置步骤如下：

```
[FLOOR3]port-group 1                              // 创建并进入端口组 1 的接口视图
[FLOOR3-port-group-1]group-member g0/0/2 to g0/0/16 // 把 G0/0/2 至 G0/0/16 接口加入
                                                  // 端口组 1 中
[FLOOR3-port-group-1]port link-type access        // 配置端口组 1 中的所有接口链路
                                                  // 模式为接入
[FLOOR3-port-group-1]port default vlan 2          // 配置端口组 1 中的所有接口的 PVID
                                                  // 为 VLAN2
[FLOOR3-port-group-1]quit                         // 返回到系统视图
[FLOOR3]port-group 2                              // 创建并进入端口组 2 的接口视图
[FLOOR3-port-group-2]group-member g0/0/17 to g0/0/24// 把 G0/0/17 至 G0/0/24 接口加
                                                  // 入端口组 2 中
[FLOOR3-port-group-2]port link-type access        // 配置端口组 2 中的所有接口链路
                                                  // 模式为接入
[FLOOR3-port-group-2]port default vlan 4          // 配置端口组 2 中的所有接口的
                                                  // PVID 为 VLAN4
[FLOOR3-port-group-2]quit                         // 返回到系统视图
```

③ 中继接口的配置。根据规划，交换机 FLOOR3 的 G0/0/1 接口需要允许 VLAN2、VLAN4 与 VLAN99 通过，因此需要将其配置为中继接口，具体配置步骤如下：

```
[FLOOR3]interface g0/0/1                          // 进入 G0/0/1 接口的接口视图
[FLOOR3-GigabitEthernet0/0/1]port link-type trunk // 配置 G0/0/1 接口链路模式为中继
[FLOOR3-GigabitEthernet0/0/1]port trunk allow-pass vlan 2 4 99
// 配置 G0/0/1 接口允许 VLAN2、VLAN4 与 VLAN99 通过
[FLOOR3-GigabitEthernet0/0/1]quit                 // 返回到系统视图
```

（5）交换机 FLOOR4 的 VLAN 配置

① 创建 VLAN。根据表 2.7，以华为 S5700-28C-HI 交换机为例，在交换机 FLOOR4 上创建 VLAN 步骤如下：

```
[FLOOR4]vlan 10                              // 进入 VLAN10 视图
[FLOOR4-vlan10]description leader            // 配置 VLAN10 的描述为 "leader"
[FLOOR4-vlan10]vlan 20                       // 进入 VLAN20 视图
[FLOOR4-vlan20]description  HR               // 配置 VLAN4 的描述为 "HR"
```

124 企业网络互联技术（双语）

② 接入接口的配置。根据表 2.7，交换机 FLOOR4 的 G0/0/2 至 G0/0/8 接口属于 VLAN10；G0/0/9 至 G0/0/24 接口属于 VLAN20；配置时采用将交换机 FLOOR4 的 G0/0/2 至 G0/0/8 接口添加到端口组 1 中，G0/0/9 至 G0/0/24 接口添加到端口组 2 中，进行接入接口的配置步骤如下：

```
[FLOOR4]port-group 1                              // 创建并进入端口组 1 的接口视图
[FLOOR4-port-group-1]group-member g0/0/2 to g0/0/8   // 把 G0/0/2 至 G0/0/8 接口加入
                                                  // 端口组 1 中
[FLOOR4-port-group-1]port link-type access        // 配置端口组 1 中的所有接口链路
                                                  // 模式为接入
[FLOOR4-port-group-1]port default vlan 10          // 配置端口组 1 中的所有接口的
                                                  // PVID 为 VLAN10
[FLOOR4-port-group-1]quit                          // 返回到系统视图
[FLOOR4]port-group 2                              // 创建并进入端口组 2 的接口视图
[FLOOR4-port-group-2]group-member g0/0/9 to g0/0/24   // 把 G0/0/9 至 G0/0/24 接口加入
                                                  // 端口组 2 中
[FLOOR4-port-group-2]port link-type access        // 配置端口组 2 中的所有接口链路
                                                  // 模式为接入
[FLOOR4-port-group-2]port default vlan 20          // 配置端口组 2 中的所有接口的
                                                  // PVID 为 VLAN20
[FLOOR4-port-group-2]quit                          // 返回到系统视图
```

③ 中继接口的配置。根据规划，交换机 FLOOR4 的 G0/0/1 接口需要允许 VLAN10 与 VLAN20 通过，因此需要将其配置为中继接口，具体配置步骤如下：

```
[FLOOR4]interface g0/0/1                           // 进入 G0/0/1 接口的接口视图
[FLOOR4-GigabitEthernet0/0/1]port link-type trunk  // 配置 G0/0/1 接口链路模式为中继
[FLOOR4-GigabitEthernet0/0/1]port trunk allow-pass vlan 10 20
// 配置 G0/0/1 接口允许 VLAN10 与 VLAN20 通过
[FLOOR4-GigabitEthernet0/0/1]quit                  // 返回到系统视图
```

（6）VLAN 的查看与验证

在交换机 CORE、FLOOR1、FLOOR2、FLOOR3 与 FLOOR4 上完成 VLAN 的创建、接入接口的配置与中继接口的配置后，可使用"display"命令查看与验证 VLAN 是否配置正确。

① 在交换机 CORE、FLOOR1、FLOOR2 和 FLOOR3 上使用"display vlan"命令，查看创建的 VLAN 信息，接口加入的 VLAN 是否携带标签等信息。图 2.45 ～图 2.49 分别为交换机 CORE、FLOOR1、FLOOR2、FLOOR3 和 FLOOR4 上使用"display vlan"命令后的显示结果。

项目 2　小型企业网络互联项目　**125**

```
[CORE]display vlan
The total number of vlans is : 7
--------------------------------------------------------------------------------
U: Up;         D: Down;         TG: Tagged;         UT: Untagged;
MP: Vlan-mapping;              ST: Vlan-stacking;
#: ProtocolTransparent-vlan;   *: Management-vlan;
--------------------------------------------------------------------------------
VID    Type       Ports
--------------------------------------------------------------------------------
1      common   UT:GE0/0/1(U)      GE0/0/2(U)       GE0/0/3(U)       GE0/0/4(D)
                 GE0/0/5(D)         GE0/0/6(D)       GE0/0/7(D)       GE0/0/8(D)
                 GE0/0/9(D)         GE0/0/10(D)      GE0/0/11(D)      GE0/0/12(D)
                 GE0/0/13(D)        GE0/0/14(D)      GE0/0/15(D)      GE0/0/16(D)
                 GE0/0/17(D)        GE0/0/18(D)      GE0/0/19(D)      GE0/0/20(D)
2      common   TG:GE0/0/2(U)      GE0/0/3(U)
3      common   TG:GE0/0/1(U)      GE0/0/2(U)
4      common   TG:GE0/0/1(U)      GE0/0/2(U)       GE0/0/3(U)
99     common   TG:GE0/0/1(U)      GE0/0/2(U)       GE0/0/3(U)
100    common   UT:GE0/0/21(D)     GE0/0/22(D)      GE0/0/23(D)
101    common   UT:GE0/0/24(D)
VID    Status     Property    MAC-LRN Statistics Description
--------------------------------------------------------------------------------
1      enable    default     enable     disable   VLAN 0001
2      enable    default     enable     disable   RD
3      enable    default     enable     disable   Sales
4      enable    default     enable     disable   Service
99     enable    default     enable     disable   Manage
100    enable    default     enable     disable   Server
101    enable    default     enable     disable   linktorouter
```

图 2.45　CORE 上使用"display vlan"命令后的显示结果

```
[FLOOR1]display vlan
The total number of vlans is : 4
--------------------------------------------------------------------------------
U: Up;         D: Down;         TG: Tagged;         UT: Untagged;
MP: Vlan-mapping;              ST: Vlan-stacking;
#: ProtocolTransparent-vlan;   *: Management-vlan;
--------------------------------------------------------------------------------
VID Type  Ports
--------------------------------------------------------------------------------
1   common  UT:GE0/0/1(U)
3   common  UT:GE0/0/2(D)      GE0/0/3(U)       GE0/0/4(D)       GE0/0/5(D)
                GE0/0/6(D)         GE0/0/7(D)       GE0/0/8(D)       GE0/0/9(D)
                GE0/0/10(D)        GE0/0/11(D)      GE0/0/12(D)      GE0/0/13(D)
                GE0/0/14(D)
            TG:GE0/0/1(U)
4   common  UT:GE0/0/15(D)     GE0/0/16(D)      GE0/0/17(D)      GE0/0/18(D)
                GE0/0/19(D)        GE0/0/20(D)      GE0/0/21(D)      GE0/0/22(D)
                GE0/0/23(D)        GE0/0/24(D)
            TG:GE0/0/1(U)
99  common  TG:GE0/0/1(U)
VID Status Property    MAC-LRN Statistics Description
--------------------------------------------------------------------------------
1   enable default     enable     disable   VLAN 0001
3   enable default     enable     disable   Sales
4   enable default     enable     disable   Service
99  enable default     enable     disable   Manage
```

图 2.46　FLOOR1 上使用"display vlan"命令后的显示结果

```
[FLOOR2]display vlan
The total number of vlans is : 5
--------------------------------------------------------------------------------
U: Up;        D: Down;        TG: Tagged;        UT: Untagged;
MP: Vlan-mapping;             ST: Vlan-stacking;
#: ProtocolTransparent-vlan;  *: Management-vlan;
--------------------------------------------------------------------------------

VID  Type   Ports
--------------------------------------------------------------------------------
1    common UT:GE0/0/1(U)

2    common UT:GE0/0/2(D)     GE0/0/3(D)      GE0/0/4(D)       GE0/0/5(D)
                GE0/0/6(U)     GE0/0/7(D)      GE0/0/8(D)       GE0/0/9(D)
                GE0/0/10(D)
             TG:GE0/0/1(U)

3    common UT:GE0/0/11(D)    GE0/0/12(D)     GE0/0/13(D)      GE0/0/14(D)
                GE0/0/15(D)    GE0/0/16(D)     GE0/0/17(D)      GE0/0/18(D)
                GE0/0/19(D)    GE0/0/20(D)
             TG:GE0/0/1(U)

4    common UT:GE0/0/21(D)    GE0/0/22(D)     GE0/0/23(D)      GE0/0/24(D)
             TG:GE0/0/1(U)

99   common TG:GE0/0/1(U)


VID  Status Property     MAC-LRN Statistics Description
--------------------------------------------------------------------------------
1    enable default      enable   disable   VLAN 0001
2    enable default      enable   disable   RD
3    enable default      enable   disable   Sales
4    enable default      enable   disable   Service
99   enable default      enable   disable   Manage
```

图 2.47　FLOOR2 上使用"display vlan"命令后的显示结果

```
[FLOOR3]display vlan
The total number of vlans is : 4
--------------------------------------------------------------------------------
U: Up;        D: Down;        TG: Tagged;        UT: Untagged;
MP: Vlan-mapping;             ST: Vlan-stacking;
#: ProtocolTransparent-vlan;  *: Management-vlan;
--------------------------------------------------------------------------------

VID  Type   Ports
--------------------------------------------------------------------------------
1    common UT:GE0/0/1(U)

2    common UT:GE0/0/2(D)     GE0/0/3(D)      GE0/0/4(D)       GE0/0/5(D)
                GE0/0/6(D)     GE0/0/7(D)      GE0/0/8(D)       GE0/0/9(D)
                GE0/0/10(D)    GE0/0/11(D)     GE0/0/12(D)      GE0/0/13(D)
                GE0/0/14(D)    GE0/0/15(D)     GE0/0/16(D)
             TG:GE0/0/1(U)

4    common UT:GE0/0/17(D)    GE0/0/18(D)     GE0/0/19(D)      GE0/0/20(D)
                GE0/0/21(D)    GE0/0/22(U)     GE0/0/23(D)      GE0/0/24(D)
             TG:GE0/0/1(U)

99   common TG:GE0/0/1(U)

VID  Status Property     MAC-LRN Statistics Description
--------------------------------------------------------------------------------
1    enable default      enable   disable   VLAN 0001
2    enable default      enable   disable   RD
4    enable default      enable   disable   Service
99   enable default      enable   disable   Manage
```

图 2.48　FLOOR3 上使用"display vlan"命令后的显示结果

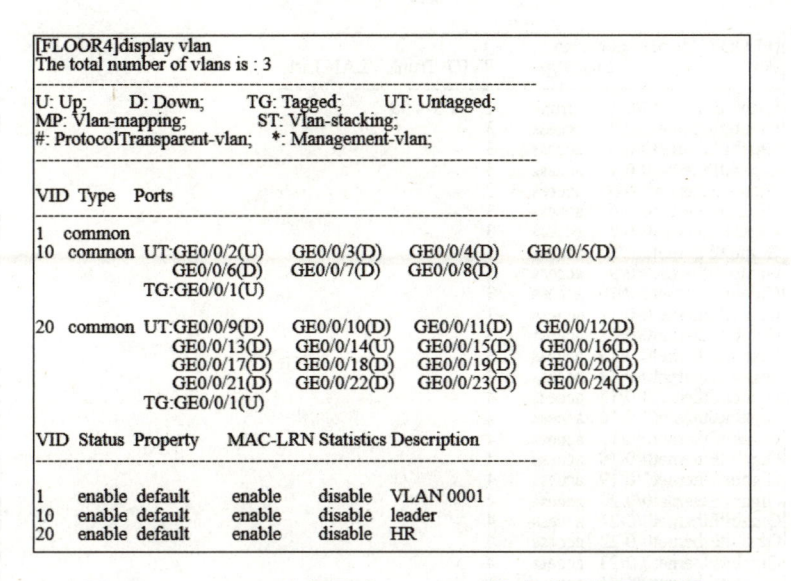

图 2.49　FLOOR4 上使用 "display vlan" 命令后的显示结果

　　② 在交换机 CORE、FLOOR1、FLOOR2 和 FLOOR3 上使用 "display port vlan" 命令查看接口的 VLAN 信息。图 2.50 ～图 2.54 分别为交换机 CORE、FLOOR1、FLOOR2、FLOOR3 和 FLOOR4 上使用 "display port vlan" 命令后的显示结果。

　　注：观察接口的链路类型、PVID 以及 Trunk 允许的 VLAN 列表是否与规划配置一致。

```
[CORE]display port vlan
Port                    Link Type   PVID  Trunk VLAN List
--------------------------------------------------------------
GigabitEthernet0/0/1    trunk       1     1 3-4 99
GigabitEthernet0/0/2    trunk       1     1-4 99
GigabitEthernet0/0/3    trunk       1     1-2 4 99
GigabitEthernet0/0/4    hybrid      1     -
GigabitEthernet0/0/5    hybrid      1     -
GigabitEthernet0/0/6    hybrid      1     -
GigabitEthernet0/0/7    hybrid      1     -
GigabitEthernet0/0/8    hybrid      1     -
GigabitEthernet0/0/9    hybrid      1     -
GigabitEthernet0/0/10   hybrid      1     -
GigabitEthernet0/0/11   hybrid      1     -
GigabitEthernet0/0/12   hybrid      1     -
GigabitEthernet0/0/13   hybrid      1     -
GigabitEthernet0/0/14   hybrid      1     -
GigabitEthernet0/0/15   hybrid      1     -
GigabitEthernet0/0/16   hybrid      1     -
GigabitEthernet0/0/17   hybrid      1     -
GigabitEthernet0/0/18   hybrid      1     -
GigabitEthernet0/0/19   hybrid      1     -
GigabitEthernet0/0/20   hybrid      1     -
GigabitEthernet0/0/21   access      100   -
GigabitEthernet0/0/22   access      100   -
GigabitEthernet0/0/23   access      100   -
GigabitEthernet0/0/24   access      101   -
```

图 2.50　CORE 上使用 "display port vlan" 命令后的显示结果

```
[FLOOR1]display port vlan
Port            Link Type   PVID  Trunk VLAN List
-------------------------------------------------------------------
GigabitEthernet0/0/1     trunk    1    1 3-4 99
GigabitEthernet0/0/2     access   3    -
GigabitEthernet0/0/3     access   3    -
GigabitEthernet0/0/4     access   3    -
GigabitEthernet0/0/5     access   3    -
GigabitEthernet0/0/6     access   3    -
GigabitEthernet0/0/7     access   3    -
GigabitEthernet0/0/8     access   3    -
GigabitEthernet0/0/9     access   3    -
GigabitEthernet0/0/10    access   3    -
GigabitEthernet0/0/11    access   3    -
GigabitEthernet0/0/12    access   3    -
GigabitEthernet0/0/13    access   3    -
GigabitEthernet0/0/14    access   3    -
GigabitEthernet0/0/15    access   4    -
GigabitEthernet0/0/16    access   4    -
GigabitEthernet0/0/17    access   4    -
GigabitEthernet0/0/18    access   4    -
GigabitEthernet0/0/19    access   4    -
GigabitEthernet0/0/20    access   4    -
GigabitEthernet0/0/21    access   4    -
GigabitEthernet0/0/22    access   4    -
GigabitEthernet0/0/23    access   4    -
GigabitEthernet0/0/24    access   4    -
```

图 2.51　FLOOR1 上使用"display port vlan"命令后的显示结果

```
[FLOOR2]display port vlan
Port            Link Type   PVID  Trunk VLAN List
-------------------------------------------------------------------
GigabitEthernet0/0/1     trunk    1    1-4 99
GigabitEthernet0/0/2     access   2    -
GigabitEthernet0/0/3     access   2    -
GigabitEthernet0/0/4     access   2    -
GigabitEthernet0/0/5     access   2    -
GigabitEthernet0/0/6     access   2    -
GigabitEthernet0/0/7     access   2    -
GigabitEthernet0/0/8     access   2    -
GigabitEthernet0/0/9     access   2    -
GigabitEthernet0/0/10    access   2    -
GigabitEthernet0/0/11    access   3    -
GigabitEthernet0/0/12    access   3    -
GigabitEthernet0/0/13    access   3    -
GigabitEthernet0/0/14    access   3    -
GigabitEthernet0/0/15    access   3    -
GigabitEthernet0/0/16    access   3    -
GigabitEthernet0/0/17    access   3    -
GigabitEthernet0/0/18    access   3    -
GigabitEthernet0/0/19    access   3    -
GigabitEthernet0/0/20    access   3    -
GigabitEthernet0/0/21    access   4    -
GigabitEthernet0/0/22    access   4    -
GigabitEthernet0/0/23    access   4    -
GigabitEthernet0/0/24    access   4    -
```

图 2.52　FLOOR2 上使用"display port vlan"命令后的显示结果

```
[FLOOR3]display port vlan
Port                    Link Type   PVID   Trunk VLAN List
-------------------------------------------------------------------
GigabitEthernet0/0/1    trunk       1      1-2 4 99
GigabitEthernet0/0/2    access      2      -
GigabitEthernet0/0/3    access      2      -
GigabitEthernet0/0/4    access      2      -
GigabitEthernet0/0/5    access      2      -
GigabitEthernet0/0/6    access      2      -
GigabitEthernet0/0/7    access      2      -
GigabitEthernet0/0/8    access      2      -
GigabitEthernet0/0/9    access      2      -
GigabitEthernet0/0/10   access      2      -
GigabitEthernet0/0/11   access      2      -
GigabitEthernet0/0/12   access      2      -
GigabitEthernet0/0/13   access      2      -
GigabitEthernet0/0/14   access      2      -
GigabitEthernet0/0/15   access      2      -
GigabitEthernet0/0/16   access      2      -
GigabitEthernet0/0/17   access      4      -
GigabitEthernet0/0/18   access      4      -
GigabitEthernet0/0/19   access      4      -
GigabitEthernet0/0/20   access      4      -
GigabitEthernet0/0/21   access      4      -
GigabitEthernet0/0/22   access      4      -
GigabitEthernet0/0/23   access      4      -
GigabitEthernet0/0/24   access      4      -
```

图 2.53　FLOOR3 上使用"display port vlan"命令后的显示结果

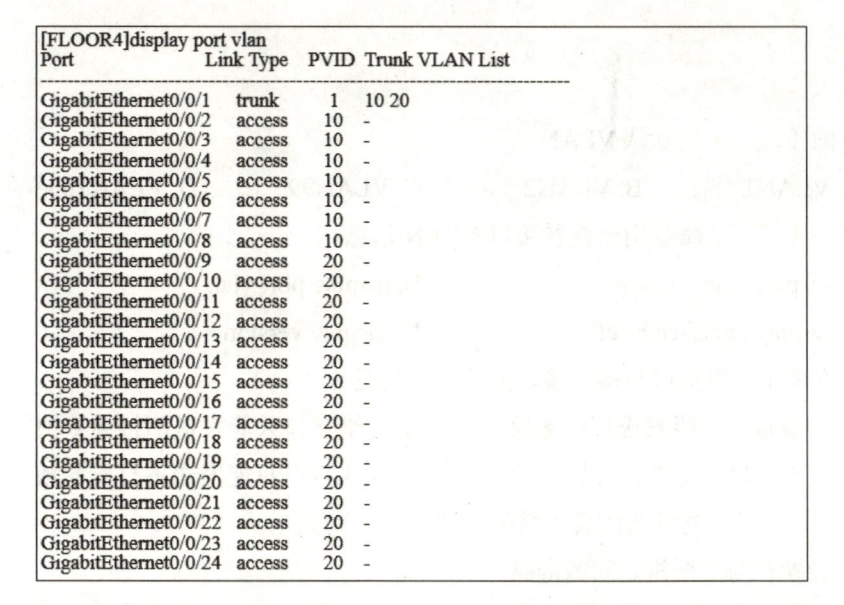

```
[FLOOR4]display port vlan
Port                    Link Type   PVID   Trunk VLAN List
-------------------------------------------------------------------
GigabitEthernet0/0/1    trunk       1      10 20
GigabitEthernet0/0/2    access      10     -
GigabitEthernet0/0/3    access      10     -
GigabitEthernet0/0/4    access      10     -
GigabitEthernet0/0/5    access      10     -
GigabitEthernet0/0/6    access      10     -
GigabitEthernet0/0/7    access      10     -
GigabitEthernet0/0/8    access      10     -
GigabitEthernet0/0/9    access      20     -
GigabitEthernet0/0/10   access      20     -
GigabitEthernet0/0/11   access      20     -
GigabitEthernet0/0/12   access      20     -
GigabitEthernet0/0/13   access      20     -
GigabitEthernet0/0/14   access      20     -
GigabitEthernet0/0/15   access      20     -
GigabitEthernet0/0/16   access      20     -
GigabitEthernet0/0/17   access      20     -
GigabitEthernet0/0/18   access      20     -
GigabitEthernet0/0/19   access      20     -
GigabitEthernet0/0/20   access      20     -
GigabitEthernet0/0/21   access      20     -
GigabitEthernet0/0/22   access      20     -
GigabitEthernet0/0/23   access      20     -
GigabitEthernet0/0/24   access      20     -
```

图 2.54　FLOOR4 上使用"display port vlan"命令后的显示结果

③ 如果使用"diplay vlan"与"display port vlan"命令检查出交换机端口的 VLAN 信息与规划配置不同，在交换机的接口视图下，执行"display this"命令查看交换机该接口的配置信息。图 2.55 所示为查看交换机 CORE 的 G0/0/1 接口的配置信息示例。图 2.56 所示为查看交换机 FLOOR1 的 G0/0/3 接口的配置信息的示例。

130 企业网络互联技术（双语）

```
[CORE-GigabitEthernet0/0/1]display this
#
interface GigabitEthernet0/0/1
 port link-type trunk
 port trunk allow-pass vlan 3 to 4 99
```

图 2.55　查看交换机 CORE 的 G0/0/1 接口的配置信息示例

```
[FLOOR1-GigabitEthernet0/0/3]display this
#
interface GigabitEthernet0/0/3
 port link-type access
 port default vlan 3
#
return
```

图 2.56　查看交换机 FLOOR1 的 G0/0/3 接口的配置信息的示例

　　如果交换机的 VLAN、交换机端口 VLAN 的配置有误，需要在交换机系统视图下先删除错误的 VLAN，再创建正确的 VLAN。在交换机接口的接口视图下删除接口的 VLAN 配置，并重新进行接口 VLAN 的配置。

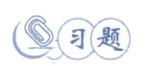 习题

选择题

（1）下面（　　）是默认 VLAN。

　　A. VLAN1　　　　　B. VLAN2　　　　　C. VLAN99　　　　　D. VLAN100

（2）下面（　　）命令用于查看接口 VLAN 信息。

　　A. display vlan　　　　　　　　　B. display port vlan

　　C. display interface brief　　　　D. display version

（3）下面关于中继接口与接入接口描述正确的是（　　）。

　　A. 中继接口只能发送打标签帧　　　B. 中继接口只能发送不打标签帧

　　C. 接入接口只能发送打标签帧　　　D. 接入接口只能发送不打标签帧

（4）下面（　　）通常被配置为接入链路。

　　A. 交换机与交换机之间的链路

　　B. 单臂路由中路由器与交换机之间的连接链路

　　C. 交换机连接主机之间的链路

（5）下面（　　）命令用于显示交换机的 VLAN 信息。

　　A. display vlan　　　　　　　　　B. display port vlan

　　C. display interface brief　　　　D. display version

项目 2　小型企业网络互联项目　131

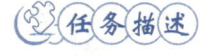　VLAN 之间通信的规划与配置

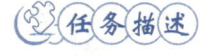 任务描述

在完成本项目任务 1 和任务 2 的基础上，根据图 2.1 所示的网络拓扑，继续完成以下任务：

① 在路由器 AR1 上完成子接口的配置，实现企业领导（VLAN10）与人力资源（VLAN20）之间的通信。

② 在路由器 AR1 上完成 DHCP 服务的配置，为企业领导（VLAN10）与人力资源（VLAN20）的主机提供自动获得 IP 地址的服务。

③ 在核心交换机 CORE 上完成 Vlanif 接口的配置，实现销售（VLAN3）、售后（VLAN4）、接入层交换机管理 VLAN（VLAN99）、企业内网服务器（VLAN100）与研发（VLAN2）之间的相互通信。

④ 在 CORE 上完成 DHCP 服务的配置，为销售（VLAN3）、售后（VLAN4）与研发（VLAN2）部门的主机提供自动获得 IP 地址的服务。

⑤ 为网络中所有的静态分配 IP 地址的设备完成 IP 的配置。

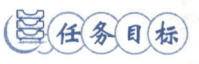 任务目标

- 理解逻辑子接口概念，掌握单臂路由的规划与配置。
- 理解三层交换机的功能，掌握使用三层交换机实现 VLAN 之间通信的规划与配置。

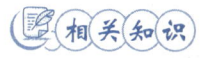 相关知识

1. 单臂路由

使用一条物理线路将实现 VLAN 之间通信的外部路由器与交换网络中的某一个交换机相连实现 VLAN 之间的通信，这种方式称为单臂路由。由于路由器只有一个接口与交换网络相连，因此需要采用 Dot1q 终结子接口，由子接口充当 VLAN 的网关。子接口（sub-interface）是物理接口中的一个逻辑接口，它是通过协议和技术将一个物理接口虚拟出来的。路由器的一个物理接口可以有多个逻辑子接口，逻辑子接口的标识是在原来的物理接口后加"."再加上一个数字，例如，"G0/0/0.2"表示物理接口"G0/0/0"的一个逻辑子接口。子接口同物理接口一样也可进行三层转发，还可以终结携带 VLAN Tag 的数据帧。

如图 2.57 所示，使用路由器 AR 实现交换网络中 VLAN2 与 VLAN3 之间的通信，其中连接路由器 AR 的交换机 SwitchA 的 G0/0/1 接口配置为 Trunk，在路由器 AR 上创建两个逻辑子接口，逻辑子接口 G0/0/0.2 终结的 VLAN 为 VLAN2，该子接口配置的 IP 地址即为 VLAN2 中主机的网关地址；逻辑子接口 G0/0/0.3 终结的 VLAN 为 VLAN3，该子接口配置的 IP 地址即为 VLAN3 中主机的网关地址。

132　企业网络互联技术（双语）

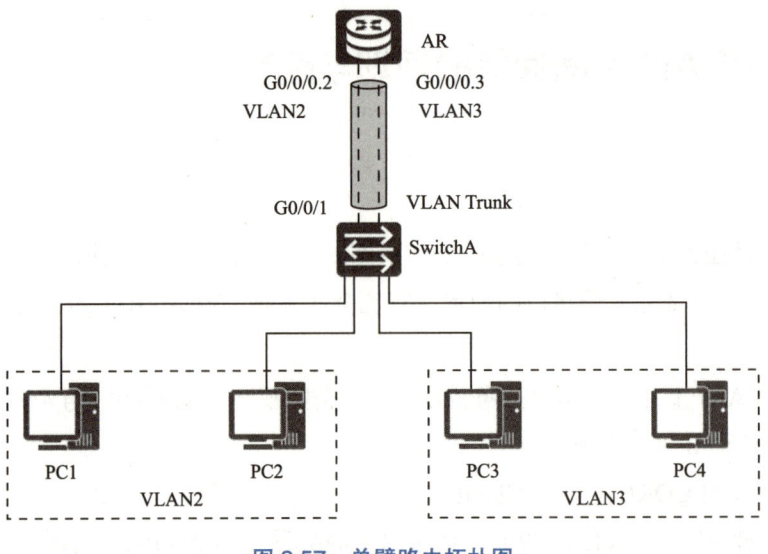

图 2.57　单臂路由拓扑图

注：逻辑子接口所属的物理接口不能配置 IP 地址。如果逻辑子接口所属的物理接口已经配置了 IP 地址，需要使用"undo ip address"命令将物理接口所配置的 IP 地址删除掉。

使用单臂路由实现 VLAN 之间通信的配置一般需要三个步骤：

① 将与路由器相连的交换机接口配置为中继接口。

② 删除与交换机相连的路由器物理接口的 IP 地址，如果路由器该物理接口禁用，还需启用该物理接口。

③ 在路由器上根据规划为每个逻辑子接口配置子接口终结的 VLAN、IP 地址与子网掩码、使能子接口的地址解析协议（address resolution protocol, ARP）广播功能。

2. Vlanif 接口

通过三层交换机的 Vlanif 接口也可以实现 VLAN 之间的通信。Vlanif 接口是一种三层的逻辑接口，每个 VLAN 对应一个 Vlanif 接口，Vlanif 接口编号与所对应的 VLAN ID 相同，如 VLAN2 对应 Vlanif2。在为 Vlanif 接口配置 IP 地址后，该接口的 IP 地址即为本 VLAN 中主机的网关。如图 2.58 所示，在三层交换机 Switch2 上创建 Vlanif2 接口和 Vlanif3 接口，Vlanif2 接口的 IP 地址为 VLAN2 中主机的网关地址，Vlanif3 接口的 IP 地址为 VLAN3 中主机的网关地址，Vlanif 接口配置简单，是实现 VLAN 间互访最常用的一种技术。

注：Vlanif1 接口是交换机自动创建的，不能删除。其他 VLAN 的 Vlanif 在系统视图下使用"interface Vlanif"命令创建，可以使用"undo interface Vlanif"命令删除 Vlanif 接口。在创建 Vlanif 接口时，要注意对应的 VLAN 在交换机中是否已存在。如果不存在，则 Vlanif 接口链路协议不能启用（Up）。

项目 2　小型企业网络互联项目　　133

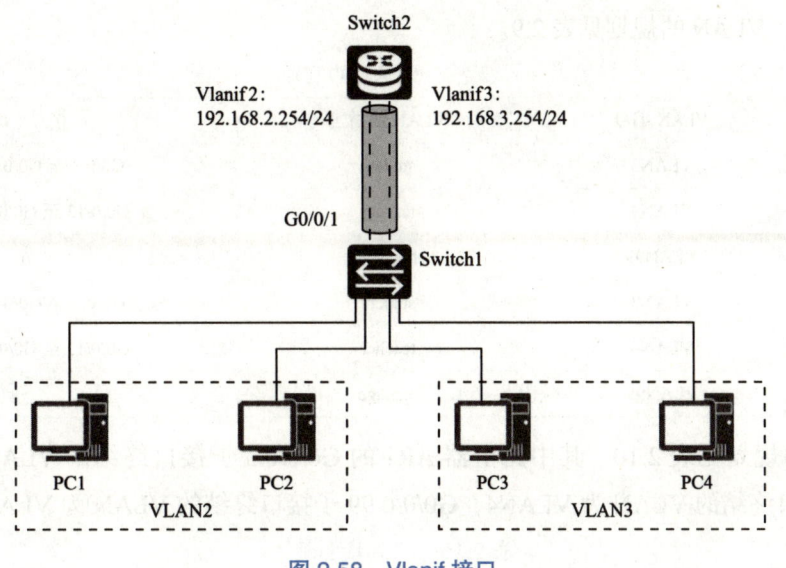

图 2.58　Vlanif 接口

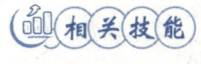

1. 使用单臂路由实现 VLAN 之间的通信

搭建如图 2.59 所示的网络拓扑图，根据表 2.8 所示的 VLAN 规划在交换机上完成 VLAN 配置，使用外部路由器 AR1 以单臂路由方式实现不同 VLAN 之间的通信，实现所有主机之间的互联互通。

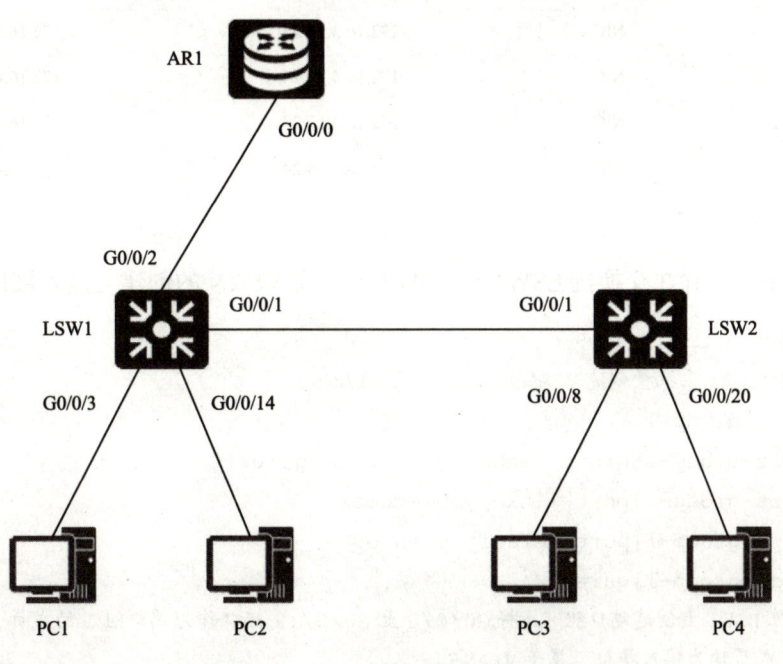

图 2.59　单臂路由网络拓扑图

交换网络共划分了三个 VLAN：VLAN3、VLAN4 与 VLAN99。其中，VLAN99 为交换机

的管理 VLAN。VLAN 的规划见表 2.9。

表 2.9　VLAN 的规划

设　　备	VLAN 编号	VLAN 的描述	接　　口
LSW1	VLAN3	student	G0/0/3 至 G0/0/12 接口
	VLAN4	teacher	G0/0/13 至 G0/0/24 接口
	VLAN99	manage	
LSW2	VLAN3	student	G0/0/2 至 G0/0/12 接口
	VLAN4	teacher	G0/0/13 至 G0/0/24 接口
	VLAN99	manage	

IP 地址的规划见表 2.10，其中路由器 AR1 的 G0/0/0.3 子接口终结的 VLAN 为 VLAN3，G0/0/0.4 子接口终结的 VLAN 为 VLAN4，G0/0/0.99 子接口终结的 VLAN 为 VLAN99。

表 2.10　IP 地址的规划

设　　备	接　　口	IP 地址与掩码	网　　关
路由器 AR1	G0/0/0.3	172.16.3.254/24	NA
	G0/0/0.4	172.16.4.254/24	NA
	G0/0/0.99	172.16.99.254/24	NA
LSW1	VLAN99	172.16.99.1/24	172.16.99.254
LSW2	VLAN99	172.16.99.2/24	172.16.99.254
PC1	NIC	172.16.3.11/24	172.16.3.254
PC2	NIC	172.16.4.12/24	172.16.4.254
PC3	NIC	172.16.3.13/24	172.16.3.254
PC4	NIC	172.16.4.14/24	172.16.4.254

（1）交换机 LSW1 与 LSW2 上 VLAN 的配置

① 参考以下的步骤在交换机 LSW1 与 LSW2 上完成 VLAN 的创建与接入接口的配置。

```
[LSW1]vlan batch 3 4 99
// 在交换机 LSW1 上批量创建 VLAN3、VLAN4 与 VLAN99
[LSW1]port-group 1
[LSW1-port-group-1]group-member g0/0/3 to g0/0/12
[LSW1-port-group-1]port link-type access
[LSW1-port-group-1]port default vlan 3
[LSW1-port-group-1]quit
// 在交换机 LSW1 上创建端口组 1，将 G0/0/3 至 G0/0/12 接口作为端口组 1 的成员，并配置端口组
//1 中所有成员均为接入接口，属于 VLAN3
[LSW1]port-group 2
[LSW1-port-group-2]group-member g0/0/13 to g0/0/24
```

```
[LSW1-port-group-2]port link-type access
[LSW1-port-group-2]port default vlan 4
// 在交换机 LSW1 上创建组端口 2，将 G0/0/13 至 G0/0/24 接口作为端口组 2 的成员，并配置端口组
//2 中所有成员均为接入接口，属于 VLAN4
[LSW2]vlan batch 3 4 99
// 在交换机 LSW2 上批量创建 VLAN3、VLAN4 与 VLAN99
[LSW2]port-group 1
[LSW2-port-group-1]group-member g0/0/2 to g0/0/12
[LSW2-port-group-1]port link-type access
[LSW2-port-group-1]port default vlan 3
[LSW2-port-group-1]quit
// 在交换机 LSW2 上创建端口组 1，将 G0/0/2 至 G0/0/12 接口作为端口组 1 的成员，并配置端口组
//1 中所有成员均为接入接口，属于 VLAN3
[LSW2]port-group 2
[LSW2-port-group-2]group-member g0/0/13 to g0/0/24
[LSW2-port-group-2]port link-type access
[LSW2-port-group-2]port default vlan 4
// 在交换机 LSW2 上创建端口组 2，将 G0/0/13 至 G0/0/24 接口作为端口组 2 的成员，并配置端口组
//2 中所有成员均为接入接口，属于 VLAN4
```

② 配置中继接口。根据图 2.59 所示的网络拓扑图以及表 2.9 所示的 VLAN 规划，图中交换机 LSW1 与交换机 LSW2 之间、交换机 LSW1 与路由器 AR1 之间的链路需要配置为中继链路，允许 VLAN3、VLAN4 与 VLAN99 通过。以华为 S5700-28C-HI 交换机为例，配置交换机 LSW1 的 G0/0/1 与 G0/0/2 接口为中继接口、LSW2 的 G0/0/1 接口为中继接口，允许 VLAN3、VLAN4 与 VLAN99 打标签通过，不允许 VLAN1 通过，配置步骤如下：

```
[LSW1]interface g0/0/1
[LSW1-GigabitEthernet0/0/1]port link-type trunk
[LSW1-GigabitEthernet0/0/1]port trunk allow-pass vlan  3 4 99
[LSW1-GigabitEthernet0/0/1]undo port trunk allow-pass vlan 1
[LSW1-GigabitEthernet0/0/1]quit
[LSW1]interface g0/0/2
[LSW1-GigabitEthernet0/0/2]port link-type trunk
[LSW1-GigabitEthernet0/0/2]port trunk allow-pass vlan 3 4 99
[LSW1-GigabitEthernet0/0/2]undo port trunk allow-pass vlan 1
[LSW1-GigabitEthernet0/0/2]quit
[LSW2]interface g0/0/1
[LSW2-GigabitEthernet0/0/1]port link-type trunk
[LSW2-GigabitEthernet0/0/1]port trunk allow-pass vlan 3 4 99
[LSW2-GigabitEthernet0/0/1]undo port trunk allow-pass vlan 1
```

```
[LSW2-GigabitEthernet0/0/1]quit
```

（2）路由器子接口的配置

在路由器 AR1 上根据表 2.9 的规划完成子接口的配置。配置步骤如下：

```
[AR1]interface g0/0/0                          // 进入 G0/0/0 接口的接口视图
[AR1-GigabitEthernet0/0/0]undo ip address      // 删除 G0/0/1 接口的 IP 地址，如果 G0/0/0 接口原来
                                               // 没有配置任何 IP 地址，则可以不配置该命令
[AR1-GigabitEthernet0/0/0]quit                 // 返回到用户视图
[AR1]interface g0/0/0.3                         // 进入 G0/0/0.3 子接口的接口视图
[AR1-GigabitEthernet0/0/0.3]dot1q termination vid 3 // 配置子接口终结的 VLAN 为 VLAN3
[AR1-GigabitEthernet0/0/0.3]ip address 172.16.3.254 24  // 配置子接口的 IP 地址与子网掩码
                                                        // 为 "172.16.3.254/24"
[AR1-GigabitEthernet0/0/0.3]arp broadcast enable // 使能子接口的 ARP 广播功能
[AR1-GigabitEthernet0/0/0.3]quit
[AR1]interface g0/0/0.4                         // 进入 G0/0/0.4 子接口的端口视图
[AR1-GigabitEthernet0/0/0.4]dot1q termination vid 4 // 配置子接口终结的 VLAN 为 VLAN4
[AR1-GigabitEthernet0/0/0.4]ip address 172.16.4.254 24  // 配置子接口的 IP 地址与子网掩码
                                                        // 为 "172.16.4.254/24"
[AR1-GigabitEthernet0/0/0.4]arp broadcast enable // 使能子端口的 ARP 广播功能
[AR1-GigabitEthernet0/0/0.4]quit
[AR1]interface g0/0/0.99                        // 进入 G0/0/0.99 子接口的端口视图
[AR1-GigabitEthernet0/0/0.99]dot1q termination vid 99 // 将子接口配置属于 VLAN3
[AR1-GigabitEthernet0/0/0.99]ip address 172.16.99.254 24// 配置子接口的 IP 地址与子网掩码
                                                        // 为 "172.16.99.254/24"
[AR1-GigabitEthernet0/0/0.99]arp broadcast enable // 启用子接口的 ARP 广播功能
[AR1-GigabitEthernet0/0/0.99]quit
```

（3）单臂路由验证与网络测试

① 在路由器 AR1 上使用"display ip interface brief"命令查看所有接口 IP 简要信息，注意观察子接口的 IP、物理与协议状态，如图 2.60 所示。

② 在路由器 AR1 上使用"display ip routing-table"命令查看路由表，如图 2.61 所示，注意观察"172.16.3.0/24"、"172.16.4.0/24"与"172.16.99.0/24"三个直连网段是否在路由表中。

③ 参考表 2.9 规划的参数完成测试主机 PC1、PC2、PC3 与 PC4 的 IP 地址、子网掩码、默认网关配置，配置完成后在 DOS 命令行下使用"ipconfig /all"命令查看其 IP 配置信息是否正确。

④ 参考表 2.9 规划的参数完成交换机 LSW1 与 LSW2 的管理 IP 地址与默认网关的配置，配置步骤如下：

```
[AR1]display ip interface brief
*down: administratively down
^down: standby
(l): loopback
(s): spoofing
The number of interface that is UP in Physical is 5
The number of interface that is DOWN in Physical is 2
The number of interface that is UP in Protocol is 4
The number of interface that is DOWN in Protocol is 3

Interface                  IP Address/Mask      Physical   Protocol
GigabitEthernet0/0/0       unassigned           up         down
GigabitEthernet0/0/0.3     172.16.3.254/24      up         up
GigabitEthernet0/0/0.4     172.16.4.254/24      up         up
GigabitEthernet0/0/0.99    172.16.99.254/24     up         up
GigabitEthernet0/0/1       unassigned           down       down
GigabitEthernet0/0/2       unassigned           down       down
NULL0                      unassigned           up         up(s)
```

图 2.60　使用"display ip interface brief"命令查看所有接口 IP 简要信息及状态

```
[AR1]display ip routing-table
Route Flags: R - relay, D - download to fib
------------------------------------------------------------------------
Routing Tables: Public
         Destinations : 13      Routes : 13

Destination/Mask      Proto    Pre   Cost   Flags   NextHop        Interface

      127.0.0.0/8     Direct    0     0       D     127.0.0.1      InLoopBack0
      127.0.0.1/32    Direct    0     0       D     127.0.0.1      InLoopBack0
127.255.255.255/32    Direct    0     0       D     127.0.0.1      InLoopBack0
     172.16.3.0/24    Direct    0     0       D     172.16.3.254   GigabitEthernet0/0/0.3
   172.16.3.254/32    Direct    0     0       D     127.0.0.1      GigabitEthernet0/0/0.3
   172.16.3.255/32    Direct    0     0       D     127.0.0.1      GigabitEthernet0/0/0.3
     172.16.4.0/24    Direct    0     0       D     172.16.4.254   GigabitEthernet0/0/0.4
   172.16.4.254/32    Direct    0     0       D     127.0.0.1      GigabitEthernet0/0/0.4
   172.16.4.255/32    Direct    0     0       D     127.0.0.1      GigabitEthernet0/0/0.4
    172.16.99.0/24    Direct    0     0       D     172.16.99.254  GigabitEthernet0/0/0.99
  172.16.99.254/32    Direct    0     0       D     127.0.0.1      GigabitEthernet0/0/0.99
  172.16.99.255/32    Direct    0     0       D     127.0.0.1      GigabitEthernet0/0/0.99
255.255.255.255/32    Direct    0     0       D     127.0.0.1      InLoopBack0
```

图 2.61　查看路由器 AR1 的路由表

```
[LSW1]interface Vlanif 99
[LSW1-Vlanif99]ip address 172.16.99.1 24
[LSW1-Vlanif99]quit
// 配置交换机 LSW1 的管理 IP 地址为"172.16.99.1/24"
[LSW1]ip route-static 0.0.0.0 0.0.0.0 172.16.99.254
// 配置交换机 LSW1 的默认网关为"172.16.99.254"
[LSW2]interface Vlanif 99
```

```
[LSW2-Vlanif99]ip address 172.16.99.2 24
[LSW2-Vlanif99]quit
// 配置交换机 LSW2 的管理 IP 地址为"172.16.99.2/24"
[LSW2]ip route-static 0.0.0.0 0.0.0.0 172.16.99.254
// 配置交换机 LSW2 的默认网关为"172.16.99.254"
```

完成交换机 LSW1 与 LSW2 的管理 IP 地址与默认网关配置后，使用"display ip interface brief"命令查看管理 VLAN 接口 IP 信息及状态。图 2.62 为在 LSW1 交换机上查看管理 VLAN 接口 IP 信息及状态的示例。使用"display ip routing-table |include Static"命令查看默认网关是否正确与生效。图 2.63 所示为查看交换机 LSW1 的默认网关。

```
[LSW1]display ip interface brief
*down: administratively down
^down: standby
(l): loopback
(s): spoofing
The number of interface that is UP in Physical is 2
The number of interface that is DOWN in Physical is 2
The number of interface that is UP in Protocol is 2
The number of interface that is DOWN in Protocol is 2

Interface             IP Address/Mask     Physical  Protocol
MEth0/0/1             unassigned          down      down
NULL0                unassigned          up        up(s)
Vlanif1              unassigned          down      down
Vlanif99             172.16.99.1/24      up        up
```

图 2.62　使用"display ip interface brief"命令查看管理 VLAN 接口 IP 信息及状态的示例

```
[LSW1]display ip routing-table | include Static
Route Flags: R - relay, D - download to fib
------------------------------------------------------------------------
Routing Tables: Public
        Destinations : 5      Routes : 5

Destination/Mask   Proto  Pre  Cost     Flags NextHop       Interface

    0.0.0.0/0   Static 60   0        RD   172.16.99.254  Vlanif99
```

图 2.63　查看交换机 LSW1 的默认网关

⑤ 在网络中的任何一台主机上，进入 DOS 命令行，使用 ping 实用程序测试到其他目标主机与交换机之间的 IP 连通性。图 2.64 所示为主机 PC1 可以 ping 通交换机 LSW1 的 IP 地址，实现了不同 VLAN 之间的通信。

项目 2　小型企业网络互联项目　**139**

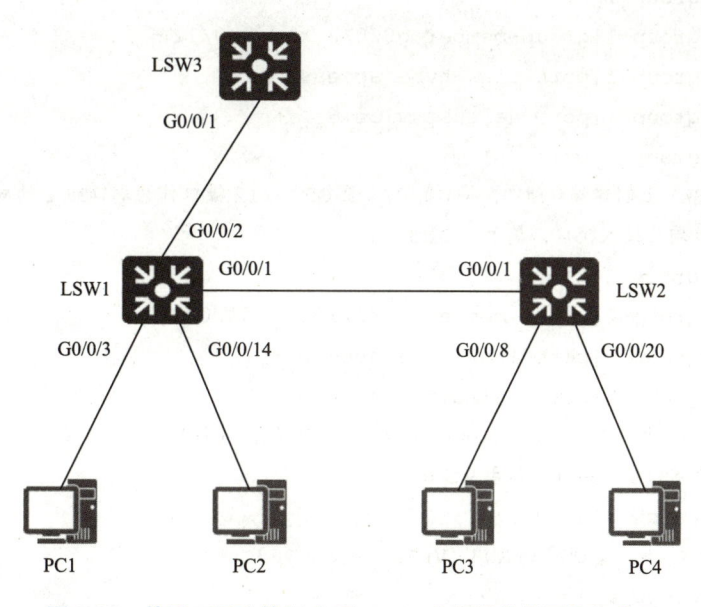

图 2.64　主机 PC1 可以 ping 通交换机 LSW1 的 IP 地址

2. 使用三层交换机实现 VLAN 之间的通信

搭建如图 2.65 所示的网络拓扑图。根据表 2.8 所示的 VLAN 规划在交换机 LSW1 与 LSW2 上完成 VLAN 配置，然后使用三层交换机 LSW3 实现 VLAN 之间的通信，实现所有主机之间的互联互通。

图 2.65　使用三层交换机实现 VLAN 之间通信的网络拓扑图

交换网络共划分了三个 VLAN：VLAN3、VLAN4 与 VLAN99，其中 VLAN99 为交换机的管理 VLAN，VLAN 的规划见表 2.9。

IP 地址的规划见表 2.11。

表 2.11　IP 地址的规划

设　　备	接　　口	IP 地址与掩码	网　　关
LSW3	Vlanif3	172.16.3.254/24	
	Vlanif4	172.16.4.254/24	
	Vlanif99	172.16.99.254/24	

续表

设　备	接　口	IP 地址与掩码	网　关
LSW1	Vlanif99	172.16.99.1/24	172.16.99.254
LSW2	Vlanif99	172.16.99.2/24	172.16.99.254
PC1	NIC	172.16.3.11/24	172.16.3.254
PC2	NIC	172.16.4.12/24	172.16.4.254
PC3	NIC	172.16.3.13/24	172.16.3.254
PC4	NIC	172.16.4.14/24	172.16.4.254

（1）交换机 LSW1 与 LSW2 上 VLAN 的配置

① 参考以下的步骤在交换机 LSW1 与 LSW2 上完成 VLAN 的创建与接入端口的配置。

```
[LSW1]vlan batch 3 4 99
// 在交换机 LSW1 上批量创建 VLAN3、VLAN4 与 VLAN99
[LSW1]port-group 1
[LSW1-port-group-1]group-member g0/0/3 to g0/0/12
[LSW1-port-group-1]port link-type access
[LSW1-port-group-1]port default vlan 3
[LSW1-port-group-1]quit
// 在交换机 LSW1 上创建端口组 1，将 G0/0/3 至 G0/0/12 接口作为端口组 1 的成员，并配置端口组
//1 中所有成员均为接入接口，属于 VLAN3
[LSW1]port-group 2
[LSW1-port-group-2]group-member g0/0/13 to g0/0/24
[LSW1-port-group-2]port link-type access
[LSW1-port-group-2]port default vlan 4
// 在交换机 LSW1 上创建端口组 2，将 G0/0/13 至 G0/0/24 接口作为端口组 2 的成员，并配置端口组
//2 中所有成员均为接入接口，属于 VLAN4
[LSW2]vlan batch 3 4 99
// 在交换机 LSW2 上批量创建 VLAN3、VLAN4 与 VLAN99
[LSW2]port-group 1
[LSW2-port-group-1]group-member g0/0/2 to g0/0/12
[LSW2-port-group-1]port link-type access
[LSW2-port-group-1]port default vlan 3
[LSW2-port-group-1]quit
// 在交换机 LSW2 上创建端口组 1，将 G0/0/2 至 G0/0/12 接口作为端口组 1 的成员，并配置端口组
//1 中所有成员均为接入接口，属于 VLAN3
[LSW2]port-group 2
[LSW2-port-group-2]group-member g0/0/13 to g0/0/24
[LSW2-port-group-2]port link-type access
[LSW2-port-group-2]port default vlan 4
```

// 在交换机 LSW2 上创建端口组 2，将 G0/0/13 至 G0/0/24 接口作为端口组 2 的成员，并配置端口组
//2 中所有成员均为接入接口，属于 VLAN4

② 配置中继接口。根据图 2.65 所示的网络拓扑图以及表 2.8 所示的 VLAN 规划，图中交换机 LSW1 与交换机 LSW2 之间、交换机 LSW1 与交换机 LSW3 之间的链路需要配置为中继链路，允许 VLAN3、VLAN4 与 VLAN99 通过。以华为 S5700-28C-HI 交换机为例，配置交换机 LSW1 的 G0/0/1 与 G0/0/2 接口为中继接口、LSW2 的 G0/0/1 接口为中继接口，允许 VLAN3、VLAN4 与 VLAN99 打标签通过，不允许 VLAN1 通过，配置步骤如下：

```
[LSW1]interface g0/0/1
[LSW1-GigabitEthernet0/0/1]port link-type trunk
[LSW1-GigabitEthernet0/0/1]port trunk allow-pass vlan  3 4 99
[LSW1-GigabitEthernet0/0/1]undo port trunk allow-pass vlan 1
[LSW1-GigabitEthernet0/0/1]quit
[LSW1]interface g0/0/2
[LSW1-GigabitEthernet0/0/2]port link-type trunk
[LSW1-GigabitEthernet0/0/2]port trunk allow-pass vlan 3 4 99
[LSW1-GigabitEthernet0/0/24]undo port trunk allow-pass vlan 1
[LSW1-GigabitEthernet0/0/2]4quit
[LSW2]interface g0/0/1
[LSW2-GigabitEthernet0/0/1]port link-type trunk
[LSW2-GigabitEthernet0/0/1]port trunk allow-pass vlan 3 4 99
[LSW2-GigabitEthernet0/0/1]undo port trunk allow-pass vlan 1
[LSW2-GigabitEthernet0/0/1]quit
```

（2）三层交换机 LSW3 的配置

① 需要在三层交换机 LSW3 创建规划的 VLAN，批量创建 VLAN3、VLAN4 与 VLAN99 的步骤如下：

```
[LSW3]vlan batch 3 4 99          // 在交换机 LSW3 上批量创建 VLAN3、VLAN4 与 VLAN99
```

② 配置中继接口。根据图 2.65 所示的网络拓扑图以及表 2.9 所示的 VLAN 规划，LSW1 与 LSW3 之间的链路为中继链路，需配置 LSW3 的 G0/0/1 接口为中继接口，允许 VLAN3、VLAN4 与 VLAN99 打标签通过，不允许 VLAN1 通过，配置步骤如下：

```
[LSW3]interface g0/0/1
[LSW3-GigabitEthernet0/0/1]port link-type trunk
[LSW3-GigabitEthernet0/0/1]port trunk allow-pass vlan  3 4 99
[LSW3-GigabitEthernet0/0/1]undo port trunk allow-pass vlan 1
[LSW3-GigabitEthernet0/0/1]quit
```

③ 配置 Vlanif2、Vlanif3 与 Vlanif99 接口。在 LSW3 上创建 Vlanif3、Vlanif4 与 Vlanif99 接

口，并配置表 2.10 规划的 IP 地址与掩码的步骤如下：

```
[LSW3]interface Vlanif 3              // 创建 Vlanif3 接口，并进入 Vlanif3 接口的接口视图
[LSW3-Vlanif3]ip address 172.16.3.254 24 // 配置接口的 IP 地址与子网掩码为"172.16.3.254/24"
[LSW3-Vlanif3]quit
[LSW3]interface Vlanif 4              // 创建 Vlanif4 接口，并进入 Vlanif4 接口的接口视图
[LSW3-Vlanif4]ip address 172.16.4.254 24 // 配置接口的 IP 地址与子网掩码为"172.16.3.254/24"
[LSW3-Vlanif4]quit
[LSW3]interface Vlanif 99             // 创建 Vlanif 99 接口，并进入 Vlanif 99 接口的
                                      // 接口视图
[LSW3-Vlanif99]ip address 172.16.99.254 24 // 配置接口的 IP 地址与子网掩码为"172.16.99.254/24"
[LSW3-Vlanif99]quit
```

（3）验证与网络测试

① 在三层交换机 LSW3 上使用"display ip interface brief"命令查看所有接口 IP 简要信息，注意观察 Vlanif3、Vlanif4 与 Vlanif99 接口的 IP 信息、物理与协议状态，如图 2.66 所示。

```
[LSW3]display ip interface brief
*down: administratively down
^down: standby
(l): loopback
(s): spoofing
The number of interface that is UP in Physical is 4
The number of interface that is DOWN in Physical is 2
The number of interface that is UP in Protocol is 4
The number of interface that is DOWN in Protocol is 2

Interface          IP Address/Mask    Physical   Protocol
MEth0/0/1          unassigned         down       down
NULL0              unassigned         up         up(s)
Vlanif1            unassigned         down       down
Vlanif3            172.16.3.254/24    up         up
Vlanif4            172.16.4.254/24    up         up
Vlanif99           172.16.99.254/24   up         up
```

图 2.66　使用"display ip interface brief"命令查看接口 IP 简要信息及状态

② 在三层交换机上使用"display ip routing-table"命令查看路由表，如图 2.67 所示，注意观察"172.16.3.0/24"、"172.16.4.0/24"与"172.16.99.0/24"三个直连网段是否在路由表中。

```
[LSW3]display ip routing-table
Route Flags: R - relay, D - download to fib
------------------------------------------------------------------
Routing Tables: Public
        Destinations : 8      Routes : 8

Destination/Mask    Proto  Pre  Cost   Flags  NextHop        Interface

       127.0.0.0/8   Direct  0    0      D     127.0.0.1      InLoopBack0
      127.0.0.1/32   Direct  0    0      D     127.0.0.1      InLoopBack0
      172.16.3.0/24  Direct  0    0      D     172.16.3.254   Vlanif3
    172.16.3.254/32  Direct  0    0      D     127.0.0.1      Vlanif3
      172.16.4.0/24  Direct  0    0      D     172.16.4.254   Vlanif4
    172.16.4.254/32  Direct  0    0      D     127.0.0.1      Vlanif4
     172.16.99.0/24  Direct  0    0      D     172.16.99.254  Vlanif99
   172.16.99.254/32  Direct  0    0      D     127.0.0.1      Vlanif99
```

图 2.67　查看三层交换机 LSW3 的路由表

③ 参考表 2.10 规划的参数完成测试主机 PC1、PC2、PC3 与 PC4 的 IP 地址、子网掩码、默认网关配置，配置完成后在 DOS 命令行下使用 "ipconfig /all" 命令查看其 IP 配置信息是否正确。

④ 参考表 2.10 规划的参数完成交换机 LSW1 与 LSW2 的管理 IP 地址与默认网关的配置，配置步骤如下：

```
[LSW1]interface Vlanif 99
[LSW1-Vlanif99]ip address 172.16.99.1 24
[LSW1-Vlanif99]quit
// 配置交换机 LSW1 的管理 IP 地址为 "172.16.99.1/24"
[LSW1]ip route-static 0.0.0.0 0.0.0.0 172.16.99.254
// 配置交换机 LSW1 的默认网关为 "172.16.99.254"
[LSW2]interface Vlanif 99
[LSW2-Vlanif99]ip address 172.16.99.2 24
[LSW2-Vlanif99]quit
// 配置交换机 LSW2 的管理 IP 地址为 "172.16.99.2/24"
[LSW2]ip route-static 0.0.0.0 0.0.0.0 172.16.99.254
// 配置交换机 LSW2 的默认网关为 "172.16.99.254"
```

完成交换机 LSW1 与 LSW2 的管理 IP 地址与默认网关配置后，使用 "display ip interface brief" 命令查看管理 VLAN 接口 IP 信息及状态。图 2.68 所示为在交换机 LSW1 上查看管理 VLAN 接口 IP 信息及状态的示例。使用 "display ip routing-table |include Static" 命令查看默认网关是否正确与生效。图 2.69 所示为查看交换机 LSW1 上的默认网关。

⑤ 在网络中的任何一台主机上，进入 DOS 命令行，使用 ping 实用程序测试到其他目标主机与交换机之间的 IP 连通性。图 2.70 所示为主机 PC1 可以 ping 通交换机 LSW1 的 IP 地址，实现了不同 VLAN 之间的通信。

```
[LSW1]display ip interface brief
*down: administratively down
^down: standby
(l): loopback
(s): spoofing
The number of interface that is UP in Physical is 2
The number of interface that is DOWN in Physical is 2
The number of interface that is UP in Protocol is 2
The number of interface that is DOWN in Protocol is 2

Interface              IP Address/Mask    Physical  Protocol
MEth0/0/1              unassigned         down      down
NULL0                 unassigned         up        up(s)
Vlanif1               unassigned         down      down
Vlanif99              172.16.99.1/24     up        up
```

图 2.68　在交换机 LSW1 上查看管理 VLAN 接口 IP 信息及状态的示例

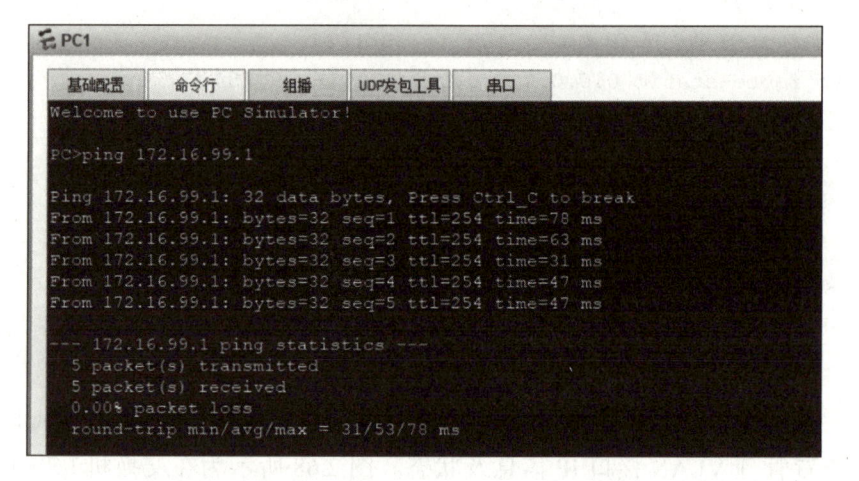

```
[LSW1]display ip routing-table | include Static
Route Flags: R - relay, D - download to fib
------------------------------------------------------------------------------
Routing Tables: Public
      Destinations : 5        Routes : 5

Destination/Mask   Proto  Pre  Cost     Flags NextHop        Interface

      0.0.0.0/0   Static  60   0          RD   172.16.99.254   Vlanif99
```

图 2.69　查看交换机 LSW1 上的默认网关

```
PC1
基础配置 | 命令行 | 组播 | UDP发包工具 | 串口
Welcome to use PC Simulator!

PC>ping 172.16.99.1

Ping 172.16.99.1: 32 data bytes, Press Ctrl_C to break
From 172.16.99.1: bytes=32 seq=1 ttl=254 time=78 ms
From 172.16.99.1: bytes=32 seq=2 ttl=254 time=63 ms
From 172.16.99.1: bytes=32 seq=3 ttl=254 time=31 ms
From 172.16.99.1: bytes=32 seq=4 ttl=254 time=47 ms
From 172.16.99.1: bytes=32 seq=5 ttl=254 time=47 ms

--- 172.16.99.1 ping statistics ---
  5 packet(s) transmitted
  5 packet(s) received
  0.00% packet loss
  round-trip min/avg/max = 31/53/78 ms
```

图 2.70　主机 PC1 可以 ping 通交换机 LSW1 的 IP 地址

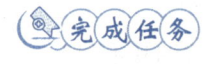

完成任务

1. 任务规划

根据任务描述和项目 2 的建网需求，表 2.12 给出了 VLAN 之间通信的 IP 地址参考规划。其中，路由器 AR1 的 G0/0/0.10 子接口终结的 VLAN 为 VLAN10，G0/0/0.20 子接口终结的 VLAN 为 VLAN20。

表 2.12　IP 地址的规划

设　备	接　口	IP 地址与掩码	网　关
路由器 AR1	G0/0/0	172.16.254.1/30	NA
	G0/0/1.10	172.16.10.254/24	NA
	G0/0/1.20	172.16.20.254/24	NA
	S4/0/0	125.71.28.93/24	NA
路由器 ISP	S4/0/0	125.71.28.1/24	NA
	G0/0/0	210.33.44.254/24	NA

续表

设　备	接　口	IP 地址与掩码	网　关
CORE	Vlanif2	172.16.2.254/24	NA
	Vlanif3	172.16.3.254/24	NA
	Vlanif4	172.16.4.254/24	NA
	Vlanif99	172.16.99.254/24	NA
	Vlanif100	172.16.100.254/24	NA
	Vlanif101	172.16.254.2/30	NA
FLOOR1	Vlanif99	172.16.99.1/24	172.16.99.254
FLOOR2	Vlanif99	172.16.99.2/24	172.16.99.254
FLOOR3	Vlanif99	172.16.99.3/24	172.16.99.254
Web 服务器	NIC	172.16.100.1/24	172.16.100.254
FTP 服务器	NIC	172.16.100.2/24	172.16.100.254
DNS 服务器	NIC	172.16.100.3/24	172.16.100.254
WWW	NIC	210.33.44.1/24	210.33.44.254
CLIENT	NIC	210.33.44.2/24	210.33.44.254

路由器 AR1 与交换机 CORE 上的 DHCP 地址池的规划见表 2.13。

表 2.13　DHCP 地址池的规划

设　备	地　址　池	网　络　号	网　关	DNS 地址	地址租期	排除地址
AR1	vlan10	172.16.10.0/24	172.16.10.254	172.16.100.3	8 h	172.16.10.50 ～ 172.16.10.70
	vlan20	172.16.20.0/24	172.16.20.254	172.16.100.3	8 h	
CORE	vlan2	172.16.2.0/24	172.16.2.254	172.16.100.3	8 h	
	vlan3	172.16.3.0/24	172.16.3.254	172.16.100.3	8 h	
	vlan4	172.16.4.0/24	172.16.4.254	172.16.100.3	8 h	

2. 任务实施

（1）路由器 AR1 的配置

① 根据表 2.12 所示的 IP 地址规划完成路由器 AR1 的子接口与接口 IP 的配置步骤如下：

```
[AR1]interface g0/0/1
[AR1-GigabitEthernet0/0/1]undo ip address
// 删除 G0/0/0 接口的 IP 地址
[AR1-GigabitEthernet0/0/1]quit
[AR1]interface g0/0/1.10
[AR1-GigabitEthernet0/0/1.10]dot1q termination vid 10
[AR1-GigabitEthernet0/0/1.10]ip address 172.16.10.254 24
```

```
[AR1-GigabitEthernet0/0/1.10]arp broadcast enable
[AR1-GigabitEthernet0/0/1.10]quit
// 创建子接口 G0/0/1.10, 配置子接口终结的 VLAN 为 VLAN10, IP 地址与子网掩码为 "172.16.
//10.254/24", 并使能子接口的 ARP 广播功能
[AR1]interface g0/0/1.20
[AR1-GigabitEthernet0/0/1.20]dot1q termination vid 20
[AR1-GigabitEthernet0/0/1.20]ip address 172.16.20.254 24
[AR1-GigabitEthernet0/0/1.20]arp broadcast enable
[AR1-GigabitEthernet0/0/1.20]quit
// 创建子接口 G0/0/1.20, 配置子接口终结的 VLAN 为 VLAN20, IP 地址与子网掩码为 "172.16.
//20.254/24", 并使能子接口的 ARP 广播功能
[AR1]interface g0/0/0
[AR1-GigabitEthernet0/0/0]ip address 172.16.254.1 30
[AR1-GigabitEthernet0/0/0]quit
// 配置 G0/0/0 接口的 IP 地址与子网掩码为 "172.16.254.1/30"
[AR1]interface s4/0/0
[AR1-Serial4/0/0]ip address 125.71.28.93 24
[AR1-Serial4/0/0]quit
// 配置 S4/0/0 接口的 IP 地址与子网掩码为 "125.71.28.93/24"
```

② DHCP 的配置。根据表 2.13, 在路由器 AR1 上配置 DHCP 的步骤如下:

```
[AR1]dhcp enable
// 启用 DHCP 服务
[AR1]ip pool vlan10
[AR1-ip-pool-vlan10]network 172.16.10.0 mask 24
[AR1-ip-pool-vlan10]gateway-list 172.16.10.254
[AR1-ip-pool-vlan10]dns-list 172.16.100.3
[AR1-ip-pool-vlan10]excluded-ip-address 172.16.10.50 172.16.10.70
[AR1-ip-pool-vlan10]lease day 0 hour 8
[AR1-ip-pool-vlan10]quit
// 配置 DHCP 地址池 "vlan10", 网络号为 "172.16.10.0/24"; 网关为 "172.16.10.254";
//DNS 地址为 "172.16.100.3", 地址租期为 8 h, 排除地址为 "172.16.10.50~172.16.10.70"
[AR1]interface g0/0/1.10
[AR1-GigabitEthernet0/0/1.10]dhcp select global
[AR1-GigabitEthernet0/0/1.10]quit
// 开启 G0/0/1.10 子接口采用全局地址池的 DHCP Server 功能
[AR1]ip pool vlan20
[AR1-ip-pool-vlan20]network 172.16.20.0 mask 24
[AR1-ip-pool-vlan20]gateway-list 172.16.20.254
[AR1-ip-pool-vlan20]dns-list 172.16.100.3
```

```
[AR1-ip-pool-vlan20]lease day 0 hour 8

[AR1-ip-pool-vlan20]quit

// 配置 DHCP 地址池 "vlan20", 网络号为 "172.16.20.0/24"; 网关为 "172.16.20.254";

//DNS 为 "172.16.100.3", 地址租期为 8 h

[AR1]interface g0/0/1.20

[AR1-GigabitEthernet0/0/1.20]dhcp select global

[AR1-GigabitEthernet0/0/1.20]quit

// 开启 G0/0/1.20 子接口采用全局地址池的 DHCP Server 功能
```

（2）交换机 CORE 的配置

① 在交换机 CORE 上创建 Vlanif2、Vlanif3、Vlanif4、Vlanif99、Vlanif100 和 Vlanif101 接口，并配置表 2.12 规划的 IP 地址与掩码的步骤如下：

```
[CORE]interface Vlanif 2

[CORE-Vlanif2]ip address 172.16.2.254 24

[CORE-Vlanif2]quit

// 创建并进入 Vlanif2 接口, 配置 Vlanif2 接口的 IP 地址与子网掩码为 "172.16.2.254/24"

[CORE]interface Vlanif 3

[CORE-Vlanif3]ip address 172.16.3.254 24

[CORE-Vlanif3]quit

// 创建并进入 Vlanif3 接口, 配置 Vlanif3 接口的 IP 地址与子网掩码为 "172.16.3.254/24"

[CORE]interface Vlanif 4

[CORE-Vlanif4]ip address 172.16.4.254 24

[CORE-Vlanif4]quit

// 创建并进入 Vlanif4 接口, 配置 Vlanif4 接口的 IP 地址与子网掩码为 "172.16.4.254/24"

[CORE]interface Vlanif 99

[CORE-Vlanif99]ip address 172.16.99.254 24

[CORE-Vlanif99]quit

// 创建并进入 Vlanif99 接口, 配置 Vlanif99 接口的 IP 地址与子网掩码为 "172.16.99.254/24"

[CORE]interface Vlanif 100

[CORE-Vlanif100]ip address 172.16.100.254 24

[CORE-Vlanif100]quit

// 创建并进入 Vlanif100 接口, 配置 Vlanif100 接口的 IP 地址与子网掩码为 "172.16.100.254/24"

[CORE]interface Vlanif 101

[CORE-Vlanif101]ip address 172.16.254.2 30

[CORE-Vlanif101]quit

// 创建并进入 Vlanif101 接口, 配置 Vlanif101 接口的 IP 地址与子网掩码为 "172.16.254.2/30"
```

148 企业网络互联技术（双语）

② DHCP 的配置。根据表 2.13，在交换机 CORE 上配置 DHCP 的步骤如下：

```
[CORE]dhcp enable
// 启用 DHCP 服务
[CORE]ip pool vlan2
[CORE-ip-pool-vlan2]network 172.16.2.0 mask 24
[CORE-ip-pool-vlan2]gateway-list 172.16.2.254
[CORE-ip-pool-vlan2]dns-list 172.16.100.3
[CORE-ip-pool-vlan2]lease day 0 hour 8
[CORE-ip-pool-vlan2]quit
// 配置 DHCP 地址池 "vlan2"，网络号为 "172.16.2.0/24"；网关为 "172.16.2.254"；DNS
// 地址为 "172.16.100.3"，地址租期为 8 h
[CORE]interface Vlanif 2
[CORE-Vlanif2]dhcp select global
[CORE-Vlanif2]quit
// 开启 Vlanif2 接口采用全局地址池的 DHCP Server 功能
[CORE]ip pool vlan3
[CORE-ip-pool-vlan3]network 172.16.3.0 mask 24
[CORE-ip-pool-vlan3]gateway-list 172.16.3.254
[CORE-ip-pool-vlan3]dns-list 172.16.100.3
[CORE-ip-pool-vlan3]lease day 0 hour 8
[CORE-ip-pool-vlan3]quit
// 配置 DHCP 地址池 "vlan3"，网络号为 "172.16.3.0/24"；网关为 "172.16.3.254"；DNS
// 地址为 "172.16.100.3"，地址租期为 8 h
[CORE]interface Vlanif 3
[CORE-Vlanif3]dhcp select global
[CORE-Vlanif3]quit
// 开启 Vlanif3 接口采用全局地址池的 DHCP Server 功能
[CORE]ip pool vlan4
[CORE-ip-pool-vlan4]network 172.16.4.0 mask 24
[CORE-ip-pool-vlan4]gateway-list 172.16.4.254
[CORE-ip-pool-vlan4]dns-list 172.16.100.3
[CORE-ip-pool-vlan4]lease day 0 hour 8
[CORE-ip-pool-vlan4]quit
// 配置 DHCP 地址池 "vlan4"，网络号为 "172.16.4.0/24"；网关为 "172.16.4.254"；DNS
// 地址为 "172.16.100.3"，地址租期为 8 h
[CORE]interface Vlanif 4
[CORE-Vlanif4]dhcp select global
[CORE-Vlanif4]quit
// 开启 Vlanif4 接口采用全局地址池的 DHCP Server 功能
```

（3）接入层交换机管理 IP 地址的配置

根据表 2.12，在接入层交换机 FLOOR1、FLOOR2 与 FLOOR3 上参考以下的步骤完成管理 IP 与默认网关的配置。

```
[FLOOR1]interface Vlanif 99
[FLOOR1-Vlanif99]ip address 172.16.99.1 24
[FLOOR1]quit
[FLOOR1]ip route-static 0.0.0.0 0.0.0.0 172.16.99.254
// 配置交换机 FLOOR1 的管理 IP 地址为"172.16.99.1/24"，默认网关为"172.16.99.254"
[FLOOR2]interface Vlanif 99
[FLOOR2-Vlanif99]ip address 172.16.99.2 24
[FLOOR2]quit
[FLOOR2]ip route-static 0.0.0.0 0.0.0.0 172.16.99.254
// 配置交换机 FLOOR2 的管理 IP 地址为"172.16.99.2/24"，默认网关为"172.16.99.254"
[FLOOR3]interface Vlanif 99
[FLOOR3-Vlanif99]ip address 172.16.99.3 24
[FLOOR3]quit
[FLOOR3]ip route-static 0.0.0.0 0.0.0.0 172.16.99.254
// 配置交换机 FLOOR3 的管理 IP 地址为"172.16.99.3/24"，默认网关为"172.16.99.254"
```

（4）验证与网络测试

在完成任务配置后，理论上企业领导（VLAN10）、人力资源（VLAN20）、研发（VLAN2）、销售（VLAN3）和售后（VLAN4）部门的测试主机均可以从 DHCP 服务器获得 IP 地址。企业领导（VLAN10）和人力资源（VLAN20）的测试主机（PC1 与 PC2）相互之间 ping 通；研发（VLAN2）、销售（VLAN3）、售后（VLAN4）部门的测试主机（PC3、PC4、PC5）、企业内网服务器（VLAN100）、FLOOR1 的管理地址、FLOOR2 的管理地址与 FLOOR3 的管理地址之间可以 ping 通。

① 使用"display ip interface brief"命令查看接口 IP 简要信息及状态，注意观察路由器配置的接口与子接口的 IP 地址与子网掩码、物理接口与链路协议的状态，如图 2.71 所示。

② 进入路由器 AR1 子接口，查看子接口的配置。图 2.72 所示为路由器 AR1 子接口 G0/0/1.10 的配置。

③ 将测试主机 PC1 与 PC2 的 IP 地址与 DNS 地址配置为自动获取，然后在命令行下使用"ipconfig"命令查看其获得的 IP 信息，并使用 ping 命令测试 PC1 与 PC2 之间的 IP 连通性。图 2.73 所示为在 PC1 上查看 IP 信息并 ping 通了 PC2 的 IP 地址的示例。

④ 在路由器上使用"display ip pool name vlan10 used"命令查看地址池"vlan10"中已使用的地址情况，如图 2.74 所示。图中显示"172.16.10.253"已动态分配给 MAC 地址为"5489-9872-427a"的主机（即测试主机 PC1）。

150 企业网络互联技术（双语）

```
[AR1]display ip interface brief
*down: administratively down
^down: standby
(l): loopback
(s): spoofing
The number of interface that is UP in Physical is 6
The number of interface that is DOWN in Physical is 2
The number of interface that is UP in Protocol is 5
The number of interface that is DOWN in Protocol is 3

Interface                    IP Address/Mask      Physical    Protocol
GigabitEthernet0/0/0         172.16.254.1/30      up          up
GigabitEthernet0/0/1         unassigned           up          down
GigabitEthernet0/0/1.10      172.16.10.254/24     up          up
GigabitEthernet0/0/1.20      172.16.20.254/24     up          up
GigabitEthernet0/0/2         unassigned           down        down
NULL0                        unassigned           up          up(s)
Serial4/0/0                  125.71.28.93/24      up          up
Serial4/0/1                  unassigned           down        down
```

图 2.71　使用"display ip interface brief"命令查看接口 IP 简要信息及状态

```
[AR1-GigabitEthernet0/0/1.10]display this
[V200R003C00]
#
interface GigabitEthernet0/0/1.10
 dot1q termination vid 10
 ip address 172.16.10.254 255.255.255.0
 arp broadcast enable
 dhcp select global
#
return
[AR1-GigabitEthernet0/0/1.10]
```

图 2.72　查看路由器 AR1 子接口 G0/0/1.10 的配置

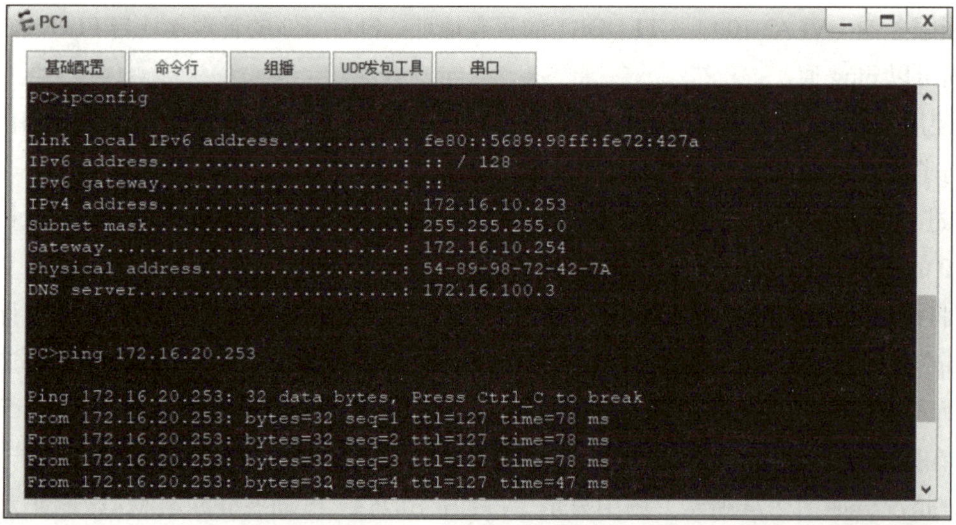

图 2.73　在 PC1 上查看 IP 信息并 ping 通了 PC2 的 IP 地址的示例

项目 2　小型企业网络互联项目　151

```
[AR1]display ip pool name vlan10 used
  Pool-name    : vlan10
  Pool-No     : 0
  Lease      : 0 Days 8 Hours 0 Minutes
  Domain-name   : -
  DNS-server0   : 172.16.100.3
  NBNS-server0  : -
  Netbios-type  : -
  Position    : Local    Status     : Unlocked
  Gateway-0    : 172.16.10.254
  Mask       : 255.255.255.0
  VPN instance  : --

   Start      End    Total Used Idle(Expired) Conflict Disable
  ------------------------------------------------------------------
   172.16.10.1 172.16.10.254 253   1     231(0)     0     21
  ------------------------------------------------------------------

  Network section :
  ------------------------------------------------------------------
  Index      IP      MAC     Lease  Status
  ------------------------------------------------------------------
   252  172.16.10.253  5489-9872-427a     163  Used
```

图 2.74　查看地址池 "vlan10" 中已使用的地址情况

⑤ 在交换机 CORE 上使用 "display ip interface brief" 命令查看接口 IP 简要信息，观察 Vlanif 接口的 IP 地址与子网掩码、物理接口与链路协议的状态，如图 2.75 所示。

```
[CORE]display ip interface brief
*down: administratively down
^down: standby
(l): loopback
(s): spoofing
The number of interface that is UP in Physical is 8
The number of interface that is DOWN in Physical is 1
The number of interface that is UP in Protocol is 7
The number of interface that is DOWN in Protocol is 2

Interface         IP Address/Mask    Physical  Protocol
MEth0/0/1         unassigned        down    down
NULL0           unassigned        up     up(s)
Vlanif1          unassigned        up     down
Vlanif2          172.16.2.254/24     up     up
Vlanif3          172.16.3.254/24     up     up
Vlanif4          172.16.4.254/24     up     up
Vlanif99         172.16.99.254/24    up     up
Vlanif100         172.16.100.254/24   up     up
Vlanif101         172.16.254.2/30     up     up
```

图 2.75　使用 "display ip interface brief" 命令查看接口 IP 简要信息及状态

⑥ 将测试主机 PC3、PC4 与 PC5 的 IP 地址与 DNS 地址配置为自动获取，然后在命令行下使用 "ipconfig" 命令查看其是否正确获得 IP 信息。图 2.76 所示为测试主机 PC3 所获得的 IP 地址信息。

152　企业网络互联技术（双语）

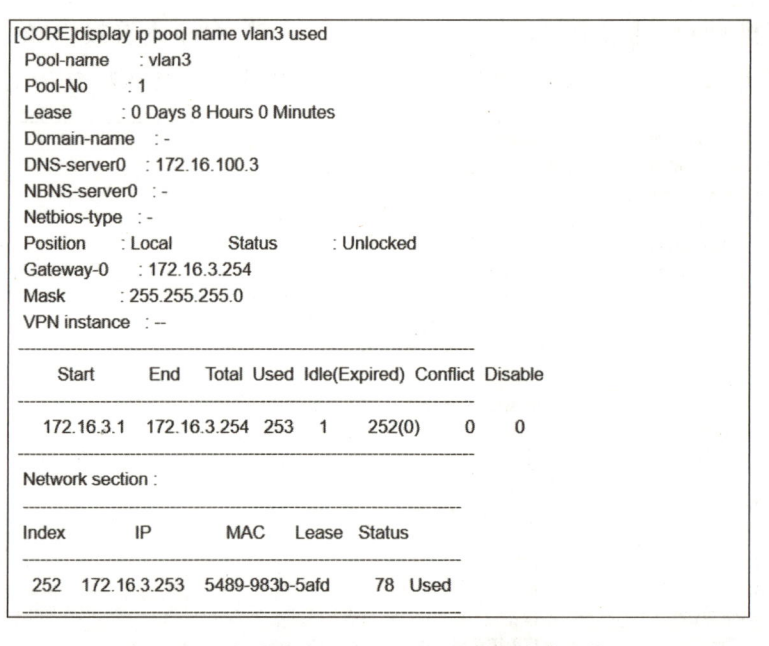

图 2.76　测试主机 PC3 所获得的 IP 地址信息

⑦ 测试主机 PC3 自动获得 IP 地址后，在交换机 CORE 上使用"display ip pool name vlan3 used"命令查看地址池"vlan3"中已使用的地址情况，如图 2.77 所示。图 2.77 中显示"172.16.3.253"已动态分配给 MAC 地址为"5489-983b-5afd"的主机（即测试主机 PC3）。同样，在测试主机 PC4 自动获得 IP 地址后，使用"display ip pool name vlan2 used"命令查看地址池"vlan2"中已使用的地址情况；在测试主机 PC5 自动获得 IP 地址后，使用"display ip pool name vlan4 used"命令查看地址池"vlan4"中已使用的地址情况。

```
[CORE]display ip pool name vlan3 used
 Pool-name    : vlan3
 Pool-No      : 1
 Lease        : 0 Days 8 Hours 0 Minutes
 Domain-name  : -
 DNS-server0  : 172.16.100.3
 NBNS-server0 : -
 Netbios-type : -
 Position     : Local      Status       : Unlocked
 Gateway-0    : 172.16.3.254
 Mask         : 255.255.255.0
 VPN instance : --
 _____

   Start      End     Total Used Idle(Expired) Conflict Disable
 _____

 172.16.3.1  172.16.3.254  253   1    252(0)       0       0

 Network section :
 _____

 Index     IP         MAC      Lease  Status
 _____

 252  172.16.3.253  5489-983b-5afd   78  Used
```

图 2.77　查看地址池"vlan3"已使用的地址情况

⑧ 根据表 2.12，设置 DNS 服务器、Web 服务器与 FTP 服务器的 IP 地址、子网掩码、默认网关以及 DNS 地址，并在 DOS 命令行下使用"ipconfig"命令查看设置的 IP 信息是否生效正确。

⑨ 在 PC3 上，使用 ping 命令测试到 PC4、PC5、FLOOR1、FLOOR2、FLOOR3 以及与服务器之间的 IP 连通性。图 2.78 所示为主机 PC3 成功 ping 通交换机 FLOOR2。

项目 2　小型企业网络互联项目　153

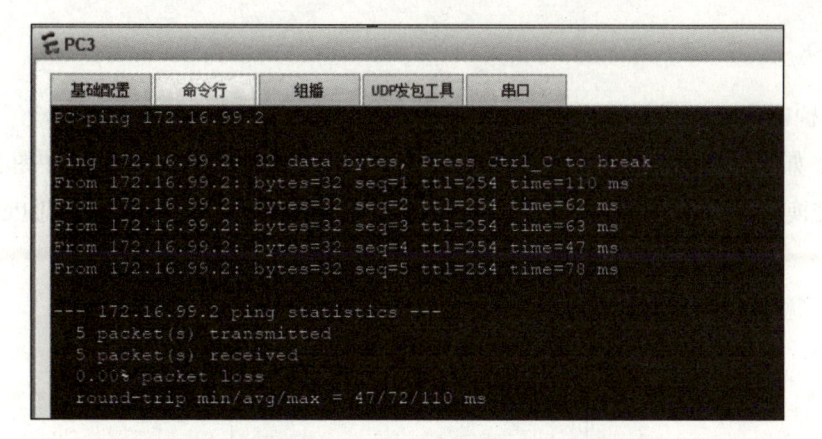

图 2.78　主机 PC3 成功 ping 通交换机 FLOOR2

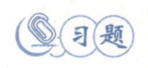

选择题

（1）下面（　　　）是路由器 G0/0/0 接口的子接口。

A. G0/0/0. 1　　　　B. G0/0/0-1　　　　C. G0/0/1. 0　　　　D. G0/0/1-1

（2）单臂路由配置路由器的子接口时，下面（　　　）不能进行配置。

A. 配置子接口终结的 VLAN　　　　B. 配置子接口的 IP 地址与子网掩码

C. 使能子接口的 ARP 广播功能　　　　D. 为子接口所属物理接口配置 IP 地址

（3）接入层交换机的管理地址配置在（　　　）接口上。

A. Vlanif1　　　　B. Vlanif99　　　　C. Vlanif100　　　　D. 管理 VLAN

任务 4　生成树的规划与配置

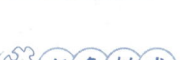

在完成本项目任务 1 到任务 3 的基础上，根据图 2.1 所示的网络拓扑，在企业内网交换机上部署生成树协议，以避免企业内网出现二层环路。

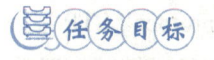

- 理解 STP 的基本概念与工作原理。
- 掌握 STP 的配置。
- 掌握快速生成树的配置。

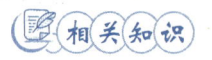

1. 生成树概念

网络中，如果网络设备之间的链路没有冗余，就会出现单点故障。例如在图 2.79 所示的网络中，如果交换机 LSW1 与交换机 LSW2 的链路出现故障，则主机 PC1 与主机 PC2 不再具有网络的连通性。

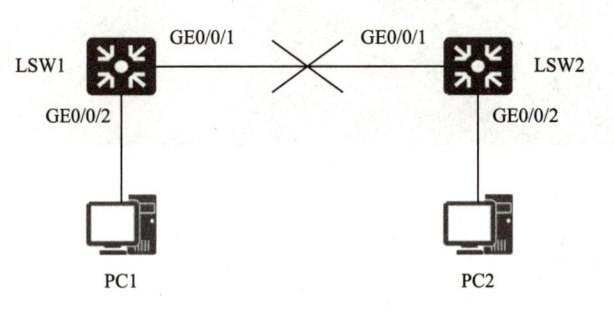

图 2.79　无冗余的网络拓扑示例

为了消除由于单点故障引起的网络中断，可以采用冗余链路来加强网络的可用性。具有冗余路径和冗余设备的网络会拥有更长的网络正常运行时间。在图 2.80 所示的具有冗余链路的网络中，当交换机 LSW1 的 G0/0/1 接口与交换机 LSW2 的 G0/0/1 接口相连的链路出现故障时，PC1 到 PC2 的流量仍然会通过图 2.80 箭头所示的冗余链路进行通信，从而保证网络的可靠性。

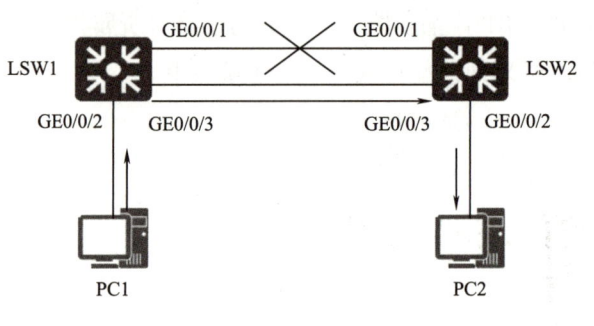

图 2.80　有冗余的网络拓扑示例

链路冗余是交换网络设计中非常重要的组成部分。通过冗余设计，可以有效保证网络的可用性，但是根据交换机的工作原理，会引起第二层环路问题、广播风暴问题、交换机 MAC 地址表不稳定问题和重复单播帧问题。

为了解决交换网络中的环路问题，IEEE 提出了生成树协议（spanning tree protocol, STP），这里是一个广义概念，包括 STP、RSTP 和 MSTP。通过阻塞交换机某些端口流量，打破二层数据环路。生成树协议具有很多类型或变体，其发展历程经历了以下三代：

（1）STP

IEEE 于 1990 年以 IEEE 802.1D 标准形式发布首个 STP 技术。STP 为单生成树协议，即交换网络中所有 VLAN 都共享同一个生成树实例，不能实现二层流量的负载均衡。

（2）RSTP（rapid spanning tree protocol，快速生成树协议）

RSTP 也是单生成树协议，IEEE 标准为 802.1W，是 STP 的升级版本，与 STP 兼容。在 IEEE 802.1D 标准基础上进行了改进，采取减少端口状态、增加端口角色、改变配置 BPDU（网桥协议数据单元）的发送方式等措施，实现当网络拓扑发生变化时，网络可以更为快速的收敛。

（3）MSTP（multiple spanning tree protocol，多生成树协议）

由于生成树概念的提出早于 VLAN 概念的提出，因此在 STP 的实现过程中并没有考虑 VLAN 的因素。RSTP 只是对 STP 的收敛机制进行改进，同样也没有考虑到 VLAN 的因素，所以 STP 与 RSTP 都属于单生成树协议，即在交换网络中所有 VLAN 共享一棵相同的生成树，这样就可能导致网络宽带的浪费。

IEEE 提出的 802.1S 标准定义的 MSTP 基于生成树实例（instance）计算出多棵生成树。在一个 MSTP 域中，多个 VLAN 可映射到同一个生成树实例中，但一个 VLAN 只能映射到一个生成树实例中，在交换机上通过对多个生成树实例的配置就可以实现链路的负载均衡。如图 2.81 所示，交换机 LSW1、LSW2 和 LSW3 为同一个 MST 域（MST region），该域中 VLAN1 与 VLAN2 映射到生成树实例（multiple spanning tree instance, MSTI）1，VLAN3 和 VLAN4 映射到实例 2，实例 1 的树根为 LSW1，交换机 LSW3 的 G0/0/1 接口被阻塞，而实例 2 的树根为 LSW2，交换机 LSW3 的 G0/0/2 接口被阻塞。这样，VLAN1 和 VLAN2 的数据从 LSW3 的 G0/0/2 接口转发，而 VLAN3 和 VLAN4 的数据从 LSW3 的 G0/0/1 接口转发，从而实现负载均衡。

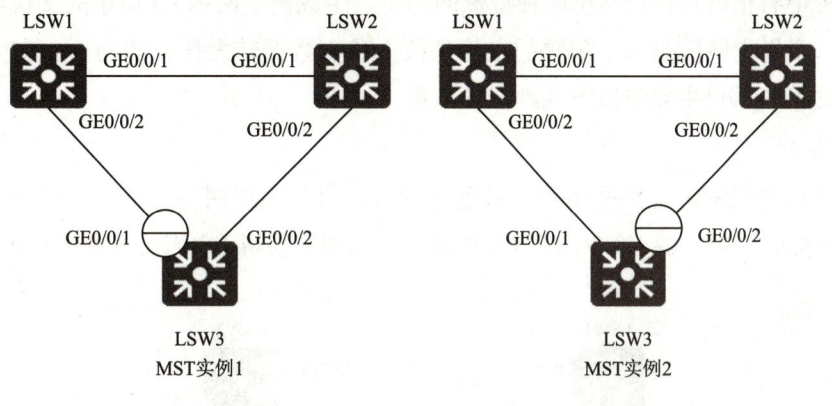

图 2.81　MSTP 示例

2. STP 的基本概念

（1）BPDU

网桥协议数据单元（bridge protocol data unit, BPDU）是运行 STP 在网桥之间交换的消息帧。STP 通过在设备间传递 BPDU 报文来确定网络的拓扑结构，并完成生成树的计算。BPDU 报文被封装在以太网数据帧中，目的 MAC 是组播地址"01-80-C2-00-00-00"。STP 协议的 BPDU 报文分为配置 BPDU 和拓扑改变通知 BPDU（topology change notification BPDU, TCN BPDU）。配置 BPDU 用来进行生成树计算和维护生成树拓扑的报文，而拓扑改变通知 BPDU 用于在网络拓

扑发生变化时，通知相关设备的报文。

（2）网桥 ID

网桥 ID（bridge ID, BID）用于在 STP 中唯一地标识网桥或交换机。网桥 ID 由两部分组成，长度共 8 字节，其中高 16 位为网桥优先级，低 48 位为网桥的 MAC 地址，如图 2.82 所示。网桥优先级的默认值为 32 768，网络管理员可以手工修改网桥的优先级，由于网桥的 MAC 地址唯一，因此网桥 ID 在网络中也是唯一的。STP 在执行生成树算法时，根据网络中网桥 ID 进行生成树树根的选择。网桥 ID 最小的交换机被选择为生成树的根桥。

网桥ID	
网桥优先级	网桥的MAC地址
高16位	低48位

图 2.82　网桥 ID 的组成

（3）路径开销

路径开销用于衡量交换机与交换机之间路径的优劣。根路径开销（root path cost）是指交换机到根桥的路径上所有链路开销的总和。默认情况下，端口开销由端口的运行速度决定。网络管理员可以通过配置交换机端口的端口开销修改其开销值，以便灵活控制到根桥的生成树路径。

3. 根与端口的角色

在生成树中有根桥与指定网桥两种特殊的网桥。生成树中网桥 ID 最小的交换机被选择为生成树的根桥。而指定网桥是一个单独的在物理段上负责数据转发任务的网桥。在生成树工作过程中，网桥/交换机端口主要有以下几种端口角色。

（1）根端口

根端口是指非根网桥上离根桥距离最近的端口。每个非根网桥只有一个根端口。如图 2.83 所示，非根交换机 LSW2 的 G0/0/1 接口为根端口，非根交换机 LSW3 的 G0/0/2 接口为根端口。

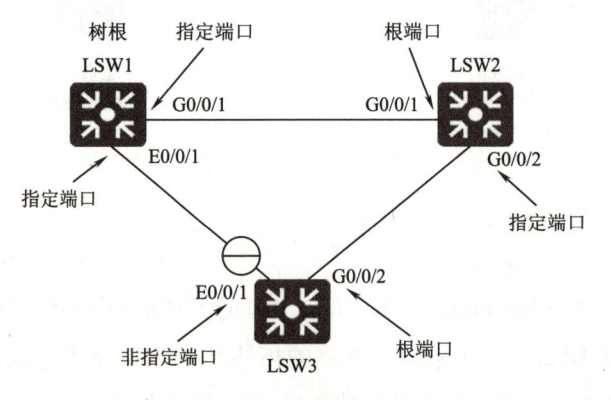

图 2.83　STP 端口角色示例

（2）指定端口

交换网络中每一个网段都需要指定一个端口用于转发该网段的数据，这个端口即为指定端口。指定端口是在 STP 中除了根端口之外可以转发流量的端口。根桥的所有端口均为指定端口。在图 2.91 中，交换机 LSW1 为根桥，因此它的所有端口（G0/0/1 与 E0/0/1 接口）均为指定端口。而对于交换机 LSW2 与交换机 LSW3 相连的网段，交换机 LSW2 的 G0/0/2 接口为指定端口。

（3）非指定端口

由于 STP 防止二层环路的需要而被设置成阻塞状态的端口称为非指定端口。在图 2.83 中，交换机 LSW3 的 E0/0/1 接口为非指定端口，该端口因处于阻塞状态而不能转发数据帧。

根端口与指定端口最后处于转发状态，而非指定交换机最终会变为阻塞状态。在 STP 中定义了禁用（disabled）、监听（listening）、学习（learning）、转发（forwarding）、阻塞（blocking）和五种端口状态，见表 2.14。

表 2.14　STP 端口状态

端 口 状 态	说　　明
disabled（禁用）	端口状态为 Down，不处理 BPDU 报文，也不转发用户流量
listening（监听）	过渡状态，开始生成树计算，端口可以接收和发送 BPDU，但不转发用户流量
learning（学习）	过渡状态，建立无环的 MAC 地址转发表，不转发用户流量
forwarding（转发）	端口可以接收和发送 BPDU，也转发用户流量。只有根端口或指定端口才能进入 Forwarding 状态
blocking（阻塞）	端口仅仅接收并处理 BPDU，不转发用户流量

4. STP 的工作过程

STP 的工作过程主要包含三个任务：一是选举根桥；二是为所有非根网桥 / 交换机选择根端口；三是为每个网段选择指定端口。STP 选举根桥、根端口和指定端口主要使用配置 BPDU 中报文中的根桥 ID、根路径开销、发送设备 BID 和发送端口 PID（端口 ID）。

（1）选举根桥

在初始化状态时，网络中的所有交换机都还没有收到其他交换机发送的配置 BPDU，因此都会先假设自己是网络中的根桥，交换机只要完成启动过程就立即开始发送以自己为根的配置 BPDU。当交换机从邻居交换机收到配置 BPDU 时，将收到 BPDU 中的根 ID 与本交换机的根 ID 进行比较，并将 BPDU 中根 ID 字段的值设置为数值小的根 ID。最终选择 BID 最小的交换机成为根桥最小 BID。

（2）选择根端口和指定端口

根端口和指定端口根据最小根路径开销、最小发送者 BID、最小发送者端口 ID 以及最小接收者端口 ID 四个条件进行决策，具体过程如下：

① 比较非根交换机上每个端口到根桥的根路径开销，具有最小根路径开销值的端口被选择为根端口。

② 如果非根交换机上有多个端口到根桥的根路径开销值相同并且为最小，则比较端口收到的配置 BPDU 中的 BID，收到具有最小发送 BID 的 BPDU 的端口为根端口。

③ 如果非根交换机上有多个端口到根桥的根路径开销值相同并且为最小，而且同时具有相同的最小发送 BID，则比较发送者端口 ID，收到具有最小端口 ID 的本网桥端口被选择为根端口。

④ 如果非根交换机上有多个端口到根桥的根路径开销值相同且最小，并且同时具有相同的最小发送 BID 和最小发送者端口 ID，则比较接收该 BPDU 的端口 ID，具有最小接收者端口 ID 的端口被选择为根端口。

（3）选择指定端口

指定端口的选择与根端口的选择类似，同样根据最小根路径开销、最小发送者 BID、最小发送者端口 ID 以及最小接收者端口 ID 四个条件进行决策。

相关技能

1. STP 的规划与配置

搭建如图 2.84 所示的网络拓扑图，根据表 2.15 的 IP 与 VLAN 规划完成 VLAN 的配置；在交换机 LSW1、LSW2 和 LSW3 上部署 STP，使 LSW1 为根桥，LSW2 为备份根桥，交换机 LSW3 的 G0/0/2 接口被阻塞。

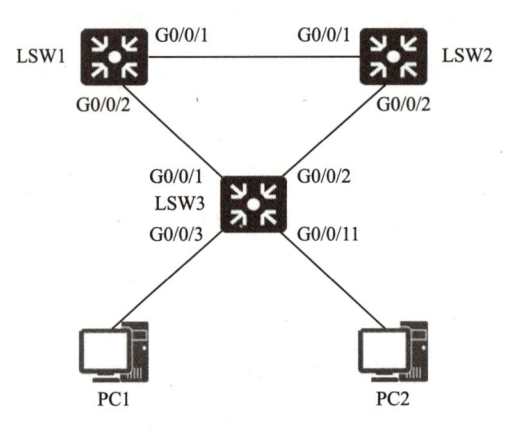

图 2.84　STP 配置的网络拓扑图

表 2.15　IP 与 VLAN 的规划

设　　备	接口或 VLAN	VLAN 的描述	二层 / 三层规划	说　　明
LSW1	VLAN2	teacher		教师
	VLAN3	student		学生
LSW2	VLAN2	teacher		教师
	VLAN3	student		学生
LSW3	VLAN2	teacher	G0/0/3 至 G0/0/11 接口	教师
	VLAN3	student	G0/0/12 至 G0/0/24 接口	学生

续表

设 备	接口或 VLAN	VLAN 的描述	二层 / 三层规划	说 明
PC1			192.168.2.11/24	
PC2			192.168.2.12/24	

（1）STP 的规划

① 所有交换机生成树均使用 STP 协议。

② 交换机 LSW1 的生成树优先级值为 4096，为 STP 的根。

③ 交换机 LSW2 的生成树优先级值为 8192，为 STP 的备份根。

④ 交换机 LSW3 的生成树优先级值为默认值。

（2）VLAN 配置

以华为 S5700-28C-HI 交换机为例，完成表 2.15 所示的 VLAN 配置步骤如下：

```
// 在 LSW1 上创建 VLAN2 与 VLAN3，配置 G0/0/1 与 G0/0/2 接口为中继接口
[LSW1]vlan 2
[LSW1-vlan2]description teacher
[LSW1-vlan2]vlan 3
[LSW1-vlan3]description student
[LSW1]port-group 1
[LSW1-port-group-1]group-member g0/0/1 to g0/0/2
[LSW1-port-group-1]port link-type trunk
[LSW1-port-group-1]port trunk allow-pass vlan  2 3
// 在 LSW2 上创建 VLAN2 与 VLAN3，配置 G0/0/1 与 G0/0/2 接口为中继接口
[LSW2]vlan 2
[LSW2-vlan2]description teacher
[LSW2-vlan2]vlan 3
[LSW2-vlan3]description student
[LSW2]port-group 1
[LSW2-port-group-1]group-member g0/0/1 to g0/0/2
[LSW2-port-group-1]port link-type trunk
[LSW2-port-group-1]port trunk allow-pass vlan  2 3
// 在 LSW3 上创建 VLAN2 与 VLAN3，配置 G0/0/1 与 G0/0/2 接口为中继接口，配置 G0/0/3 至
//G0/0/11 接口属于 VLAN2，G0/0/12 至 G0/0/24 接口属于 VLAN3
[LSW3]vlan 2
[LSW3-vlan2]description teacher
[LSW3-vlan2]vlan 3
[LSW3-vlan3]description student
[LSW3]port-group 1
[LSW3-port-group-1]group-member g0/0/1 to g0/0/2
```

```
[LSW3-port-group-1]port link-type trunk
[LSW3-port-group-1]port trunk allow-pass vlan  2 3
[LSW3-port-group-1]quit
[LSW3]port-group 2
[LSW3-port-group-2]group-member g0/0/3 to g0/0/11
[LSW3-port-group-2]port link-type access
[LSW3-port-group-2]port default vlan 2
[LSW3-port-group-2]quit
[LSW3]port-group 3
[LSW3-port-group-3]group-member g0/0/12 to g0/0/24
[LSW3-port-group-3]port link-type access
[LSW3-port-group-3]port default vlan 3
```

（3）STP 的配置

在交换机上进行 STP 的配置通常需要：

① 启用 STP。

② 配置交换机生成树模式为 STP。

③ 根据规划配置交换机的生成树优先级，使其成为根桥或者备份根桥。

以华为 S5700-28C-HI 交换机为例，在交换机 LSW1、LSW2 与 LSW3 上完成 STP 的配置步骤如下：

```
[LSW1]stp enable              // 启用 STP
[LSW1]stp mode stp            // 配置生成树模式为 STP
[LSW1]stp priority 4096       // 配置 LSW1 生成树优先级为 4096
[LSW2]stp enable              // 启用 STP
[LSW2]stp mode stp            // 配置生成树模式为 STP
[LSW2]stp priority 8192       // 配置 LSW1 生成树优先级为 8192
[LSW3]stp enable              // 启用 STP
[LSW3]stp mode stp            // 配置生成树模式为 STP
```

（4）STP 的查看与验证

在交换机 LSW1、LSW2 和 LSW3 上完成了 STP 配置后，等 STP 计算稳定后，进行 STP 的查看与验证。

① 在 LSW1、LSW2 与 LSW3 上执行 "display stp" 命令查看生成树信息。图 2.85 为在交换机 LSW1 上执行 "display stp" 命令的显示结果。结果显示交换机 LSW1 的生成树模式为 STP，交换机 LSW1 的 BID 为 "4096.4c1f-cc7c-26e4"，根桥的 BID 为 "4096.4c1f-cc7c-26e4"，根桥的 ID 与交换机 LSW1 的 BID 相同，因此，交换机 LSW1 为生成树的树根，选择结果与规划结果一致。

项目 2　小型企业网络互联项目　　161

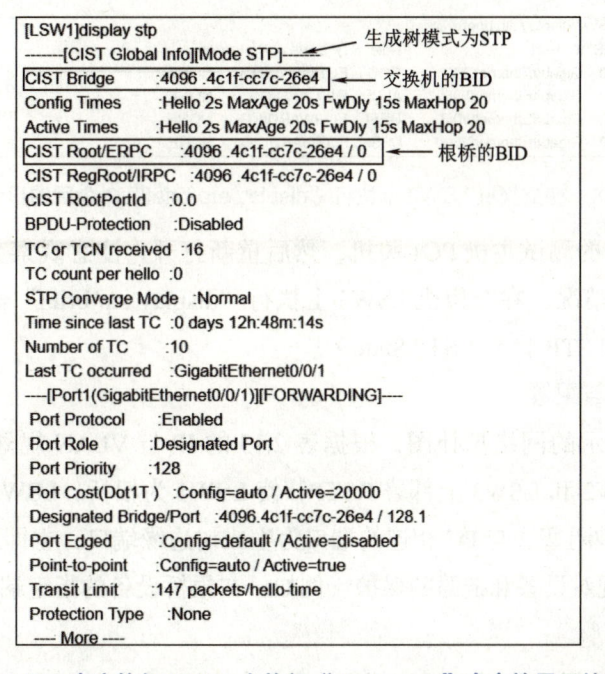

```
[LSW1]display stp
-------[CIST Global Info][Mode STP]----          ← 生成树模式为STP
CIST Bridge      :4096 .4c1f-cc7c-26e4           ← 交换机的BID
Config Times     :Hello 2s MaxAge 20s FwDly 15s MaxHop 20
Active Times     :Hello 2s MaxAge 20s FwDly 15s MaxHop 20
CIST Root/ERPC   :4096 .4c1f-cc7c-26e4 / 0       ← 根桥的BID
CIST RegRoot/IRPC :4096 .4c1f-cc7c-26e4 / 0
CIST RootPortId  :0.0
BPDU-Protection  :Disabled
TC or TCN received :16
TC count per hello :0
STP Converge Mode  :Normal
Time since last TC :0 days 12h:48m:14s
Number of TC     :10
Last TC occurred :GigabitEthernet0/0/1
----[Port1(GigabitEthernet0/0/1)][FORWARDING]----
Port Protocol    :Enabled
Port Role        :Designated Port
Port Priority    :128
Port Cost(Dot1T ) :Config=auto / Active=20000
Designated Bridge/Port :4096.4c1f-cc7c-26e4 / 128.1
Port Edged       :Config=default / Active=disabled
Point-to-point   :Config=auto / Active=true
Transit Limit    :147 packets/hello-time
Protection Type  :None
---- More ----
```

图 2.85　在交换机 LSW1 上执行 "display stp" 命令的显示结果

② 在 LSW1、LSW2 与 LSW3 上执行 "display stp brief" 命令查看生成树简要信息，注意观察交换机端口角色与端口状态。如图 2.86 所示，交换机 LSW1 的所有端口角色均为指定端口（DESI），端口的生成树状态为转发状态（FORWARDING），该交换机即为树根。如图 2.87 所示，交换机 LSW2 的 GigabitEthernet0/0/1 端口角色为根端口（ROOT），状态为转发状态（FORWARDING）；GigabitEthernet0/0/2 端口角色为指定端口（DESI），状态为转发状态（FORWARDING）。如图 2.88 所示，交换机 LSW3 的 GigabitEthernet0/0/1 端口角色为根端口（ROOT），状态为转发状态（FORWARDING）；GigabitEthernet0/0/3 与 GigabitEthernet0/0/11 端口角色为指定端口（DESI），状态为转发状态（FORWARDING）；GigabitEthernet0/0/2 端口角色为其他端口（ALTE），状态为抛弃状态（DISCARDING）。

```
[LSW1]display stp brief
MSTID  Port                      Role   STP State    Protection
 0   GigabitEthernet0/0/1        DESI   FORWARDING   NONE
 0   GigabitEthernet0/0/2        DESI   FORWARDING   NONE
```

图 2.86　在交换机 LSW1 上执行 "display stp brief" 命令后的显示结果

```
[LSW2]display stp brief
MSTID  Port                      Role   STP State    Protection
 0   GigabitEthernet0/0/1        ROOT   FORWARDING   NONE
 0   GigabitEthernet0/0/2        DESI   FORWARDING   NONE
```

图 2.87　在交换机 LSW2 上执行 "display stp brief" 命令后的显示结果

162 企业网络互联技术（双语）

```
[LSW3]display stp brief
MSTID  Port                    Role  STP State    Protection
  0    GigabitEthernet0/0/1    ROOT  FORWARDING   NONE
  0    GigabitEthernet0/0/2    ALTE  DISCARDING   NONE
  0    GigabitEthernet0/0/3    DESI  FORWARDING   NONE
  0    GigabitEthernet0/0/11   DESI  FORWARDING   NONE
```

图 2.88　在交换机 LSW3 上执行"display stp brief"命令后的显示结果

③ 将网络拓扑中的测试主机 PC1 关机，然后重新打开，注意观察交换机 LSW3 连接 PC1 的接口 G0/0/3 信号灯情况。在交换机 LSW3 上执行"display stp brief"命令，不停观察 G0/0/3 接口的角色（Role）和 STP 状态（STP State）。

2．RSTP 的规划与配置

搭建如图 2.84 所示的网络拓扑图，根据表 2.13 的 IP 与 VLAN 规划完成 VLAN 的配置；在交换机 LSW1、LSW2 和 LSW3 上部署 RSTP，使 LSW1 为根桥，LSW2 为备份根桥，交换机 LSW3 的 G0/0/2 接口被阻塞，与 PC 相连的端口为生成树边缘端口，立即进入转发状态；配置生成树的保护功能，实现对设备和链路的保护。例如：在根桥设备的指定端口配置根保护功能。

（1）RSTP 的规划

① 所有交换机生成树均使用 RSTP 协议。

② 交换机 LSW1 为树根。

③ 交换机 LSW2 为备份根桥。

④ 交换机 LSW1 的 G0/0/1 与 G0/0/2 接口配置根保护功能。

⑤ 交换机 LSW3 的 G0/0/3 与 G0/0/24 接口配置为边缘端口，并启用 BPDU 保护。

（2）VLAN 配置

参考前文"STP 的规划与配置"中"VLAN 配置"的步骤完成 VLAN 的配置。

（3）RSTP 的配置

根据规划，完成 RSTP 的配置需要完成以下配置任务：

① 启用 STP。

② 配置交换机生成树模式为 RSTP。

③ 配置交换机 LSW1 为根桥，并启用根保护。

④ 配置交换机 LSW2 为备份根桥。

⑤ 配置交换机 LSW3 的 G0/0/3 至 G0/0/24 接口为边缘端口，并启用 BPDU 保护。

以华为 S5700-28C-HI 交换机为例，在交换机 LSW1、LSW2 与 LSW3 上完成 RSTP 的配置步骤如下：

```
[LSW1]stp enable            // 启用 STP
[LSW1]stp mode rstp         // 配置生成树模式为 RSTP
[LSW1]stp root primary      // 配置 LSW1 为生成树的根桥
[LSW1]interface g0/0/1
```

```
[LSW1-GigabitEthernet0/0/1]stp root-protection        // 启用根保护
[LSW1-GigabitEthernet0/0/1]quit
[LSW1]interface g0/0/2
[LSW1-GigabitEthernet0/0/2]stp root-protection        // 启用根保护
[LSW1-GigabitEthernet0/0/2]quit
[LSW2]stp enable                                       // 启用 STP
[LSW2]stp mode rstp                                    // 配置生成树模式为 RSTP
[LSW2]stp root secondary                               // 配置 LSW1 为生成树的备份根桥
[LSW3]stp enable                                       // 启用 STP
[LSW3]stp mode rstp                                    // 配置生成树模式为 RSTP
[LSW3]port-group 2
[LSW3-port-group-2]group-member g0/0/3 to g0/0/24
[LSW3-port-group-2]stp edged-port enable               // 配置为边缘端口
[LSW3-port-group-2]quit
[LSW3]stp bpdu-protection                               // 配置交换机的 BPDU 保护功能
```

（4）RSTP 的查看与验证

在交换机 LSW1、LSW2 和 LSW3 上完成了 RSTP 配置后，等 RSTP 计算稳定后，进行 RSTP 的查看与验证。

① 在 LSW1、LSW2 与 LSW3 上执行 "display stp" 命令查看生成树信息，图 2.89 所示为在交换机 LSW1 上执行 "display stp" 命令的显示结果。结果显示交换机 LSW1 的生成树模式为 RSTP，交换机 LSW1 的 BID 为 "0.4c1f-cc7c-26e4"，根桥的 BID 为 "0.4c1f-cc7c-26e4"，根桥的 ID 与交换机 LSW1 的 BID 相同，因此，交换机 LSW1 为生成树的树根，选择结果与规划结果一致。

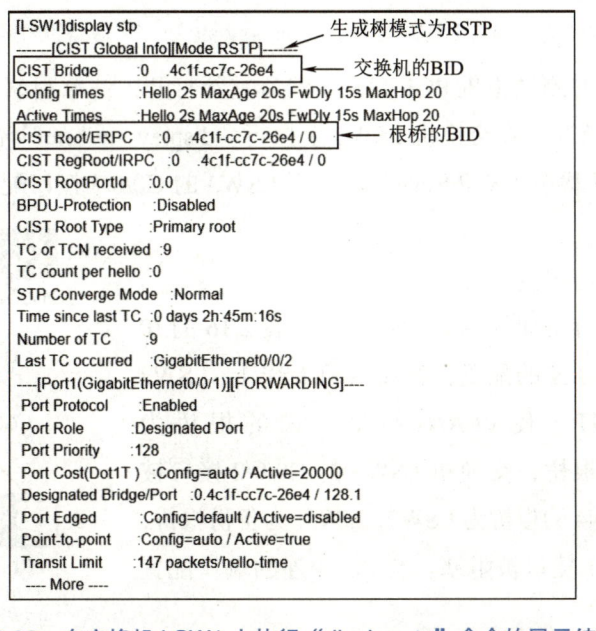

图 2.89　在交换机 LSW1 上执行 "display stp" 命令的显示结果

② 在 LSW1、LSW2 与 LSW3 上执行 "display stp brief" 命令查看生成树简要信息，注意观察交换机端口角色与端口状态。如图 2.90 所示，交换机 LSW1 的所有端口角色均为指定端口（DESI），端口的生成树状态为转发状态（FORWARDING），该交换机即为树根。如图 2.91 所示，交换机 LSW2 的 GigabitEthernet0/0/1 端口角色为根端口（ROOT），状态为转发状态（FORWARDING）；GigabitEthernet0/0/2 端口角色为指定端口（DESI），状态为转发状态（FORWARDING）。如图 2.92 所示，交换机 LSW3 的 GigabitEthernet0/0/1 端口角色为根端口（ROOT），状态为转发状态（FORWARDING）；GigabitEthernet0/0/3 与 GigabitEthernet0/0/11 端口角色为指定端口（DESI），状态为转发状态（FORWARDING）；GigabitEthernet0/0/2 端口角色为其他端口（ALTE），状态为抛弃状态（DISCARDING）。

```
[LSW1]display stp brief
MSTID Port                      Role   STP State    Protection
 0    GigabitEthernet0/0/1      DESI FORWARDING     NONE
 0    GigabitEthernet0/0/2      DESI FORWARDING     NONE
```

图 2.90　在交换机 LSW1 上执行 "display stp brief" 命令后的显示结果

```
[LSW2]display stp brief
MSTID Port                      Role   STP State    Protection
 0    GigabitEthernet0/0/1      ROOT FORWARDING     NONE
 0    GigabitEthernet0/0/2      DESI FORWARDING     NONE
```

图 2.91　在交换机 LSW2 上执行 "display stp brief" 命令后的显示结果

```
[LSW3]display stp brief
MSTID Port                      Role STP State    Protection
 0    GigabitEthernet0/0/1      ROOT FORWARDING    NONE
 0    GigabitEthernet0/0/2      ALTE DISCARDING    NONE
 0    GigabitEthernet0/0/3      DESI FORWARDING    BPDU
 0    GigabitEthernet0/0/11     DESI FORWARDING    BPDU
```

图 2.92　在交换机 LSW3 上执行 "display stp brief" 命令后的显示结果

③ 将网络拓扑中的测试主机 PC1 关机，然后重新打开，注意观察交换机 LSW3 连接 PC1 的接口 G0/0/3 信号灯情况。在交换机 LSW3 上执行 "display stp brief" 命令，观察 G0/0/3 接口的角色（Role）和 STP 状态（STP State）。由于 LSW3 的 G0/0/3 接口配置为边缘端口，因此立即进入转发状态。

3. MSTP 的规划与配置

搭建如图 2.93 所示的网络拓扑图，根据表 2.16 的 IP 与 VLAN 规划完成 VLAN 的配置；在交换机 LSW1、LSW2 和 LSW3 上部署 MSTP，使 VLAN1 与 VLAN2 的根桥为 LSW1，LSW2 为备份根桥，交换机 LSW3 的 G0/0/2 接口被阻塞。VLAN3 与 VLAN4 的根桥为 LSW2，LSW1 为备份根桥，交换机 LSW3 的 G0/0/1 接口被阻塞。与 PC 相连的端口配置为生成树边缘端口。

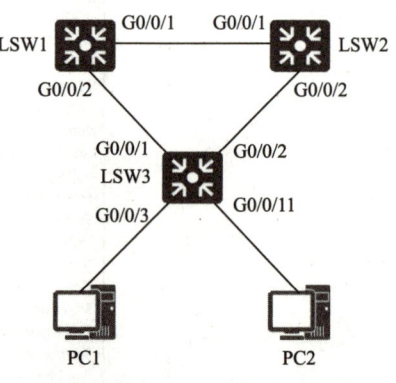

图 2.93　MSTP 配置的网络拓扑图

项目 2　小型企业网络互联项目　165

表 2.16　IP 与 VLAN 的规划

设　备	接口或 VLAN	VLAN 的描述	二层 / 三层规划	说　明
LSW1	VLAN2	teacher		教师
	VLAN3	student		学生
	VLAN4	jwc		教务处
LSW2	VLAN2	teacher		教师
	VLAN3	student		学生
	VLAN4	jwc		教务处
LSW3	VLAN2	teacher	G0/0/3 至 G0/0/10 接口	教师
	VLAN3	student	G0/0/11 至 G0/0/19 接口	学生
	VLAN4	jwc	G0/0/20 至 G0/0/24 接口	
PC1			192.168.2.11/24	
PC2			192.168.3.12/24	

（1）MSTP 的规划

① 所有交换机生成树均使用 MSTP 协议。

② MSTP 域名为 cdp，修订级别为 1，实例 1 包括 VLAN1 与 VLAN2，实例 2 包括 VLAN3 与 VLAN4。

③ 交换机 LSW1 为实例 0 和 1 的主根，实例 2 的备份根。

④ 交换机 LSW2 为实例 0 和 1 的备份根，实例 2 的主根。

⑤ 交换机 LSW3 的 G0/0/3 至 G0/0/24 接口配置为边缘端口，并启用 BPDU 保护。

（2）VLAN 配置

参考前文 "STP 的规划与配置" 中 "VLAN 配置" 的步骤完成 VLAN 的配置。

（3）MSTP 的配置

根据规划，完成 MSTP 的配置需要完成以下配置任务：

① 启用 STP。

② 配置交换机生成树模式为 MSTP。

③ 配置 MSTP 域名、修订版本、实例与 VLAN 的映射关系以及激活 MSTP 域。

④ 配置交换机 LSW1 为实例 1 根桥，实例 2 的备份根桥。

⑤ 配置交换机 LSW2 为实例 1 的备份根桥，实例 2 的根桥。

⑥ 配置交换机 LSW3 的 G0/0/3 至 G0/0/24 接口为边缘端口，并启用 BPDU 保护。

以华为 S5700-28C-HI 交换机为例，在交换机 LSW1、LSW2 与 LSW3 上完成 MSTP 的配置步骤如下：

```
[LSW1]stp enable              // 启用 STP
[LSW1]stp mode mstp           // 配置生成树模式为 MSTP
```

```
[LSW1]stp region-configuration                          // 进入 MSTP 域视图
[LSW1-mst-region]region-name cdp                        // 配置 MSTP 域名为 "cdp"
[LSW1-mst-region]revision-level 1                       // 配置修订级别为 1
[LSW1-mst-region]instance 1 vlan 1 2                    // 配置实例 1 包括 VLAN1 和 VLAN2
[LSW1-mst-region]instance 2 vlan 3 4                    // 配置实例 2 包括 VLAN3 和 VLAN4
[LSW1-mst-region]active region-configuration           // 激活 MSTP 域
[LSW1]stp instance 0 root primary                       // 配置 LSW1 为实例 0 的根桥
[LSW1]stp instance 1 root primary                       // 配置 LSW1 为实例 1 的根桥
[LSW1]stp instance 2 root secondary                     // 配置 LSW1 为实例 2 的备份根桥
[LSW2]stp enable                                        // 启用 STP
[LSW2]stp mode mstp                                     // 配置生成树模式为 MSTP
[LSW2]stp region-configuration                          // 进入 MSTP 域视图
[LSW2-mst-region]region-name cdp                        // 配置 MSTP 域名为 "cdp"
[LSW2-mst-region]revision-level 1                       // 配置修订级别为 1
[LSW2-mst-region]instance 1 vlan 1 2                    // 配置实例 1 包括 VLAN1 和 VLAN2
[LSW2-mst-region]instance 2 vlan 3 4                    // 配置实例 2 包括 VLAN3 和 VLAN4
[LSW2-mst-region]active region-configuration           // 激活 MSTP 域
[LSW2]stp instance 0 root secondary                     // 配置 LSW2 为实例 0 的备份根桥
[LSW2]stp instance 1 root secondary                     // 配置 LSW1 为实例 1 的备份根桥
[LSW2]stp instance 2 root primary                       // 配置 LSW1 为实例 2 的根桥
[LSW3]stp enable                                        // 启用 STP
[LSW3]stp mode mstp                                     // 配置生成树模式为 MSTP
[LSW3]stp region-configuration                          // 进入 MSTP 域视图
[LSW3-mst-region]region-name cdp                        // 配置 MSTP 域名为 "cdp"
[LSW3-mst-region]revision-level 1                       // 配置修订级别为 1
[LSW3-mst-region]instance 1 vlan 1 2                    // 配置实例 1 包括 VLAN1 和 VLAN2
[LSW3-mst-region]instance 2 vlan 3 4                    // 配置实例 2 包括 VLAN3 和 VLAN4
[LSW3-mst-region]active region-configuration           // 激活 MSTP 域
[LSW3]port-group 5
[LSW3-port-group-2]group-member g0/0/3 to g0/0/24
[LSW3-port-group-2]stp edged-port enable               // 配置 G0/0/3 至 G0/0/24 为边缘端口
[LSW3-port-group-2]quit
[LSW3]stp bpdu-protection                               // 配置交换机的 BPDU 保护功能
```

（4）MSTP 的查看与验证

在交换机 LSW1、LSW2 和 LSW3 上完成了 MSTP 配置后，等 MSTP 计算稳定后，进行 MSTP 的查看与验证。

① 在 LSW1、LSW2 与 LSW3 上执行 "display stp region-configuration" 命令查看 MSTP 域配置信息。图 2.94 所示为 LSW1 上 MSTP 域的配置信息。注意观察生成树域名、修订级别、实例包含的 VLAN 是否与规划配置一致。图 2.94 中的实例 0 为默认实例，未配置指定属于哪个实例的所有 VLAN 均属于实例 0。

项目 2　小型企业网络互联项目　　167

```
[LSW1]display stp region-configuration
Oper configuration
 Format selector   :0
 Region name       :cdp
 Revision level    :1

 Instance   VLANs Mapped
    0       5 to 4094
    1       1 to 2
    2       3 to 4
```

图 2.94　LSW1 上 MSTP 域的配置信息

② 在 LSW1 上执行"display stp brief"命令查看生成树简要信息，注意观察交换机端口角色与端口状态。如图 2.95 所示，交换机 LSW1 的生成树实例 0 和实例 1 的所有端口角色均为指定端口（DESI），生成树状态为转发状态（FORWARDING），该交换机为实例 0 和实例 1 的树根；交换机 LSW1 的生成树实例 2 的 GigabitEthernet0/0/1 端口为根端口，GigabitEthernet0/0/2 端口为指定端口，均为转发状态（FORWARDING）。

```
[LSW1]display stp brief
MSTID  Port                 Role   STP State    Protection
 0     GigabitEthernet0/0/1  DESI   FORWARDING   NONE
 0     GigabitEthernet0/0/2  DESI   FORWARDING   NONE
 1     GigabitEthernet0/0/1  DESI   FORWARDING   NONE
 1     GigabitEthernet0/0/2  DESI   FORWARDING   NONE
 2     GigabitEthernet0/0/1  ROOT   FORWARDING   NONE
 2     GigabitEthernet0/0/2  DESI   FORWARDING   NONE
```

图 2.95　在交换机 LSW1 上执行"display stp brief"命令后的显示结果

③ 在 LSW2 上执行"display stp brief"命令查看生成树简要信息。如图 2.96 所示，交换机 LSW2 的生成树实例 2 的所有端口角色均为指定端口（DESI），生成树状态为转发状态（FORWARDING），该交换机为实例 2 的树根；交换机 LSW2 的生成树实例 0 和实例 1 的 GigabitEthernet0/0/1 端口为根端口，GigabitEthernet0/0/2 端口为指定端口，均为转发状态（FORWARDING）

```
[LSW2]display stp brief
MSTID  Port                 Role   STP State    Protection
 0     GigabitEthernet0/0/1  ROOT   FORWARDING   NONE
 0     GigabitEthernet0/0/2  DESI   FORWARDING   NONE
 1     GigabitEthernet0/0/1  ROOT   FORWARDING   NONE
 1     GigabitEthernet0/0/2  DESI   FORWARDING   NONE
 2     GigabitEthernet0/0/1  DESI   FORWARDING   NONE
 2     GigabitEthernet0/0/2  DESI   FORWARDING   NONE
```

图 2.96　在交换机 LSW2 上执行"display stp brief"命令后的显示结果

④ 在 LSW3 上执行"display stp brief"命令查看生成树简要信息。如图 2.97 所示，交换机 LSW3 的生成树实例 0 和实例 1 的 GigabitEthernet0/0/2 端口角色为其他端口（ALTE），状态为抛弃状态（DISCARDING），而对实例 2 则是 GigabitEthernet0/0/1 端口角色为其他端口（ALTE），状态为抛弃状态（DISCARDING）。

168 企业网络互联技术（双语）

```
[LSW3]display stp brief
MSTID  Port                    Role    STP State      Protection
  0    GigabitEthernet0/0/1    ROOT    FORWARDING     NONE
  0    GigabitEthernet0/0/2    ALTE    DISCARDING     NONE
  0    GigabitEthernet0/0/3    DESI    FORWARDING     BPDU
  0    GigabitEthernet0/0/11   DESI    FORWARDING     BPDU
  1    GigabitEthernet0/0/1    ROOT    FORWARDING     NONE
  1    GigabitEthernet0/0/2    ALTE    DISCARDING     NONE
  1    GigabitEthernet0/0/3    DESI    FORWARDING     BPDU
  2    GigabitEthernet0/0/1    ALTE    DISCARDING     NONE
  2    GigabitEthernet0/0/2    ROOT    FORWARDING     NONE
  2    GigabitEthernet0/0/11   DESI    FORWARDING     BPDU
```

图 2.97　在交换机 LSW3 上执行"display stp brief"命令后的显示结果

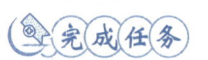

1. 任务规划

根据任务描述和图 2.1 所示的网络拓扑，企业内网采用 MSTP 协议打破二层环路。MSTP 的规划如下：

① 企业内网所有交换机生成树均使用 MSTP 协议。

② 交换机 CORE、FLOOR1、FLOOR2 和 FLOOR3 属于同一个 MSTP 域，MSTP 域名为 cdp，修订级别为 1，所有 VLAN 映射到默认实例 0，实例 0 的根桥为交换机 CORE。

③ 交换机 FLOOR4 属于另一个 MSTP 域，域名为 leader，修订级别为 1，所有 VLAN 映射到默认实例 0，实例 0 的根桥为交换机 FLOOR4。

④ 在核心交换机 CORE 的所有指定端口启用根保护功能。

⑤ 核心交换机 CORE 连接路由器 AR1 的 G0/0/24 接口不参与生成树。

⑥ 所有接入层交换机接主机的端口配置为边缘端口，并启用 BPDU 保护。

2. 任务实施

（1）MSTP 域"cdp"中的交换机生成树配置

① 根据生成树的规划，核心交换机 CORE 生成树的配置步骤如下：

```
[CORE]stp enable
[CORE]stp mode mstp
// 启用生成树与配置生成树模式为 MSTP
[CORE]stp region-configuration
[CORE-mst-region]region-name cdp
[CORE-mst-region]revision-level 1
[CORE-mst-region]active region-configuration
[CORE-mst-region]quit
[CORE]
// 配置 MSTP 域，并激活域，MSTP 域名为"cdp"，修订级别为 1，所有 VLAN 均映射到默认实例 0 中
[CORE]stp instance 1 root primary
```

项目 2　小型企业网络互联项目　169

```
// 配置交换机 CORE 为 cdp 域中实例 0 的根桥
[CORE]interface g0/0/24
[CORE-GigabitEthernet0/0/24]stp disable
[CORE-GigabitEthernet0/0/24]quit
[CORE]
// 交换机 CORE 的 G0/0/24 接口不参与生成树选举
[CORE]port-group 1
[CORE-port-group-1]group-member g0/0/21 to g0/0/24
[CORE-port-group-1]stp edged-port enable
[CORE-port-group-1]quit
[CORE]stp bpdu-protection
// 配置交换机 CORE 连接服务器的 G0/0/21 至 G0/0/24 接口为边缘端口，并启用 BPDU 保护
[CORE]port-group 2
[CORE-port-group-2]group-member g0/0/1 to g0/0/3
[CORE-port-group-2]stp root-protection
// 交换机 CORE 的 G0/0/1 至 G0/0/3 接口启用根保护
```

② 根据生成树的规划，交换机 FLOOR1 生成树的配置步骤如下：

```
[FLOOR1]stp enable
[FLOOR1]stp mode mstp
[FLOOR1]// 启用生成树与配置生成树模式为 MSTP
[FLOOR1]stp region-configuration
[FLOOR1-mst-region]region-name cdp
[FLOOR1-mst-region]revision-level  1
[FLOOR1-mst-region]active region-configuration
[FLOOR1-mst-region]quit
// 配置 MSTP 域，并激活域，MSTP 域名为 "cdp"，修订级别为 1，所有 VLAN 均映射到默认实例 0 中
[FLOOR1]port-group 5
[FLOOR1-port-group-5]group-member g0/0/2 to g0/0/24
[FLOOR1-port-group-5]stp edged-port enable
[FLOOR1-port-group-5]quit
[FLOOR1]stp bpdu-protection
// 配置交换机 FLOOR1 的 G0/0/2 至 G0/0/24 接口为边缘端口，并启用 BPDU 保护
```

③ 根据生成树的规划，交换机 FLOOR2 生成树的配置步骤如下：

```
[FLOOR2]stp enable
[FLOOR2]stp mode mstp
[FLOOR2]
// 启用生成树与配置生成树模式为 MSTP
[FLOOR2]stp region-configuration
```

```
[FLOOR2-mst-region]region-name cdp
[FLOOR2-mst-region]revision-level  1
[FLOOR2-mst-region]active region-configuration
[FLOOR2-mst-region]quit
// 配置 MSTP 域，并激活域，MSTP 域名为"cdp"，修订级别为 1，所有 VLAN 均映射到默认实例 0 中
[FLOOR2]port-group 5
[FLOOR2-port-group-5]group-member g0/0/2 to g0/0/24
[FLOOR2-port-group-5]stp edged-port enable
[FLOOR2-port-group-5]quit
[FLOOR2]stp bpdu-protection
// 配置交换机 FLOOR2 的 G0/0/2 至 G0/0/24 接口为边缘端口，并启用 BPDU 保护
```

④根据生成树的规划，交换机 FLOOR3 生成树的配置步骤如下：

```
[FLOOR3]stp enable
[FLOOR3]stp mode mstp
[FLOOR3]
// 启用生成树与配置生成树模式为 MSTP
[FLOOR3]stp region-configuration
[FLOOR3-mst-region]region-name cdp
[FLOOR3-mst-region]revision-level  1
[FLOOR3-mst-region]active region-configuration
[FLOOR3-mst-region]quit
// 配置 MSTP 域，并激活域，MSTP 域名为"cdp"，修订级别为 1，所有 VLAN 均映射到默认实例 0 中
[FLOOR3]port-group 5
[FLOOR3-port-group-5]group-member g0/0/2 to g0/0/24
[FLOOR3-port-group-5]stp edged-port enable
[FLOOR3-port-group-5]quit
[FLOOR3]stp bpdu-protection
// 配置交换机 FLOOR3 的 G0/0/2 至 G0/0/24 接口为边缘端口，并启用 BPDU 保护
```

（2）MSTP 域"leader"中的交换机生成树配置

根据生成树的规划，交换机 FLOOR4 生成树的配置步骤如下：

```
[FLOOR4]stp enable
[FLOOR4]stp mode mstp
[FLOOR4]
// 启用生成树与配置生成树模式为 MSTP
[FLOOR4]stp region-configuration
[FLOOR4-mst-region]region-name leader
[FLOOR4-mst-region]revision-level  1
[FLOOR2-mst-region]active region-configuration
```

```
[FLOOR2-mst-region]quit
// 配置 MSTP 域，并激活域，MSTP 域名为"leader"，修订级别为 1，所有 VLAN 均映射到默认实例 0 中
[FLOOR4]stp instance 0 root primary
// 配置交换机 FLOOR4 为 leader 域中实例 0 的根桥
[FLOOR4]port-group 5
[FLOOR4-port-group-5]group-member g0/0/2 to g0/0/24
[FLOOR4-port-group-5]stp edged-port enable
[FLOOR4-port-group-5]quit
[FLOOR4]stp bpdu-protection
// 配置交换机 FLOOR4 的 G0/0/2 至 G0/0/24 接口为边缘端口，并启用 BPDU 保护
```

（3）验证与网络测试

在企业内网中的所有交换机上完成了 MSTP 配置后，等 MSTP 计算稳定后，进行 MSTP 的查看与验证。

① 在交换机 CORE、FLOOR1、FLOOR2 与 FLOOR3 上执行"display stp region-configuration"命令查看 MSTP 域配置信息。图 2.98 所示为交换机 CORE 生成树 MSTP 域配置显示结果。

注：根据规划，交换机 CORE、FLOOR1、FLOOR2 与 FLOOR3 的 MSTP 域信息（域名、修订级别、实例与 VLAN 的映射关系）需要完全相同。

```
[CORE]display stp region-configuration
Oper configuration
 Format selector    :0
 Region name        :cdp
 Revision level     :1

 Instance   VLANs Mapped
   0       1 to 4094
```

图 2.98　交换机 CORE 生成树 MSTP 域配置显示结果

② 在交换机 CORE、FLOOR1、FLOOR2 与 FLOOR3 上执行"display stp brief"命令查看生成树简要信息，显示结果如图 2.99～图 2.102 所示。注意观察交换机端口角色与端口状态。

注：根据规划，CORE 为根，其实例 0 中所有端口均为指定端口，均为转发状态。接入层交换机 FLOOR1、FLOOR2、FLOOR3 与 CORE 相连的端口为根端口，连主机的端口为指定端口，也均处于转发状态。

```
[CORE]display stp brief
MSTID  Port                   Role   STP State    Protection
 0     GigabitEthernet0/0/1   DESI   FORWARDING   NONE
 0     GigabitEthernet0/0/2   DESI   FORWARDING   NONE
 0     GigabitEthernet0/0/3   DESI   FORWARDING   NONE
 0     GigabitEthernet0/0/21  DESI   FORWARDING   BPDU
 0     GigabitEthernet0/0/22  DESI   FORWARDING   BPDU
 0     GigabitEthernet0/0/23  DESI   FORWARDING   BPDU
```

图 2.99　在交换机 CORE 上执行"display stp brief"命令后的显示结果

```
[FLOOR1]display stp brief
MSTID  Port                       Role    STP State     Protection
  0    GigabitEthernet0/0/1      ROOT  FORWARDING     NONE
  0    GigabitEthernet0/0/3      DESI  FORWARDING     BPDU
```

图 2.100　在交换机 FLOOR1 上执行 "display stp brief" 命令后的显示结果

```
[FLOOR2]display stp brief
MSTID  Port                       Role    STP State     Protection
  0    GigabitEthernet0/0/1      ROOT  FORWARDING     NONE
  0    GigabitEthernet0/0/6      DESI   FORWARDING     BPDU
```

图 2.101　在交换机 FLOOR2 上执行 "display stp brief" 命令后的显示结果

```
[FLOOR3]display stp brief
MSTID  Port                       Role    STP State     Protection
  0    GigabitEthernet0/0/1      ROOT  FORWARDING     NONE
  0    GigabitEthernet0/0/22     DESI   FORWARDING     BPDU
```

图 2.102　在交换机 FLOOR3 上执行 "display stp brief" 命令后的显示结果

③ 在交换机 FLOOR4 上执行 "display stp region-configuration" 命令，查看 MSTP 域配置信息，如图 2.103 所示。

```
[FLOOR4]display stp region-configuration
 Oper configuration
  Format selector   :0
  Region name       :leader
  Revision level    :1

 Instance  VLANs Mapped
    0      1 to 4094
```

图 2.103　在交换机 FLOOR4 上执行 "display stp region-configuration" 命令后的显示结果

④ 在交换机 FLOOR4 上执行 "display stp brief" 命令查看生成树简要信息，显示结果如图 2.104 所示。注意观察交换机端口角色与端口状态。根据规划，FLOOR4 为 leader 域的根，其实例 0 中所有端口均为指定端口，均为转发状态。

```
[FLOOR4]display stp brief
MSTID  Port                       Role    STP State     Protection
  0    GigabitEthernet0/0/1      DESI  FORWARDING     NONE
  0    GigabitEthernet0/0/2      DESI  FORWARDING     BPDU
  0    GigabitEthernet0/0/14     DESI  FORWARDING     BPDU
```

图 2.104　在交换机 FLOOR4 上执行 "display stp brief" 命令后的显示结果

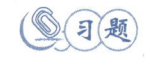 习题

选择题

（1）STP 在选择根桥时，下面（　　　）会被选择成为树根。

　　A. 具有最小 MAC 地址的交换机　　　　B. 具有最大 MAC 地址的交换机

项目 2　小型企业网络互联项目　　173

　　C. 具有最小 BID 的交换机　　　　　D. 具有最大 BID 的交换机

（2）下面（　　　）可把数据帧的源 MAC 地址学习到 MAC 地址表中，但不转发该数据帧。

　　A. 转发状态　　　　B. 学习状态　　　　C. 阻塞状态　　　　D. 侦听状态

（3）STP 中，（　　　）的端口不会发送配置 BPDU。

　　A. 转发状态　　　　B. 学习状态　　　　C. 阻塞状态　　　　D. 侦听状态

（4）在交换机中使用"stp edged-port enable"命令配置交换机端口的作用是（　　　）。

　　A. 用来快速消除二层环路

　　B. 该命令在连接到其他交换机的中继接口上配置以减少 STP 收敛时间

　　C. 如果该命令配置在接入端口上，接入接口会立即从阻塞状态进入转发状态

（5）下面（　　　）为链路聚合技术正式标准。

　　A. IEEE 802. 1Q　　B. IEEE 802. 3AD　C. IEEE 802. 1W　　D. IEEE 802. 1D

（6）下面（　　　）为 MSTP 正式标准。

　　A. IEEE 802. 1S　　B. IEEE 802. 3AD　C. IEEE 802. 1W　　D. IEEE 802. 1D

任务 5　PPP 的规划与配置

　　在完成本项目任务 1 到任务 4 的基础上，在图 2.1 所示的网络中，完成使用 PPP 认证功能实现企业边界路由器 AR1 与运营商路由器 ISP 之间双向身份认证的规划与配置任务。

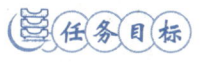

● 理解 PPP 的工作过程。

● 理解 PAP 的工作过程，掌握 PAP 的配置。

● 理解 CHAP 的工作过程，掌握 CHAP 的配置。

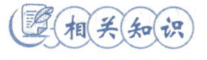

1. PPP 概述

　　PPP（point-to-point protocol，点到点协议）是一种常见的广域网数据链路层协议，是在 SLIP（serial line internet protocol，串行线 IP）协议的基础上发展起来的，可用在同步和异步专线、异步拨号链路和同步拨号链路上。

　　PPP 是由一系列协议构成的协议簇，PPP 协议的分层结构如图 2.105 所示。PPP 提供 LCP（link control protocol，链路控制协议），用于各种链路层参数的协商，例如最大接收单元、认

证模式等，提供 NCP（network control protocol，网络控制协议），如 IPCP（IP control protocol，IP 控制协议），用于各网络层参数的协商，更好地支持了网络层协议。

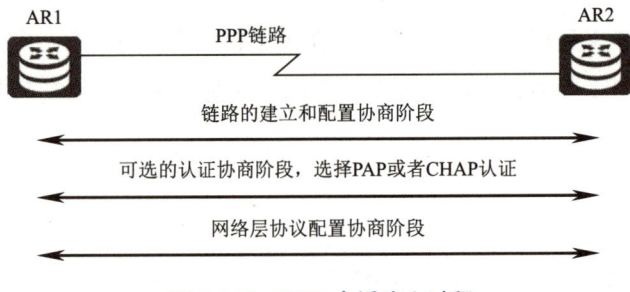

图 2.105　PPP 协议的分层结构

2. PPP 会话的建立

PPP 会话的建立分为链路层协商、认证协商（可选）和网络层协商三个阶段，如图 2.106 所示。

（1）链路层协商

通过 LCP 报文进行链路参数协商，如建立链路层连接。

（2）认证协商（可选）

通过链路建立阶段协商的认证方式进行链路认证。

（3）网络层协商

通过 NCP 协商来选择和配置一个网络层协议并进行网络层参数协商。

图 2.106　PPP 会话建立过程

3. PPP 认证

PPP 提供了 PAP（password authentication protocol，密码认证协议）和 CHAP（challenge handshake authentication protocol，挑战握手认证协议）。

（1）PAP

PAP 认证协议为两次握手认证协议，密码以明文方式在链路上发送，如图 2.107 所示。认证过程如下：

① 被认证方将配置的用户名和密码信息以明文方式发送给认证方。

② 认证方收到被认证方发送的用户名和密码信息之后，根据本地配置的用户名和密码数据库检查用户名和密码信息是否匹配，如果匹配，则发送 ACK 消息，通告对方通过认证，允许进

入下一阶段；如果不匹配，则发送 NAK 消息，通告对方认证失败。

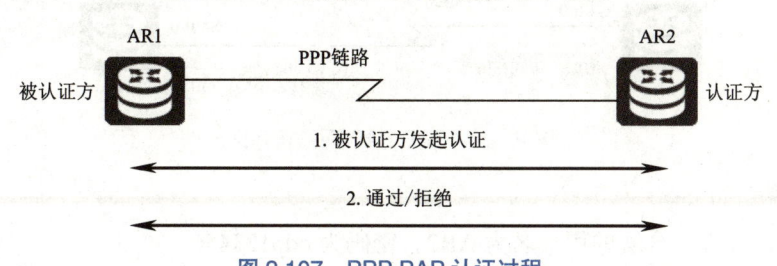

图 2.107　PPP PAP 认证过程

（2）CHAP

CHAP 认证使用三次握手认证，协商报文被加密后再在链路上传输，如图 2.108 所示。认证过程如下：

① 认证方主动发起认证请求，认证方向被认证方发送 Challenge 报文，报文内包含随机数（Random）和 ID。

② 被认证方收到此 Challenge 报文之后，进行一次加密运算，得到一个摘要信息，然后将此摘要信息和端口上配置的 CHAP 用户名一起封装在 Response 报文中发回认证方。

③ 认证方接收到被认证方发送的 Response 报文之后，按照其中的用户名在本地查找相应的密码信息，得到密码信息之后，进行一次加密运算，运算方式和被认证方的加密运算方式相同；然后将加密运算得到的摘要信息和 Response 报文中封装的摘要信息做比较，相同则认证成功，不相同则认证失败。

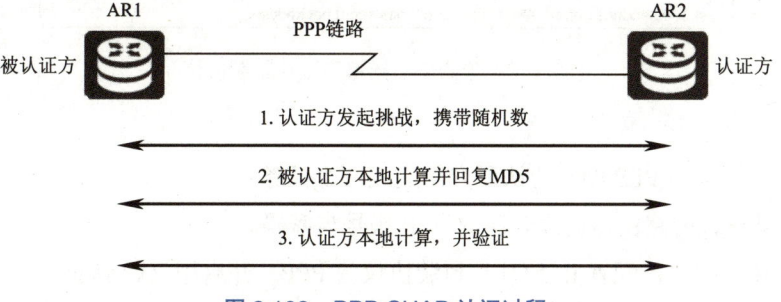

图 2.108　PPP CHAP 认证过程

使用 CHAP 认证方式时，被认证方的密码是被加密后再进行传输的，这样就极大地提高了安全性。

相关技能

1. PPP PAP 本地认证的规划与配置

搭建如图 2.109 所示的网络拓扑图，配置路由器 AR1 与 AR2 的串口，使用 PPP PAP 进行单向身份认证，其中路由器 AR1 为认证方，路由器 AR2 为被认证方，并进行测试与查看。

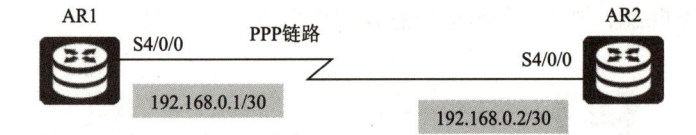

图 2.109　PPP PAP 网络拓扑图

PPP PAP 规划如下；

①AR1 为认证方，合法的用户名为 AR2，密码为 cdp123456

②AR2 为被认证方。

（1）观察华为串口默认数据链路层协议

分别在路由器 AR1 与 AR2 上使用"display interface s4/0/0"命令显示关于接口 S4/0/0 的 IP 信息。观察图 2.110 所示的信息。

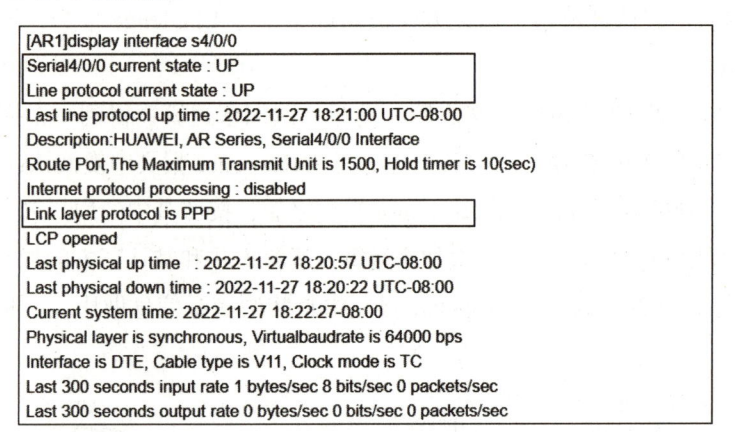

图 2.110　华为路由器串口默认封装协议

（2）认证方路由器 AR1 的配置

在 AR1 上配置成为 PPP PAP 的认证方需完成两个任务：

①为远程被认证的路由器创建合法的 PPP 账号与密码。

②进入 S4/0/0 接口，配置 IP 地址与封装协议为 PPP，并启用 PAP 认证。

配置的步骤如下：

```
[AR1]aaa
[AR1-aaa]local-user AR2 password cipher cdp123456
[AR1-aaa]local-user AR2 service-type ppp
[AR1-aaa]quit
// 创建合法的 PPP 账号与密码,账号服务类型为 PPP
[AR1]interface s4/0/0
[AR1-Serial4/0/0]ip address 192.168.0.1 30
[AR1-Serial4/0/0]link-protocol ppp
[AR1-Serial4/0/0]ppp authentication-mode pap
```

```
[AR1-Serial4/0/0]quit
// 配置 S4/0/0 接口协议为 PPP 协议，启用 PAP 认证
```

（3）被认证方路由器 AR2 的配置

在 AR2 上配置成为 PPP 的被认证方只需要进入串口，配置封装协议为 PPP（华为路由器串口默认协议为 PPP），发送 PPP PAP 的用户名与密码（该用户名与密码必须能与认证方配置的合法用户名与密码匹配）。配置步骤如下：

```
[AR2]interface s4/0/0
[AR2-Serial4/0/0]ip address 192.168.0.2 30
[AR2-Serial4/0/0]link-protocol ppp
[AR2-Serial4/0/0]ppp pap local-user AR2 password cipher cdp123456
[AR2-Serial4/0/0]
```

（4）验证与测试

① 在路由器 AR1 或者 AR2 上将 S4/0/0 接口禁用，然后再启用接口后，执行"display interface serial 4/0/0"命令查看接口的配置信息，接口的物理和链路协议状态都是"UP"，并且 PPP 的 LCP 和 IPCP 都是"opened"状态，说明链路的 PPP 协商已经成功，如图 2.111 所示。

```
[AR2-Serial4/0/0]shutdown          // 禁用接口
[AR2-Serial4/0/0]undo shutdown     // 启用接口
```

```
[AR1]display interface s4/0/0
Serial4/0/0 current state : UP
Line protocol current state : UP
Last line protocol up time : 2023-02-27 16:14:24 UTC-08:00
Description:HUAWEI, AR Series, Serial4/0/0 Interface
Route Port,The Maximum Transmit Unit is 1500, Hold timer is 10(sec)
Internet Address is 125.71.28.93/24
Link layer protocol is PPP
LCP opened, IPCP opened
Last physical up time   : 2023-02-27 16:12:47 UTC-08:00
Last physical down time : 2023-02-27 16:12:46 UTC-08:00
Current system time: 2023-02-27 16:15:02-08:00
Physical layer is synchronous, Virtualbaudrate is 64000 bps
Interface is DTE, Cable type is V11, Clock mode is TC
Last 300 seconds input rate 8 bytes/sec 64 bits/sec 0 packets/sec
Last 300 seconds output rate 3 bytes/sec 24 bits/sec 0 packets/sec

Input: 1810 packets, 57398 bytes
 Broadcast:        0, Multicast:        0
 Errors:           0, Runts:            0
 Giants:           0, CRC:              0

 Alignments:       0, Overruns:         0
 Dribbles:         0, Aborts:           0
 No Buffers:       0, Frame Error:      0
 ---- More ----
```

图 2.111　PPP 协商成功

② 在路由器 AR1 和 AR2 上可以互相 ping 通对方，如图 2.112 所示，路由器 AR1 可以 ping 通 AR2 的 S4/0/0 接口 IP 地址。

```
[AR1]ping 192.168.0.2
 PING 192.168.0.2: 56  data bytes, press CTRL_C to break
  Reply from 192.168.0.2: bytes=56 Sequence=1 ttl=255 time=30 ms
  Reply from 192.168.0.2: bytes=56 Sequence=2 ttl=255 time=20 ms
  Reply from 192.168.0.2: bytes=56 Sequence=3 ttl=255 time=20 ms
  Reply from 192.168.0.2: bytes=56 Sequence=4 ttl=255 time=30 ms
  Reply from 192.168.0.2: bytes=56 Sequence=5 ttl=255 time=20 ms
```

图 2.112　路由器 AR1 ping 通路由器 AR2 的 S4/0/0 接口

2. PPP CHAP 本地认证的规划与配置

搭建如图 2.113 所示的网络拓扑图，配置路由器 AR1 与 AR2 的串口，使用 PPP CHAP 进行单向身份认证，其中路由器 AR1 为认证方，路由器 AR2 为被认证方，并进行测试与查看。

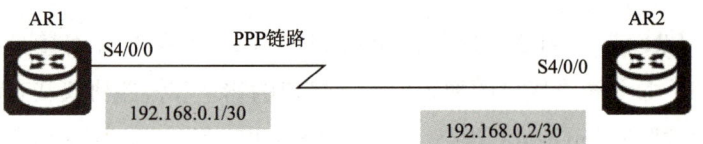

图 2.113　PPP CHAP 网络拓扑图

PPP CHAP 规划如下：

① AR1 为认证方，合法的用户名为"AR2"，密码为"cdp123456"。

② AR2 为被认证方。

（1）认证方路由器 AR1 的配置

在 AR1 上配置成为 PPP CHAP 的认证方需完成两个任务：

① 为远程被认证的路由器创建合法的 PPP 账号与密码。

② 进入 S4/0/0 接口，配置 IP 地址与封装协议为 PPP，并启用 CHAP 认证。

配置的步骤如下：

```
[AR1]aaa
[AR1-aaa]local-user AR2 password cipher cdp123456
[AR1-aaa]local-user AR2 service-type ppp
[AR1-aaa]quit
// 创建合法的 PPP 账号与密码，账号服务类型为 PPP
[AR1]interface s4/0/0
[AR1-Serial4/0/0]ip address 192.168.0.1 30
[AR1-Serial4/0/0]link-protocol ppp
[AR1-Serial4/0/0]ppp authentication-mode chap
[AR1-Serial4/0/0]quit
```

```
// 配置 S4/0/0 接口协议为 PPP 协议，启用 CPAP 认证
```

（2）被认证方路由器 AR2 的配置

在 AR2 上配置成为 PPP 的被认证方只需要进入串口，配置封装协议为 PPP，发送 PPP CHAP 的用户名与密码（该用户名与密码必须能与认证方配置的合法用户名与密码匹配）。配置的步骤如下：

```
[AR2]interface s4/0/0
[AR2-Serial4/0/0]ip address 192.168.0.2 30
[AR2-Serial4/0/0]link-protocol ppp
[AR2-Serial4/0/0]ppp chap user AR2
[AR2-Serial4/0/0]ppp chap password cipher cdp123456
[AR2-Serial4/0/0]
```

（3）验证与测试

① 在路由器 AR1 或者 AR2 上将 S4/0/0 接口禁用，然后再启用接口后，执行 "display interface serial 4/0/0" 命令查看接口的配置信息，接口的物理和链路协议的状态都是 "UP"，并且 PPP 的 LCP 和 IPCP 都是 "opened" 状态，说明链路的 PPP 协商已经成功，如图 2.114 所示。

```
[AR2-Serial4/0/0]shutdown            // 禁用接口
[AR2-Serial4/0/0]undo shutdown       // 启用接口
```

```
[AR1]display interface s4/0/0
Serial4/0/0 current state : UP
Line protocol current state : UP
Last line protocol up time : 2023-03-30 14:47:28 UTC-08:00
Description:HUAWEI, AR Series, Serial4/0/0 Interface
Route Port,The Maximum Transmit Unit is 1500, Hold timer is 10(sec)
Internet Address is 192.168.0.1/30
Link layer protocol is PPP
LCP opened, IPCP opened
Last physical up time   : 2023-03-30 14:47:25 UTC-08:00
Last physical down time : 2023-03-30 14:47:20 UTC-08:00
Current system time: 2023-03-30 14:47:36-08:00
Physical layer is synchronous, Virtualbaudrate is 64000 bps
Interface is DTE, Cable type is V11, Clock mode is TC
Last 300 seconds input rate 6 bytes/sec 48 bits/sec 0 packets/sec
Last 300 seconds output rate 2 bytes/sec 16 bits/sec 0 packets/sec
```

图 2.114　PPP 协商成功

② 在路由器 AR1 和 AR2 上可以互相 ping 通对方，如图 2.115 所示，路由器 AR1 ping 通路由器 AR2 的 S4/0/0 接口。

180 企业网络互联技术（双语）

```
[AR1]ping 192.168.0.2
  PING 192.168.0.2: 56  data bytes, press CTRL_C to break
    Reply from 192.168.0.2: bytes=56 Sequence=1 ttl=255 time=30 ms
    Reply from 192.168.0.2: bytes=56 Sequence=2 ttl=255 time=20 ms
    Reply from 192.168.0.2: bytes=56 Sequence=3 ttl=255 time=20 ms
    Reply from 192.168.0.2: bytes=56 Sequence=4 ttl=255 time=30 ms
    Reply from 192.168.0.2: bytes=56 Sequence=5 ttl=255 time=20 ms
```

图 2.115　路由器 AR1 ping 通路由器 AR2 的 S4/0/0 接口

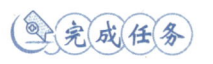 完成任务

1. 任务规划

根据任务描述和图 2.1 所示的拓扑结构，企业边界路由器 AR1 与运营商路由器 ISP 之间使用 PPP CHAP 实现双向认证，认证密码为"cdp123456"，IP 规划见表 2.12。

2. 任务实施

（1）PPP CHAP 本地认证的配置

① 根据 PPP CHAP 的规划，路由器 AR1 配置步骤如下：

```
[AR1]aaa
[AR1-aaa]local-user ISP password cipher cdp123456
[AR1-aaa]local-user ISP service-type ppp
[AR1-aaa]quit
// 创建合法的 PPP 账号与密码，账号服务类型为 PPP
[AR1]interface s4/0/0
[AR1-Serial4/0/0]link-protocol ppp
[AR1-Serial4/0/0]ip address 125.71.28.93 24
[AR1-Serial4/0/0]ppp authentication-mode chap
[AR1-Serial4/0/0]ppp chap user AR1
[AR1-Serial4/0/0]ppp chap password cipher cdp123456
[AR1-Serial4/0/0]quit
// 配置 S4/0/0 接口协议为 PPP 协议，启用 CPAP 认证，CHAP 认证的用户名为"AR1"，认证密码为
// "cdp123456"
```

② 根据 PPP CHAP 的规划，路由器 ISP 配置步骤如下：

```
[ISP]aaa
[ISP-aaa]local-user AR1 password cipher cdp123456
[ISP-aaa]local-user AR1 service-type ppp
[ISP-aaa]quit
// 创建合法的 PPP 账号与密码，账号服务类型为 PPP
[ISP]interface s4/0/0
[ISP-Serial4/0/0]link-protocol ppp
```

```
[ISP-Serial4/0/0]ip address 125.71.28.1 24
[ISP-Serial4/0/0]ppp authentication-mode chap
[ISP-Serial4/0/0]ppp chap user ISP
[ISP-Serial4/0/0]ppp chap password cipher cdp123456
[ISP-Serial4/0/0]quit
```
// 配置 S4/0/0 接口协议为 PPP 协议，启用 CPAP 认证，CHAP 认证的用户名为 "ISP"，认证密码为
// "cdp123456"

（2）验证与测试

① 将路由器上配置了 PPP CHAP 认证的接口禁用，再启用接口，执行 "display interface serial 4/0/0" 命令查看接口的配置信息，接口的物理和链路协议的状态都是 "UP"，并且 PPP 的 LCP 和 IPCP 都是 "opened" 状态，说明链路的 PPP 协商已经成功，如图 2.116 所示。

```
[AR1-Serial4/0/0]shutdown        // 禁用接口
[AR1-Serial4/0/0]undo shutdown   // 启用接口
```

```
[AR1]display interface s4/0/0
Serial4/0/0 current state : UP
Line protocol current state : UP
Last line protocol up time : 2023-02-27 16:14:24 UTC-08:00
Description:HUAWEI, AR Series, Serial4/0/0 Interface
Route Port,The Maximum Transmit Unit is 1500, Hold timer is 10(sec)
Internet Address is 125.71.28.93/24
Link layer protocol is PPP
LCP opened, IPCP opened
Last physical up time   : 2023-02-27 16:12:47 UTC-08:00
Last physical down time : 2023-02-27 16:12:46 UTC-08:00
Current system time: 2023-02-27 16:15:02-08:00
Physical layer is synchronous, Virtualbaudrate is 64000 bps
Interface is DTE, Cable type is V11, Clock mode is TC
Last 300 seconds input rate 8 bytes/sec 64 bits/sec 0 packets/sec
Last 300 seconds output rate 3 bytes/sec 24 bits/sec 0 packets/sec

Input: 1810 packets, 57398 bytes
 Broadcast:        0, Multicast:        0
 Errors:           0, Runts:            0
 Giants:           0, CRC:              0

 Alignments:       0, Overruns:         0
 Dribbles:         0, Aborts:           0
 No Buffers:       0, Frame Error:      0
---- More ----
```

图 2.116 PPP 协商成功

② 在路由器 AR1 和 ISP 上可以互相 ping 通对方，如图 2.117 所示，路由器 AR1 ping 通路由器 ISP 的 S4/0/0 接口。

企业网络互联技术（双语）

```
[AR1]ping 125.71.28.1
 PING 125.71.28.1: 56  data bytes, press CTRL_C to break
  Reply from 125.71.28.1: bytes=56 Sequence=1 ttl=255 time=40 ms
  Reply from 125.71.28.1: bytes=56 Sequence=2 ttl=255 time=30 ms
  Reply from 125.71.28.1: bytes=56 Sequence=3 ttl=255 time=30 ms
  Reply from 125.71.28.1: bytes=56 Sequence=4 ttl=255 time=20 ms
  Reply from 125.71.28.1: bytes=56 Sequence=5 ttl=255 time=20 ms

 --- 125.71.28.1 ping statistics ---
  5 packet(s) transmitted
  5 packet(s) received
  0.00% packet loss
  round-trip min/avg/max = 20/28/40 ms
```

图 2.117　路由器 AR1 ping 通路由器 ISP 的 S4/0/0 接口

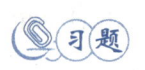 习题

选择题

（1）PPP PAP 认证采用（　　　）次握手。

　　A. 一　　　　　　　B. 二　　　　　　　C. 三　　　　　　　D. 四

（2）PPP CHAP 认证采用（　　　）次握手。

　　A. 一　　　　　　　B. 二　　　　　　　C. 三　　　　　　　D. 四

（3）下面（　　　）不是 PPP 协议的组成。

　　A. LCP　　　　　　B. NCP　　　　　　C. CHAP　　　　　　D. IP

任务6　RIP 的规划与配置

 任务描述

　　在完成本项目任务 1 到任务 5 的基础上，在图 2.1 所示的网络中，完成使用默认路由与动态路由协议 RIP 实现企业内网互联互通，企业内网可访问 Internet 的路由规划与配置任务。

 任务目标

- 理解动态路由与静态路由的区别。
- 理解 RIPv1 的工作原理与特点。
- 掌握 RIPv1 的配置、查看与排障。
- 理解 RIPv2 的工作原理与特点。
- 掌握 RIPv2 的配置、查看与排障。

相关知识

1. 动态路由与路由协议

静态路由由网络管理员负责选路，并手工进行配置，路由的开销少，但不灵活。当网络互连规模增大时，静态路由手工方式生成和维护一个路由表会非常复杂，且不能及时适应网络状态的变化。因此，当网络互连规模较大或网络中的不稳定因素较多时，通常使用动态路由的方式生成和维护路由表。在路由器上运行路由协议并进行相应的路由协议配置即可保证路由器自动生成并维护路由表信息。

路由协议有多种分类方法：

① 按适用范围的不同可分为外部网关协议（EGP）和内部网关协议（IGP）。

② 按路由算法的不同，路由协议分为距离矢量路由协议和链路状态路由协议。

IPv4 的路由协议有：RIPv1、RIPv2、OSPF、IS-IS 和 BGP 等，IPv6 的路由协议有：RIPng、OSPFv3、BGPv4 等，如图 2.118 所示。

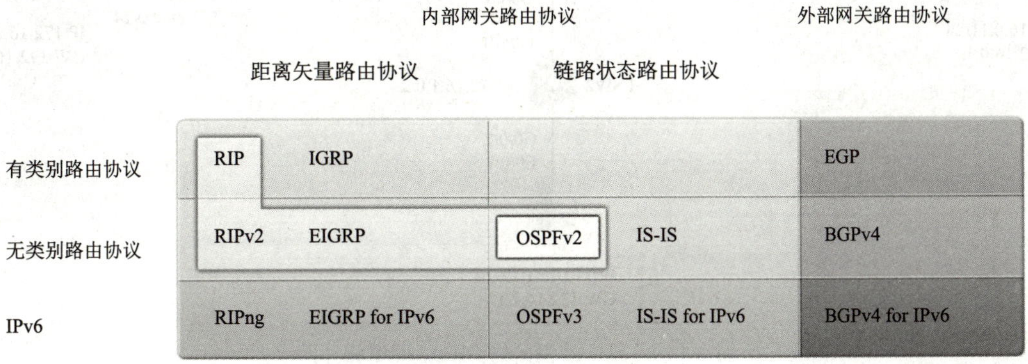

图 2.118 路由协议的分类

使用路由协议动态构建的路由表能更好地适应网络状态的变化，如网络拓扑或网络流量的变化等，同时也减少了人工生成与维护路由表的工作量。但为此付出的代价则是用于路由信息更新的资源耗费，包括网络带宽、路由器 CPU 和存储资源。

2. RIP 协议

RIP 为 routing information protocol（路由信息协议）的简称，它是一种内部网关协议（interior gateway protocol）。RIP 是基于距离矢量（distance vector, D-V）算法的协议，使用跳数（hop count）作为度量来衡量到达目标网络的距离。跳数是指到达目标网络所需要经过的路由器数目。RIP 通过 UDP 报文进行路由信息的交换，使用的端口号为 520。

RIP 虽然实现及配置都非常简单，但该协议所使用的距离矢量算法可能会在网络中形成路由环路。为此，RIP 引入了水平分割、抑制更新定时器、规定最大跳数、毒化路由等机制来减少或防止路由环路的出现。RIP 的最大跳数规定为 15，凡是到目标网络超出了 15 跳，RIP 则认为目标网络不可达。运行 RIP 协议的邻居路由器之间采用定期更新的方式进行路由信息的交换。

RIP 有两个版本，即 RIPv1 和 RIPv2。RIPv1 采用广播发送路由更新信息，且 RIPv1 的协议标准是早于 VLSM 与 CIDR 之前提出的，因此，RIPv1 不支持 VLSM 与 CIDR，在 VLSM 与 CIDR 网络互连环境中无法启用 RIPv1。在 RIPv1 的路由更新信息中，不包括子网掩码信息。RIPv2 继承了 RIPv1 的所有功能，并在此基础上进行改进，改进的主要内容包括对 VLSM 和认证的支持，RIPv2 采用组播发送路由更新信息，组播地址为 224.0.0.9。

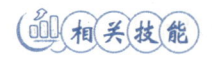 相关技能

1. 基于路由器的 RIPv2 规划与配置

搭建如图 2.119 所示的网络拓扑图与 IP 地址分配，在路由器 AR1、AR2 和 AR3 上使用 RIPv2 路由协议实现网络的互联互通，并进行路由查看与测试。

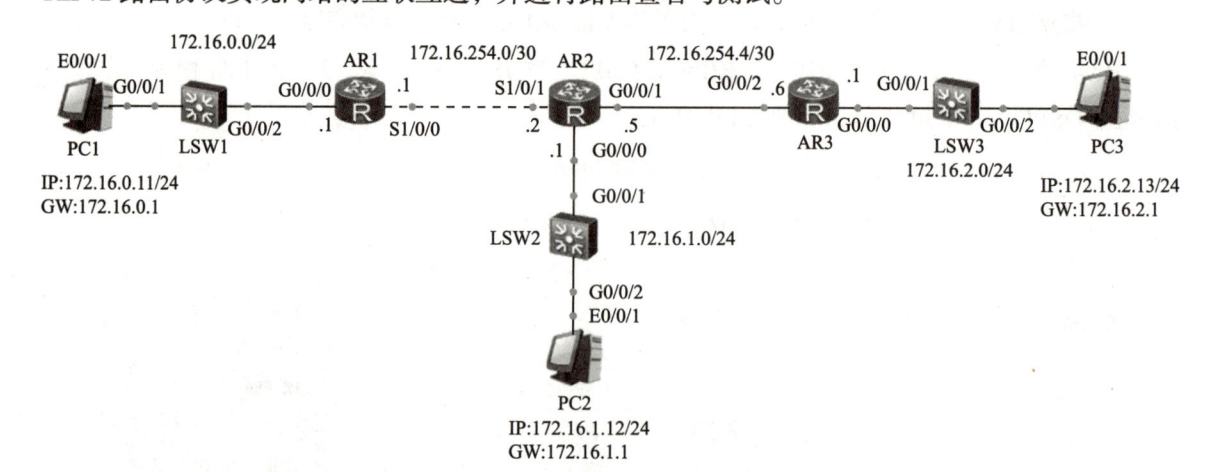

图 2.119　基于路由器的 RIPv2 网络拓扑图

RIPv2 的规划见表 2.17。

表 2.17　RIPv2 的规划

设　　备	RIP 进程号	参与 RIP 更新的网络	静默接口	是否禁用自动汇总
AR1	1	172.16.0.0	G0/0/0	是
AR2	1	172.16.0.0	G0/0/0	是
AR3	1	172.16.0.0	G0/0/0	是

（1）路由器接口的配置

① 在路由器 AR1、AR2 与 AR3 上根据图 2.119 规划的 IP 完成接口的配置，步骤如下：

```
[AR1]interface g0/0/0
[AR1-GigabitEthernet0/0/0]ip address 172.16.0.1 24
[AR1-GigabitEthernet0/0/0]quit
// 配置路由器 AR1 的 G0/0/0 接口 IP 地址与掩码为 "172.16.0.1/24"
[AR1]interface s1/0/0
```

```
[AR1-Serial1/0/0]ip address 172.16.254.1 30
[AR1-Serial1/0/0]quit
// 配置路由器 AR1 的 S1/0/0 接口 IP 地址与掩码为 "172.16.254.1/30"
[AR2]interface g0/0/0
[AR2-GigabitEthernet0/0/0]ip address 172.16.1.1 24
[AR2-GigabitEthernet0/0/0]quit
// 配置路由器 AR2 的 G0/0/0 接口 IP 地址与掩码为 "172.16.1.1/24"
[AR2]interface s1/0/1
[AR2-Serial1/0/1]ip address 172.16.254.2 30
[AR2-Serial1/0/1]quit
// 配置路由器 AR2 的 S1/0/1 接口 IP 地址与掩码为 "172.16.254.2/30"
[AR2]interface g0/0/1
[AR2-GigabitEthernet0/0/1]ip address 172.16.254.5 30
[AR2-GigabitEthernet0/0/1]quit
// 配置路由器 AR2 的 G0/0/1 接口 IP 地址与掩码为 "172.16.254.5/30"
[AR3]interface g0/0/0
[AR3-GigabitEthernet0/0/0]ip address 172.16.2.1 24
[AR3-GigabitEthernet0/0/0]quit
// 配置路由器 AR3 的 G0/0/0 接口 IP 地址与掩码为 "172.16.2.1/24"
[AR3]interface g0/0/2
[AR3-GigabitEthernet0/0/2]ip address 172.16.254.6 30
[AR3-GigabitEthernet0/0/2]quit
// 配置路由器 AR3 的 G0/0/2 接口 IP 地址与掩码为 "172.16.254.6/30"
```

② 在路由器 AR1、AR2 和 AR3 上使用 "display ip interface brief" 命令检查路由器接口 IP 地址是否正确、接口的物理与协议状态是否均为 "UP"。

（2）RIPv2 的配置

RIPv2 的配置需要完成以下任务：

① 创建并运行 RIP 进程，指定 RIP 进程号。

② 指定 RIP 的版本为 "2"。

③ 使用 "network" 命令指定哪些网络参与 RIP 的路由更新。

"network" 命令包含两层含义：一方面用来指定本机上哪些直连的网络参与 RIP 更新，另一方面用来指定哪些接口可以收发 RIP 更新。

④ 配置静默接口（可选配置），配置为静默接口的端口不发送 RIP 更新，只能接收 RIP 更新。

⑤ 配置禁用自动汇总（可选配置）。

配置的步骤如下：

```
[AR1]rip 1                    // 启用 RIP 协议，进程号为 1
[AR1-rip-1]version 2          // 指定 RIP 版本为 RIPv2
```

```
[AR1-rip-1]network 172.16.0.0    // 网络 "172.16.0.0" 参与 RIPv2 的路由更新，注：RIP 配
                                 // 置的网络号为有类别网络号
[AR1-rip-1]silent-interface g0/0/0   // 配置 G0/0/0 接口为静默接口
[AR1-rip-1]undo summary          // 禁用自动汇总
[AR2]rip 1                       // 启用 RIP 协议，进程号为 1
[AR2-rip-1]version 2             // 指定 RIP 版本为 RIPv2
[AR2-rip-1]network 172.16.0.0    // 网络 "172.16.0.0" 参与 RIPv2 的路由更新，注：
                                 // RIP 指定的网络号为有类别网络号
[AR2-rip-1]silent-interface g0/0/0   // 配置 G0/0/0 接口为静默接口
[AR2-rip-1]undo summary          // 禁用自动汇总
[AR3]rip 1                       // 启用 RIP 协议，进程号为 1
[AR3-rip-1]version 2             // 指定 RIP 版本为 RIPv2
[AR3-rip-1]network 172.16.0.0    // 网络 "172.16.0.0" 参与 RIPv2 的路由更新，注：
                                 // RIP 指定的网络号为有类别网络号
[AR3-rip-1]silent-interface g0/0/0   // 配置 G0/0/0 接口为静默接口
[AR3-rip-1]undo summary          // 禁用自动汇总
```

（3）验证与测试

① 在路由器 AR1、AR2 与 AR3 上分别执行 "display ip routing-table | include RIP" 命令查看路由器路由表中协议为 RIP 的路由条目。图 2.120、图 2.121 和图 2.122 分别为路由器 AR1、AR2 和 AR3 上执行 "display ip routing-table | include RIP" 命令后的显示结果。

注：路由条目中的 "RIP" 表示该路由条目通过 RIP 协议学习到，RIP 的默认路由优先级值为 "100"，Cost 值为到目标网络的跳数。

② 在测试主机 PC1、PC2 和 PC3 上根据图 2.119 规划的 IP 完成 IP 地址、子网掩码和默认网关的配置，然后在命令行中使用 "ipconfig /all" 命令验证 IP 地址、子网掩码和默认网关的正确性。使用 ping 命令进行全网的连通性测试。图 2.123 所示为测试主机 PC1 成功 PING 通了 PC2 和 PC3。

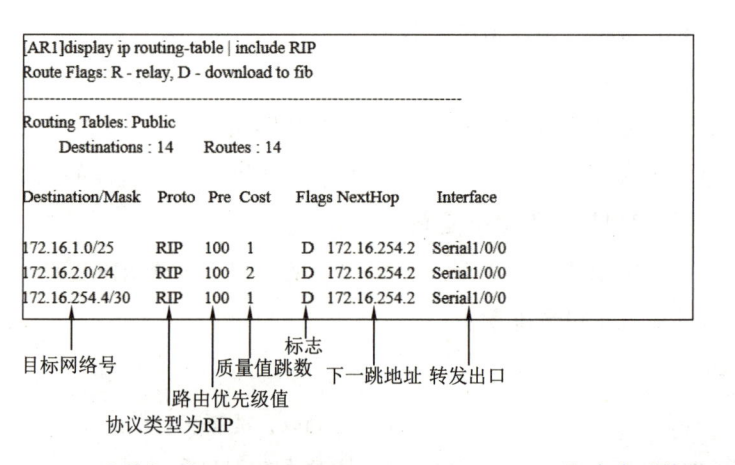

图 2.120　AR1 上执行 "display ip routing-table | include RIP" 命令后的显示结果

```
[AR2]display ip routing-table | include RIP
Route Flags: R - relay, D - download to fib
------------------------------------------------------------------------
Routing Tables: Public
         Destinations : 16        Routes : 16

Destination/Mask   Proto  Pre  Cost    Flags NextHop      Interface

172.16.0.0/24  RIP    100  1        D  172.16.254.1   Serial1/0/1
172.16.2.0/24  RIP    100  1        D  172.16.254.6   GigabitEthernet0/0/1
```

图 2.121　AR2 上执行"display ip routing-table | include RIP"命令后的显示结果

```
[AR3]display ip routing-table | include RIP
Route Flags: R - relay, D - download to fib
------------------------------------------------------------------------
Routing Tables: Public
         Destinations : 13        Routes : 13

Destination/Mask   Proto   Pre  Cost    Flags NextHop      Interface

172.16.0.0/24      RIP    100  2        D  172.16.254.5   GigabitEthernet0/0/2
172.16.1.0/25      RIP    100  1        D  172.16.254.5   GigabitEthernet0/0/2
172.16.254.0/30    RIP    100  1        D  172.16.254.5   GigabitEthernet0/0/2
```

图 2.122　AR3 上执行"display ip routing-table | include RIP"命令后的显示结果

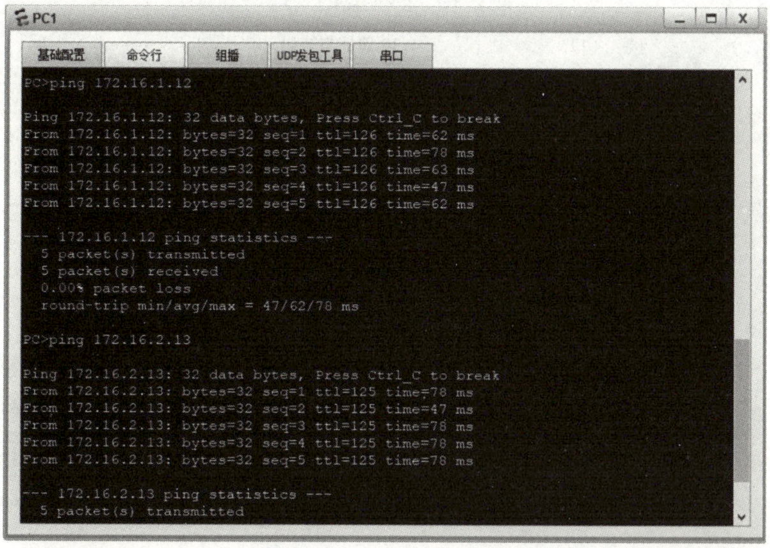

图 2.123　测试主机 PC1 成功 ping 通了 PC2 和 PC3

2. 基于三层交换机的 RIPv2 规划与配置

搭建如图 2.124 所示的网络拓扑图（**注：交换机 LSW1 与 LSW2 为支持 RIP 协议的三层交换机**）。VLAN 的规划见表 2.18，IP 地址分配见表 2.19，使用 RIPv2 路由协议实现网络的互联互通，并进行路由查看与测试。

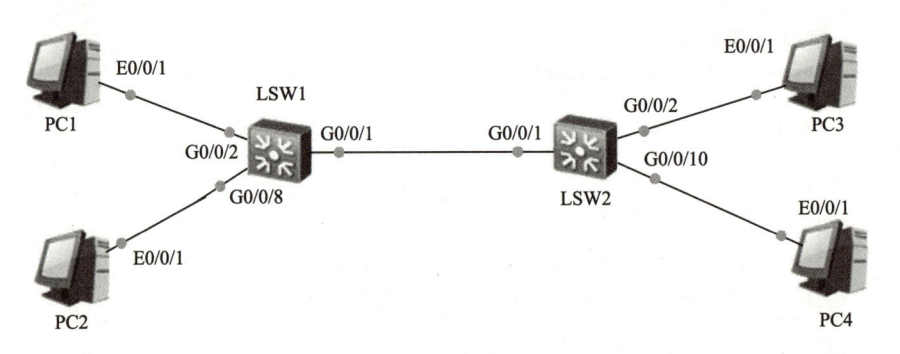

图 2.124　基于三层交换机的 RIPv2 拓扑图

表 2.18　VLAN 的规划

设　　备	接口或 VLAN	VLAN 的描述	接　　口	说　　明
LSW1	VLAN2	student	G0/0/2	
	VLAN3	teacher	G0/0/8	
	VLAN100	LinktoLSW2	G0/0/1	用于连接到 LSW2
LSW2	VLAN20	jwc	G0/0/2 端口	
	VLAN30	cw	G0/0/10 端口	
	VLAN100	LinktoLSW1	G0/0/1	用于连接到 LSW1

表 2.19　IP 地址的规划

设　　备	接　　口	IP 地址与掩码	网　　关
LSW1	Vlanif 2	192.168.0.1/25	
	Vlanif 3	192.168.0.129/25	
	Vlanif100	192.168.254.1/30	
LSW2	Vlanif20	192.168.1.1/25	
	Vlanif30	192.168.1.129/25	
	Vlanif100	192.168.254.2/30	
PC1	NIC	192.168.0.2/25	192.168.0.1
PC2	NIC	192.168.0.130/25	192.168.0.129
PC3	NIC	192.168.1.2/25	192.168.1.1
PC4	NIC	192.168.1.130/25	192.168.1.129

RIPv2 的规划见表 2.20。

表 2.20　RIPv2 的规划

设　　备	RIP 进程号	参与 RIP 更新的网络	静默接口	是否禁用自动汇总
LSW1	1	192.168.0.0，192.168.254.0	VLAN2，VLAN3	是
LSW2	1	192.168.1.0，192.168.254.0	VLAN20，VLAN30	是

（1）VLAN 与 IP 的配置

① 在三层交换机 LSW1 上根据表 2.18 的 VLAN 规划和表 2.19 的 IP 规划完成 VLAN 和 IP 的配置，步骤如下：

```
[LSW1]vlan 2
[LSW1-vlan2]description student
[LSW1-vlan2]vlan 3
[LSW1-vlan3]description teacher
[LSW1-vlan3]vlan 100
[LSW1-vlan100]description linktoLSW2
[LSW1-vlan100]quit
[LSW1]interface g0/0/2
[LSW1-GigabitEthernet0/0/2]port link-type access
[LSW1-GigabitEthernet0/0/2]port default vlan 2
[LSW1-GigabitEthernet0/0/2]quit
[LSW1]interface g0/0/8
[LSW1-GigabitEthernet0/0/8]port link-type access
[LSW1-GigabitEthernet0/0/8]port default vlan 3
[LSW1-GigabitEthernet0/0/8]quit
[LSW1]interface g0/0/1
[LSW1-GigabitEthernet0/0/1]port link-type access
[LSW1-GigabitEthernet0/0/1]port default vlan 100
[LSW1-GigabitEthernet0/0/1]quit
[LSW1]interface Vlanif 2
[LSW1-Vlanif2]ip address 192.168.0.1 25
[LSW1-Vlanif2]quit
[LSW1]interface Vlanif 3
[LSW1-Vlanif3]ip address 192.168.0.129 25
[LSW1-Vlanif3]quit
[LSW1]interface Vlanif 100
[LSW1-Vlanif100]ip address 192.168.254.1 30
[LSW1-Vlanif100]quit
```

② 在三层交换机 LSW2 上根据表 2.18 的 VLAN 规划和表 2.19 的 IP 规划完成 VLAN 和 IP 的配置，步骤如下：

```
[LSW2]vlan 20
[LSW2-vlan20]description jwc
[LSW2-vlan20]vlan 30
[LSW2-vlan30]description cw
[LSW2-vlan30]vlan 100
```

```
[LSW2-vlan100]description linktoLSW1
[LSW2-vlan100]quit
[LSW2]interface g0/0/2
[LSW2-GigabitEthernet0/0/2]port link-type access
[LSW2-GigabitEthernet0/0/2]port default vlan 20
[LSW2-GigabitEthernet0/0/2]quit
[LSW2]interface g0/0/10
[LSW2-GigabitEthernet0/0/10]port link-type access
[LSW2-GigabitEthernet0/0/10]port default vlan 30
[LSW2-GigabitEthernet0/0/10]quit
[LSW2]interface g0/0/1
[LSW2-GigabitEthernet0/0/1]port link-type access
[LSW2-GigabitEthernet0/0/1]port default vlan 100
[LSW2-GigabitEthernet0/0/1]quit
[LSW2]interface Vlanif 20
[LSW2-Vlanif20]ip address 192.168.1.1 25
[LSW2-Vlanif20]quit
[LSW2]interface Vlanif 30
[LSW2-Vlanif30]ip address 192.168.1.129 25
[LSW2-Vlanif30]quit
[LSW2]interface Vlanif 100
[LSW2-Vlanif100]ip address 192.168.254.2 30
[LSW2-Vlanif100]quit
```

③ 在三层交换机 LSW1 和 LSW2 上使用"display ip interface brief"命令检查三层交换机的 Vlanif 接口 IP 地址与子网掩码是否正确、接口的物理与协议状态是否均为"UP"。

（2）RIPv2 的配置

LSW1 与 LSW2 的 RIP 配置步骤如下：

```
[LSW1]rip 1                              // 启用 RIP 协议, 进程号为 1
[LSW1-rip-1]version 2                    // 指定 RIP 版本为 RIPv2
[LSW1-rip-1]network 192.168.0.0          // 网络"192.168.0.0"参与 RIPv2 的路由更新, 注:
                                         // 子网"192.168.0.0/25"和"192.168.0.128/25"
                                         // 的有类别网络号均为"192.168.0.0"
[LSW1-rip-1]network 192.168.254.0        // 网络"192.168.254.0"参与 RIPv2 的路由更新
[LSW1-rip-1]silent-interface Vlanif 2    // 配置 Vlanif2 接口为静默接口
[LSW1-rip-1]silent-interface Vlanif 3    // 配置 Vlanif3 接口为静默接口
[LSW1-rip-1]undo summary                 // 禁用自动汇总
[LSW1-rip-1]quit
[LSW2]rip 1                              // 启用 RIP 协议, 进程号为 1
```

```
[LSW2-rip-1]version 2                    // 指定 RIP 版本为 RIPv2
[LSW2-rip-1]network 192.168.1.0          // 网络"192.168.1.0"参与 RIPv2 的路由更新，注:
                                         // 子网"192.168.1.0/25"和"192.168.1.128/25"
                                         // 的有类别网络号均为"192.168.1.0"
[LSW2-rip-1]network 192.168.254.0        // 网络"192.168.254.0"参与 RIPv2 的路由更新
[LSW2-rip-1]silent-interface Vlanif 20   // 配置 Vlanif20 接口为静默接口
[LSW2-rip-1]silent-interface Vlanif 30   // 配置 Vlanif30 接口为静默接口
[LSW2-rip-1]undo summary                 // 禁用自动汇总
[LSW2-rip-1]quit
```

（3）验证与测试

① 在路由器 AR1、AR2 与 AR3 上分别执行 "display ip routing-table" 命令显示路由表。图 2.125、图 2.126 分别为在三层交换机 LSW1 和 LSW2 上执行 "display ip routing-table" 命令后的显示结果。观察路由协议为 "RIP" 的路由条目中的 Cost 值、下一跳和转发出口。

```
[LSW1]display ip routing-table
Route Flags: R - relay, D - download to fib
------------------------------------------------------------
Routing Tables: Public
       Destinations : 10      Routes : 10

Destination/Mask    Proto   Pre  Cost    Flags NextHop        Interface

        127.0.0.0/8     Direct  0    0       D     127.0.0.1      InLoopBack0
        127.0.0.1/32    Direct  0    0       D     127.0.0.1      InLoopBack0
      192.168.0.0/25    Direct  0    0       D     192.168.0.1    Vlanif2
      192.168.0.1/32    Direct  0    0       D     127.0.0.1      Vlanif2
    192.168.0.128/25    Direct  0    0       D     192.168.0.129  Vlanif3
    192.168.0.129/32    Direct  0    0       D     127.0.0.1      Vlanif3
      192.168.1.0/25    RIP     100  1       D     192.168.254.2  Vlanif100
    192.168.1.128/25    RIP     100  1       D     192.168.254.2  Vlanif100
    192.168.254.0/30    Direct  0    0       D     192.168.254.1  Vlanif100
    192.168.254.1/32    Direct  0    0       D     127.0.0.1      Vlanif100
```

图 2.125 在 LSW1 上执行 "display ip routing-table" 命令后的显示结果

```
[LSW2]display ip routing-table
Route Flags: R - relay, D - download to fib
------------------------------------------------------------
Routing Tables: Public
       Destinations : 10      Routes : 10

Destination/Mask    Proto   Pre  Cost    Flags NextHop        Interface

        127.0.0.0/8     Direct  0    0       D     127.0.0.1      InLoopBack0
        127.0.0.1/32    Direct  0    0       D     127.0.0.1      InLoopBack0
      192.168.0.0/25    RIP     100  1       D     192.168.254.1  Vlanif100
    192.168.0.128/25    RIP     100  1       D     192.168.254.1  Vlanif100
      192.168.1.0/25    Direct  0    0       D     192.168.1.1    Vlanif20
      192.168.1.1/32    Direct  0    0       D     127.0.0.1      Vlanif20
    192.168.1.128/25    Direct  0    0       D     192.168.1.129  Vlanif30
    192.168.1.129/32    Direct  0    0       D     127.0.0.1      Vlanif30
    192.168.254.0/30    Direct  0    0       D     192.168.254.2  Vlanif100
    192.168.254.2/32    Direct  0    0       D     127.0.0.1      Vlanif100
```

图 2.126 在 LSW2 上执行 "display ip routing-table" 命令后的显示结果

② 在测试主机 PC1、PC2、PC3 和 PC4 上根据表 2.19 规划的 IP 完成 IP 地址、子网掩码和默认网关的配置。然后在命令行中使用 "ipconfig /all" 命令验证 IP 地址、子网掩码和默认网关的正确性。使用 ping 命令进行全网的连通性测试，图 2.127 所示为测试主机 PC1 成功 ping 通了 PC2 和 PC3。

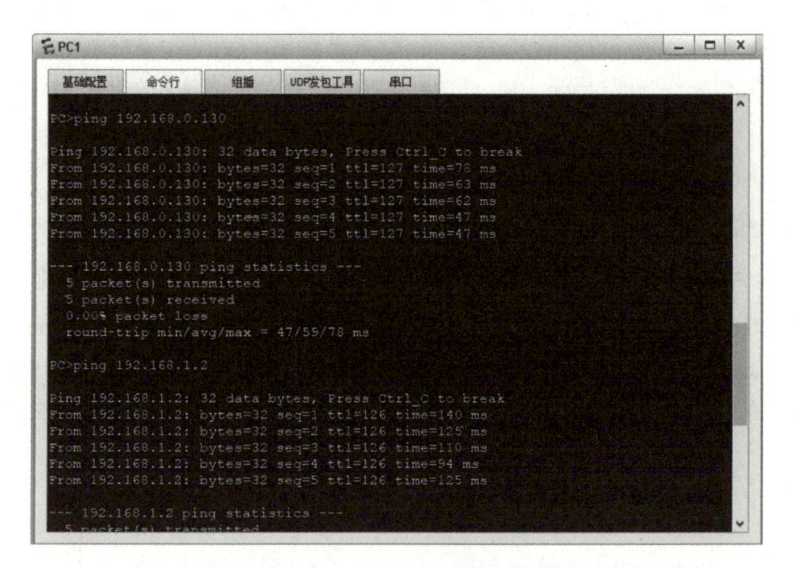

图 2.127　测试主机 PC1 成功 ping 通了 PC2 和 PC3

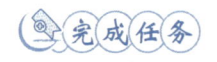

1. 任务规划

根据任务描述、图 2.1 所示的网络拓扑、表 2.12 的 IP 规划和项目 2 建网需求，本任务路由的具体要求如下：

① 在企业边界路由器 AR1 上配置到 Internet 的默认路由。

② 企业内部使用 RIPv2 协议实现内网所有网段的互联互通。

③ 在 AR1 上把到 Internet 的默认路由引入 RIPv2 路由协议中。

④ 实现路由的优化。

路由的规划如下：

① 在路由器 AR1 上配置到 Internet 的默认路由，转发出口为 S4/0/0。

② 在路由器 AR1 与三层交换机 CORE 上配置 RIPv2，实现企业内部网络所有网段的互联互通，RIP 的进程号为 1。

③ 路由器 AR1 上有类别网络号 "172.16.0.0" 参与 RIPv2 的路由更新；三层交换机 CORE 上有类别网络号 "172.16.0.0" 参与 RIPv2 的路由更新。

注： AR1 与内网直连的网络 172.16.254.1/30、172.16.10.254/24 与 172.16.20.254/24 的有类别网络号均为 "172.16.0.0"；三层交换机 CORE 直连的网络 172.16.2.254/24、172.16.3.254/24、

172.16.4.254/24、172.16.99.254/24、172.16.100.254/24和172.16.254.2/30的有类别网络号均为"172.16.0.0"。

④ 在路由器 AR1 上把默认路由引入 RIP 中。

⑤ AR1 的 G0/0/1.10 和 G0/0/1.20 配置为静默接口，CORE 的 VLAN2、VLAN3、VLAN4、VLAN99 和 VLAN100 接口配置为静默接口。

2. 任务实施

（1）AR1 上路由的配置

```
[AR1]ip route-static 0.0.0.0 0.0.0.0 s4/0/0
// 配置到 Internet 的默认路由，转发出口为 S4/0/0
[AR1]rip 1
[AR1-rip-1]version 2
[AR1-rip-1]network 172.16.0.0
[AR1-rip-1]silent-interface g0/0/1.10
[AR1-rip-1]silent-interface g0/0/1.20
[AR1-rip-1]undo summary
// 启用 RIP 协议，进程号为 1，RIP 的版本为 RIPv2，网络"172.16.0.0"参与 RIPv2 的路由更新，
//G0/0/1.10 和 G0/0/1.20 和为静默接口，禁用自动汇总功能
[AR1-rip-1]default-route-originate
// 把默认路由引入 RIP 路由协议中
```

（2）三层交换机 CORE 上路由的配置

```
[CORE]rip 1
[CORE-rip-1]version 2
[CORE-rip-1]network 172.16.0.0
[CORE-rip-1]silent-interface Vlanif 2
[CORE-rip-1]silent-interface Vlanif 3
[CORE-rip-1]silent-interface Vlanif 4
[CORE-rip-1]silent-interface Vlanif 99
[CORE-rip-1]silent-interface Vlanif 100
[CORE-rip-1]quit
// 启用 RIP 协议，进程号为 1，RIP 的版本为 RIPv2，网络"172.16.0.0"参与 RIPv2 的路由更新，
//Vlanif2、Vlanif3、Vlanif4、Vlanif99 和 Vlanif100 为静默接口，禁用自动汇总功能
```

（3）验证与测试

① 在路由器 AR1 与三层交换机 CORE 上分别执行"display ip routing-table"命令查看路由表，观察路由表默认路由选路与通过 RIPv2 学习到的远程目标网络路由条目选路。

② 在路由器 AR1 上执行"display ip routing-table | include Static"命令显示路由表中静态路由条目，如图 2.128 所示。再执行"display ip routing-table | include RIP"命令显示路由表中所有通过 RIP 学习到的路由条目，如图 2.129 所示。

194　企业网络互联技术（双语）

```
[AR1]display ip routing-table | include Static
Route Flags: R - relay, D - download to fib
------------------------------------------------------------------------------
Routing Tables: Public
        Destinations : 23        Routes : 23

Destination/Mask   Proto  Pre  Cost    Flags NextHop       Interface

0.0.0.0/0  Static 60  0          D   125.71.28.93   Serial4/0/0
```

图 2.128　查看路由器 AR1 路由表中静态路由条目

```
[AR1]display ip routing-table | include RIP
Route Flags: R - relay, D - download to fib
------------------------------------------------------------------------------
Routing Tables: Public
        Destinations : 23        Routes : 23

Destination/Mask   Proto  Pre  Cost    Flags NextHop       Interface

    172.16.2.0/24    RIP   100  1       D   172.16.254.2   GigabitEthernet0/0/0
    172.16.3.0/24    RIP   100  1       D   172.16.254.2   GigabitEthernet0/0/0
    172.16.4.0/24    RIP   100  1       D   172.16.254.2   GigabitEthernet0/0/0
    172.16.99.0/24   RIP   100  1       D   172.16.254.2   GigabitEthernet0/0/0
    172.16.100.0/24  RIP   100  1       D   172.16.254.2   GigabitEthernet0/0/0
```

图 2.129　查看路由器 AR1 上路由表中 RIP 学习到的路由条目

③ 在三层交换机 CORE 上使用 "display ip routing-table | include RIP" 命令显示路由表中所有通过 RIP 学习到的路由条目，如图 2.130 所示。

注： CORE 上路由表中显示默认路由条目是通过 RIP 学习到的。

```
[CORE]display ip routing-table | include RIP
Route Flags: R - relay, D - download to fib
------------------------------------------------------------------------------
Routing Tables: Public
        Destinations : 17        Routes : 17

Destination/Mask    Proto  Pre  Cost    Flags NextHop        Interface

     0.0.0.0/0   RIP  100   1       D       172.16.254.1   Vlanif101
 172.16.10.0/24  RIP  100   1       D       172.16.254.1   Vlanif101
 172.16.20.0/24  RIP  100   1       D       172.16.254.1   Vlanif101
```

图 2.130　查看 CORE 上路由表中 RIP 学习到的路由条目

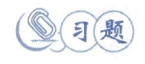 习题

选择题

（1）RIP 协议默认（　　　）发送一次更新。

　　A. 20 s　　　　　　B. 30 s　　　　　　C. 60 s　　　　　　D. 90 s

（2）RIPv1 发送更新时，其目标 IP 地址为（　　　）。

　　A. 255. 255. 255. 255　　　　　　　B. 255. 255. 255. 0

项目 2　小型企业网络互联项目　195

C. 224. 0. 0. 9　　　　　　　　　　　D. 224. 0. 0. 10

（3）RIPv2 发送更新时，其目标 IP 地址为（　　　）。

A. 255. 255. 255. 255　　　　　　　　B. 255. 255. 255. 0

C. 224. 0. 0. 9　　　　　　　　　　　D. 224. 0. 0. 10

（4）在华为的 VRP 平台上，RIP 协议默认路由优先级值为（　　　）。

A. 10　　　　　B. 60　　　　　C. 100　　　　　D. 120

（5）RIP 协议根据（　　　）计算到目标网络的最佳路径。

A. 开销　　　　B. 延迟　　　　C. 跳数　　　　D. 负载

任务 7　NAT 的规划与配置

任务描述

在完成本项目任务 1 到任务 6 的基础上，在图 2.1 所示的网络中，完成 NAT 的规划与配置任务，实现项目 2 "企业内网中的主机可以访问 Internet、Internet 可以访问企业的 Web 服务器对外的 Web 服务"网络部署需求。

任务目标

- 理解私有 IP 地址的作用。
- 理解 NAT SERVER 和 EASY NAT 的应用场景和工作过程。
- 掌握 NAT SERVER 和 EASY NAT 的配置。

相关知识

1. NAT 的工作原理

RFC1918 规定的私有地址是专门为私有或内部网络使用而保留的地址，包括 10.0.0.0/8、172.16.0.0/12 和 192.168.0.0/16 三个 IP 地址块。私有地址只能用于组建内部 IP 网络，但使用私有地址的主机不能与外部因特网上使用公有地址的 IP 主机进行通信。NAT 一般部署在网络出口设备上，例如路由器或防火墙上。

网络地址转换（network address translation, NAT）是一种用来将一个地址转换为另一个地址的技术，典型的应用是将私有地址转换为可以在公网上被路由的公有 IP 地址，以便于使用私有地址的主机可以访问 Internet。提供 NAT 功能的设备，一般运行在末节（stub）区域的边界上，当末节区域内部的一台主机想要向外部的主机进行数据传输时，它先将数据包发到 NAT 设备。NAT进程查看所收到的 IP 分组头部，如果合适，就用一个本地全局地址（通常为公有地址）替换掉"源

地址"字段中的内部私有地址。当外部目标主机发送回应分组时，NAT 进程将接收它，通过查看网络地址转换表，将"目标地址"地址字段中的本地全局地址替换成原来的私有地址。图 2.131 所示为 NAT 工作过程的实例。

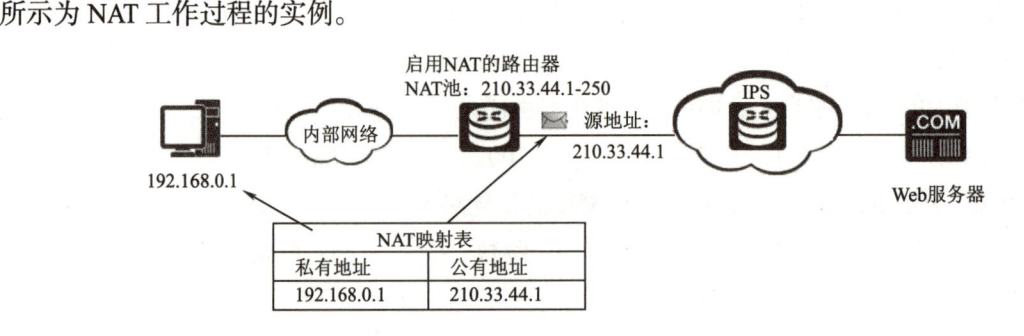

图 2.131　NAT 工作过程的实例

2. NAT 的类型

（1）静态 NAT

静态 NAT 是指管理员通过手工配置实现内部本地地址与内部全局地址之间的一对一转换，如图 2.132 所示。除非管理员重新进行静态 NAT 配置，否则静态配置的 NAT 映射将一直保持不变。静态 NAT 支持双向访问：内部网络的主机可以经过 NAT 访问外网；外部网络的主机可以通过转换过后的全局地址（通常为公有地址）访问内部网络的主机。

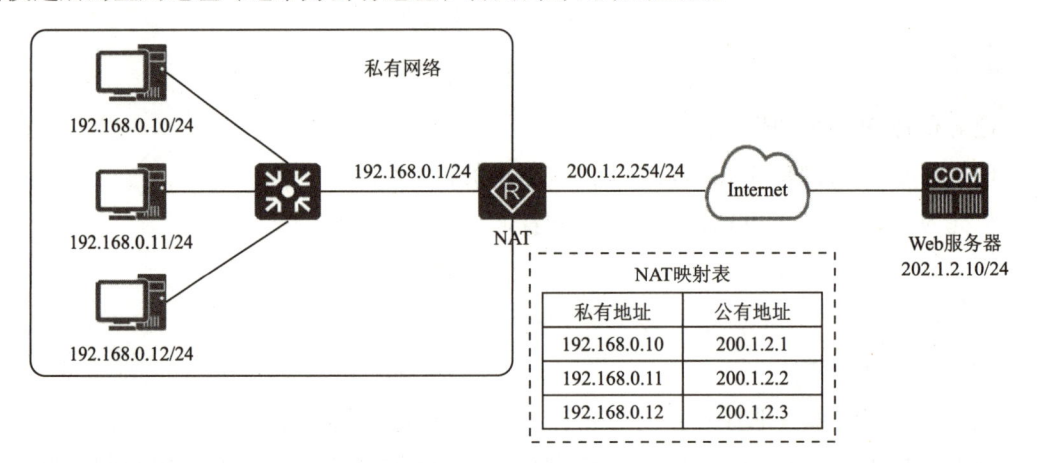

图 2.132　静态 NAT 工作过程的实例

（2）动态 NAT

动态 NAT 通过定义本地全局地址池，以先来先服务的方式动态实现内部本地地址与本地全局地址之间的一对一映射。首先，需要在 NAT 设备上定义一个本地全局地址池。当内部主机的数据经过 NAT 设备时，动态 NAT 从内部全局地址池中选择一个未被其他主机占用的本地全局地址进行地址转换，并将动态获得的地址转换或映射记录写入网络地址转换表。一旦本地全局地址池中没有足够的地址可供转换时，就会出现 NAT 转换失败，如图 2.133 所示。

项目 2　小型企业网络互联项目　　197

动态 NAT 在选择地址池中的地址进行地址转换时不会转换端口号，即 No-PAT（No-Port Address Translation，非端口地址转换），公有地址与私有地址还是 1:1 的映射关系，无法提高公有地址利用率。

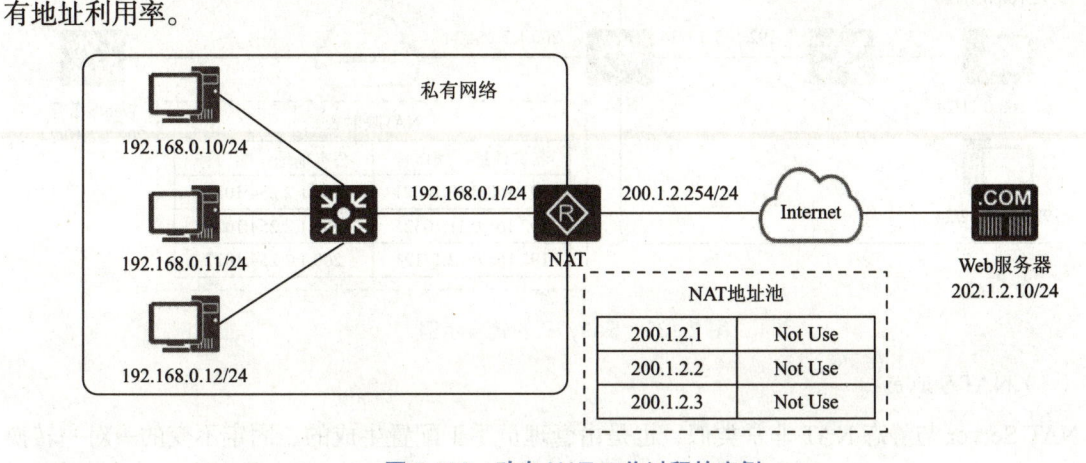

图 2.133　动态 NAT 工作过程的实例

（3）NAPT

NAPT（network address and port translation，网络地址端口转换）在从地址池中选择地址进行地址转换时不仅转换 IP 地址，同时也会对端口号进行转换。如图 2.134 所示，NAPT 借助端口可以实现一个公有地址同时对应多个私有地址。由于 NAPT 同时对 IP 地址和传输层端口进行转换，能够实现不同私有地址（不同的私有地址，不同的源端口）映射到同一个公有地址（相同的公有地址，不同的源端口），有效提高公有地址利用率。

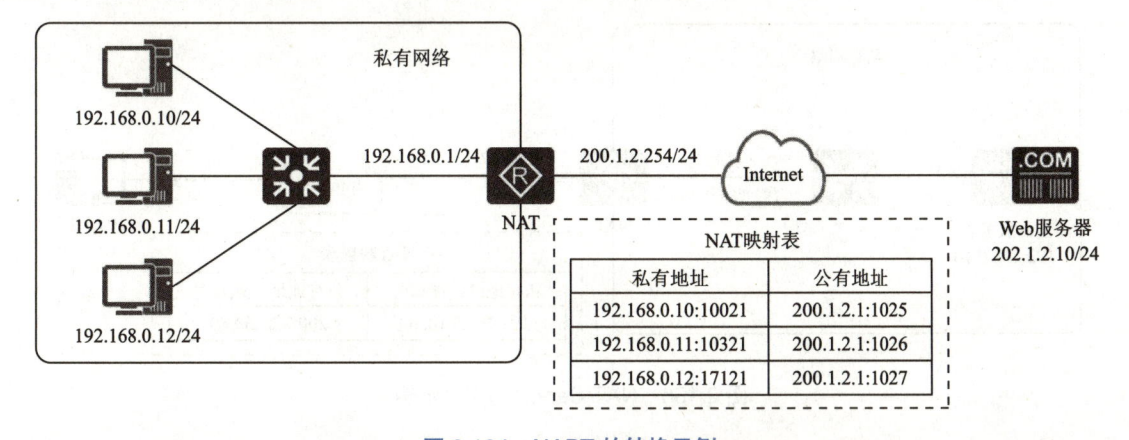

图 2.134　NAPT 的转换示例

（4）Easy IP

Easy IP 的实现原理和 NAPT 相同，在进行转换时，即转换 IP 地址，也同时转换传输层端口。但 Easy IP 没有地址池的概念，它使用 NAT 设备出口的接口 IP 地址作为地址转换后的公有地址。

Easy IP 通常适用于不具备固定公网 IP 地址的场景。如通过 DHCP、PPPoE 拨号获取地址的私有网络出口，可以直接使用获取到的动态地址进行转换。图 2.135 所示为 Easy IP 的转换示例。

198 企业网络互联技术（双语）

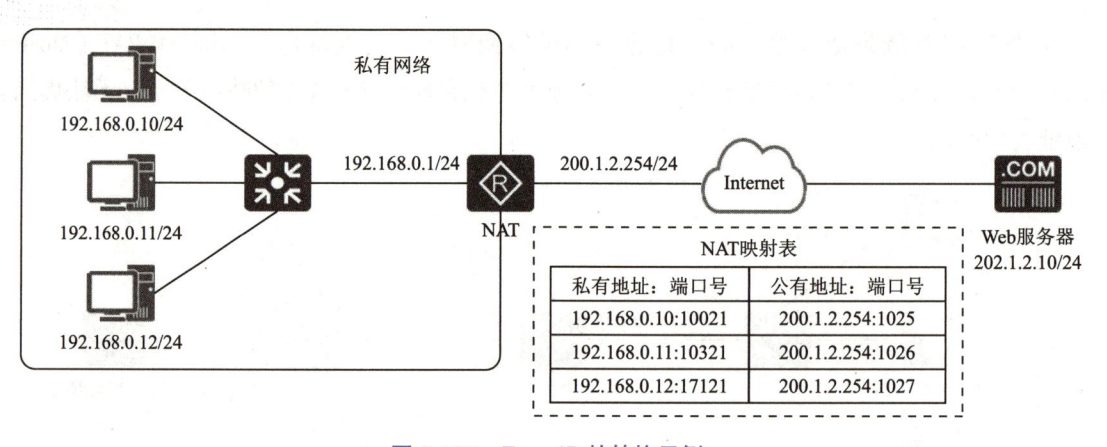

图 2.135　Easy IP 的转换示例

（5）NAT Server

　　NAT Server 与静态 NAT 非常类似，也是由管理员手工配置生成的、固定不变的一对一转换映射。除非管理员删除或重新配置，否则 NAT Server 转换映射会一直保持不变。但与静态地址转换不同的是，NAT Server 转换映射除了定义内部本地地址与内部全局地址外，还需要指定传输层的协议和端口信息，它是 [公有地址 : 端口] 与 [私有地址 : 端口] 的一对一映射关系。NAT Server 通常用于企业网内部采用私有地址进行寻址，但需要对外提供访问服务的服务器，如企业对外的 Web 与 E-mail 服务器等。通过 NAT Server，外部网络的主机可以使用转换后的全局公有地址及对应的协议端口访问这些服务器所提供的相应服务。图 2.136 所示为 NAT Server 的转换示例。

图 2.136　NAT Server 的转换示例

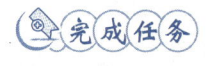　完成任务

　　1. 任务规划

　　根据任务描述、图 2.1 所示的网络拓扑、表 2.12 的 IP 规划和项目 2 网络部署需求，可在出口路由器 AR1 上配置 Easy IP 与 NAT Server，实现企业内网主机可以同时访问 Internet，Internet 上的测试主机可以访问企业对外提供的 WWW 和 DNS 服务的网络需求。NAT Server 的规划见表 2.21。

项目 2　小型企业网络互联项目　199

表 2.21　NAT Server 的规划

服 务 器	协　　议	内 部 地 址	内部端口号	接　　口	全局端口号
Web	TCP	172.16.100.1/24	80	AR1 S4/0/0 当前接口	80
DNS	TCP 和 UDP	172.16.100.3/24	53	AR1 S4/0/0 当前接口	53

Easy IP 的规划：

ACL 列 表 号 为 2000，允 许 172.16.2.0/24、172.16.3.0/24、172.16.4.0/24、172.16.99.0/24、172.16.100.0/24、172.16.10.0/24 和 172.16.20.0/24 网段的地址进行 Easy IP 转换

2. 任务实施

（1）NAT Server 的配置

在企业出口路由器 AR1 上，根据表 2.21 规划的 NAT Server 完成配置，步骤如下：

```
[AR1]interface s4/0/0
[AR1-Serial4/0/0]nat server protocol tcp global current-interface 80 inside
172.16.100.1 80
[AR1-Serial4/0/0]nat server protocol tcp global current-interface 53 inside
172.16.100.3 53
[AR1-Serial4/0/0]nat server protocol udp global current-interface 53 inside
172.16.100.3 53
```

（2）Easy IP 的配置

在企业出口路由器 AR1 上，Easy IP 配置需完成的任务包括：

① 使用 ACL 配置允许进行转换的内网地址段。

② 在连接到 Internet 的出口上配置 Easy IP 使 ACL 定义的内网地址可以使用出口的 IP 地址访问 Internet。

```
[AR1]acl 2000
[AR1-acl-basic-2000]rule permit source 172.16.3.0 0.0.0.255
[AR1-acl-basic-2000]rule permit source 172.16.4.0 0.0.0.255
[AR1-acl-basic-2000]rule permit source 172.16.99.0 0.0.0.255
[AR1-acl-basic-2000]rule permit source 172.16.100.0 0.0.0.255
[AR1-acl-basic-2000]rule permit source 172.16.10.0 0.0.0.255
[AR1-acl-basic-2000]rule permit source 172.16.20.0 0.0.0.255
[AR1-acl-basic-2000]quit
// 配置允许进行 Easy IP 转换的内网地址段，ACL 列表号为 2000，允许进行 Easy IP 转换的内网地
// 址段包括：172.16.2.0/24、172.16.3.0/24、172.16.4.0/24、172.16.99.0/24、172.16.100.0/
//24、172.16.10.0/24 和 172.16.20.0/24
[AR1]interface s4/0/0
[AR1-Serial4/0/0]nat outbound 2000
// 配置 Easy IP，让 ACL2000 定义的网段地址主机通过 S4/0/0 接口的公有 IP 地址访问 Internet
```

（3）NAT 的查看与验证

完成 NAT 的配置后，可使用 display 命令进行 NAT 查看与验证。

① 执行"display nat server"命令，查看 NAT server 信息。图 2.137 所示为在路由器 AR1 上查看 NAT server 信息的示例。

```
[AR1]display nat server

Nat Server Information:
Interface  : Serial4/0/0
  Global IP/Port    : current-interface/80(www) (Real IP : 125.71.28.93)
  Inside IP/Port    : 172.16.100.1/80(www)
  Protocol : 6(tcp)
  VPN instance-name  : ----
  Acl number         : ----
  Description : ----

  Global IP/Port    : current-interface/53(domain) (Real IP : 125.71.28.93)
  Inside IP/Port    : 172.16.100.3/53(domain)
  Protocol : 6(tcp)
  VPN instance-name  : ----
  Acl number         : ----
  Description : ----

  Global IP/Port    : current-interface/53(dns) (Real IP : 125.71.28.93)
  Inside IP/Port    : 172.16.100.3/53(dns)
  Protocol : 17(udp)
  VPN instance-name  : ----
  Acl number         : ----
  Description : ----

Total :  3
```

图 2.137　在路由器 AR1 上查看 NAT server 信息的示例

② 在路由器 AR1 的 S4/0/0 接口视图下执行"display this"命令，可查看该接口下的配置，注意观察 NAT 的配置信息是否与规划一致，如图 2.138 所示。

```
[AR1-Serial4/0/0]display this
[V200R003C00]
#
interface Serial4/0/0
 link-protocol ppp
 ppp authentication-mode chap
 ppp chap user AR1
 ppp chap password cipher %$%$WJ(-3^`dfCH0^ZT!~%d&,+Jl%$%$
 ip address 125.71.28.93 255.255.255.0
 nat server protocol tcp global current-interface www inside 172.16.100.1 www
 nat server protocol tcp global current-interface domain inside 172.16.100.3 dom
ain
 nat server protocol udp global current-interface dns inside 172.16.100.3 dns
 nat outbound 2000
#
return
```

图 2.138　查看路由器 AR1 的 S4/0/0 接口下的配置

③ 在 Internet 测试主机 Client1 上，打开浏览器，在网址中输入企业 Web 服务器转换后的公有地址（即路由器 AR1 的 S4/0/0 接口的 IP 地址），可访问企业 Web 服务器提供的 Web 服务，如图 2.139 所示。

④ 在企业内网的多台主机上，同时使用不间断 ping 命令，ping Internet 上的任何测试主机，可以访问。图 2.140 所示为企业内网主机 PC3 与 PC4 同时 ping 通 Internet 上测试主机 Client1。

项目 2　小型企业网络互联项目　201

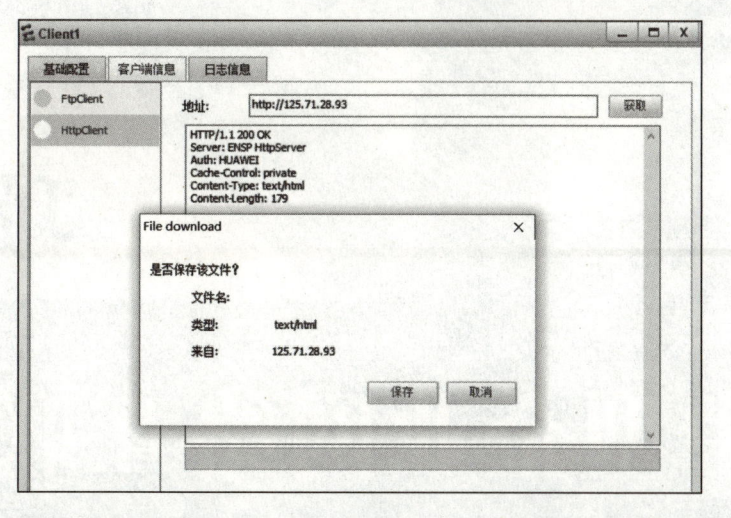

图 2.139　Internet 上的测试主机访问企业 Web 服务器对外提供的 Web 服务

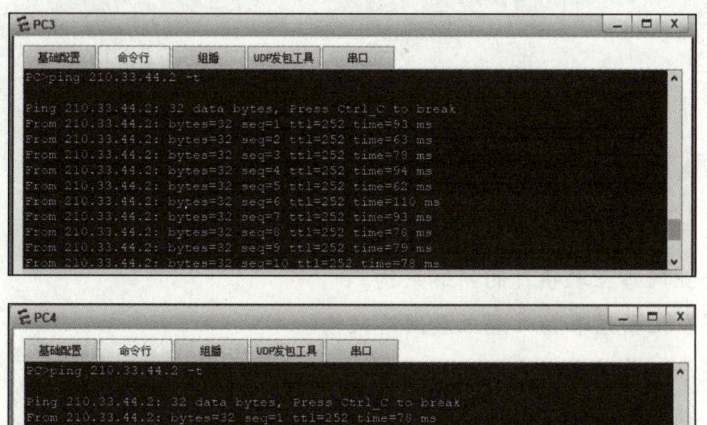

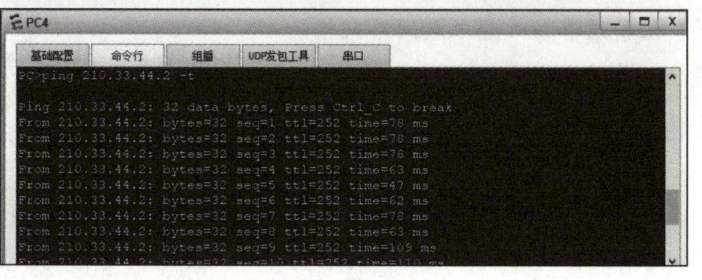

图 2.140　PC3 与 PC4 同时 ping 通 Internet 上测试主机 Client1

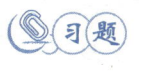

选择题

（1）（　　　）可以使企业内网的多台主机共享 NAT 设备出口的 IP 地址访问因特网。

　　A. 静态 NAT　　　　B. 动态 NAT　　　　C. Easy IP　　　　　　　　D. NAPT

（2）（　　　）不允许外网的主机主动发起到内网主机的访问。

　　A. NAT server　　　B. NAPT　　　　　　C. 静态 NAT

（3）在配置 Easy IP 时，使用（　　　）定义内网地址可以进行地址转换。

　　A. 地址池　　　　　B. ACL　　　　　　　C. 配置 NAT 的设备接口　　D. NAT 表

项目 3

中型企业网络互联项目

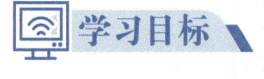 **学习目标**

知识目标

◎ 了解中型企业网络互联项目的网络架构。

◎ 理解接入层、汇聚层和核心层三层网络架构。

◎ 理解链路聚合工作过程。

◎ 理解 OSPF 概念及工作过程。

能力目标

◎ 具有独立部署、实施一个中型企业网络互联工程项目的能力。包括：IP 与 VLAN 的配置、路由器交换机配置、链路聚合的配置、生成树的配置、VLAN 之间通信的配置、OSPF 的配置和 NAT 的配置等。

素质目标

◎ 具有团队协作能力。

◎ 具有分析问题与解决问题的能力。

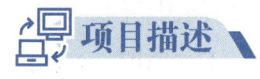 **项目描述**

　　某中型企业拟组建如图 3.1 所示的网络，网络采用接入层、汇聚层和核心层的三层网络架构。行政楼的部门有保卫处、学生处、财务处、人事处。教学楼有教室和办公室；实训楼有软件实训室和网络实训室。每幢楼均安装了数字摄像头用于监控。企业申请到五个公有地址 125.71.28.65 ～ 125.71.28.69。现需要进行该企业网络的规划与部署，实现企业内网各网段相

项目 3　中型企业网络互联项目　203

互连通，企业内的主机可以访问 Internet，Internet 上的主机可以访问企业对外提供的 WWW 和 DNS 服务。

企业网络部署需求如下：

◎企业不同部门间业务数据需完全隔离。

◎管理员可以使用 SSH 远程登录到企业内部所有的交换机上。

◎实训楼汇聚层交换机 SXL-AGG 与接入层交换机 SXL-FLOOR2 之间的链路使用链路聚合技术提高链路带宽，增加链路的高可靠性。

◎实训楼汇聚层交换机 SXL-AGG 与核心交换机 CORE 之间的链路使用链路聚合技术提高链路带宽，增加链路的高可靠性。

◎企业内网需互联互通。

◎企业内网采用私有地址寻址，在出口路由器上部署 NAPT 和 NAT SERVER，实现企业内网所有主机均可访问 Internet，Internet 上的主机可以访问企业对外的网络服务。

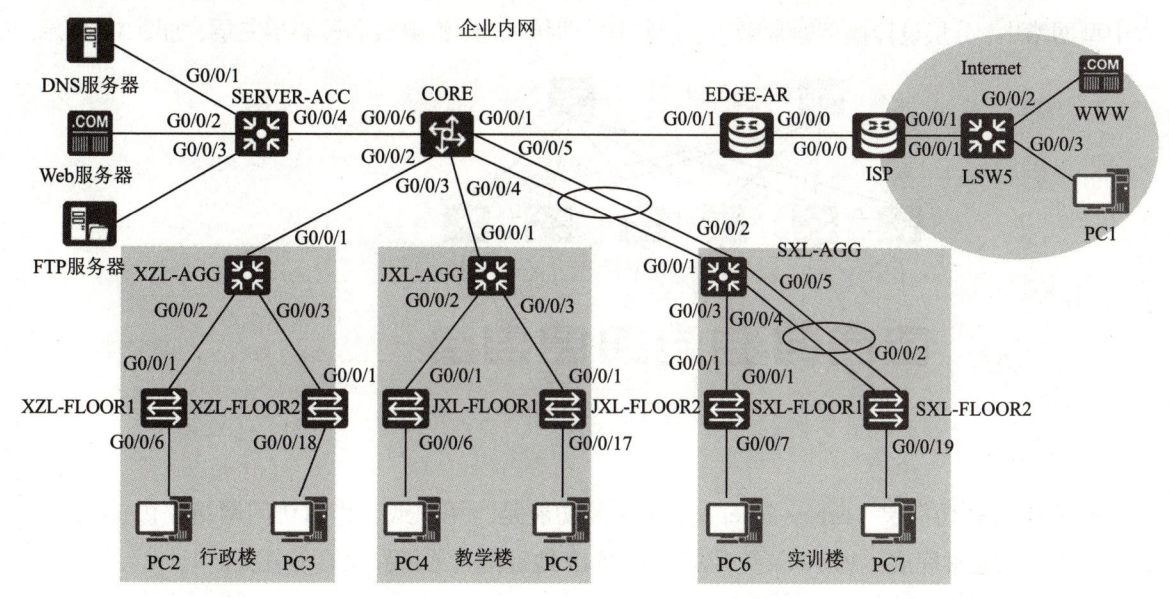

图 3.1　中型企业网络互联项目网络拓扑图

任务 1　IP、VLAN 与 NAT 的规划与配置

任务描述

使用华为 ENSP 仿真环境或者使用华为的路由器交换机搭建如图 3.1 所示的网络拓扑。根据项目 3 的网络部署需求，完成该网络 IP、VLAN、NAPT 和 NAT Server 的规划与配置任务，并进行验证与测试。

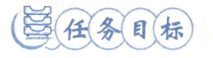任务目标

- 具有中型企业网络互联项目 VLAN 的部署实施能力。
- 具有中型企业网络互联项目 IP 的部署实施能力。
- 掌握中型网络互联项目 Easy IP 的规划与配置方法。
- 掌握中型网络互联项目 NAT Server 的配置方法。

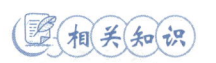相关知识

1. 中型企业网络设计的层次化方法

在构建中型企业需求的网络时，为了使建立的网络更容易管理、更灵活和更具有扩展性，通常采用分层设计模型。分层设计模型又称结构化设计模型，通过对网络的功能进行较明确的划分，采用分层的思想将整个网络结构分成若干个层次，从而使各层的设计变得相对简单。在大中型网络中，分层设计模型通常采用三层结构，即接入层、汇聚层和核心层三层，如图 3.2 所示。

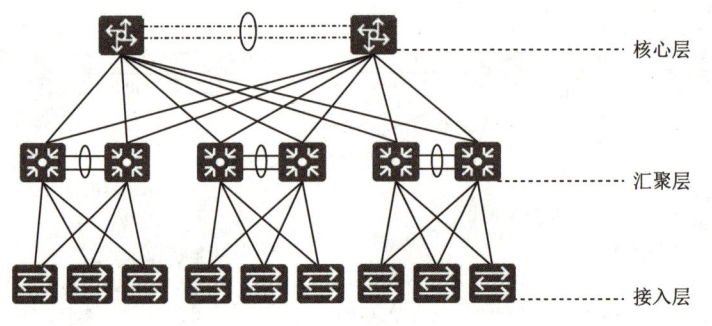

图 3.2　三层结构

（1）接入层

接入层又称访问层（access layer），其主要功能是为用户或工作组访问网络提供接入。企业网的接入层通常完成：VLAN、生成树、管理 IP 地址配置或链路聚合等任务。

（2）汇聚层

汇聚层又称分布层（distribution layer），是接入层和核心层之间的分界点。汇聚层通过提供基于策略的连接管理，实现接入层对核心层的可控制的传输。汇聚层主要实现 VLAN 之间的路由等功能。

（3）核心层

核心层（core layer）是企业网的高速主干，其主要功能是为相互通信的节点间提供高速优化的带宽传输，因此核心层需要保持高可用性和高冗余性，避免访问控制列表和数据包过滤之类的功能。

注：在实际的网络设计中，不是所有的网络都必须具有全部的三层。当网络的规模较小或网络的功能较少时，三层设计往往可以简化为二层的设计（接入层＋核心层），甚至是一层的设计。

2. 分层交换机网络中的交换机

与上述分层结构的交换网络相对应，企业网中的交换机被分成接入层交换机、汇聚层交换机和核心层交换机。以图 3.1 所示的企业网拓扑结构为例，图中的交换机 CORE 为核心层交换机，交换机 XZL-AGG、JXL-AGG 和 SXL-AGG 为汇聚层交换机，交换机 XZL-FLOOR1、XZL-FLOOR2 等为接入层交换机。

① 接入层交换机通常需要具有 VLAN 功能和一定的安全功能，通常可以选用二层交换机。

② 汇聚层交换机需要实现 VLAN 之间的通信，需具有路由的功能，因此需选用三层交换机。汇聚层交换机需提供路由功能，因此汇聚层交换机与核心层交换机之间相连的链路通常配置为三层链路。市场上大部分三层交换机的端口可以直接配置为路由端口，如华三、思科等三层交换机，也可以把该三层链路的物理端口加入某个单独的 VLAN，三层的 IP 地址配置在该 VLAN 端口上，通常会在该物理端口上禁用生成树功能。

③ 核心层交换机主要是为了实现数据包在主干链路上的快速转发，因此也需要选用三层交换机，且对于交换转发性能的要求相对于汇聚层交换机更高。核心层交换机上的基本配置为路由协议。此外，若企业网的企业级服务器连接在核心交换机上，那么核心层交换机还需要完成 VLAN 的创建、将交换机的端口指派给相应的 VLAN 和 VLAN 之间的通信等配置工作。

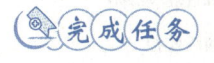 完成任务

1. 任务规划

① 表 3.1 所示的 IP 与 VLAN 规划是根据三层网络设计模型原则设计的。请根据图 3.1 与表 3.1 完成 VLAN 配置。在接入层交换机、汇聚层交换机、核心层交换机、出口路由器 EDGE-AR、运营商路由器 ISP 上和服务器以及 Internet 上的测试主机 PC1 上完成 IP 地址的配置，并在汇聚层交换机上配置 DHCP 服务，为该楼层的终端接入设备提供 IP 地址服务。DHCP 服务的规划见表 3.2。

② 根据图 3.1、表 3.3，在出口路由器 EDGE-AR 上完成 NAT Server 与 NAPT 的配置，以实现企业内网主机可以同时访问 Internet，Internet 上的测试主机可以访问企业对外提供的 WWW 和 DNS 服务。

注：接入层交换机的管理 IP 地址为静态分配，监控设备的 IP 地址也为静态分配，因此不需要为其进行 DHCP 服务的配置。

表 3.1　IP 与 VLAN 的规划

设　　备	接口或 VLAN	VLAN 的描述	二层或三层规划	说　　明
XZL-FLOOR1	VLAN2	baoan	G0/0/2 至 G0/0/12 接口	保卫处
	VLAN3	student	G0/0/13 至 G0/0/18 接口	学生处
	VLAN90	manage	10.0.90.1/24	交换机管理 VLAN
	VLAN91	monitor	G0/0/19 至 G0/0/24 接口	网络监控

设备	接口或VLAN	VLAN的描述	二层或三层结构	说 明
XZL-FLOOR2	VLAN4	caiwu	G0/0/2 至 G0/0/12 接口	财务处
	VLAN5	hr	G0/0/13 至 G0/0/18 接口	人事处
	VLAN90	manage	10.0.90.2/24	交换机管理VLAN
	VLAN91	monitor	G0/0/19 至 G0/0/24 接口	监控
XZL-AGG	VLAN2	baoan	10.0.2.254/24	
	VLAN3	student	10.0.3.254/24	
	VLAN4	caiwu	10.0.4.254/24	
	VLAN5	hr	10.0.5.254/24	
	VLAN90	manage	10.0.90.254/24	
	VLAN91	monitor	10.0.91.254/24	
	VLAN99	linkcore	10.254.254.2/30	到CORE的连接网段VLAN及IP 相比
JXL-FLOOR1	VLAN12	jiaoshi	G0/0/2 至 G0/0/15 接口	教室
	VLAN13	office	G0/0/16 至 G0/0/18 接口	办公室
	VLAN92	manage	10.1.92.1/24	交换机管理VLAN
	VLAN93	monitor	G0/0/19 至 G0/0/24 接口	网络监控
JXL-FLOOR2	VLAN12	jiaoshi	G0/0/2 至 G0/0/15 接口	教室
	VLAN13	office	G0/0/16 至 G0/0/18 接口	办公室
	VLAN92	manage	10.1.92.2/24	交换机管理VLAN
	VLAN93	monitor	G0/0/19 至 G0/0/24 接口	网络监控
JXL-AGG	VLAN12	jiaoshi	10.1.12.254/24	
	VLAN13	office	10.1.13.254/24	
	VLAN92	manage	10.1.92.254/24	
	VLAN93	monitor	10.1.93.254/24	
	VLAN99	linkcore	10.254.254.6/30	到CORE的连接网段VLAN及IP 相比
SXL-FLOOR1	VLAN22	sxs-rj	G0/0/2 至 G0/0/10 接口	软件实训室
	VLAN23	sxs-wl	G0/0/11 至 G0/0/18 接口	网络实训室
	VLAN94	manage	10.2.94.1/24	交换机管理VLAN
	VLAN95	monitor	G0/0/19 至 G0/0/24 接口	网络监控
SXL-FLOOR2	VLAN22	sxs-rj	G0/0/3 至 G0/0/10 接口	软件实训室
	VLAN23	sxs-wl	G0/0/11 至 G0/0/18 接口	网络实训室
	VLAN94	manage	10.2.94.2/24	交换机管理VLAN
	VLAN95	monitor	G0/0/19 至 G0/0/24 接口	网络监控

续表

表 3.2 DHCP 服务的规划

设备	DHCP 地址池名	IP 地址网络号	网关地址	DNS 地址	地址租期
XZL-AGG	VLAN2	10.0.2.0/24	10.0.2.254	10.254.253.2	8 h
	VLAN3	10.0.3.0/24	10.0.3.254	10.254.253.2	8 h
	VLAN4	10.0.4.0/24	10.0.4.254	10.254.253.2	8 h
	VLAN5	10.0.5.0/24	10.0.5.254	10.254.253.2	8 h
JXL-AGG	VLAN12	10.1.12.0/24	10.1.12.254	10.254.253.2	8 h
	VLAN13	10.1.13.0/24	10.1.13.254	10.254.253.2	8 h
SXL-AGG	VLAN22	10.2.22.0/24	10.2.22.254	10.254.253.2	8 h
	VLAN23	10.2.23.0/24	10.2.23.254	10.254.253.2	8 h

设备	接口或 VLAN	VLAN 的描述	二层或三层接口	说 明
SXL-AGG	VLAN22	rj-sxs	10.2.22.254/24	
	VLAN23	wl-sxs	10.2.23.254/24	
	VLAN94	manage	10.2.94.254/24	
	VLAN95	monitor	10.2.95.254/24	
CORE	VLAN199	linktocore	10.254.254.10/30	到交换机 CORE 的互连接网段 VLAN
	VLAN199	linkto-sxl-agg	10.254.254.9/30	到交换机 SXL-AGG 的互连接网段 VLAN
	VLAN200	linkto-jxl-agg	10.254.254.5/30	到交换机 JXL-AGG 的互连接网段 VLAN
	VLAN201	linkto-xzl-agg	10.254.254.1/30	到交换机 XZL-AGG 的互连接网段 VLAN
	VLAN202	SERVER	10.254.253.254/24	服务器网段
	VLAN203	linkto-edge-ar	10.254.254.13/30	到 CORE 的互连接网段 VLAN 及 IP 地址
EDGE-AR	G0/0/1	NA	10.254.254.14/30	
	G0/0/0	NA	125.71.28.65/24	
	G0/0/1	NA	202.1.1.254/24	
ISP	G0/0/0	NA	125.71.28.1/24	
Web	F0/0	NA	10.254.253.1/24	
DNS	F0/0	NA	10.254.253.2/24	
FTP	F0/0	NA	10.254.253.3/24	
WWW	F0/0	NA	202.1.1.1/24	
PC1	F0/0	NA	202.1.1.10/24	

续表

表 3.3 NAT Server 的规划

服 务 器	协 议	内 部 地 址	内部端口号	外 部 地 址	全局端口号
Web	TCP	10.254.253.1/24	80	125.71.28.66	80
DNS	TCP 和 UDP	10.254.253.2/24	53	125.71.28.66	53

注：

① 每幢楼的汇聚层交换机实现该楼宇的 VLAN 之间的通信，即汇聚层交换机 XZL-AGG 实现行政楼 VLAN 之间的通信，JXL-AGG 实现教学楼 VLAN 之间的通信，SXL-AGG 实现实训楼 VLAN 之间的通信。

② 每幢楼汇聚层交换机与核心交换机 CORE 之间采用三层连接。

NAPT 的规划：

① ACL 列表号为 2000，允许 10.0.2.0/24、10.0.3.0/24、10.0.4.0/24、10.0.5.0/24、10.1.12.0/24、10.1.13.0/24、10.2.22.0/24、10.2.23.0/24 网段的地址进行 Easy IP 转换。

② 转换后的公有地址池 ID 为"1"，地址池范围为 125.71.28.67 至 125.71.28.69。

2. 任务实施

（1）接入层交换机 VLAN 的配置

根据表 3.1 的规划，接入层交换机需完成的 VLAN 配置任务包括：创建 VLAN，配置 VLAN 描述；配置到汇聚层交换机的链路为中继链路；配置接入链路。

① 在接入层交换机 XZL-FLOOR1 上完成 VLAN 配置，步骤如下：

```
[XZL-FLOOR1]vlan 2
[XZL-FLOOR1-vlan2]name baoan
[XZL-FLOOR1-vlan2]description baoan
[XZL-FLOOR1-vlan2]vlan 3
[XZL-FLOOR1-vlan3]description student
[XZL-FLOOR1-vlan3]vlan 90
[XZL-FLOOR1-vlan90]description manage
[XZL-FLOOR1-vlan90]vlan 91
[XZL-FLOOR1-vlan91]description monitor
[XZL-FLOOR1-vlan91]quit
// 创建 VLAN2、VLAN3、VLAN90、VLAN91，并配置 VLAN 描述
[XZL-FLOOR1]port-group 1
[XZL-FLOOR1-port-group-1]group-member g0/0/2 to g0/0/12
[XZL-FLOOR1-port-group-1]port link-type access
[XZL-FLOOR1-port-group-1]port default vlan 2
// 配置 G0/0/2 至 G0/0/12 接口为接入接口，属于 VLAN2
[XZL-FLOOR1]port-group 2
[XZL-FLOOR1-port-group-2]group-member g0/0/13 to g0/0/18
```

```
[XZL-FLOOR1-port-group-2]port link-type access
[XZL-FLOOR1-port-group-2]port default vlan 3
[XZL-FLOOR1-port-group-2]quit
// 配置 G0/0/13 至 G0/0/18 接口为接入接口，属于 VLAN3
[XZL-FLOOR1]port-group 3
[XZL-FLOOR1-port-group-3]group-member g0/0/19 to g0/0/24
[XZL-FLOOR1-port-group-3]port link-type access
[XZL-FLOOR1-port-group-3]port default vlan 91
// 配置 G0/0/19 至 G0/0/24 接口为接入接口，属于 VLAN91
[XZL-FLOOR1]interface g0/0/1
[XZL-FLOOR1-GigabitEthernet0/0/1]port link-type trunk
[XZL-FLOOR1-GigabitEthernet0/0/1]port trunk allow-pass vlan 2 3 90 91
[XZL-FLOOR1-GigabitEthernet0/0/1]quit
// 配置 G0/0/1 端口为中继端口，允许 VLAN2、VLAN3、VLAN90 和 VLAN91 通过
```

② 在接入层交换机 XZL-FLOOR2 上完成 VLAN 的配置，步骤如下：

```
[XZL-FLOOR2]vlan 4
[XZL-FLOOR2-vlan4]description caiwu
[XZL-FLOOR2-vlan4]vlan 5
[XZL-FLOOR2-vlan5]description HR
[XZL-FLOOR2-vlan5]vlan 90
[XZL-FLOOR2-vlan90]description manage
[XZL-FLOOR2-vlan90]vlan 91
[XZL-FLOOR2-vlan91]description monitor
[XZL-FLOOR2-vlan91]quit
[XZL-FLOOR2]
// 创建 VLAN4、VLAN5、VLAN90、VLAN91，并配置 VLAN 描述
[XZL-FLOOR2]port-group 1
[XZL-FLOOR2-port-group-1]group-member g0/0/2 to g0/0/12
[XZL-FLOOR2-port-group-1]port link-type access
[XZL-FLOOR2-port-group-1]port default vlan 4
// 配置 G0/0/2 至 G0/0/12 接口为接入接口，属于 VLAN4
[XZL-FLOOR2]port-group 2
[XZL-FLOOR2-port-group-2]group-member g0/0/13 to g0/0/18
[XZL-FLOOR2-port-group-2]port link-type access
[XZL-FLOOR2-port-group-2]port default vlan 5
[XZL-FLOOR2-port-group-2]quit
// 配置 G0/0/13 至 G0/0/18 接口为接入端口，属于 VLAN5
[XZL-FLOOR2]port-group 3
[XZL-FLOOR2-port-group-3]group-member g0/0/19 to g0/0/24
```

```
[XZL-FLOOR2-port-group-3]port link-type access
[XZL-FLOOR2-port-group-3]port default vlan 91
[XZL-FLOOR2-port-group-3]quit
// 配置 G0/0/19 至 G0/0/24 接口为接入接口，属于 VLAN91
[XZL-FLOOR2]interface g0/0/1
[XZL-FLOOR2-GigabitEthernet0/0/1]port link-type trunk
[XZL-FLOOR2-GigabitEthernet0/0/1]port trunk allow-pass vlan 4 5 90 91
[XZL-FLOOR2-GigabitEthernet0/0/1]quit
// 配置 G0/0/1 接口为中继接口，允许 VLAN4、VLAN5、VLAN90 和 VLAN91 通过
```

③ 在接入层交换机 JXL-FLOOR1 上完成 VLAN 的配置，步骤如下：

```
[JXL-FLOOR1]vlan 12
[JXL-FLOOR1-vlan12]name jiaoshi
[JXL-FLOOR1-vlan12]description jiaoshi
[JXL-FLOOR1-vlan12]vlan 13
[JXL-FLOOR1-vlan13]description office
[JXL-FLOOR1-vlan13]vlan 92
[JXL-FLOOR1-vlan92]description manage
[JXL-FLOOR1-vlan92]vlan 93
[JXL-FLOOR1-vlan93]description monitor
[JXL-FLOOR1-vlan93]quit
// 创建 VLAN12、VLAN13、VLAN92、VLAN93，并配置 VLAN 描述
[JXL-FLOOR1]port-group 1
[JXL-FLOOR1-port-group-1]group-member g0/0/2 to g0/0/15
[JXL-FLOOR1-port-group-1]port link-type access
[JXL-FLOOR1-port-group-1]port default vlan 12
// 配置 G0/0/2 至 G0/0/15 接口为接入接口，属于 VLAN12
[JXL-FLOOR1]port-group 2
[JXL-FLOOR1-port-group-2]group-member g0/0/16 to g0/0/18
[JXL-FLOOR1-port-group-2]port link-type access
[JXL-FLOOR1-port-group-2]port default vlan 13
[JXL-FLOOR1-port-group-2]quit
// 配置 G0/0/16 至 G0/0/18 接口为接入接口，属于 VLAN13
[JXL-FLOOR1]port-group 3
[JXL-FLOOR1-port-group-3]group-member g0/0/19 to g0/0/24
[JXL-FLOOR1-port-group-3]port link-type access
[JXL-FLOOR1-port-group-3]port default vlan 93
[JXL-FLOOR1-port-group-3]quit
// 配置 G0/0/19 至 G0/0/24 接口为接入接口，属于 VLAN93
[JXL-FLOOR1]interface g0/0/1
```

项目 3　中型企业网络互联项目　211

```
[JXL-FLOOR1-GigabitEthernet0/0/1]port link-type trunk
[JXL-FLOOR1-GigabitEthernet0/0/1]port trunk allow-pass vlan 12 13 92 93
[JXL-FLOOR1-GigabitEthernet0/0/1]quit
// 配置 G0/0/1 接口为中继接口，允许 VLAN12、VLAN13、VLAN92 和 VLAN93 通过
```

④ 在接入层交换机 JXL-FLOOR2 上完成 VLAN 的配置，步骤如下：

```
[JXL-FLOOR2]vlan 12
[JXL-FLOOR2-vlan12]description jiaoshi
[JXL-FLOOR2-vlan12]vlan 13
[JXL-FLOOR2-vlan13]description office
[JXL-FLOOR2-vlan13]vlan 92
[JXL-FLOOR2-vlan92]description manage
[JXL-FLOOR2-vlan92]vlan 93
[JXL-FLOOR2-vlan93]description monitor
[JXL-FLOOR2-vlan93]quit
// 创建 VLAN12、VLAN13、VLAN92、VLAN93，并配置 VLAN 描述
[JXL-FLOOR2]port-group 1
[JXL-FLOOR2-port-group-1]group-member g0/0/2 to g0/0/15
[JXL-FLOOR2-port-group-1]port link-type access
[JXL-FLOOR2-port-group-1]port default vlan 12
[JXL-FLOOR2-port-group-1]quit
// 配置 G0/0/2 至 G0/0/15 接口为接入接口，属于 VLAN12
[JXL-FLOOR2]port-group 2
[JXL-FLOOR2-port-group-2]group-member g0/0/16 to g0/0/18
[JXL-FLOOR2-port-group-2]port link-type access
[JXL-FLOOR2-port-group-2]port default vlan 13
[JXL-FLOOR2-port-group-2]quit
// 配置 G0/0/16 至 G0/0/18 接口为接入接口，属于 VLAN13
[JXL-FLOOR2]port-group 3
[JXL-FLOOR2-port-group-3]group-member g0/0/19 to g0/0/24
[JXL-FLOOR2-port-group-3]port link-type access
[JXL-FLOOR2-port-group-3]port default vlan 93
[JXL-FLOOR2-port-group-3]quit
// 配置 G0/0/19 至 G0/0/24 接口为接入接口，属于 VLAN93
[JXL-FLOOR2]interface g0/0/1
[JXL-FLOOR2-GigabitEthernet0/0/1]port link-type trunk
[JXL-FLOOR2-GigabitEthernet0/0/1]port trunk allow-pass vlan 12 13 92 93
[JXL-FLOOR2-GigabitEthernet0/0/1]quit
// 配置 G0/0/1 接口为中继接口，允许 VLAN12、VLAN13、VLAN92 和 VLAN93 通过
```

⑤ 在接入层交换机 SXL-FLOOR1 上完成 VLAN 的配置，步骤如下：

```
[SXL-FLOOR1]vlan 22
[SXL-FLOOR1-vlan22]description rj-sxs
[SXL-FLOOR1-vlan22]vlan 23
[SXL-FLOOR1-vlan23]description wl-sxs
[SXL-FLOOR1-vlan23]vlan 94
[SXL-FLOOR1-vlan94]description manage
[SXL-FLOOR1-vlan94]vlan 95
[SXL-FLOOR1-vlan95]description monitor
[SXL-FLOOR1-vlan95]quit
[SXL-FLOOR1]
// 创建 VLAN22、VLAN23、VLAN94、VLAN95, 并配置 VLAN 描述
[SXL-FLOOR1]port-group 1
[SXL-FLOOR1-port-group-1]group-member g0/0/2 to g0/0/10
[SXL-FLOOR1-port-group-1]port link-type access
[SXL-FLOOR1-port-group-1]port default vlan 22
[SXL-FLOOR1-port-group-1]quit
// 配置 G0/0/2 至 G0/0/10 接口为接入接口, 属于 VLAN22
[SXL-FLOOR1]port-group 2
[SXL-FLOOR1-port-group-2]group-member g0/0/11 to g0/0/18
[SXL-FLOOR1-port-group-2]port link-type access
[SXL-FLOOR1-port-group-2]port default vlan 23
[SXL-FLOOR1-port-group-2]quit
// 配置 G0/0/11 至 G0/0/18 接口为接入接口, 属于 VLAN23
[SXL-FLOOR1]port-group 3
[SXL-FLOOR1-port-group-3]group-member g0/0/19 to g0/0/24
[SXL-FLOOR1-port-group-3]port link-type access
[SXL-FLOOR1-port-group-3]port default vlan 95
[SXL-FLOOR1-port-group-3]quit
// 配置 G0/0/19 至 G0/0/24 接口为接入接口, 属于 VLAN95
[SXL-FLOOR1]interface g0/0/1
[SXL-FLOOR1-GigabitEthernet0/0/1]port link-type trunk
[SXL-FLOOR1-GigabitEthernet0/0/1]port trunk allow-pass vlan 22 23 94 95
[SXL-FLOOR1-GigabitEthernet0/0/1]quit
// 配置 G0/0/1 接口为中继接口, 允许 VLAN22、VLAN23、VLAN94 和 VLAN95 通过
```

⑥ 在接入层交换机 SXL-FLOOR2 上完成 VLAN 的配置, 步骤如下:

```
[SXL-FLOOR2]vlan 22
[SXL-FLOOR2-vlan22]description rj-sxs
[SXL-FLOOR2-vlan22]vlan 23
[SXL-FLOOR2-vlan23]description wl-sxs
```

```
[SXL-FLOOR2-vlan23]vlan 94
[SXL-FLOOR2-vlan94]description manage
[SXL-FLOOR2-vlan94]vlan 95
[SXL-FLOOR2-vlan95]description monitor
[SXL-FLOOR2-vlan95]quit
[SXL-FLOOR2]
// 创建 VLAN22、VLAN23、VLAN94、VLAN95，并配置 VLAN 描述
[SXL-FLOOR2]port-group 1
[SXL-FLOOR2-port-group-1]group-member g0/0/3 to g0/0/10
[SXL-FLOOR2-port-group-1]port link-type access
[SXL-FLOOR2-port-group-1]port default vlan 22
[SXL-FLOOR2-port-group-1]quit
// 配置 G0/0/3 至 G0/0/10 接口为接入接口，属于 VLAN22
[SXL-FLOOR2]port-group 2
[SXL-FLOOR2-port-group-2]group-member g0/0/11 to g0/0/18
[SXL-FLOOR2-port-group-2]port link-type access
[SXL-FLOOR2-port-group-2]port default vlan 23
[SXL-FLOOR2-port-group-2]quit
// 配置 G0/0/11 至 G0/0/18 端口为接入接口，属于 VLAN23
[SXL-FLOOR2]port-group 3
[SXL-FLOOR2-port-group-3]group-member g0/0/19 to g0/0/24
[SXL-FLOOR2-port-group-3]port link-type access
[SXL-FLOOR2-port-group-3]port default vlan 95
[SXL-FLOOR2-port-group-3]quit
// 配置 G0/0/19 至 G0/0/24 端口为接入接口，属于 VLAN95
```

注：实训楼接入层交换机 SXL-FLOOR2 与实训楼汇聚层交换机 SXL-AGG 之间的两条链路使用链路聚合技术提升链路带宽，提高链路高可靠性。因此，实训楼接入层交换机 SXL-FLOOR2 的这两条链路接口（G0/0/1 和 G0/0/2）先不做 VLAN 的配置。

（2）汇聚层交换机 VLAN 的配置

每一幢楼的汇聚层交换机需实现该楼宇不同 VLAN 之间的通信，根据表 3.1 的规划，汇聚层交换机需完成的 VLAN 配置任务包括：创建 VLAN，配置 VLAN 描述；配置到接入层交换机的中继链路；配置到 CORE 的链路为接入链路。

① 在汇聚层交换机 XZL-AGG 上完成 VLAN 的配置，步骤如下：

```
[XZL-AGG]vlan 2
[XZL-AGG-vlan2]description baoan
[XZL-AGG-vlan2]vlan 3
[XZL-AGG-vlan3]description student
[XZL-AGG-vlan3]vlan 4
```

```
[XZL-AGG-vlan4]description caiwu
[XZL-AGG-vlan4]vlan 5
[XZL-AGG-vlan5]description hr
[XZL-AGG-vlan5]vlan 90
[XZL-AGG-vlan99]description manage
[XZL-AGG-vlan99]vlan 91
[XZL-AGG-vlan91]description monitor
[XZL-AGG-vlan91]vlan 199
[XZL-AGG-vlan199]description linktocore
[XZL-AGG-vlan199]quit
[XZL-AGG]
// 创建 VLAN2、VLAN3、VLAN4、VLAN5、VLAN90、VLAN91 和 VLAN199，并配置 VLAN 描述
[XZL-AGG]interface g0/0/2
[XZL-AGG-GigabitEthernet0/0/2]port link-type trunk
[XZL-AGG-GigabitEthernet0/0/2]port trunk allow-pass vlan 2 3 90 91
[XZL-AGG-GigabitEthernet0/0/2]quit
// 配置 G0/0/2 接口为中继接口，允许 VLAN2、VLAN3、VLAN90 和 VLAN91 通过
[XZL-AGG]interface g0/0/3
[XZL-AGG-GigabitEthernet0/0/3]port link-type trunk
[XZL-AGG-GigabitEthernet0/0/3]port trunk allow-pass vlan 4 5 90 91
[XZL-AGG-GigabitEthernet0/0/3]quit
// 配置 G0/0/3 接口为中继接口，允许 VLAN4、VLAN5、VLAN90 和 VLAN91 通过
[XZL-AGG]interface g0/0/1
[XZL-AGG-GigabitEthernet0/0/1]port link-type access
[XZL-AGG-GigabitEthernet0/0/1]port default vlan 199
[XZL-AGG-GigabitEthernet0/0/1]quit
// 配置 G0/0/1 接口为接入接口，属于 VLAN199
```

② 在汇聚层交换机 JXL-AGG 上完成 VLAN 的配置，步骤如下：

```
[JXL-AGG]vlan 12
[JXL-AGG-vlan12]description jiaoshi
[JXL-AGG-vlan12]vlan 13
[JXL-AGG-vlan13]description office
[JXL-AGG-vlan13]vlan 92
[JXL-AGG-vlan92]description manage
[JXL-AGG-vlan92]vlan 93
[JXL-AGG-vlan93]description monitor
[JXL-AGG-vlan93]vlan 199
[JXL-AGG-vlan199]description linktocore
[JXL-AGG-vlan199]quit
```

```
// 创建 VLAN12、VLAN13、VLAN92、VLAN93 和 VLAN199，并配置 VLAN 描述
[JXL-AGG]port-group 1
[JXL-AGG-port-group-1]group-member g0/0/2 to g0/0/3
[JXL-AGG-port-group-1]port link-type trunk
[JXL-AGG-port-group-1]port trunk allow-pass vlan 12 13 92 93
[JXL-AGG-port-group-1]quit
// 配置 G0/0/2 接口和 G0/0/3 接口为中继接口，允许 VLAN12、VLAN13、VLAN92 和 VLAN93 通过
[JXL-AGG]interface g0/0/1
[JXL-AGG-GigabitEthernet0/0/1]port link-type access
[JXL-AGG-GigabitEthernet0/0/1]port default vlan 199
[JXL-AGG-GigabitEthernet0/0/1]quit
// 配置 G0/0/1 接口为接入接口，属于 VLAN199
```

③ 在汇聚层交换机 SXL-AGG 上完成 VLAN 的配置，步骤如下：

```
[SXL-AGG]vlan 22
[SXL-AGG-vlan22]description rj-sxs
[SXL-AGG-vlan22]vlan 23
[SXL-AGG-vlan23]description wl-sxs
[SXL-AGG-vlan23]vlan 94
[SXL-AGG-vlan94]description manage
[SXL-AGG-vlan94]vlan 95
[SXL-AGG-vlan95]description monitor
[SXL-AGG-vlan95]vlan 199
[SXL-AGG-vlan199]description linktocore
[SXL-AGG-vlan199]quit
// 创建 VLAN22、VLAN23、VLAN94、VLAN95 和 VLAN199，并配置 VLAN 描述
[SXL-AGG]interface g0/0/3
[SXL-AGG-GigabitEthernet0/0/3]port link-type trunk
[SXL-AGG-GigabitEthernet0/0/3]port trunk allow-pass vlan 22 23 94 95
[SXL-AGG-GigabitEthernet0/0/3]quit
// 配置 G0/0/3 接口为中继接口，允许 VLAN22、VLAN23、VLAN94 和 VLAN95 通过
```

注：实训楼汇聚层交换机 SXL-AGG 与核心层交换机 CORE 之间相连的两条链路、实训楼汇聚层交换机 SXL-AGG 与实训楼接入层交换机 SXL-FLOOR2 相连的两条链路都要使用链路聚合技术提升链路带宽，提高链路高可靠性。因此，实训楼汇聚层交换机 SXL-AGG 的这四条链路接口（G0/0/1、G0/0/2、G0/0/4 和 G0/0/5）先不做 VLAN 的配置。

（3）核心层交换机的 VLAN 的配置

根据表 3.1 的规划，核心层交换机需完成的 VLAN 配置任务包括：创建 VLAN，配置 VLAN 描述；配置到汇聚层交换机的链路为接入链路；配置到服务器接入交换机 SERVER-ACC

216 企业网络互联技术（双语）

的链路为接入链路。步骤如下：

```
[CORE]vlan 199
[CORE-vlan199]description linkto-sxl-agg
[CORE-vlan199]vlan 200
[CORE-vlan200]description linkto-jxl-agg
[CORE-vlan200]vlan 201
[CORE-vlan201]description linkto-xzl-agg
[CORE-vlan201]vlan 202
[CORE-vlan202]description server
[CORE-vlan202]vlan 203
[CORE-vlan203]description linkto-edge-ar
[CORE-vlan203]quit
// 创建 VLAN199、VLAN200、VLAN201、VLAN202 和 VLAN203，并配置 VLAN 描述
[CORE]interface g0/0/1
[CORE-GigabitEthernet0/0/1]port link-type access
[CORE-GigabitEthernet0/0/1]port default vlan 203
[CORE-GigabitEthernet0/0/1]quit
// 配置 G0/0/1 接口为接入接口，属于 VLAN203。注：该接口连接到出口路由器 EDGE-AR
[CORE]interface g0/0/2
[CORE-GigabitEthernet0/0/2]port link-type access
[CORE-GigabitEthernet0/0/2]port default vlan 201
[CORE-GigabitEthernet0/0/2]quit
// 配置 G0/0/2 接口为接入接口，属于 VLAN201。注：该接口连接到行政楼汇聚层交换机 XZL-AGG
[CORE]interface g0/0/3
[CORE-GigabitEthernet0/0/3]port link-type access
[CORE-GigabitEthernet0/0/3]port default vlan 200
[CORE-GigabitEthernet0/0/3]quit
// 配置 G0/0/3 接口为接入接口，属于 VLAN200。注：该接口连接到教学楼汇聚层交换机 JXL-AGG
[CORE]interface g0/0/6
[CORE-GigabitEthernet0/0/6]port link-type access
[CORE-GigabitEthernet0/0/6]port default vlan 202
[CORE-GigabitEthernet0/0/6]quit
// 配置 G0/0/6 接口为接入接口，属于 VLAN202。注：该接口连接到服务器接入层交换机 SERVER-ACC
```

（4）VLAN 验证

在所有交换机上执行 "display vlan" 命令和 "display port vlan" 命令，观察交换机显示的
VLAN 信息和端口的 VLAN 信息是否与表 3.1 的规划一致。图 3.3 为核心层交换机 CORE 上端
口 VLAN 显示结果。图 3.4 为汇聚层交换机 XZL-AGG 上端口 VLAN 显示结果。图 3.5 为接入

层交换机 XZL-FLOOR1 上端口 VLAN 显示结果。注意观察图中加框中显示结果是否与表 3.1 的
规划一致。

```
[CORE]display port vlan
Port            Link Type    PVID  Trunk VLAN List
----------------------------------------------------------------
GigabitEthernet0/0/1    access      203  -
GigabitEthernet0/0/2    access      201  -
GigabitEthernet0/0/3    access      200  -
GigabitEthernet0/0/4    hybrid       1   -
GigabitEthernet0/0/5    hybrid       1   -
GigabitEthernet0/0/6    access      202  -
GigabitEthernet0/0/7    hybrid       1   -
GigabitEthernet0/0/8    hybrid       1   -
GigabitEthernet0/0/9    hybrid       1   -
GigabitEthernet0/0/10   hybrid       1   -
GigabitEthernet0/0/11   hybrid       1   -
GigabitEthernet0/0/12   hybrid       1   -
GigabitEthernet0/0/13   hybrid       1   -
GigabitEthernet0/0/14   hybrid       1   -
GigabitEthernet0/0/15   hybrid       1   -
```

图 3.3   核心层交换机 CORE 上端口 VLAN 显示结果

```
[XZL-AGG]display port vlan
Port            Link Type    PVID  Trunk VLAN List
----------------------------------------------------------------
GigabitEthernet0/0/1    access      199  -
GigabitEthernet0/0/2    trunk        1   1-3 90-91
GigabitEthernet0/0/3    trunk        1   1 4-5 90-91
GigabitEthernet0/0/4    hybrid       1   -
GigabitEthernet0/0/5    hybrid       1   -
GigabitEthernet0/0/6    hybrid       1   -
GigabitEthernet0/0/7    hybrid       1   -
GigabitEthernet0/0/8    hybrid       1   -
GigabitEthernet0/0/9    hybrid       1   -
GigabitEthernet0/0/10   hybrid       1   -
GigabitEthernet0/0/11   hybrid       1   -
GigabitEthernet0/0/12   hybrid       1   -
GigabitEthernet0/0/13   hybrid       1   -
GigabitEthernet0/0/14   hybrid       1   -
GigabitEthernet0/0/15   hybrid       1   -
```

图 3.4   汇聚层交换机 XZL-AGG 上端口 VLAN 显示结果

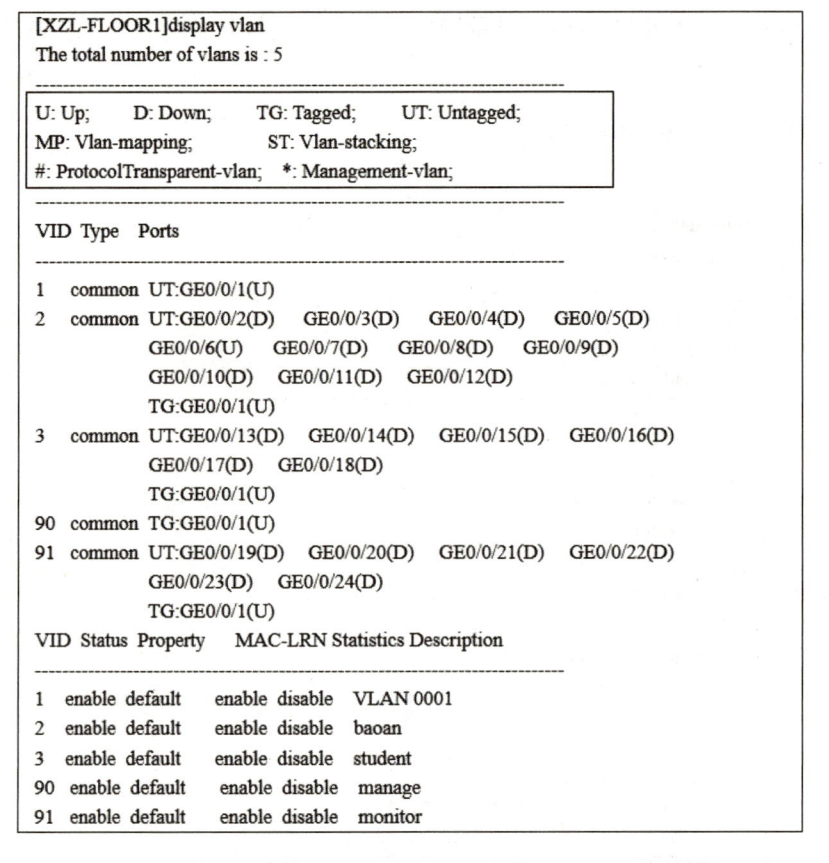

图 3.5　接入层交换机 XZL-FLOOR1 上端口 VLAN 显示结果

（5）接入层交换机管理 IP 的配置

接入层交换机管理 IP 的配置任务有：进入管理 VLAN，配置 IP 地址与掩码；配置接入层交换机的默认网关地址。

① 在接入层交换机 XZL-FLOOR1 上完成管理 IP 的配置，步骤如下：

```
[XZL-FLOOR1]interface Vlanif 90
[XZL-FLOOR1-Vlanif90]ip address 10.0.90.1 24
[XZL-FLOOR1-Vlanif90]quit
[XZL-FLOOR1]ip route-static 0.0.0.0 0.0.0.0 10.0.90.254
// 配置交换机 XZL-FLOOR1 的管理 IP 地址为 10.0.90.1/24，默认网关为 10.0.90.254
```

② 在接入层交换机 XZL-FLOOR2 上完成管理 IP 的配置，步骤如下：

```
[XZL-FLOOR2]interface Vlanif 90
[XZL-FLOOR2-Vlanif90]ip address 10.0.90.2 24
[XZL-FLOOR2-Vlanif90]quit
[XZL-FLOOR2]ip route-static 0.0.0.0 0.0.0.0 10.0.90.254
// 配置交换机 XZL-FLOOR2 的管理 IP 地址为 10.0.90.2/24，默认网关为 10.0.90.254
```

③ 在接入层交换机 JXL-FLOOR1 上完成管理 IP 的配置，步骤如下：

```
[JXL-FLOOR1]interface Vlanif 92
[JXL-FLOOR1-Vlanif92]ip address 10.1.92.1 24
[JXL-FLOOR1-Vlanif92]quit
[JXL-FLOOR1]ip route-static 0.0.0.0 0.0.0.0 10.1.92.254
// 配置交换机 JXL-FLOOR1 的管理 IP 地址为 10.1.92.1/24，默认网关为 10.0.92.254
```

④ 在接入层交换机 JXL-FLOOR2 上完成管理 IP 的配置，步骤如下：

```
[JXL-FLOOR2]interface Vlanif 92
[JXL-FLOOR2-Vlanif92]ip address 10.1.92.2 24
[JXL-FLOOR2-Vlanif92]quit
[JXL-FLOOR2]ip route-static 0.0.0.0 0.0.0.0 10.1.92.254
// 配置交换机 JXL-FLOOR2 的管理 IP 地址为 10.1.92.2/24，默认网关为 10.0.92.254
```

⑤ 在接入层交换机 SXL-FLOOR1 上完成管理 IP 的配置，步骤如下：

```
[SXL-FLOOR1]interface Vlanif 94
[SXL-FLOOR1-Vlanif94]ip address 10.2.94.1 24
[SXL-FLOOR1-Vlanif94]quit
[SXL-FLOOR1]ip route-static 0.0.0.0 0.0.0.0 10.2.94.254
[SXL-FLOOR1]quit
// 配置交换机 SXL-FLOOR1 的管理 IP 地址为 10.1.94.1/24，默认网关为 10.0.94.254
```

⑥ 在接入层交换机 SXL-FLOOR2 上完成管理 IP 的配置，步骤如下：

```
[SXL-FLOOR2]interface Vlanif 94
[SXL-FLOOR2-Vlanif94]ip address 10.2.94.2 24
[SXL-FLOOR2-Vlanif94]quit
[SXL-FLOOR2]ip route-static 0.0.0.0 0.0.0.0 10.2.94.254
// 配置交换机 SXL-FLOOR2 的管理 IP 地址为 10.1.94.2/24，默认网关为 10.0.94.254
```

（6）汇聚层交换机 IP 地址与 DHCP 服务的配置

汇聚层交换机需要为楼宇的终端设备动态分配 IP 地址，因此除了完成表 3.1 规划的接口 IP 地址配置以外，还需要进行 DHCP 服务配置。

① 在汇聚层交换机 XZL-AGG 上完成表 3.1 规划的 IP 配置和 DHCP 服务的配置，步骤如下：

```
[XZL-AGG]dhcp enable
// 启用 DHCP 服务
[XZL-AGG]ip pool vlan2
[XZL-AGG-ip-pool-vlan2]network 10.0.2.0 mask 24
[XZL-AGG-ip-pool-vlan2]gateway-list 10.0.2.254
[XZL-AGG-ip-pool-vlan2]dns-list 10.254.253.2
```

```
[XZL-AGG-ip-pool-vlan2]lease day 0 hour 8
[XZL-AGG-ip-pool-vlan2]quit
```
// 配置 VLAN2 的 DHCP 地址池, 地址池名为 vlan2, 网络号为 10.0.2.0/24, 网关地址为
//10.0.2.254, DNS 服务器地址为 10.254.253.2, 地址租期为 8 h
```
[XZL-AGG]ip pool vlan3
[XZL-AGG-ip-pool-vlan3]network 10.0.3.0 mask 24
[XZL-AGG-ip-pool-vlan3]gateway-list 10.0.3.254
[XZL-AGG-ip-pool-vlan3]dns-list 10.254.253.2
[XZL-AGG-ip-pool-vlan3]lease day 0 hour 8
[XZL-AGG-ip-pool-vlan3]quit
```
// 配置 VLAN3 的 DHCP 地址池, 地址池名为 vlan3, 网络号为 10.0.3.0/24, 网关地址为 10.0.
//3.254, DNS 服务器地址为 10.254.253.2, 地址租期为 8 h
```
[XZL-AGG]ip pool vlan4
[XZL-AGG-ip-pool-vlan4]network 10.0.4.0 mask 24
[XZL-AGG-ip-pool-vlan4]gateway-list 10.0.4.254
[XZL-AGG-ip-pool-vlan4]dns-list 10.254.253.2
[XZL-AGG-ip-pool-vlan4]lease day 0 hour 8
[XZL-AGG-ip-pool-vlan4]quit
```
// 配置 VLAN4 的 DHCP 地址池, 地址池名为 vlan4, 网络号为 10.0.4.0/24, 网关地址为 10.0.
//4.254, DNS 服务器地址为 10.254.253.2, 地址租期为 8 h
```
[XZL-AGG]ip pool vlan5
[XZL-AGG-ip-pool-vlan5]network 10.0.5.0 mask 24
[XZL-AGG-ip-pool-vlan5]gateway-list 10.0.5.254
[XZL-AGG-ip-pool-vlan5]dns-list 10.254.253.2
[XZL-AGG-ip-pool-vlan5]lease day 0 hour 8
[XZL-AGG-ip-pool-vlan5]quit
```
// 配置 VLAN5 的 DHCP 地址池, 地址池名为 vlan5, 网络号为 10.0.5.0/24, 网关地址为 10.0.5.
//254, DNS 服务器地址为 10.254.253.2, 地址租期为 8 h
```
[XZL-AGG]interface Vlanif 2
[XZL-AGG-Vlanif2]ip address 10.0.2.254 24
[XZL-AGG-Vlanif2]dhcp select global
[XZL-AGG-Vlanif2]quit
[XZL-AGG]interface Vlanif 3
[XZL-AGG-Vlanif3]ip address 10.0.3.254 24
[XZL-AGG-Vlanif3]dhcp select global
[XZL-AGG-Vlanif3]quit
[XZL-AGG]interface Vlanif 4
[XZL-AGG-Vlanif4]ip address 10.0.4.254 24
[XZL-AGG-Vlanif4]dhcp select global
[XZL-AGG-Vlanif4]quit
```

```
[XZL-AGG]interface Vlanif 5
[XZL-AGG-Vlanif5]ip address 10.0.5.254 24
[XZL-AGG-Vlanif5]dhcp select global
[XZL-AGG-Vlanif5]quit
[XZL-AGG]interface Vlanif 90
[XZL-AGG-Vlanif91]ip address 10.0.9.254 24
[XZL-AGG-Vlanif91]quit
[XZL-AGG]interface Vlanif 91
[XZL-AGG-Vlanif91]ip address 10.0.91.254 24
[XZL-AGG-Vlanif91]quit
[XZL-AGG]interface Vlanif 199
[XZL-AGG-Vlanif199]ip address 10.254.254.2 30
[XZL-AGG-Vlanif199]quit
// 配置交换机 SXL-AGG 的 Vlanif2、Vlanif3、Vlanif4、Vlanif5、Vlanif90、Vlanif
//91 和 Vlanif199 接口的 IP 地址, 开启 Vlanif2、Vlanif3、Vlanif4 和 Vlanif5 接口
// 采用全局地址池的 DHCP Server 功能
```

② 在汇聚层交换机 JXL-AGG 上完成表 3.1 规划的 IP 配置和 DHCP 服务的配置, 步骤如下:

```
[JXL-AGG]dhcp enable
// 启用 DHCP 服务
[JXL-AGG]ip pool vlan12
[JXL-AGG-ip-pool-vlan12]network 10.1.12.0 mask 24
[JXL-AGG-ip-pool-vlan12]gateway-list 10.1.12.254
[JXL-AGG-ip-pool-vlan12]dns-list 10.254.253.2
[JXL-AGG-ip-pool-vlan12]lease day 0 hour 8
[JXL-AGG-ip-pool-vlan12]quit
// 配置 VLAN12 的 DHCP 地址池, 地址池名为 vlan12, 网络号为 10.1.12.0/24, 网关地址为
//10.1.12.254, DNS 服务器地址为 10.254.253.2, 地址租期为 8 h
[JXL-AGG]ip pool vlan13
[JXL-AGG-ip-pool-vlan13]network 10.1.13.0 mask 24
[JXL-AGG-ip-pool-vlan13]gateway-list 10.1.13.254
[JXL-AGG-ip-pool-vlan13]dns-list 10.254.253.2
[JXL-AGG-ip-pool-vlan13]lease day 0 hour 8
[JXL-AGG-ip-pool-vlan13]quit
// 配置 VLAN13 的 DHCP 地址池, 地址池名为 vlan13, 网络号为 10.1.13.0/24, 网关地址为
//10.1.13.254, DNS 服务器地址为 10.254.253.2, 地址租期为 8 h
[JXL-AGG]interface Vlanif 12
[JXL-AGG-Vlanif12]ip address 10.1.12.254 24
[JXL-AGG-Vlanif12]dhcp select global
[JXL-AGG-Vlanif12]quit
```

```
[JXL-AGG]interface Vlanif 13
[JXL-AGG-Vlanif13]ip address 10.1.13.254 24
[JXL-AGG-Vlanif13]dhcp select global
[JXL-AGG-Vlanif13]quit
[JXL-AGG]interface Vlanif 92
[JXL-AGG-Vlanif92]ip address 10.1.92.254 24
[JXL-AGG-Vlanif92]quit
[JXL-AGG]interface Vlanif 93
[JXL-AGG-Vlanif93]ip address 10.1.93.254 24
[JXL-AGG-Vlanif93]quit
[JXL-AGG]interface Vlanif 199
[JXL-AGG-Vlanif199]ip add
[JXL-AGG-Vlanif199]ip address 10.254.254.6 30
[JXL-AGG-Vlanif199]quit
// 配置交换机 JXL-AGG 的 Vlanif12、Vlanif13、Vlanif92、Vlanif93 和 Vlanif199
// 接口的 IP 地址，开启 Vlanif12 和 Vlanif13 接口采用全局地址池的 DHCP Server 功能
```

③ 在汇聚层交换机 SXL-AGG 上完成表 3.1 规划的 IP 配置和 DHCP 服务的配置，步骤如下：

```
[SXL-AGG]dhcp enable
// 启用 DHCP 服务
[SXL-AGG]ip pool vlan22
[SXL-AGG-ip-pool-vlan22]network 10.2.22.0 mask 24
[SXL-AGG-ip-pool-vlan22]gateway-list 10.2.22.254
[SXL-AGG-ip-pool-vlan22]dns-list 10.254.253.2
[SXL-AGG-ip-pool-vlan22]lease day 0 hour 8
[SXL-AGG-ip-pool-vlan22]quit
// 配置 VLAN22 的 DHCP 地址池，地址池名为 vlan22，网络号为 10.2.22.0/24，网关地址为
//10.2.22.254，DNS 服务器地址为 10.254.253.2，地址租期为 8 h
[SXL-AGG]ip pool vlan23
[SXL-AGG-ip-pool-vlan23]network 10.2.23.0 mask 24
[SXL-AGG-ip-pool-vlan23]gateway-list 10.2.23.254
[SXL-AGG-ip-pool-vlan23]dns-list 10.254.253.2
[SXL-AGG-ip-pool-vlan23]lease day 0 hour 8
[SXL-AGG-ip-pool-vlan23]quit
// 配置 VLAN23 的 DHCP 地址池，地址池名为 vlan23，网络号为 10.2.23.0/24，网关地址为
//10.2.23.254，DNS 服务器地址为 10.254.253.2，地址租期为 8 h
[SXL-AGG]interface Vlanif 22
[SXL-AGG-Vlanif22]ip address 10.2.22.254 24
[SXL-AGG-Vlanif22]dhcp select global
[SXL-AGG-Vlanif22]quit
```

```
[SXL-AGG]interface Vlanif 23
[SXL-AGG-Vlanif23]ip address 10.2.23.254 24
[SXL-AGG-Vlanif23]dhcp select global
[SXL-AGG-Vlanif23]quit
[SXL-AGG]interface Vlanif 94
[SXL-AGG-Vlanif94]ip address 10.2.94.254 24
[SXL-AGG-Vlanif94]quit
[SXL-AGG]interface Vlanif 95
[SXL-AGG-Vlanif95]ip address 10.2.95.254 24
[SXL-AGG-Vlanif95]quit
[SXL-AGG]interface Vlanif 199
[SXL-AGG-Vlanif199]ip address 10.254.254.10 30
[SXL-AGG-Vlanif199]quit
// 配置交换机 SXL-AGG 的 Vlanif22、Vlanif23、Vlanif94、Vlanif95 和 Vlanif199
// 接口的 IP 地址，开启 Vlanif22 和 Vlanif23 接口采用全局地址池的 DHCP Server 功能．
```

（7）核心层交换机 IP 地址的配置

根据表 3.1 的 IP 规划，配置核心层交换机 CORE 的 Vlanif199、Vlanif200、Vlanif201、Vlanif202 和 Vlanif203 接口 IP 地址，步骤如下：

```
[CORE]interface Vlanif 199
[CORE-Vlanif199]ip address 10.254.254.9 30
[CORE-Vlanif199]quit
[CORE]interface Vlanif 200
[CORE-Vlanif200]ip address 10.254.254.5 30
[CORE-Vlanif200]quit
[CORE]interface Vlanif 201
[CORE-Vlanif201]ip address 10.254.254.1 30
[CORE-Vlanif201]quit
[CORE]interface Vlanif 202
[CORE-Vlanif202]ip address 10.254.253.254 24
[CORE-Vlanif202]quit
[CORE]interface Vlanif 203
[CORE-Vlanif203]ip address 10.254.254.13 30
[CORE-Vlanif203]quit
```

（8）出口路由器和运营商路由器接口 IP 配置

① 根据表 3.1 的 IP 规划，配置出口路由器 EDGE-AR 的接口 IP 地址的步骤如下：

```
[EDGE-AR]interface g0/0/1
[EDGE-AR-GigabitEthernet0/0/1]ip address 10.254.254.14 30
[EDGE-AR-GigabitEthernet0/0/1]undo shutdown
```

```
[EDGE-AR-GigabitEthernet0/0/1]quit
[EDGE-AR]interface g0/0/0
[EDGE-AR-GigabitEthernet0/0/0]ip address 125.71.28.65 24
[EDGE-AR-GigabitEthernet0/0/0]undo shutdown
[EDGE-AR-GigabitEthernet0/0/0]quit
```

② 根据表 3.1 的 IP 规划，配置运营路由器 ISP 的接口 IP 地址的步骤如下：

```
[ISP]interface g0/0/1
[ISP-GigabitEthernet0/0/1]ip address 202.1.1.254 24
[ISP-GigabitEthernet0/0/1]undo shutdown
[ISP-GigabitEthernet0/0/1]quit
[ISP]interface g0/0/0
[ISP-GigabitEthernet0/0/0]ip address 125.71.28.1 24
[ISP-GigabitEthernet0/0/0]undo shutdown
[ISP-GigabitEthernet0/0/0]quit
```

（9）NAT Server 的配置

在企业出口路由器 EDGE-AR 上，根据表 3.3 规划的 **NAT Server** 完成配置，步骤如下：

```
[EDGE-AR]interface g0/0/0
[EDGE-AR-GigabitEthernet0/0/0]nat server protocol tcp global 125.71.28.66
www inside 10.254.253.1 www
    [EDGE-AR-GigabitEthernet0/0/0]nat server protocol tcp global 125.71.28.66 53
inside 10.254.253.2 53
    [EDGE-AR-GigabitEthernet0/0/0]nat server protocol udp global 125.71.28.66 53
inside 10.254.253.2 53
```

（10）NAPT 配置

在企业出口路由器 EDGE-AR 上，NAPT 配置需完成的任务包括：使用 ACL 配置允许进行转换的内网地址段；配置 NAT 公有地址池。

```
[EDGE-AR]acl 2000
[EDGE-AR-acl-basic-2000]rule permit source 10.0.2.0 0.0.0.255
[EDGE-AR-acl-basic-2000]rule permit source 10.0.3.0 0.0.0.255
[EDGE-AR-acl-basic-2000]rule permit source 10.0.4.0 0.0.0.255
[EDGE-AR-acl-basic-2000]rule permit source 10.0.5.0 0.0.0.255
[EDGE-AR-acl-basic-2000]rule permit source 10.1.12.0 0.0.0.255
[EDGE-AR-acl-basic-2000]rule permit source 10.1.13.0 0.0.0.255
[EDGE-AR-acl-basic-2000]rule permit source 10.2.22.0 0.0.0.255
[EDGE-AR-acl-basic-2000]rule permit source 10.2.23.0 0.0.0.255
[EDGE-AR-acl-basic-2000]rule permit source 10.0.90.0 0.0.0.255
[EDGE-AR-acl-basic-2000]rule permit source 10.0.91.0 0.0.0.255
```

```
[EDGE-AR-acl-basic-2000]rule permit source 10.1.92.0 0.0.0.255
[EDGE-AR-acl-basic-2000]rule permit source 10.1.93.0 0.0.0.255
[EDGE-AR-acl-basic-2000]rule permit source 10.2.94.0 0.0.0.255
[EDGE-AR-acl-basic-2000]rule permit source 10.2.95.0 0.0.0.255
[EDGE-AR-acl-basic-2000]rule permit source 10.254.253.0 0.0.0.255
[EDGE-AR-acl-basic-2000]rule permit source 10.254.254.0 0.0.0.255
// 配置允许进行 NAT 转换的内网地址段，ACL 列表号为 2000，允许进行 NAT 转换的内网地址段包括：
//10.0.2.0/24、10.0.3.0/24、10.0.4.0/24、10.0.5.0/24、10.1.12.0/24、
//10.1.13.0/24、10.2.22.0/24、10.2.23.0/24、10.0.90.0/24、10.0.91.0/
//24、10.1.92.0/24、10.1.93.0/24、10.2.94.0/24、10.2.95.0/24、10.254.253.0/24
// 和 10.254.254.0/24
[EDGE-AR]nat address-group 1 125.71.28.67 125.71.28.69
// 配置 NAT 公有地址池，地址池编号为 1，地址池范围为 125.71.28.67~125.71.28.69
[EDGE-AR-GigabitEthernet0/0/0]nat outbound 2000 address-group 1
// 在出口 G0/0/0 接口下关联 ACL，与地址池进行 NAPT 转换
```

（11）服务器与 PC1 的 IP 地址规划

根据表 3.1 的 IP 地址规划，企业内网服务器、Internet 上服务器与测试主机 PC1 的 IP 地址为静态分配，请按规划配置 IP 地址、子网掩码、默认网关和 DNS 服务器 IP 地址，并在命令行中使用 ipconfig 命令查看配置的 IP 信息是否生效。

（12）验证与测试

完成 IP 地址与 DHCP 服务的配置后，使用以下方法进行验证：

① 在楼宇的所有接入层交换机、汇聚层交换、核心层交换机以及路由器上使用"display ip interface brief"命令显示接口 IP 简要信息，检查管理 VLAN 端口、管理 IP 地址、物理和协议状态是否为 UP。图 3.6 所示为接入层交换机 XZL-FLOOR1 上使用"display ip interface brief"命令后的显示结果。图 3.7 所示为汇聚层交换机 XZL-AGG 上使用"display ip interface brief"命令后的显示结果。图 3.8 所示为核心层交换机 CORE 上使用"display ip interface brief"命令后的显示结果。

```
[XZL-FLOOR1]display ip interface brief
*down: administratively down
^down: standby
(l): loopback
(s): spoofing
The number of interface that is UP in Physical is 3
The number of interface that is DOWN in Physical is 1
The number of interface that is UP in Protocol is 2
The number of interface that is DOWN in Protocol is 2


Interface           IP Address/Mask    Physical  Protocol
MEth0/0/1           unassigned         down      down
NULL0               unassigned         up        up(s)
Vlanif1             unassigned         up        down
Vlanif90            10.0.90.1/24       up        up
```

图 3.6　接入层交换机 XZL-FLOOR1 上使用"display ip interface brief"命令后的显示结果

```
[XZL-AGG]display ip interface brief
*down: administratively down
^down: standby
(l): loopback
(s): spoofing
The number of interface that is UP in Physical is 9
The number of interface that is DOWN in Physical is 1
The number of interface that is UP in Protocol is 8
The number of interface that is DOWN in Protocol is 2

Interface              IP Address/Mask     Physical    Protocol
MEth0/0/1              unassigned          down        down
NULL0                 unassigned          up          up(s)
Vlanif1               unassigned          up          down
Vlanif2               10.0.2.254/24       up          up
Vlanif3               10.0.3.254/24       up          up
Vlanif4               10.0.4.254/24       up          up
Vlanif5               10.0.5.254/24       up          up
Vlanif90              10.0.90.254/24      up          up
Vlanif91              10.0.91.254/24      up          up
Vlanif199             10.254.254.2/30     up          up
```

图 3.7　汇聚层交换机 XZL-AGG 上使用"display ip interface brief"命令后的显示结果

```
[CORE]display ip interface brief
*down: administratively down
^down: standby
(l): loopback
(s): spoofing
The number of interface that is UP in Physical is 6
The number of interface that is DOWN in Physical is 2
The number of interface that is UP in Protocol is 5
The number of interface that is DOWN in Protocol is 3

Interface              IP Address/Mask     Physical    Protocol
MEth0/0/1              unassigned          down        down
NULL0                 unassigned          up          up(s)
Vlanif1               unassigned          up          down
Vlanif199             10.254.254.9/30     down        down
Vlanif200             10.254.254.5/30     up          up
Vlanif201             10.254.254.1/30     up          up
Vlanif202             10.254.253.254/24   up          up
Vlanif203             10.254.254.13/30    up          up
```

图 3.8　核心层交换机 CORE 上使用"display ip interface brief"命令后的显示结果

注：VLAN199 端口为连接到实训楼汇聚层交换机 SXL-AGG 的连接端口，由于目前还没有配置任何物理接口属于 VLAN199，因此 Vlanif199 接口的物理和协议状态均为 down。

② 在楼宇的所有接入层交换机上使用"display ip routing-table"命令显示路由表内容，查看是否存在默认路由，默认路由的下一跳地址是否为规划的网关地址。图 3.9 所示为在接入层交换机 XZL-FLOOR1 上使用"display ip routing-table"命令后的显示结果。默认路由的下一跳地址为10.0.90.254，与规划的网关地址相同。

```
[XZL-FLOOR1]display ip routing-table
Route Flags: R - relay, D - download to fib
------------------------------------------------------------------------------
Routing Tables: Public
        Destinations : 5      Routes : 5

Destination/Mask   Proto  Pre  Cost    Flags NextHop       Interface

   0.0.0.0/0    Static 60   0        RD  10.0.90.254    Vlanif90
   10.0.90.0/24  Direct 0    0        D   10.0.90.1      Vlanif90
   10.0.90.1/32  Direct 0    0        D   127.0.0.1      Vlanif90
   127.0.0.0/8   Direct 0    0        D   127.0.0.1      InLoopBack0
   127.0.0.1/32  Direct 0    0        D   127.0.0.1      InLoopBack0
```

图 3.9　在接入层交换机 XZL-FLOOR1 上使用 "display ip routing-table" 命令后的显示结果

注：在每个接入层上查看路由表，观察是否存在默认路由，下一跳地址是否为规划的网关地址

③ 在楼宇的所有接入层交换机上使用 ping 命令，测试到默认网关的连通性。图 3.10 所示为接入层交换机 XZL-FLOOR1 成功 ping 通其配置的默认网关。

```
[XZL-FLOOR1]ping 10.0.90.254
 PING 10.0.90.254: 56  data bytes, press CTRL_C to break
   Reply from 10.0.90.254: bytes=56 Sequence=1 ttl=255 time=10 ms
   Reply from 10.0.90.254: bytes=56 Sequence=2 ttl=255 time=20 ms
   Reply from 10.0.90.254: bytes=56 Sequence=3 ttl=255 time=50 ms
   Reply from 10.0.90.254: bytes=56 Sequence=4 ttl=255 time=30 ms
   Reply from 10.0.90.254: bytes=56 Sequence=5 ttl=255 time=50 ms

 --- 10.0.90.254 ping statistics ---
   5 packet(s) transmitted
   5 packet(s) received
   0.00% packet loss
   round-trip min/avg/max = 10/32/50 ms
```

图 3.10　接入层交换机 XZL-FLOOR1 成功 ping 通其配置的默认网关

④ 在楼宇的所有接入层交换机任何端口接一台测试主机，将其 IP 地址与 DNS 服务器地址配置为自动获取，在测试主机上使用 ipconfig 命令查看是否正常获取 IP 地址，再使用 ping 命令测试到默认网关的 IP 连通性。图 3.11 为测试主机 PC2 成功从 DHCP 服务器获取 IP 地址、子网掩码和默认网关，并成功 ping 通其默认网关地址。

使用以下 display 命令进行 NAT 查看与验证。

① 使用 "display nat server" 命令，可查看 NAT Server 信息。图 3.12 所示为在路由器 EDGE-AR 上查看 NAT server 信息的示例。

② 在路由器 EDGE-AR 的 G0/0/0 接口视图下使用 "display this" 命令，可查看该接口下的配置，注意观察 NAT 的配置信息是否与规划一致，如图 3.13 所示。

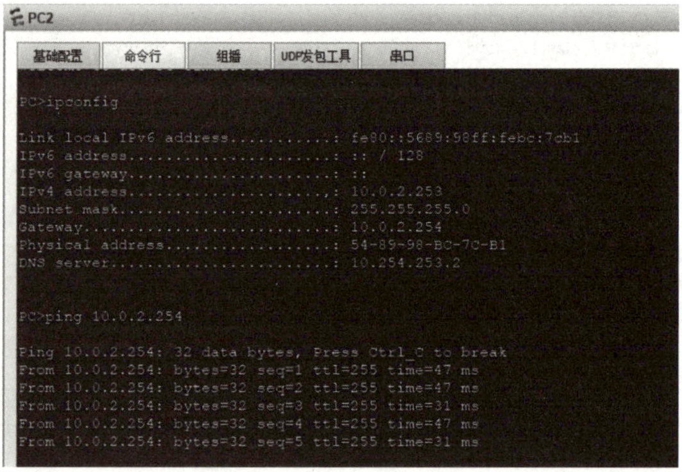

228　企业网络互联技术（双语）

图 3.11　测试主机 PC2 成功从 DHCP 服务器获取 IP 地址

```
[EDGE-AR]display nat server
 Nat Server Information:
 Interface : GigabitEthernet0/0/0
  Global IP/Port   : 125.71.28.66/80(www)
  Inside IP/Port   : 10.254.253.1/80(www)
  Protocol : 6(tcp)
  VPN instance-name  : ----
  Acl number       : ----
  Description : ----
  Global IP/Port   : 125.71.88.66/53(domain)
  Inside IP/Port   : 10.254.253.2/53(domain)
  Protocol : 6(tcp)
  VPN instance-name  : ----
  Acl number       : ----
  Description : ----
  Global IP/Port   : 125.71.88.66/53(dns)
  Inside IP/Port   : 10.254.253.2/53(dns)
  Protocol : 17(udp)
  VPN instance-name  : ----
  Acl number       : ----
  Description : ----

 Total :   3
```

图 3.12　查看 NAT Server 信息的示例

```
[EDGE-AR-GigabitEthernet0/0/0]display this
[V200R003C00]
#
interface GigabitEthernet0/0/0
 ip address 125.71.28.65 255.255.255.0
 nat server protocol tcp global 125.71.28.66 www inside 10.254.253.1 www
 nat server protocol tcp global 125.71.88.66 domain inside 10.254.253.2 domain
 nat server protocol udp global 125.71.88.66 dns inside 10.254.253.2 dns
 nat outbound 2000 address-group 1 return
```

图 3.13　查看路由器 EDGE-AR 的 G0/0/0 接口下的配置

项目 3 中型企业网络互联项目 229

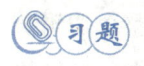

选择题

（1）下面关于企业网络层次化设计，说法错误的是（　　）。

　　A. 核心层为相互通信的节点间提供高速优化的带宽传输

　　B. 汇聚层为相互通信的节点间提供高速优化的带宽传输

　　C. 核心层设备之间、核心层设备与汇聚层设备通常采用冗余链路的光纤连接

　　D. 接入层网络用于将终端设备接入网络中

（2）下面（　　）通常为使用 DHCP 自动获取。

　　A. 路由器接口 IP 地址　　　　　　　　B. 接入层交换机的管理 IP 地址

　　C. 服务器的 IP 地址　　　　　　　　　D. 普通用户的笔记本计算机

（3）接入层交换机通常不完成（　　）。

　　A. VLAN 之间通信配置　　　　　　　B. VLAN 的配置

　　C. 管理 IP 地址的配置　　　　　　　　D. 生成树的配置

任务 2　链路高可靠性的规划与配置

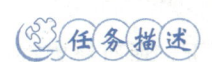

在完成本项目任务 1 的基础上，在图 3.1 所示的网络中完成链路聚合和 MSTP 的规划与配置，实现增加链路带宽与高可靠性、避免二层环路的网络部署需求。

任务目标

- 理解链路聚合的应用。
- 掌握手工模式的链路聚合配置、查看与排障。
- 掌握 LACP 模式的链路聚合配置、查看与排障。
- 具有链路聚合和 MSTP 的部署实施能力。

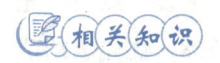

1. 链路聚合基本原理

链路聚合技术又称链路捆绑技术（bonding），它是将两台设备间的多条物理链路捆绑在一起形成一条逻辑链路，又称聚合链路。链路聚合技术正式标准为 IEEE 802 委员会制定的 IEEE 802.3ad，它适用于 1000 Mbit/s 与 10 Gbit/s 等以太网技术。链路聚合技术可以实现数据流量在构成聚合链路的所有物理链路之间的分担，以有效提高网络连接的带宽。此外，形成一条聚合

链路的各个物理链路之间彼此互为动态备份，只要存在一条正常工作的物理链路，整个逻辑链路就不会失效，因此可以有效地增加网络的可靠性。图 3.14 所示的链路聚合示例中，交换机 LSW1 与 LSW2 之间的三条物理链路聚合成了一条逻辑链路。

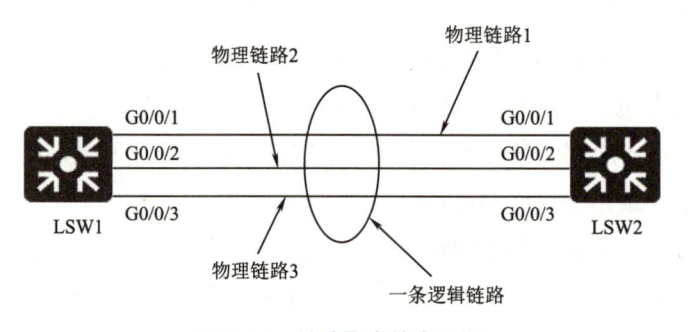

图 3.14　链路聚合技术示例

在企业网络中，通常在核心层交换机与汇聚层交换机之间、汇聚层交换机与接入层交换机之间、服务器接入层交换机与服务器之间部署多条物理链路并使用链路聚合，以提升交换机之间的链路带宽、交换机与服务器之间的带宽以及保证其可靠性。

注： 华为的链路聚合为 Eth-Trunk。

2. 链路聚合模式

链路聚合可以分为手工模式和 LACP 模式。

（1）手工模式

手工模式下，Eth-Trunk 的建立、成员接口的加入均由手工配置。正常情况下，所有链路都是活动链路，即都参与数据的转发。如果某一条活动链路发生了故障，Eth-Trunk 自动在余下的活动链路中平均分担流量。手工模式必须保证链路聚合本端接口中的所有成员接口对应的对端成员接口都是通过同一设备加入的，这个只能通过管理员人工确认。设备只能通过物理层状态判断对端接口是否正常工作。通常，当聚合链路两端设备中只要有一个不支持 LACP 协议时，使用手工模式。

（2）LACP 模式

LACP 模式是采用 LACP 协议的一种链路聚合模式。链路聚合设备间通过链路聚合控制协议数据单元（link aggregation control protocol data unit, LACPDU）进行交互，协商确保对端是同一台设备、同一个聚合接口的成员接口。LACPDU 报文中有设备优先级、MAC 地址、接口优先级、接口号等字段。LACP 模式下，通过系统 LACP 优先级（默认优先级为 32 768，优先级的值越小，优先级越高）确定主动端，另一端（被动端）根据主动端选择活动接口。LACP 模式支持配置最大活动接口数目，当成员接口数目超过最大活动接口数目时会通过比较接口优先级、接口号选举出较优的链路聚合成员接口成为活动接口，余下成员接口为非活动接口（备份端口），交换机只会从活动接口中发送、接收数据帧。当活动接口出现故障时，非活动接口则被选举为活动接口。如图 3.15 所示，交换机 LSW1 与 LSW2 之间的链路聚合有三个成员接口，但最大活

动接口数为 2，正常情况下，从活动接口 1 和 2 进行转发数据帧，接口 3 为非活动接口，当某个活动链路（如端口 1）出现故障时，则端口 3 会被选择成为活动接口。

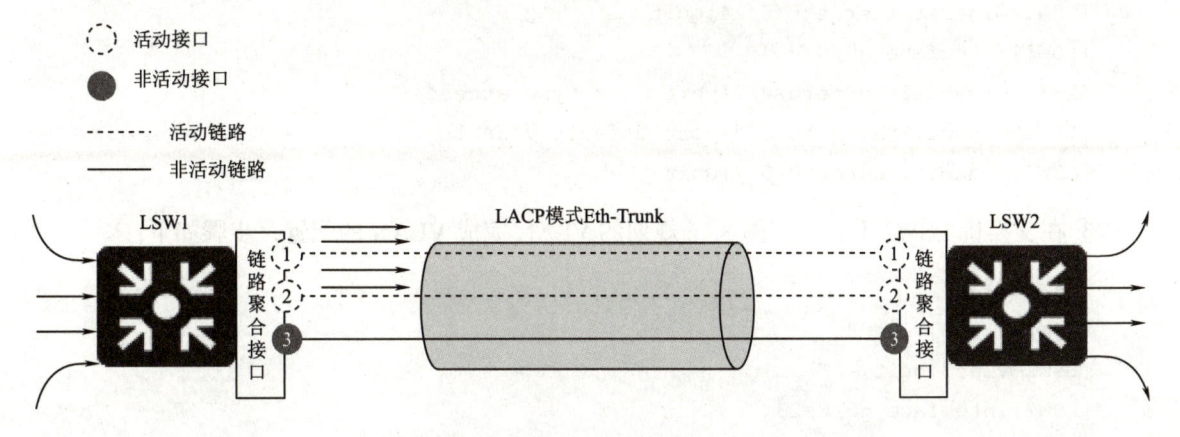

图 3.15　链路聚合最大活动接口数示例

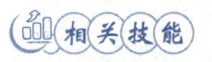

1. 手工模式链路聚合的规划与配置

搭建如图 3.16 所示的网络拓扑图。在交换机 LSW1 和 LSW2 之间使用手工模式配置链路聚合，实现 LSW1 与 LSW2 之间的三条链路同时处于数据转发状态，以充分利用这三条链路的带宽，并进行链路聚合的验证与测试。

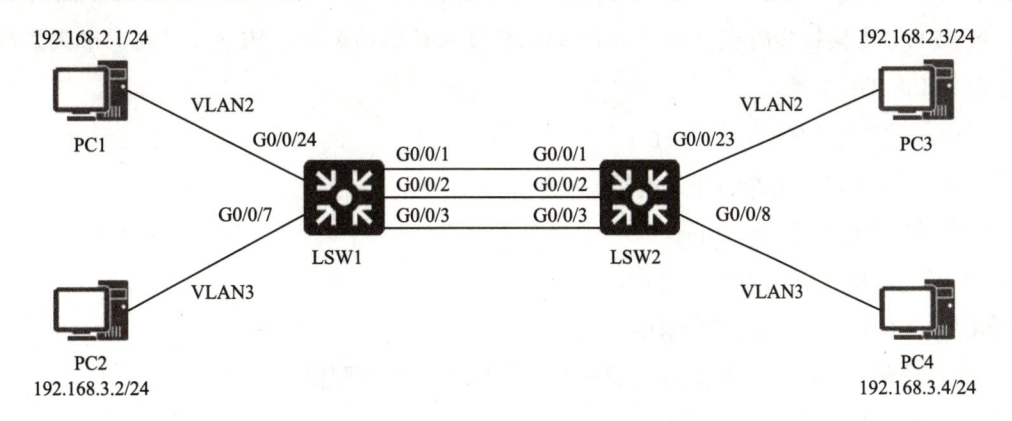

图 3.16　手工模式链路聚合网络拓扑图

（1）环境搭建与 VLAN 的配置

① 搭建如图 3.16 所示的网络拓扑图，根据图中 IP 的规划完成主机的 IP 配置。

② 在交换机 LSW1 上，根据图 3.16 规划的 VLAN 完成 VLAN 的配置，步骤如下：

```
[LSW1]vlan 2
[LSW1-vlan2]vlan 3
[LSW1]interface g0/0/24
```

```
[LSW1-GigabitEthernet0/0/24]port link-type access
[LSW1-GigabitEthernet0/0/24]port default vlan 2
[LSW1-GigabitEthernet0/0/24]quit
[LSW1]interface g0/0/7
[LSW1-GigabitEthernet0/0/7]port link-type access
[LSW1-GigabitEthernet0/0/7]port default vlan 3
[LSW1-GigabitEthernet0/0/7]quit
```

③ 在交换机 LSW2 上，根据图 3.16 规划的 VLAN 完成 VLAN 的配置，步骤如下：

```
[LSW2]vlan 2
[LSW2-vlan2]vlan 3
[LSW2-vlan3]quit
[LSW2]interface g0/0/23
[LSW2-GigabitEthernet0/0/23]port link-type access
[LSW2-GigabitEthernet0/0/23]port default vlan 2
[LSW2-GigabitEthernet0/0/23]quit
[LSW2]interface g0/0/8
[LSW2-GigabitEthernet0/0/8]port link-type access
[LSW2-GigabitEthernet0/0/8]port default vlan 3
[LSW2-GigabitEthernet0/0/8]quit
```

注：华为交换机链路聚合组的成员接口不能有任何的 VLAN 配置，因此不能对交换机
LSW1 与 LSW2 连接链路的成员接口进行 VLAN 接入链路的配置，VLAN 信息的配置需配置在
链路聚合组 Eth-Trunk 里。

（2）手工模式链路聚合的配置

手工模式链路聚合配置的任务包括：

① 创建链路聚合组 Eth-Trunk。

② 配置链路聚合模式为手工模式。

③ 把成员接口加入链路聚合组中。

④ 进入链路聚合组，为其配置二层或者三层属性（可选配置）。

⑤ 配置负载均衡方式（可选）。

配置步骤如下：

```
[LSW1]interface Eth-Trunk 1  // 创建链路聚合组 Eth-Trunk1，并进入 Eth-Trunk 接口视图
[LSW1-Eth-Trunk1]mode manual load-balance // 配置链路聚合模式为手工负载分担模式 . 注：
                                           // Eth-Trunk 的工作模式默认为手工负载分担
                                           // 模式，可以不配置该命令
[LSW1-Eth-Trunk1]trunkport g0/0/1          // 把 G0/0/1 接口加入 Eth-Trunk1 中
[LSW1-Eth-Trunk1]trunkport g0/0/2          // 把 G0/0/2 接口加入 Eth-Trunk1 中
```

```
[LSW1-Eth-Trunk1]trunkport g0/0/3              // 把 G0/0/3 接口加入 Eth-Trunk1 中
[LSW1-Eth-Trunk1]port link-type trunk         // 配置链路聚合组 Eth-Trunk1 为中继接口
[LSW1-Eth-Trunk1]port trunk allow-pass vlan 2 3 // 配置链路聚合组 Eth-Trunk1 允许 VLAN2、
                                               // VLAN3 通过
[LSW1-Eth-Trunk1]quit
[LSW2]interface Eth-Trunk 1                    // 创建链路聚合组 Eth-Trunk1，并进入
                                               // Eth-Trunk 接口视图
[LSW2-Eth-Trunk1]trunkport g0/0/1             // 把 G0/0/1 接口加入 Eth-Trunk1 中
[LSW2-Eth-Trunk1]trunkport g0/0/2             // 把 G0/0/2 接口加入 Eth-Trunk1 中
[LSW2-Eth-Trunk1]trunkport g0/0/3             // 把 G0/0/3 接口加入 Eth-Trunk1 中
[LSW2-Eth-Trunk1]port link-type trunk         // 配置链路聚合组 Eth-Trunk1 为中继端口
[LSW2-Eth-Trunk1]port trunk allow-pass vlan 2 3 // 配置链路聚合组 Eth-Trunk1 允许 VLAN2、
                                               // VLAN3 通过
[LSW2-Eth-Trunk1]quit
```

注：可在 Eth-Trunk 的端口视图下使用 trunkport 命令添加链路聚合组的成员接口，也可以在成员接口的接口视图下使用"eth-trunk"命令将其添加到链路聚合组 Eth-Trunk 中。在将成员接口加入 Eth-Trunk 时，需要注意以下问题：

① 一个以太网接口只能加入一个 Eth-Trunk 接口。

② Eth-Trunk 接口不能嵌套，即 Eth-Trunk 接口的成员接口不能是 Eth-Trunk 接口。

③ 如果本地设备使用了 Eth-Trunk，与成员接口直连的对端接口也必须捆绑为 Eth-Trunk 接口，两端才能正常通信。

④ Eth-Trunk 链路两端相连的物理接口的数量、速率、双工方式等必须一致。

（3）验证与测试

① 在交换机 LSW1 与 LSW2 上分别执行"display eth-trunk"命令，查看 Eth-Trunk 接口状态，如图 3.17、图 3.18 所示。

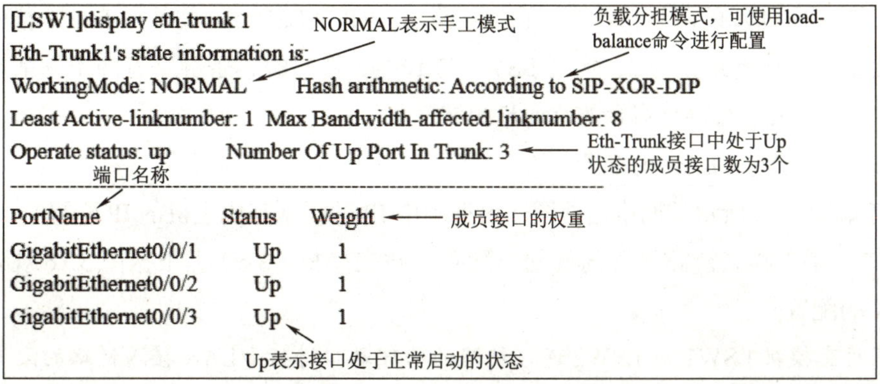

图 3.17　交换机 LSW1 的 Eth-Trunk 1 接口状态

```
[LSW2]display eth-trunk 1
Eth-Trunk1's state information is:
WorkingMode: NORMAL        Hash arithmetic: According to SIP-XOR-DIP
Least Active-linknumber: 1  Max Bandwidth-affected-linknumber: 8
Operate status: up         Number Of Up Port In Trunk: 3
--------------------------------------------------------------------------

PortName              Status    Weight
GigabitEthernet0/0/1    Up        1
GigabitEthernet0/0/2    Up        1
GigabitEthernet0/0/3    Up        1
```

图 3.18　交换机 LSW2 的 Eth-Trunk 1 接口状态

② 在交换机 LSW1 与 LSW2 上分别执行 "display port vlan" 命令。图 3.19 为 LSW1 上的显示结果，可以看出，Eth-Trunk1 接口链路类型为 trunk，允许 VLAN1 到 VLAN3 通过。

```
<LSW1>display port vlan
Port              Link Type   PVID  Trunk VLAN List
--------------------------------------------------------------------------
Eth-Trunk1          trunk       1     1-3
GigabitEthernet0/0/1    hybrid    0     -
GigabitEthernet0/0/2    hybrid    0     -
GigabitEthernet0/0/3    hybrid    0     -
GigabitEthernet0/0/4    hybrid    1     -
GigabitEthernet0/0/5    hybrid    1     -
GigabitEthernet0/0/6    hybrid    1     -
GigabitEthernet0/0/7    access    3     -
GigabitEthernet0/0/8    hybrid    1     -
GigabitEthernet0/0/9    hybrid    1     -
GigabitEthernet0/0/10   hybrid    1     -
GigabitEthernet0/0/11   hybrid    1     -
```

图 3.19　交换机 LSW1 的端口 VLAN 信息

2. LACP 模式链路聚合的规划与配置

搭建如图 3.16 所示的网络拓扑图。在交换机 LSW1 和 LSW2 之间使用 LACP 模式配置链路聚合，配置最大活动链路数为 2，实现 LSW1 与 LSW2 之间的三条链路有两条为活动链路，一条为非活动链路。并进行链路聚合的验证与测试。

（1）环境搭建与 VLAN 的配置

① 搭建如图 3.16 所示的网络拓扑图，根据图中 IP 的规划完成主机的 IP 配置。

② 参考 "手工模式链路聚合的规划与配置" 在交换机 LSW1 上根据图 3.16 规划的 VLAN 完成 VLAN 的配置。

注： 不对交换机 LSW1 与 LSW2 连接链路的物理端口进行 VLAN 接入链路的配置。

（2）LACP 模式链路聚合的配置

LACP 模式链路聚合的配置任务包括：

① 配置系统 LACP 优先级（可选配置）。

② 创建链路聚合组。

③ 配置链路聚合模式为 LACP 模式。

④ 使能 LACP 优先级抢占功能（可选配置）。

⑤ 配置最大活动链路数（可选配置）。

⑥ 配置负载均衡方式（可选）。

⑦ 把成员接口加入链路聚合组中。

⑧ 配置链路聚合端口二层或者三层属性（可选配置）。

步骤如下：

```
[LSW1]lacp priority 20000       // 配置交换机 LSW1 的 LACP 优先级为 20000
[LSW1]interface Eth-Trunk 1     // 创建链路聚合组 Eth-Trunk1，并进入 Eth-Trunk 接口视图
[LSW1-Eth-Trunk1]mode lacp-static          // 配置链路聚合模式为 LACP 模式
[LSW1-Eth-Trunk1]lacp preempt enable       // 使能 LACP 优先级抢占功能
[LSW1-Eth-Trunk1]max active-linknumber 2   // 配置最大活动链路数为 2
[LSW1-Eth-Trunk1]load-balance dst-ip       // 配置按目的 IP 进行负载均衡
[LSW1-Eth-Trunk1]trunkport g0/0/1          // 把 G0/0/1 接口加入 Eth-Trunk1 中
[LSW1-Eth-Trunk1]trunkport g0/0/2          // 把 G0/0/2 接口加入 Eth-Trunk1 中
[LSW1-Eth-Trunk1]trunkport g0/0/3          // 把 G0/0/3 接口加入 Eth-Trunk1 中
[LSW1-Eth-Trunk1]port link-type trunk
[LSW1-Eth-Trunk1]port trunk allow-pass vlan 2 3
[LSW1-Eth-Trunk1]quit
// 配置 Eth-Trunk1 端口为中继接口，允许 VLAN2、VLAN3 通过
[LSW2]interface Eth-Trunk 1     // 创建链路聚合组 Eth-Trunk1，并进入 Eth-Trunk 接口视图
[LSW2-Eth-Trunk1]mode lacp-static          // 配置链路聚合模式为 LACP 模式
[LSW2-Eth-Trunk1]lacp preempt enable       // 使能 LACP 优先级抢占功能
[LSW2-Eth-Trunk1]load-balance dst-ip       // 配置最大活动链路数为 2
[LSW2-Eth-Trunk1]trunkport g0/0/1          // 把接口 G0/0/1 加入 Eth-Trunk1 中
[LSW2-Eth-Trunk1]trunkport g0/0/2          // 把接口 G0/0/2 加入 Eth-Trunk1 中
[LSW2-Eth-Trunk1]trunkport g0/0/3          // 把接口 G0/0/3 加入 Eth-Trunk1 中
[LSW2-Eth-Trunk1]port link-type trunk
[LSW2-Eth-Trunk1]port trunk allow-pass vlan 2 3
[LSW2-Eth-Trunk1]quit
// 配置 Eth-Trunk1 端口为中继接口，允许 VLAN2、VLAN3 通过
```

（3）验证与测试

① 在交换机 LSW1 与 LSW2 上分别执行"display eth-trunk"命令查看 Eth-Trunk 接口状态。图 3.20 为在交换机 LSW1 上查看 Eth-Trunk 1 接口状态。图 3.21 为在交换机 LSW2 上查看 Eth-Trunk 1 接口状态。

```
[LSW1]display eth-trunk 1
Eth-Trunk1's state information is:          表示LACP模式
Local:    LACP优先级为20000                       使用目的IP进行负载分担
LAG ID: 1                WorkingMode: STATIC
Preempt Delay Time: 30    Hash arithmetic: According to DIP  ←
System Priority: 20000    System ID: 4c1f-ccd4-7dc2
Least Active-linknumber: 1 Max Active-linknumber: 2  ←  最大活动链路为2个
Operate status: up        Number Of Up Port In Trunk: 2
--------------------------------------------------------------------------
                      Selected表示接口为活动接口

ActorPortName       Status  PortType PortPri PortNo PortKey PortState Weight
GigabitEthernet0/0/1 Selected 1GE     32768   2      305    10111100  1
GigabitEthernet0/0/2 Selected 1GE     32768   3      305    10111100  1
GigabitEthernet0/0/3 Unselect 1GE     32768   4      305    10100000  1
Partner:
--------------------------------------------------------------------------
                      Unselected表示接口为非活动接口

ActorPortName       SysPri  SystemID      PortPri PortNo PortKey PortState
GigabitEthernet0/0/1 32768  4c1f-cc1e-578f 32768   2      305    10111100
GigabitEthernet0/0/2 32768  4c1f-cc1e-578f 32768   3      305    10111100
GigabitEthernet0/0/3 32768  4c1f-cc1e-578f 32768   4      305    10110000
```

图 3.20　在交换机 LSW1 上查看 Eth-Trunk 1 接口状态

```
[LSW2]display eth-trunk
Eth-Trunk1's state information is:
Local:
LAG ID: 1                WorkingMode: STATIC
Preempt Delay Time: 30    Hash arithmetic: According to DIP
System Priority: 32768    System ID: 4c1f-cc1e-578f
Least Active-linknumber: 1 Max Active-linknumber: 8
Operate status: up        Number Of Up Port In Trunk: 2
--------------------------------------------------------------------------

ActorPortName       Status  PortType PortPri PortNo PortKey PortState Weight
GigabitEthernet0/0/1 Selected 1GE     32768   2      305    10111100  1
GigabitEthernet0/0/2 Selected 1GE     32768   3      305    10111100  1
GigabitEthernet0/0/3 Unselect 1GE     32768   4      305    10110000  1

Partner:

--------------------------------------------------------------------------

ActorPortName       SysPri  SystemID      PortPri PortNo PortKey PortState
GigabitEthernet0/0/1 20000  4c1f-ccd4-7dc2 32768   2      305    10111100
GigabitEthernet0/0/2 20000  4c1f-ccd4-7dc2 32768   3      305    10111100
GigabitEthernet0/0/3 20000  4c1f-ccd4-7dc2 32768   4      305    10100000
```

图 3.21　在交换机 LSW2 上查看 Eth-Trunk 1 接口状态

② 在交换机 LSW1 与 LSW2 上分别执行 "display port vlan" 命令。图 3.22 所示为交换机 LSW1 的端口 VLAN 信息，可以看出，Eth-Trunk1 接口链路类型为 trunk，允许 VLAN1 到 VLAN3 通过。

③ 将交换机 LSW1 与 LSW2 之间的活动链路（如 G0/0/1 接口的链路）删除或者将 G0/0/0 接口禁用以模拟链路故障，再在交换机 LSW1 与 LSW2 上使用 "display eth-trunk" 命令查看 Eth-Trunk 接口状态。图 3.23 为 LSW1 上的显示结果，可以看出，此时原来的备份链路（G0/0/3）接口被选择为活动接口，而 G0/0/1 接口为非活动接口。

项目 3　中型企业网络互联项目　　237

```
<LSW1>display port vlan
Port            Link Type   PVID  Trunk VLAN List
------------------------------------------------------------
Eth-Trunk1          trunk      1    1-3
GigabitEthernet0/0/1    hybrid     0    -
GigabitEthernet0/0/2    hybrid     0    -
GigabitEthernet0/0/3    hybrid     0    -
GigabitEthernet0/0/4    hybrid     1    -
GigabitEthernet0/0/5    hybrid     1    -
GigabitEthernet0/0/6    hybrid     1    -
GigabitEthernet0/0/7    access     3    -
GigabitEthernet0/0/8    hybrid     1    -
GigabitEthernet0/0/9    hybrid     1    -
GigabitEthernet0/0/10   hybrid     1    -
GigabitEthernet0/0/11   hybrid     1    -
```

图 3.22　交换机 LSW1 的端口 VLAN 信息

```
[LSW1]display eth-trunk
Eth-Trunk1's state information is:
Local:
LAG ID: 1            WorkingMode: STATIC
Preempt Delay Time: 30     Hash arithmetic: According to DIP
System Priority: 20000     System ID: 4c1f-ccd4-7dc2
Least Active-linknumber: 1  Max Active-linknumber: 2
Operate status: up      Number Of Up Port In Trunk: 2
--------------------------------------------------------------
ActorPortName        Status  PortType PortPri PortNo PortKey PortState Weight
GigabitEthernet0/0/1  Unselect 1GE    32768   2      305    10100010 1
GigabitEthernet0/0/2  Selected 1GE    32768   3      305    10111100 1
GigabitEthernet0/0/3  Selected 1GE    32768   4      305    10111100 1

Partner:
--------------------------------------------------------------
ActorPortName      SysPri  SystemID       PortPri PortNo PortKey PortState
GigabitEthernet0/0/1  0    0000-0000-0000  0      0      0     10100011
GigabitEthernet0/0/2  32768  4c1f-cc1e-578f 32768  3     305    10111100
GigabitEthernet0/0/3  32768  4c1f-cc1e-578f 32768  4     305    10111100
```

图 3.23　交换机 LSW1 的 Eth-Trunk 1 接口状态

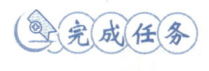

1. 任务规划

根据任务描述、图 3.1 所示的网络拓扑和项目 3 网络部署需求，接入层交换机 SXL-FLOOR2 与汇聚层交换机 SXL-AGG 之间的链路、汇聚层交换机 SXL-AGG 与核心层交换机 CORE 之间的链路使用 LACP 模式进行链路聚合的规划与配置，见表 3.4。

238　企业网络互联技术（双语）

表 3.4　使用 LACP 链路聚合的规划

设　备	聚合口 ID	成员接口	系统 LACP 优先级	最大活动链路数	描　　述
CORE	1	G0/0/4、G0/0/5	默认值	默认值	linkto-sxl-agg
SXL-AGG	1	G0/0/1、G0/0/2	20000	默认值	linkto-core
	2	G0/0/4、G0/0/5	20000	默认值	linkto-sxl-floor2
SXL-FLOOR2	1	G0/0/1、G0/0/2	默认值	默认值	linkto-sxl-agg

MSTP 的规划如下：

① 交换机生成树均使用 MSTP 协议。

② 行政楼交换机的 MSTP 域名为 xzl，修订级别为 1，实例 1 包括 VLAN2、VLAN3、VLAN4、VLAN5、VLAN90 和 VLAN91；汇聚层交换机 XZL-AGG 为实例 0 和实例 1 的主根，接入层交换机接终端的端口配置为边缘端口，并启用 BPDU 保护。

③ 教学楼交换机的 MSTP 域名为 jxl，修订级别为 1，实例 1 包括 VLAN12、VLAN13、VLAN92 和 VLAN93；汇聚层交换机 JXL-AGG 为实例 0 和实例 1 的主根，接入层交换机接终端的端口配置为边缘端口，并启用 BPDU 保护。

④ 实训楼交换机的 MSTP 域名为 sxl，修订级别为 1，实例 1 包括 VLAN22、VLAN23、VLAN94 和 VLAN95；汇聚层交换机 SXL-AGG 为实例 0 和实例 1 的主根，接入层交换机接终端的端口配置为边缘端口，并启用 BPDU 保护。

⑤ 核心层交换机 CORE 和服务器接入层交换机 SERVER-ACC 的 MSTP 域名为 server，修订级别为 1，实例 1 包括 VLAN202；核心层交换机 CORE 为实例 0 和实例 1 的主根，服务器接入层交换机接终端的端口配置为边缘端口，并启用 BPDU 保护。

⑥ 核心层交换机与汇聚层交换机之间链路端口禁用 STP 功能，核心层交换机连接边缘路由器的端口也禁用 STP 功能。

2. 任务实施

（1）链路聚合的配置

① 在核心层交换机 CORE 上根据表 3.4 的链路聚合规划和表 3.1 的 VLAN 规划（聚合口 Eth-Trunk1 为 ACCESS 模式，属于 VLAN199），完成链路聚合口 Eth-Trunk1 的配置，步骤如下：

```
[CORE]interface Eth-Trunk 1
[CORE-Eth-Trunk1]mode lacp-static
[CORE-Eth-Trunk1]trunkport g0/0/4
[CORE-Eth-Trunk1]trunkport g0/0/5
[CORE-Eth-Trunk1]port link-type access
[CORE-Eth-Trunk1]port default vlan 199
[CORE-Eth-Trunk1]description linkto-sxl-agg
[CORE-Eth-Trunk1]quit
```

// 创建聚合口 Eth-Trunk1，聚合模式为 LACP，成员接口为 G0/0/4 和 G0/0/5，链路类型为
//ACCESS，属于 VLAN199，描述为 "linkto-sxl-agg"

② 在汇聚层交换机 SXL-AGG 上根据表 3.4 的链路聚合规划和表 3.1 的 VLAN 规划（聚合口 Eth-Trunk1 为 ACCESS 模式，属于 VLAN199；聚合口 Eth-Trunk2 为 TRUNK 模式，允许 VLAN22、VLAN23、VLAN94 和 VLAN95 通过），完成链路聚合口 Eth-Trunk1 和聚合口 Eth-Trunk2 的配置，步骤如下：

```
[SXL-AGG]lacp priority 20000
// 配置 LACP 优先级为 20000
[SXL-AGG]interface Eth-Trunk 1
[SXL-AGG-Eth-Trunk1]mode lacp-static
[SXL-AGG-Eth-Trunk1]trunkport g0/0/1
[SXL-AGG-Eth-Trunk1]trunkport g0/0/2
[SXL-AGG-Eth-Trunk1]description linkto-core
[SXL-AGG-Eth-Trunk1]port link-type access
[SXL-AGG-Eth-Trunk1]port default vlan 199
[SXL-AGG-Eth-Trunk1]quit
// 创建聚合口 Eth-Trunk1，聚合模式为 LACP，成员接口为 G0/0/1 和 G0/0/2，链路类型为
//ACCESS，属于 VLAN199，描述为 "linkto-core"
[SXL-AGG]interface Eth-Trunk 2
[SXL-AGG-Eth-Trunk2]mode lacp-static
[SXL-AGG-Eth-Trunk2]trunkport g0/0/4
[SXL-AGG-Eth-Trunk2]trunkport g0/0/5
[SXL-AGG-Eth-Trunk2]port link-type trunk
[SXL-AGG-Eth-Trunk2]port trunk allow-pass vlan 22 23 94 95
[SXL-AGG-Eth-Trunk2]description linkto-sxl-floor2
[SXL-AGG-Eth-Trunk2]quit
// 创建聚合口 Eth-Trunk2，聚合模式为 LACP，成员接口为 G0/0/4 和 G0/0/5，链路类型为 TRUNK，
// 允许 VLAN22、VLAN23、VLAN94 和 VLAN95 通过，描述为 "linkto-sxl-floor2"
```

③ 在接入层交换机 SXL-FLOOR2 上根据表 3.4 的链路聚合规划和表 3.1 的 VLAN 规划（聚合口 Eth-Trunk1 为 TRUNK 模式，允许 VLAN22、VLAN23、VLAN94 和 VLAN95 通过），完成链路聚合口 Eth-Trunk1 的配置，步骤如下：

```
[SXL-FLOOR2]interface Eth-Trunk 1
[SXL-FLOOR2-Eth-Trunk1]mode lacp-static
[SXL-FLOOR2-Eth-Trunk1]trunkport g0/0/1
[SXL-FLOOR2-Eth-Trunk1]trunkport g0/0/2
[SXL-FLOOR2-Eth-Trunk1]port link-type trunk
[SXL-FLOOR2-Eth-Trunk1]port trunk allow-pass vlan 22 23 94 95
```

```
[SXL-FLOOR2-Eth-Trunk1]description linkto-sxl-agg
[SXL-FLOOR2-Eth-Trunk1]quit
// 创建聚合口 Eth-Trunk1，聚合模式为 LACP，成员接口为 G0/0/4 和 G0/0/5，链路类型为 TRUNK，
// 允许 VLAN22、VLAN23、VLAN94 和 VLAN95 通过，描述为 "linkto-sxl-agg"
```

（2）链路聚合的验证

① 完成链路聚合的配置后，在 CORE、SXL-AGG 和 SXL-FLOOR2 交换机上分别使用 "display eth-trunk" 命令查看链路聚合 Eth-Trunk 接口信息。图 3.24 为在交换机 SXL-AGG 上显示的 Eth-Trunk1 接口信息。图 3.25 为在交换机 SXL-FLOOR 2 上显示的 Eth-Trunk1 接口信息。

```
[SXL-AGG]display eth-trunk 1
Eth-Trunk1's state information is:
Local:
LAG ID: 1              WorkingMode: STATIC
Preempt Delay: Disabled      Hash arithmetic: According to SIP-XOR-DIP
System Priority: 20000      System ID: 4c1f-cc95-3bd2
Least Active-linknumber: 1  Max Active-linknumber: 8
Operate status: up        Number Of Up Port In Trunk: 2
--------------------------------------------------------------------------------
ActorPortName        Status  PortType PortPri PortNo PortKey PortState Weight
GigabitEthernet0/0/1  Selected 1GE     32768   2    305    10111100  1
GigabitEthernet0/0/2  Selected 1GE     32768   3    305    10111100  1

Partner:
--------------------------------------------------------------------------------
ActorPortName        SysPri  SystemID      PortPri PortNo PortKey PortState
GigabitEthernet0/0/1  32768   4c1f-cc89-64b2 32768   5    305    10111100
GigabitEthernet0/0/2  32768   4c1f-cc89-64b2 32768   6    305    10111100
```

图 3.24　在交换机 SXL-AGG 上显示的 Eth-Trunk 1 接口信息

```
[SXL-FLOOR2]display eth-trunk 1
Eth-Trunk1's state information is:
Local:
LAG ID: 1              WorkingMode: STATIC
Preempt Delay: Disabled      Hash arithmetic: According to SIP-XOR-DIP
System Priority: 32768      System ID: 4c1f-cc49-5291
Least Active-linknumber: 1  Max Active-linknumber: 8
Operate status: up        Number Of Up Port In Trunk: 2
--------------------------------------------------------------------------------
ActorPortName        Status  PortType PortPri PortNo PortKey PortState Weight
GigabitEthernet0/0/1  Selected 1GE     32768   2    305    10111100  1
GigabitEthernet0/0/2  Selected 1GE     32768   3    305    10111100  1

Partner:
--------------------------------------------------------------------------------
ActorPortName        SysPri  SystemID      PortPri PortNo PortKey PortState
GigabitEthernet0/0/1  20000   4c1f-cc95-3bd2 32768   5    561    10111100
GigabitEthernet0/0/2  20000   4c1f-cc95-3bd2 32768   6    561    10111100
```

图 3.25　在交换机 SXL-FLOOR2 上显示的 Eth-Trunk 1 接口信息

② 在交换机 CORE 和 SXL-AGG 上使用"display ip interface brief"命令显示接口 IP 简要信息。图 3.26 为在汇聚层交换机 SXL-AGG 上使用"display ip interface brief"命令后的显示结果，图 3.27 为在核心层交换机 CORE 上使用"display ip interface brief"命令后的显示结果。

```
[SXL-AGG]display ip interface brief
*down: administratively down
^down: standby
(l): loopback
(s): spoofing
The number of interface that is UP in Physical is 7
The number of interface that is DOWN in Physical is 1
The number of interface that is UP in Protocol is 6
The number of interface that is DOWN in Protocol is 2

Interface          IP Address/Mask      Physical  Protocol
MEth0/0/1          unassigned           down      down
NULL0              unassigned           up        up(s)
Vlanif1            unassigned           up        down
Vlanif22           10.2.22.254/24       up        up
Vlanif23           10.2.23.254/24       up        up
Vlanif94           10.2.94.254/24       up        up
Vlanif95           10.2.95.254/24       up        up
Vlanif199          10.254.254.10/30     up        up
```

图 3.26　在汇聚层交换机 SXL-AGG 上使用"display ip interface brief"命令后的显示结果

```
[CORE]display ip interface brief
*down: administratively down
^down: standby
(l): loopback
(s): spoofing
The number of interface that is UP in Physical is 6
The number of interface that is DOWN in Physical is 2
The number of interface that is UP in Protocol is 6
The number of interface that is DOWN in Protocol is 2

Interface          IP Address/Mask      Physical  Protocol
MEth0/0/1          unassigned           down      down
NULL0              unassigned           up        up(s)
Vlanif1            unassigned           down      down
Vlanif199          10.254.254.9/30      up        up
Vlanif200          10.254.254.5/30      up        up
Vlanif201          10.254.254.1/30      up        up
Vlanif202          10.254.253.254/24    up        up
Vlanif203          10.254.254.13/30     up        up
```

图 3.27　在核心层交换机 CORE 上使用"display ip interface brief"命令后的显示结果

（3）MSTP 的配置

① 在行政楼的交换机上根据 MSTP 规划完成 MSTP 的配置。所有交换机的 MSTP 域名为 xzl，修订级别为 1，实例 1 包括 VLAN2、VLAN3、VLAN4、VLAN5、VLAN90 和 VLAN91；汇聚层交换机 XZL-AGG 为实例 0 和实例 1 的主根，其 G0/0/1 接口需要禁用生成树功能；接入

242 企业网络互联技术（双语）

层交换机 XZL-FLOOR1 与 XZL-FLOOR2 接终端设备的接口（G0/0/2 至 G0/0/24），为生成树边缘端口，并启用 BPDU 保护。行政楼交换机 MSTP 的配置步骤如下：

```
[XZL-AGG]stp enable
[XZL-AGG]stp mode mstp
// 启用生成树，生成树模式为 MSTP
[XZL-AGG]stp region-configuration
[XZL-AGG-mst-region]region-name xzl
[XZL-AGG-mst-region]revision-level 1
[XZL-AGG-mst-region]instance 1 vlan 2 3 4 5 90 91
[XZL-AGG-mst-region]active region-configuration
// 配置 MSTP 域名、修订级别、实例 1 对应的 VLAN，激活 MSTP 域
[XZL-AGG]interface g0/0/1
[XZL-AGG-GigabitEthernet0/0/1]stp disable
// 交换机 XZL-AGG 的 G0/0/1 接口禁用生成树功能
[XZL-AGG]stp instance 0 root primary
[XZL-AGG]stp instance 1 root primary
// 配置交换机 XZL-AGG 为实例 0 与实例 1 的树根
[XZL-FLOOR1]stp enable
[XZL-FLOOR1]stp mode mstp
// 启用生成树，生成树模式为 MSTP
[XZL-FLOOR1]stp region-configuration
[XZL-FLOOR1-mst-region]region-name xzl
[XZL-FLOOR1-mst-region]revision-level 1
[XZL-FLOOR1-mst-region]instance 1 vlan 2 3 4 5 90 91
[XZL-FLOOR1-mst-region]active region-configuration
[XZL-FLOOR1-mst-region]quit
// 配置 MSTP 域名、修订级别、实例 1 对应的 VLAN，激活 MSTP 域
[XZL-FLOOR1]port-group 4
[XZL-FLOOR1-port-group-4]group-member g0/0/2 to g0/0/24
[XZL-FLOOR1-port-group-4]stp edged-port enable
[XZL-FLOOR1-port-group-4]quit
[XZL-FLOOR1]stp bpdu-protection
// 配置交换机 XZL-FLOOR1 的 G0/0/2 至 G0/0/24 接口为生成树边缘端口，并启用 BPDU 保护功能
[XZL-FLOOR2]stp enable
[XZL-FLOOR2]stp mode mstp
// 启用生成树，生成树模式为 MSTP
[XZL-FLOOR2]stp region-configuration
[XZL-FLOOR2-mst-region]region-name xzl
[XZL-FLOOR2-mst-region]revision-level 1
```

项目 3　中型企业网络互联项目　243

```
[XZL-FLOOR2-mst-region]instance 1 vlan 2 3 4 5 90 91
[XZL-FLOOR2-mst-region]active region-configuration
[XZL-FLOOR2-mst-region]quit
// 配置 MSTP 域名、修订级别、实例 1 对应的 VLAN，激活 MSTP 域
[XZL-FLOOR2]port-group 4
[XZL-FLOOR2-port-group-4]group-member g0/0/2 to g0/0/24
[XZL-FLOOR2-port-group-4]stp edged-port enable
[XZL-FLOOR2-port-group-4]quit
[XZL-FLOOR2]stp bpdu-protection
// 配置交换机 XZL-FLOOR2 的 G0/0/2 至 G0/0/24 接口为生成树边缘端口，并启用 BPDU 保护功能
```

② 在教学楼的交换机上根据 MSTP 规划完成 MSTP 的配置。所有交换机的 MSTP 域名为 jxl，修订级别为 1，实例 1 包括 VLAN12、VLAN13、VLAN92 和 VLAN93；汇聚层交换机 JXL-AGG 为实例 0 和实例 1 的主根，其 G0/0/1 接口禁用生成树功能；接入层交换机 JXL-FLOOR1 与 JXL-FLOOR2 接终端设备的接口（G0/0/2 至 G0/0/24）为生成树边缘端口，并启用 BPDU 保护。教学楼交换机的 MSTP 的配置步骤如下：

```
[JXL-AGG]stp enable
[JXL-AGG]stp mode mstp
// 启用生成树，生成树模式为 MSTP
[JXL-AGG]stp region-configuration
[JXL-AGG-mst-region]region-name jxl
[JXL-AGG-mst-region]revision-level 1
[JXL-AGG-mst-region]instance 1 vlan 12 13 92 93
[JXL-AGG-mst-region]active region-configuration
[JXL-AGG-mst-region]quit
// 配置 MSTP 域名、修订级别、实例 1 对应的 VLAN，激活 MSTP 域
[JXL-AGG]stp instance 0 root primary
[JXL-AGG]stp instance 1 root primary
// 配置交换机 JXL-AGG 为实例 0 与实例 1 的树根
[JXL-AGG]interface g0/0/1
[JXL-AGG-GigabitEthernet0/0/1]stp disable
[JXL-AGG-GigabitEthernet0/0/1]quit
// 交换机 JXL-AGG 的 G0/0/1 接口禁用生成树功能
[JXL-FLOOR1]stp enable
[JXL-FLOOR1]stp mode mstp
// 启用生成树，生成树模式为 MSTP
[JXL-FLOOR1]stp region-configuration
[JXL-FLOOR1-mst-region]region-name jxl
[JXL-FLOOR1-mst-region]revision-level 1
```

```
[JXL-FLOOR1-mst-region]instance 1 vlan 12 13 92 93
[JXL-FLOOR1-mst-region]active region-configuration
[JXL-FLOOR1-mst-region]quit
// 配置 MSTP 域名、修订级别、实例 1 对应的 VLAN，激活 MSTP 域
[JXL-FLOOR1]port-group 4
[JXL-FLOOR1-port-group-4]group-member g0/0/2 to g0/0/24
[JXL-FLOOR1-port-group-4]stp edged-port enable
[JXL-FLOOR1]stp bpdu-protection
// 配置教学楼接入层交换机 JXL-FLOOR1 的 MSTP
[JXL-FLOOR2]stp enable
[JXL-FLOOR2]stp mode mstp
// 启用生成树，生成树模式为 MSTP
[JXL-FLOOR2]stp region-configuration
[JXL-FLOOR2-mst-region]region-name jxl
[JXL-FLOOR2-mst-region]revision-level 1
[JXL-FLOOR2-mst-region]instance 1 vlan 12 13 92 93
[JXL-FLOOR2-mst-region]active region-configuration
[JXL-FLOOR2-mst-region]quit
// 配置 MSTP 域名、修订级别、实例 1 对应的 VLAN，激活 MSTP 域
[JXL-FLOOR2]port-group 4
[JXL-FLOOR2-port-group-4]group-member g0/0/2 to g0/0/24
[JXL-FLOOR2-port-group-4]stp edged-port enable
[JXL-FLOOR2]stp bpdu-protection
// 配置交换机 JXL-FLOOR2 的 G0/0/2 至 G0/0/24 接口为生成树边缘端口，并启用 BPDU 保护功能
```

③ 在实训楼的交换机上根据 MSTP 规划完成 MSTP 的配置。所有交换机的 MSTP 域名为 sxl，修订级别为 1，实例 1 包括 VLAN22、VLAN23、VLAN94 和 VLAN95；汇聚层交换机 SXL-AGG 为实例 0 和实例 1 的主根，其 Eth-Trunk 1 端口禁用生成树功能；接入层交换机 SXL-FLOOR1 与 JXL-FLOOR2 接终端设备的端口（SXL-FLOOR1 的 G0/0/2 至 G0/0/24 接口，SXL-FLOOR2 的 G0/0/3 至 G0/0/24 接口）为生成树边缘端口，并启用 BPDU 保护。实训楼交换机 MSTP 的配置步骤如下：

```
[SXL-AGG]stp enable
[SXL-AGG]stp mode mstp
// 启用生成树，生成树模式为 MSTP
[SXL-AGG]stp region-configuration
[SXL-AGG-mst-region]region-name sxl
[SXL-AGG-mst-region]revision-level 1
[SXL-AGG-mst-region]instance 1 vlan 22 23 94 95
[SXL-AGG-mst-region]active region-configuration
```

```
[SXL-AGG-mst-region]quit
// 配置 MSTP 域名、修订级别、实例 1 对应的 VLAN，激活 MSTP 域
[SXL-AGG]stp instance 0 root primary
[SXL-AGG]stp instance 1 root primary
// 配置交换机 SXL-AGG 为实例 0 与实例 1 的树根
[SXL-AGG]interface Eth-Trunk 1
[SXL-AGG-Eth-Trunk1]stp disable
[SXL-AGG-Eth-Trunk1]quit
// 交换机 SXL-AGG 的 Eth-Trunk 1 接口禁用生成树功能
[SXL-FLOOR1]stp enable
[SXL-FLOOR1]stp mode mstp
// 启用生成树，生成树模式为 MSTP
[SXL-FLOOR1]stp region-configuration
[SXL-FLOOR1-mst-region]region-name sxl
[SXL-FLOOR1-mst-region]revision-level 1
[SXL-FLOOR1-mst-region]instance 1 vlan 22 23 94 95
[SXL-FLOOR1-mst-region]active region-configuration
[SXL-FLOOR1-mst-region]quit
// 配置 MSTP 域名、修订级别、实例 1 对应的 VLAN，激活 MSTP 域
[SXL-FLOOR1]port-group 4
[SXL-FLOOR1-port-group-4]group-member g0/0/2 to g0/0/24
[SXL-FLOOR1-port-group-4]stp edged-port enable
[SXL-FLOOR1]stp bpdu-protection
// 配置交换机 SXL-FLOOR1 的 G0/0/2 至 G0/0/24 接口为生成树边缘端口，并启用 BPDU 保护功能
[SXL-FLOOR2]stp enable
[SXL-FLOOR2]stp mode mstp
// 启用生成树，生成树模式为 MSTP
[SXL-FLOOR2]stp region-configuration
[SXL-FLOOR2-mst-region]region-name sxl
[SXL-FLOOR2-mst-region]revision-level 1
[SXL-FLOOR2-mst-region]instance 1 vlan 22 23 94 95
[SXL-FLOOR2-mst-region]active region-configuration
[SXL-FLOOR2-mst-region]quit
// 配置 MSTP 域名、修订级别、实例 1 对应的 VLAN，激活 MSTP 域
[SXL-FLOOR2]port-group 4
[SXL-FLOOR2-port-group-4]group-member g0/0/3 to g0/0/24
[SXL-FLOOR2-port-group-4]stp edged-port enable
[SXL-FLOOR2-port-group-4]quit
[SXL-FLOOR2]stp bpdu-protection
// 配置交换机 SXL-FLOOR2 的 G0/0/3 至 G0/0/24 接口为生成树边缘端口，并启用 BPDU 保护功能
```

④ 在核心层交换机 CORE 和服务器接入层交换机 SERVER-ACC 上根据 MSTP 规划完成 MSTP 的配置，交换机 CORE 与 SERVER-ACC 的 MSTP 域名为 server，修订级别为 1，实例 1 包括 VLAN202；核心层交换机 CORE 为实例 0 和实例 1 的主根，其 Eth-Trunk 1、G0/0/1、G0/0/2 和 G0/0/3 接口禁用生成树功能（这些接口为连接到汇聚层交换机与边缘路由器 EDGE-AR 连接网段接口）；服务器接入层交换机 SERVER-ACC 接服务器的接口（G0/0/1 至 G0/0/3 接口）为生成树边缘端口，并启用 BPDU 保护，核心层交换机和服务器接入层交换机 MSTP 的配置步骤如下：

```
[CORE]stp enable
[CORE]stp mode mstp
// 启用生成树，生成树模式为 MSTP
[CORE]stp region-configuration
[CORE-mst-region]region-name server
[CORE-mst-region]revision-level 1
[CORE-mst-region]instance 1 vlan 202
[CORE-mst-region]active region-configuration
[CORE-mst-region]quit
// 配置 MSTP 域名、修订级别、实例 1 对应的 VLAN，激活 MSTP 域
[CORE]stp instance 0 root primary
[CORE]stp instance 1 root primary
// 配置交换机 CORE 为实例 0 与实例 1 的树根
[CORE]port-group 1
[CORE-port-group-1]group-member g0/0/1 to g0/0/3 Eth-Trunk 1
[CORE-port-group-1]stp disable
[CORE-port-group-1]quit
// 交换机 CORE 的 G0/0/1 至 G0/0/3 和 Eth-Trunk 1 接口禁用生成树功能
[SERVER-ACC]stp enable
[SERVER-ACC]stp mode mstp
// 启用生成树，生成树模式为 MSTP
[SERVER-ACC]stp region-configuration
[SERVER-ACC-mst-region]region-name server
[SERVER-ACC-mst-region]revision-level 1
[SERVER-ACC-mst-region]instance 1 vlan 202
[SERVER-ACC-mst-region]active region-configuration
[SERVER-ACC-mst-region]quit
// 配置 MSTP 域名、修订级别、实例 1 对应的 VLAN，激活 MSTP 域
[SERVER-ACC]port-group 1
[SERVER-ACC-port-group-1]group-member g0/0/1 to g0/0/3
[SERVER-ACC-port-group-1]stp edged-port enable
```

项目 3　中型企业网络互联项目　247

```
[SERVER-ACC-port-group-1]quit
[SERVER-ACC]stp bpdu-protection
// 配置交换机 SERVER-ACC 的 G0/0/1 至 G0/0/3 接口为生成树边缘端口，并启用 BPDU 保护功能
```

（4）验证与测试

① 在所有的交换机上分别执行 "display stp region-configuration" 命令查看 MSTP 域配置信息，同一幢楼宇的交换机 MSTP 域配置要相同。图 3.28 所示为实训楼汇聚层交换机 SXL-AGG 上查看 MSTP 域配置信息示例。图 3.29 所示为实训楼接入层交换机 SXL-FLOOR2 上查看 MSTP 域配置信息示例。

```
[SXL-AGG]display stp region-configuration
Oper configuration
 Format selector   :0
 Region name       :sxl
 Revision level    :1

 Instance   VLANs Mapped
  0    1 to 21, 24 to 93, 96 to 4094
  1    22 to 23, 94 to 95
```

图 3.28　实训楼汇聚层交换机 SXL-AGG 上查看 MSTP 域配置信息示例

```
[SXL-FLOOR2]display stp region-configuration
Oper configuration
 Format selector   :0
 Region name       :sxl
 Revision level    :1

 Instance   VLANs Mapped
  0    1 to 21, 24 to 93, 96 to 4094
  1    22 to 23, 94 to 95
```

图 3.29　实训楼接入层交换机 SXL-FLOOR2 上查看 MSTP 域配置信息示例

② 在所有的交换机上分别执行 "display stp brief" 命令查看生成树简要信息。图 3.30 所示为实训楼汇聚层交换机 SXL-AGG 上查看生成树简要信息示例。图 3.31 所示为实训楼接入层交换机 SXL-FLOOR2 上查看生成树简要信息示例。

```
[SXL-AGG]display stp brief
MSTID Port              Role STP State   Protection
  0  GigabitEthernet0/0/3  DESI FORWARDING   NONE
  0  Eth-Trunk2            DESI FORWARDING   NONE
  1  GigabitEthernet0/0/3  DESI FORWARDING   NONE
  1  Eth-Trunk2            DESI FORWARDING   NONE
```

图 3.30　实训楼汇聚层交换机 SXL-AGG 上查看生成树简要信息示例

248　企业网络互联技术（双语）

```
[SXL-FLOOR2]display stp brief
MSTID Port                    Role    STP State    Protection
  0   GigabitEthernet0/0/19   DESI    FORWARDING   BPDU
  0   Eth-Trunk1              ROOT    FORWARDING   NONE
  1   GigabitEthernet0/0/19   DESI    FORWARDING   BPDU
  1   Eth-Trunk1              ROOT    FORWARDING   NONE
```

图 3.31　实训楼接入层交换机 SXL-FLOOR2 上查看生成树简要信息示例

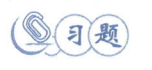 习题

选择题

（1）下面（　　　）用来查看 Eth-Trunk 接口状态。

A. display eth-trunk   B. display current-configuration

C. display this   D. display ip interface brief

（2）当采用 LACP 模式进行链路聚合时，华为交换机的默认系统优先级是（　　　）。

A. 0   B. 64   C. 4096   D. 32768

（3）图 3.17 为交换机 LSW1 的 Eth-Trunk 1 接口输出信息，如果想要删除 Eth-Trunk 1，下列命令正确的是（　　　）。

A. interface GigabitEthernet 0/0/1

　　undo eth-trunk

　　quit

　　undo interface Eth-Trunk

B. interface GigabitEthernet 0/0/1

　　undo eth-trunk

　　quit

　　interface GigabitEthernet 0/0/2

　　undo eth- trunk

　　quit

　　interface GigabitEthernet 0/0/3

　　undo eth- trunk

　　quit

　　undo interface Eth-Trunk 1

C. undo interface Eth-Trunk 1

D. inter GigabitEthernet 0/0/1

　　undo eth trunk

　　quit undo interface Eth-Trunk 1

項目 3　中型企业网络互联项目　249

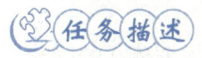

任务 3　OSPF 的规划与配置

任务描述

在完成本项目任务 1 和任务 2 的基础上，在图 3.1 所示的网络中，完成使用默认路由与动态路由协议 OSPF 实现企业内网互联互通，企业内网可访问 Internet 的路由规划与配置任务。

任务目标

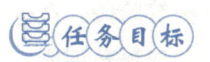

- 理解 OSPF 的基本概念。
- 理解 OSPF 的工作过程。
- 掌握单域 OSPF 的配置。
- 掌握单域 OSPF 的验证与排障。

相关知识

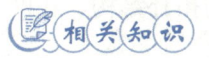

1. OSPF 概述

OSPF（open shortest path first，开放最短路径优先）是一种基于开放标准的链路状态路由选择协议。

OSPF 采用链路状态路由选择算法，每个 OSFP 路由器使用 HELLO 协议识别邻居路由器并与邻居路由器建立通信关系，并通过泛洪的方式将它们自己的链路状态和接收到的链路状态信息告知同一区域中其他所有的 OSPF 路由器。当位于同一区域的 OSPF 路由器具有了完整的链路状态数据库，即得到一张统一的网络拓扑图后，每个 OSPF 路由器即以本地路由器为根，采用最短路径优先（shortest path first，SPF）算法计算到每个目的网络的最短路径，然后根据 SPF 树，使用通向每个网络的最佳路径填充 IP 路由表，如图 3.32 所示。

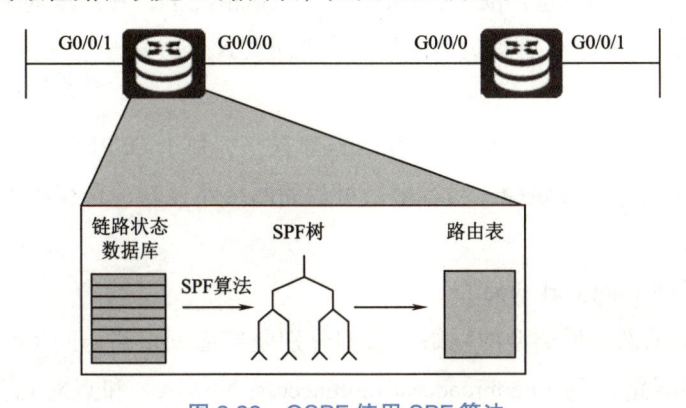

图 3.32　OSPF 使用 SPF 算法

2. OSPF 路由器工作过程

OSPF 路由器工作过程大致分为三个步骤：

首先，每台 OSPF 路由器生成描述自己接口状态的 LSA（link state advertisement，链路状态通告）。

其次，OSPF 路由器通过交换 LSA 实现 LSDB（link state database，链路状态数据库）的同步，同一区域的所有 OSPF 路由器 LSDB 相同。

最后，OSPF 路由器根据 LSDB 使用 SPF 计算出路由，并插入路由表中。

OSPF 的工作过程中包含了发现邻居、交换路由信息、计算路由以及路由维护等阶段，在这些阶段中，每个 OSPF 路由器上需要维护 OSPF 实现相关的三个基本数据结构，即邻接数据库、链路状态数据库和路由表。

① 邻接数据库（adjacencies database）：用于保存所有已经和路由器建立起双向通信关系的邻居路由器。

② 链路状态数据库（LSDB），又称拓扑结构数据库（topological database），用于保存关于 OSPF 网络中所有其他路由器的链路状态信息。该数据库显示出了全网络的拓扑结构。在网络收敛时，同一个 OSPF 区域中的所有 OSPF 路由器都有相同的链路状态数据库。

③ 路由表（route table）：在链路状态数据库上运行 SPF 算法所产生的路由被插入路由表中。

3. 与 OSPF 相关的重要概念

在 OSPF 的工作过程中涉及以下概念：

（1）链路（link）

一条由线路和传输路径组成的网络通信信道。

（2）链路状态（link-state）

两个路由器或者两个路由器接口之间链路的状态以及路由器与邻居路由器的联系。

（3）开销（cost）

给 OSPF 链路所分配的度量标准（metric）值。默认情况下，开销是基于接口上的带宽计算出来的，即"10^8/接口带宽"，管理员可以手工为链路配置开销值。

（4）SPF 算法

SPF 算法即最短路径优先算法（又称 Dijkstra 算法），每个 OSPF 路由器以自己为根节点，计算根节点到每个网络的最短路径。当有多条可达的网络路径时，具有最小开销累加值的路径被认为是最短路径。

（5）网络的类型（network type）

OSPF 接口自动识别三种类型的网络，它们分别是广播型多路访问（broadcast multiaccess, BMA）、非广播型多路访问（nonbroadcast multiaccess, NBMA）和点对点型（point to point）网络。

（6）指定路由器（designated router, DR）和备份的指定路由器（backup designated router, BDR）

在多路访问（如以太网）的 OSPF 网络环境中，由于可以有多台 OSPF 路由器，每台路由器与所有其他路由器建立毗邻关系时将产生许多额外的开销。因此，在多路访问网络中选择一台 OSPF 路由器作为 DR，一台 OSFP 路由器为 BDR，DR 与 BDR 都与该网络中的所有其他路由器建立毗邻关系。由 DR 向其他网络发送关于所有本地路由器的链路状态通告包（LSAs）（即 2 类 -LSA）。BDR 被作为 DR 的备份，以提高网络容错能力。但 BDR 不负责向其他路由器发送路由更新信息，也不发送网络 LSA。当 DR 出现故障时，BDR 就接任 DR 的角色，随后该网络再选举一个 BDR。

注：在点到点网络中不会选举 DR 或 BDR。

在 BMA、NBMA 网络中选择 DR、BDR 的原则：优先级最大的 OSPF 路由器为 DR，优先级次大的 OSPF 路由器为 BDR，如果优先级一样，则选择路由器 ID 最大的为 DR，路由器 ID 次大的为 BDR。

（7）OSPF 路由器 ID

OSPF 路由器 ID 用于唯一标识 OSPF 中的每台路由器。一个路由器 ID 其实就是一个 IP 地址。路由器的 ROUTER-ID 可以使用命令配置，也可以自动选择。

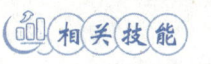

1. 基于路由器的单域 OSPF 规划与配置

搭建如图 3.33 所示的网络拓扑图，使用单域 OSPF 路由协议实现网络的互联互通，并进行路由查看与测试。

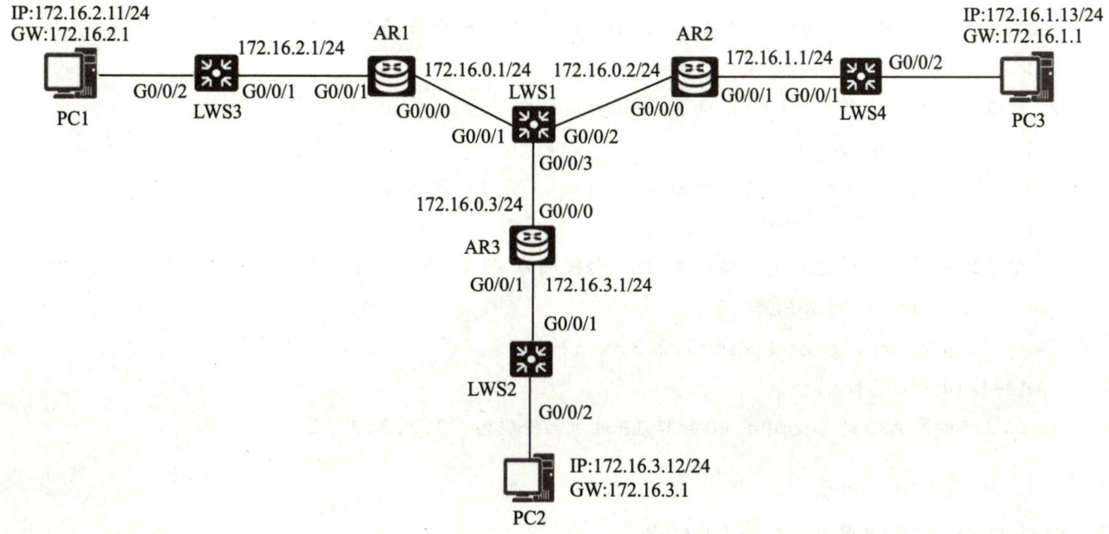

图 3.33　单域 OSPF 网络拓扑图

252 企业网络互联技术（双语）

（1）OSPF 规划

① 使用单域 OSPF，区域号为 0。

② AR1 的 ROUTER-ID 为 1.1.1.1，AR2 的 ROUTER-ID 为 2.2.2.2，AR3 的 ROUTER-ID 为 3.3.3.3。

③ AR1、AR2 与 AR3 的 G0/0/1 接口配置为静默接口，接口不发送路由更新。

④ AR1 的 G0/0/0 接口 DR 选举优先级值为 10，AR2 的 G0/0/0 接口 DR 选举优先级值为 5。

（2）路由器接口的配置

在路由器 AR1、AR2 与 AR3 上根据图 3.33 规划的 IP 完成接口的配置，步骤如下：

```
[AR1]interface g0/0/0
[AR1-GigabitEthernet0/0/0]ip address 172.16.0.1 24
[AR1-GigabitEthernet0/0/0]quit
// 配置路由器 AR1 的 G0/0/0 接口 IP 地址与掩码为 "172.16.0.1/24"
[AR1]interface g0/0/1
[AR1-GigabitEthernet0/0/1]ip address 172.16.2.1 24
[AR1-GigabitEthernet0/0/1]quit
// 配置路由器 AR1 的 G0/0/1 接口 IP 地址与掩码为 "172.16.2.1/24"
[AR2]interface g0/0/0
[AR2-GigabitEthernet0/0/0]ip address 172.16.0.2 24
[AR2-GigabitEthernet0/0/0]quit
// 配置路由器 AR2 的 G0/0/0 接口 IP 地址与掩码为 "172.16.0.2/24"
[AR2]interface g0/0/1
[AR2-GigabitEthernet0/0/1]ip address 172.16.1.1 24
[AR2-GigabitEthernet0/0/1]quit
// 配置路由器 AR2 的 G0/0/1 接口 IP 地址与掩码为 "172.16.1.1/24"
[AR3]interface g0/0/0
[AR3-GigabitEthernet0/0/0]ip address 172.16.0.3 24
[AR3-GigabitEthernet0/0/0]quit
// 配置路由器 AR3 的 G0/0/0 接口 IP 地址与掩码为 "172.16.0.3/24"
[AR3]interface g0/0/1
[AR3-GigabitEthernet0/0/1]ip address 172.16.3.1 24
[AR3-GigabitEthernet0/0/1]quit
// 配置路由器 AR3 的 G0/0/1 接口 IP 地址与掩码为 "172.16.3.1/24"
[AR3]interface LoopBack 0
[AR3-LoopBack0]ip address 3.3.3.3 32
[AR3-LoopBack0]quit
// 配置路由器 AR3 的 LoopBack0 接口 IP 地址与掩码为 "3.3.3.3/32"
```

（3）单域 OSPF 的配置

单域 OSPF 的配置需要完成以下任务：

项目 3　中型企业网络互联项目　253

① 修改路由器接口的选举 DR 时的优先级值（可选配置）。

② 创建并运行 OSPF 进程，指定 OSPF 进程号，配置路由器的 ROUTER-ID，默认 OSPF 进程号为 1，路由器的 ROUTER-ID 为可选配置。

③ 创建并进入 OSPF 区域。

④ 指定运行 OSPF 的接口。

⑤ 配置路由器的接口为静默接口（可选配置）。

配置的步骤如下：

```
[AR1]interface g0/0/0
[AR1-GigabitEthernet0/0/0]ospf dr-priority 10  //设置 G0/0/0 接口在选举 DR 时的优先级值
                                               //为 10，默认值为 1
[AR1]ospf 1 router-id 1.1.1.1 //创建并运行 OSPF 进程 1，路由器的 ROUTER-ID 配置为"1.1.1.1"
[AR1-ospf-1]area 0                             //创建并进入 OSPF 区域 0
[AR1-ospf-1-area-0.0.0.0]network 172.16.2.0 0.0.0.255 //网络 172.16.2.0/24 所在接口参
                                               //与 OSPF 进程，并运行在区域 0
[AR1-ospf-1-area-0.0.0.0]network 172.16.0.0 0.0.0.255 //网络 172.16.0.0/24 所在接口参
                                               //与 OSPF 进程，并运行在区域 0

[AR1-ospf-1-area-0.0.0.0]quit
[AR1-ospf-1]silent-interface g0/0/1            //配置接口为静默接口
[AR2]interface g0/0/0
[AR2-GigabitEthernet0/0/0]ospf dr-priority 5   //设置 G0/0/0 接口在选举 DR 时的
                                               //优先级值为 5
[AR2]ospf 1 router-id 2.2.2.2 //创建并运行 OSPF 进程 1，路由器的 ROUTER-ID 配置为"2.2.2.2"
[AR2-ospf-1]area 0.0.0.0                        //创建并进入 OSPF 区域 0
[AR2-ospf-1-area-0.0.0.0]network 172.16.1.0 0.0.0.255 //网络 172.16.1.0/24 所在接口参
                                               //与 OSPF 进程，并运行在区域 0
[AR2-ospf-1-area-0.0.0.0]network 172.16.0.0 0.0.0.255 //网络 172.16.0.0/24 所在接口参
                                               //与 OSPF 进程，并运行在区域 0

[AR2-ospf-1-area-0.0.0.0]quit
[AR2-ospf-1]silent-interface g0/0/1            //配置接口为静默接口
[AR3]ospf 1 router-id 3.3.3.3                  //创建并运行 OSPF 进程 1，路由器的
                                               //ROUTER-ID 配置为"3.3.3.3"
[AR3-ospf-1]area 0                             //创建并进入 OSPF 区域 0
[AR3-ospf-1-area-0.0.0.0]network 172.16.3.0 0.0.0.255 //网络 172.16.3.0/24 所在接口参
                                               //与 OSPF 进程，并运行在区域 0
[AR3-ospf-1-area-0.0.0.0]network 172.16.0.0 0.0.0.255 //网络 172.16.0.0/24 所在接口参
                                               //与 OSPF 进程，并运行在区域 0

[AR3-ospf-1-area-0.0.0.0]quit
[AR3-ospf-1]silent-interface g0/0/1            //配置接口为静默接口
```

254 企业网络互联技术（双语）

（4）验证与测试

① 在 AR1、AR2 与 AR3 上分别执行"display ospf peer brief"命令，查看 OSPF 邻居表，如图 3.34～图 3.36 所示。注意观察每个路由器的邻居 ID 值，状态是否为 FULL。

```
[AR1]display ospf peer brief

        OSPF Process 1 with Router ID 1.1.1.1
              Peer Statistic Information
                                                          FULL表示与邻居
-------------------------------------------------------    建立了全毗邻关系
Area Id      Interface              Neighbor id    State
0.0.0.0      GigabitEthernet0/0/0      2.2.2.2      Full  ←
0.0.0.0      GigabitEthernet0/0/0      3.3.3.3      Full

-------------------------------------------------------


OSPF区域ID          接口              邻居OSPF路由器的
                                       ROUTER ID
```

图 3.34　在 AR1 上执行"display ospf peer brief"命令后的显示结果

```
[AR2]display ospf peer brief

        OSPF Process 1 with Router ID 2.2.2.2
              Peer Statistic Information

-------------------------------------------------------
Area Id      Interface              Neighbor id    State
0.0.0.0      GigabitEthernet0/0/0      1.1.1.1      Full
0.0.0.0      GigabitEthernet0/0/0      3.3.3.3      Full
-------------------------------------------------------
```

图 3.35　在 AR2 上执行"display ospf peer brief"命令后的显示结果

```
[AR3]display ospf peer brief

        OSPF Process 1 with Router ID 3.3.3.3
              Peer Statistic Information

-------------------------------------------------------
Area Id      Interface              Neighbor id    State
0.0.0.0      GigabitEthernet0/0/0      1.1.1.1      Full
0.0.0.0      GigabitEthernet0/0/0      2.2.2.2      Full
-------------------------------------------------------
```

图 3.36　在 AR3 上执行"display ospf peer brief"命令后的显示结果

② 在 AR1、AR2 与 AR3 上执行"display ip routing-table"命令，查看路由器的路由表。图 3.37 所示为 AR1 上路由表信息。

③ 在 AR1、AR2 与 AR3 上执行"display ip routing-table | include OSPF"命令，查看路由器路由表中包含 OSPF 的路由条目。显示结果如图 3.38～图 3.40 所示。

④ 参照图 3.33 规划的参数完成测试主机 PC1、PC2 与 PC3 的 IP 地址、子网掩码与默认网关配置，配置完成后在 DOS 命令行下使用"ipconfig"命令查看其 IP 配置信息是否正确。在主机的 DOS 命令行下，使用 ping 实用程序测试到另外两台主机的 IP 连通性。图 3.41 所示为测试主机 PC1 成功 ping 通测试主机 PC2 与 PC3。

```
[AR1]display ip routing-table
Route Flags: R - relay, D - download to fib
------------------------------------------------------------------

Routing Tables: Public
        Destinations : 12     Routes : 12

Destination/Mask   Proto   Pre  Cost    Flags NextHop      Interface

      127.0.0.0/8    Direct  0    0       D   127.0.0.1    InLoopBack0
      127.0.0.1/32   Direct  0    0       D   127.0.0.1    InLoopBack0
127.255.255.255/32   Direct  0    0       D   127.0.0.1    InLoopBack0
     172.16.0.0/24   Direct  0    0       D   172.16.0.1   GigabitEthernet0/0/0
    172.16.0.1/32    Direct  0    0       D   127.0.0.1    GigabitEthernet0/0/0
   172.16.0.255/32   Direct  0    0       D   127.0.0.1    GigabitEthernet0/0/0
    172.16.1.0/24    OSPF   10    2       D   172.16.0.2   GigabitEthernet0/0/0
    172.16.2.0/24    Direct  0    0       D   172.16.2.1   GigabitEthernet0/0/1
    172.16.2.1/32    Direct  0    0       D   127.0.0.1    GigabitEthernet0/0/1
   172.16.2.255/32   Direct  0    0       D   127.0.0.1    GigabitEthernet0/0/1
    172.16.3.0/24    OSPF   10    2       D   172.16.0.3   GigabitEthernet0/0/0
255.255.255.255/32   Direct  0    0       D   127.0.0.1    InLoopBack0
```

图 3.37   AR1 上路由表信息

```
[AR1]display ip routing-table | include OSPF
Route Flags: R - relay, D - download to fib
------------------------------------------------------------------
Routing Tables: Public
        Destinations : 12     Routes : 12

Destination/Mask   Proto   Pre  Cost    Flags NextHop      Interface

172.16.1.0/24   OSPF   10    2       D   172.16.0.2   GigabitEthernet0/0/0
172.16.3.0/24   OSPF   10    2       D   172.16.0.3   GigabitEthernet0/0/0
```

图 3.38   在 AR1 上执行 "display ip routing-table | include OSPF" 命令后的显示结果

```
[AR2]display ip routing-table | include OSPF
Route Flags: R - relay, D - download to fib
------------------------------------------------------------------
Routing Tables: Public
        Destinations : 12     Routes : 12

Destination/Mask   Proto   Pre  Cost    Flags NextHop      Interface

172.16.2.0/24   OSPF   10    2       D   172.16.0.1   GigabitEthernet0/0/0
172.16.3.0/24   OSPF   10    2       D   172.16.0.3   GigabitEthernet0/0/0
```

图 3.39   在 AR2 上执行 "display ip routing-table | include OSPF" 命令后的显示结果

```
[AR3]display ip routing-table | include OSPF
Route Flags: R - relay, D - download to fib
------------------------------------------------------------------
Routing Tables: Public
        Destinations : 13     Routes : 13

Destination/Mask   Proto   Pre  Cost    Flags NextHop      Interface

172.16.1.0/24   OSPF   10    2       D   172.16.0.2   GigabitEthernet0/0/0
172.16.2.0/24   OSPF   10    2       D   172.16.0.1   GigabitEthernet0/0/0
```

图 3.40   在 AR3 上执行 "display ip routing-table | include OSPF" 命令后的显示结果

256 企业网络互联技术（双语）

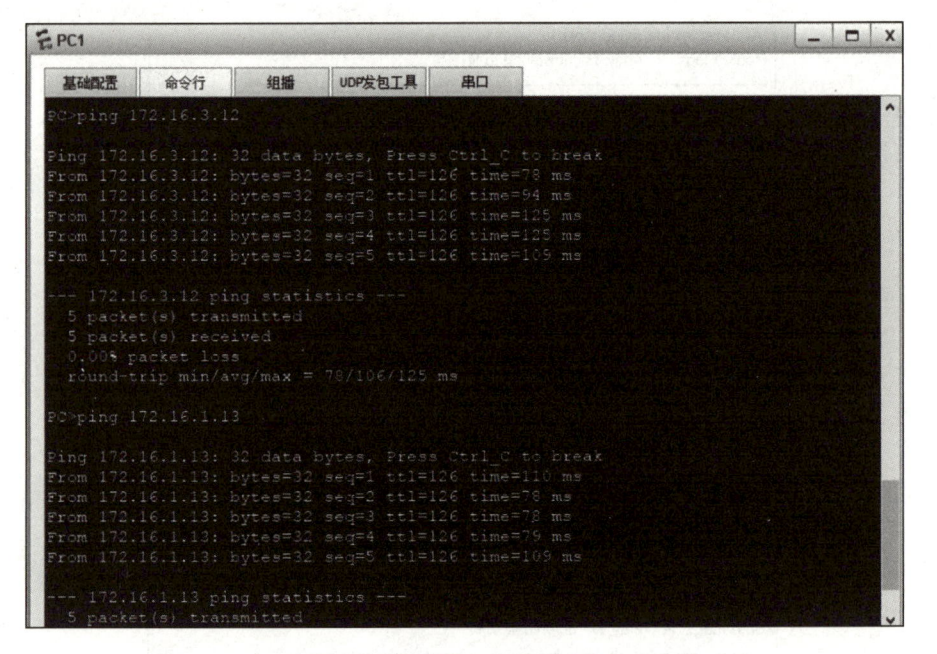

图 3.41　测试主机 PC1 成功 ping 通测试主机 PC2 与 PC3

2. 基于三层交换机的 OSPF 规划与配置

搭建如图 3.42 所示的网络拓扑图，图中交换机 CORE、AGG1 和 AGG2 均为三层交换机，CORE 与 AGG1 之间、CORE 与 AGG2 之间为三层连接网段。IP 地址的规划见表 3.5，其中，VLAN100 与 VLAN101 为交换机 CORE 与 AGG1、AGG2 连接网段 VLAN，VLAN2 为测试主机PC1 所属 VLAN，VLAN20 为测试主机 PC2 所属 VLAN。

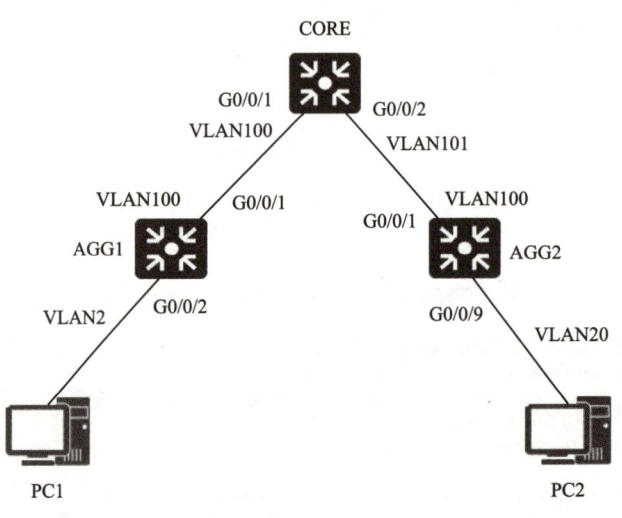

图 3.42　基于三层交换机的 OSPF 规划与配置网络拓扑图

请在 CORE、AGG1 和 AGG2 上配置单域 OSPF 路由，实现网络的互联互通，并进行路由查看与测试。

项目 3　中型企业网络互联项目　257

表 3.5　IP 地址的规划

设　备	接　口	IP 地址与掩码	网　关
CORE	Vlanif100	172.16.254.1/30	
	Vlanif101	172.16.254.5/30	
AGG1	Vlanif100	172.16.254.2/30	
	Vlanif2	172.16.2.1/24	
AGG2	Vlanif100	172.16.254.6/30	
	Vlanif20	172.16.20.1/24	
PC1	NIC	172.16.2.11/24	172.16.2.1
PC2	NIC	172.16.20.12/24	172.16.20.1

（1）OSPF 规划

①使用单域 OSPF，区域号为 0。

② CORE 的 ROUTER-ID 为 1.1.1.1，AGG1 的 ROUTER-ID 为 2.2.2.2，AGG3 的 ROUTER-ID 为 3.3.3.3。

③ AGG1 的 VLAN2 接口配置为静默接口，AGG2 的 VLAN20 接口配置为静默接口，不发送路由更新。

④ CORE 上 172.16.254.0/30 与 172.16.254.4/30 网段接口参与 OSPF 进程；AGG1 上 172.16.254.0/30 与 172.16.2.0/24 网段接口参与 OSPF 进程；AGG2 上 172.16.254.4/30 与 172.16.20.0/24 网段接口参与 OSPF 进程。

（2）IP 与 VLAN 的配置

在交换机 CORE、AGG1 与 AGG2 上，根据图 3.42 与表 3.5，完成 IP 与 VLAN 的配置，并禁用 CORE 与 AGG1、CORE 与 AGG2 连接网段接口生成树。配置步骤如下：

```
[CORE]vlan 100
[CORE-vlan100]description linktoAGG1
[CORE-vlan100]vlan 101
[CORE-vlan101]description linktoAGG2
[CORE-vlan101]quit
// 创建 VLAN，并配置 VLAN 的描述
[CORE]interface g0/0/1
[CORE-GigabitEthernet0/0/1]port link-type access
[CORE-GigabitEthernet0/0/1]port default vlan 100
[CORE-GigabitEthernet0/0/1]stp disable
[CORE-GigabitEthernet0/0/1]quit
// 配置 G0/0/1 接口为接入接口，属于 VLAN100，并在接口上禁用生成树功能
[CORE]interface g0/0/2
```

```
[CORE-GigabitEthernet0/0/2]port link-type access
[CORE-GigabitEthernet0/0/2]port default vlan 101
[CORE-GigabitEthernet0/0/2]stp disable
[CORE-GigabitEthernet0/0/2]quit
// 配置 G0/0/2 接口为接入接口，属于 VLAN101，并在接口上禁用生成树功能
[CORE]interface Vlanif 100
[CORE-Vlanif100]ip address 172.16.254.1 30
[CORE-Vlanif100]quit
[CORE]interface Vlanif 101
[CORE-Vlanif101]ip address 172.16.254.5 30
[CORE-Vlanif101]quit
// 配置 VLAN 接口的 IP 地址与掩码
[AGG1]vlan 2
[AGG1-vlan2]description linktopc1
[AGG1-vlan2]vlan 100
[AGG1-vlan100]description linktoCORE
[AGG1-vlan100]quit
// 创建 VLAN，并配置 VLAN 的描述
[AGG1]interface g0/0/1
[AGG1-GigabitEthernet0/0/1]port link-type access
[AGG1-GigabitEthernet0/0/1]port default vlan 100
[AGG1-GigabitEthernet0/0/1]stp disable
[AGG1-GigabitEthernet0/0/1]quit
// 配置 G0/0/1 接口为接入接口，属于 VLAN100，并在接口上禁用生成树功能
[AGG1]interface g0/0/2
[AGG1-GigabitEthernet0/0/2]port link-type access
[AGG1-GigabitEthernet0/0/2]port default vlan 2
[AGG1-GigabitEthernet0/0/2]quit
// 配置 G0/0/2 接口为接入接口，属于 VLAN2
[AGG1]interface Vlanif 100
[AGG1-Vlanif100]ip address 172.16.254.2 30
[AGG1-Vlanif100]quit
[AGG1]interface Vlanif 2
[AGG1-Vlanif2]ip address 172.16.2.1 24
[AGG1-Vlanif2]quit
// 配置 VLAN 接口的 IP 地址与掩码
[AGG2]vlan 20
[AGG2-vlan20]description linktopc2
[AGG2-vlan20]vlan 100
[AGG2-vlan100]description linktoCORE
```

```
[AGG2-vlan100]quit
// 创建 VLAN，并配置 VLAN 的描述
[AGG2]interface g0/0/1
[AGG2-GigabitEthernet0/0/1]port link-type access
[AGG2-GigabitEthernet0/0/1]port default vlan 100
[AGG2-GigabitEthernet0/0/1]stp disable
[AGG2-GigabitEthernet0/0/1]quit
// 配置 G0/0/1 接口为接入接口，属于 VLAN100，并在接口上禁用生成树功能
[AGG2]interface g0/0/9
[AGG2-GigabitEthernet0/0/9]port link-type access
[AGG2-GigabitEthernet0/0/9]port default vlan 20
[AGG2-GigabitEthernet0/0/9]quit
// 配置 G0/0/9 接口为接入接口，属于 VLAN20
[AGG2]interface Vlanif 100
[AGG2-Vlanif100]ip address 172.16.254.6 30
[AGG2-Vlanif100]quit
[AGG2]interface Vlanif 20
[AGG2-Vlanif20]ip address 172.16.20.1 24
[AGG2-Vlanif20]quit
// 配置 VLAN 接口的 IP 地址与掩码
```

（3）单域 OSPF 的配置

根据 OSPF 的规划，在交换机 CORE、AGG1、AGG2 上完成单域 OSPF 的配置，步骤如下：

```
[CORE]ospf 1 router-id 1.1.1.1
[CORE-ospf-1]area 0
[CORE-ospf-1-area-0.0.0.0]network 172.16.254.0 0.0.0.3
[CORE-ospf-1-area-0.0.0.0]network 172.16.254.4 0.0.0.3
[CORE-ospf-1-area-0.0.0.0]quit
[CORE-ospf-1]quit
// 创建并运行 OSPF 进程 1，ROUTER-ID 为 "1.1.1.1"，创建并进入 OSPF 区域 0，网络 172.16.
//254.0/30 与 172.16.254.4/30 所在接口参与 OSPF 进程，并运行在区域 0
[AGG1]ospf 1 router-id 2.2.2.2
[AGG1-ospf-1] area-0
[AGG1-ospf-1-area-0.0.0.0]network 172.16.2.0 0.0.0.255
[AGG1-ospf-1-area-0.0.0.0]network 172.16.254.0 0.0.0.3
[AGG1-ospf-1-area-0.0.0.0]quit
[AGG1-ospf-1]silent-interface Vlanif 2
[AGG1-ospf-1]quit
// 创建并运行 OSPF 进程 1，ROUTER-ID 为 "2.2.2.2"，创建并进入 OSPF 区域 0，网络 172.16.
//0.0/24 与 172.16.254.0/30 所在接口参与 OSPF 进程，并运行在区域 0
```

企业网络互联技术（双语）

```
[AGG2]ospf 1 router-id 3.3.3.3
[AGG2-ospf-1]area 0
[AGG2-ospf-1-area-0.0.0.0]network 172.16.20.0 0.0.0.255
[AGG2-ospf-1-area-0.0.0.0]network 172.16.254.4 0.0.0.3
[AGG2-ospf-1-area-0.0.0.0]quit
[AGG2-ospf-1]silent-interface Vlanif 20
[AGG2-ospf-1]quit
// 创建并运行 OSPF 进程 1，ROUTER-ID 为"3.3.3.3"，创建并进入 OSPF 区域 0，网络 172.16.
//20.0/24 与 172.16.254.4/30 所在接口参与 OSPF 进程，并运行在区域 0
```

（4）验证与测试

① 在交换机 CORE、AGG1 与 AGG2 上分别执行"display ospf peer brief"命令，查看 OSPF 邻居表，如图 3.43～图 3.45 所示。注意观察每个路由器的邻居 ID 值，状态是否为 FULL。

```
[CORE]display ospf peer brief

           OSPF Process 1 with Router ID 1.1.1.1
                  Peer Statistic Information
-------------------------------------------------------------
Area Id       Interface          Neighbor id     State
0.0.0.0       Vlanif100          2.2.2.2         Full
0.0.0.0       Vlanif101          3.3.3.3         Full
-------------------------------------------------------------
```

图 3.43　在 CORE 上执行"display ospf peer brief"命令后的显示结果

```
[AGG1]display ospf peer brief

           OSPF Process 1 with Router ID 2.2.2.2
                  Peer Statistic Information
-------------------------------------------------------------
Area Id       Interface          Neighbor id     State
0.0.0.0       Vlanif100          1.1.1.1         Full
-------------------------------------------------------------
```

图 3.44　在 AGG1 上执行"display ospf peer brief"命令后的显示结果

```
[AGG2]display ospf peer brief

           OSPF Process 1 with Router ID 3.3.3.3
                  Peer Statistic Information
-------------------------------------------------------------
Area Id       Interface          Neighbor id     State
0.0.0.0       Vlanif100          1.1.1.1         Full
-------------------------------------------------------------
```

图 3.45　在 AGG2 上执行"display ospf peer brief"命令后的显示结果

② 在交换机 CORE、AGG1 与 AGG2 上执行"display ip routing-table"命令，查看路由器的路由表，如图 3.46～图 3.48 所示。

```
[CORE]display ip routing-table
Route Flags: R - relay, D - download to fib
------------------------------------------------------------------------------
Routing Tables: Public
        Destinations : 8      Routes : 8

Destination/Mask   Proto   Pre  Cost     Flags NextHop       Interface

       127.0.0.0/8   Direct  0    0         D    127.0.0.1       InLoopBack0
      127.0.0.1/32   Direct  0    0         D    127.0.0.1       InLoopBack0
      172.16.2.0/24  OSPF   10   2         D    172.16.254.2    Vlanif100
     172.16.20.0/24  OSPF   10   2         D    172.16.254.6    Vlanif101
    172.16.254.0/30  Direct  0    0         D    172.16.254.1    Vlanif100
    172.16.254.1/32  Direct  0    0         D    127.0.0.1       Vlanif100
    172.16.254.4/30  Direct  0    0         D    172.16.254.5    Vlanif101
    172.16.254.5/32  Direct  0    0         D    127.0.0.1       Vlanif101
```

图 3.46   CORE 的路由表

```
[AGG1]display ip routing-table
Route Flags: R - relay, D - download to fib
------------------------------------------------------------------------------
Routing Tables: Public
        Destinations : 8      Routes : 8

Destination/Mask   Proto   Pre  Cost     Flags NextHop       Interface

       127.0.0.0/8   Direct  0    0         D    127.0.0.1       InLoopBack0
      127.0.0.1/32   Direct  0    0         D    127.0.0.1       InLoopBack0
      172.16.2.0/24  Direct  0    0         D    172.16.2.1      Vlanif2
      172.16.2.1/32  Direct  0    0         D    127.0.0.1       Vlanif2
     172.16.20.0/24  OSPF   10   3         D    172.16.254.1    Vlanif100
    172.16.254.0/30  Direct  0    0         D    172.16.254.2    Vlanif100
    172.16.254.2/32  Direct  0    0         D    127.0.0.1       Vlanif100
    172.16.254.4/30  OSPF   10   2         D    172.16.254.1    Vlanif100
```

图 3.47   AGG1 的路由表

```
[AGG2]display ip routing-table
Route Flags: R - relay, D - download to fib
------------------------------------------------------------------------------
Routing Tables: Public
        Destinations : 7      Routes : 7

Destination/Mask   Proto   Pre  Cost     Flags NextHop       Interface

       127.0.0.0/8   Direct  0    0         D    127.0.0.1       InLoopBack0
      127.0.0.1/32   Direct  0    0         D    127.0.0.1       InLoopBack0
      172.16.2.0/24  OSPF   10   2         D    172.16.254.5    Vlanif100
     172.16.20.0/24  Direct  0    0         D    172.16.20.1     Vlanif20
     172.16.20.1/32  Direct  0    0         D    127.0.0.1       Vlanif20
    172.16.254.0/30  OSPF   10   2         D    172.16.254.5    Vlanif100
    172.16.254.4/30  Direct  0    0         D    172.16.254.6    Vlanif100
    172.16.254.6/32  Direct  0    0         D    127.0.0.1       Vlanif100
```

图 3.48   AGG2 的路由表

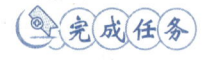 完成任务

1. 任务规划

根据任务描述、图 3.1 所示的网络拓扑、表 3.1 的 IP 规划和项目 3 网络路由需求，本任务路由的具体实现要求如下：

① 在出口路由器 EDGE-AR 上配置到 Internet 的默认路由。

② 使用单域 OSPF 实现企业内部网络所有网段的互联互通。

③ 在 EDGE-AR 上把到 Internet 的默认路由引入 OSPF 路由协议中。

④ 实现路由的优化。

OSPF 的规划见表 3.6。CORE 与路由器 EDGE-AR、汇聚层交换机 XZL-AGG、汇聚层交换机 JXL-AGG 和汇聚层交换机 SXL-AGG 之间链路的 OSPF 类型更改为点到点类型。

表 3.6　OSPF 的规划

路由设备	ROUTER-ID	进程号	区域号	参与 OSPF 进程的网段	静默接口
EDGE-AR	1.1.1.1	1	0.0.0.0	10.254.254.12/30	无
CORE	2.2.2.2	1	0.0.0.0	10.254.254.0/30、10.254.254.4/30、10.254.254.8/30、10.254.254.12/30、10.254.253.0/24	Vlanif 202
XZL-AGG	3.3.3.3	1	0.0.0.0	10.0.2.0/24、10.0.3.0/24、10.0.4.0/24、10.0.5.0/24、10.0.90.0/24、10.0.91.0/24、10.254.254.0/30	Vlanif 2、Vlanif 3、Vlanif 4、Vlanif 5、Vlanif 90 和 Vlanif 91
JXL-AGG	4.4.4.4	1	0.0.0.0	10.1.12.0/24、10.1.13.0/24、10.1.92.0/24、10.1.93.0/24 与 10.254.254.4/30	Vlanif 12、Vlanif 13、Vlanif 92 和 Vlanif 93
SXL-AGG	5.5.5.5	1	0.0.0.0	10.2.22.0/24、10.2.23.0/24、10.2.94.0/24、10.2.95.0/24 与 10.254.254.8/30	Vlanif 22、Vlanif 23、Vlanif 94 和 Vlanif 95

2. 任务实施

（1）边缘路由器 EDGE-AR1 上 OSPF 的配置

```
[EDGE-AR]ip route-static 0.0.0.0 0.0.0.0 125.71.28.1
// 配置到 Internet 的默认路由，下一跳地址为 125.71.28.1
[EDGE-AR]interface g0/0/1
[EDGE-AR-GigabitEthernet0/0/1]ospf network-type p2p
[EDGE-AR-GigabitEthernet0/0/1]quit
// 配置 G0/0/1 接口的 OSPF 网络类型为点到点类型
[EDGE-AR]ospf 1 router-id 1.1.1.1
[EDGE-AR-ospf-1]area 0.0.0.0
```

```
[EDGE-AR-ospf-1-area-0.0.0.0]network 10.254.254.12 0.0.0.3
[EDGE-AR-ospf-1-area-0.0.0.0]quit
// 启用 OSPF 进程 1,ROUTER-ID 为 1.1.1.1,10.254.254.12/30 网段参与 OSPF 进程,并运行在区域 0
[EDGE-AR-ospf-1]default-route-advertise always
// 把默认路由引入 OSPF 路由协议中,always 参数表示只要 EDGE-AR 上配置了默认路由就引入 OSPF
// 路由协议中
```

(2) 核心层与汇聚层交换机上 OSPF 的配置

① 在核心层交换机 CORE 上完成 OSPF 的配置。

```
[CORE]interface Vlanif 199
[CORE-Vlanif199]ospf network-type p2p
[CORE-Vlanif199]quit
[CORE]interface Vlanif 200
[CORE-Vlanif200]ospf network-type p2p
[CORE-Vlanif200]quit
[CORE]interface Vlanif 201
[CORE-Vlanif201]ospf network-type p2p
[CORE-Vlanif201]quit
[CORE]interface Vlanif 203
[CORE-Vlanif203]ospf network-type p2p
[CORE-Vlanif203]quit
// 配置 Vlanif199、Vlanif200、Vlanif201 和 Vlanif203 接口的 OSPF 网络类型为点到点类型
[CORE]ospf 1 router-id 2.2.2.2
[CORE-ospf-1]area 0.0.0.0
[CORE-ospf-1-area-0.0.0.0]network 10.254.254.0 0.0.0.3
[CORE-ospf-1-area-0.0.0.0]network 10.254.254.4 0.0.0.3
[CORE-ospf-1-area-0.0.0.0]network 10.254.254.8 0.0.0.3
[CORE-ospf-1-area-0.0.0.0]network 10.254.254.12 0.0.0.3
[CORE-ospf-1-area-0.0.0.0]network 10.254.253.0 0.0.0.255
[CORE-ospf-1-area-0.0.0.0]quit
// 启用 OSPF 进程 1,ROUTER-ID 为 2.2.2.2,10.254.254.0/30、10.254.254.3/30、10.254.254.8/
//30、10.254.254.12/30 与 10.254.253.0/24 网段的接口参与 OSPF 进程,并运行在区域 0
[CORE-ospf-1]silent-interface Vlanif 202
[CORE-ospf-1]quit
// 配置 Vlanif202 接口为静默接口
```

② 在汇聚层交换机 XZL-AGG 上完成 OSPF 的配置。

```
[XZL-AGG]interface Vlanif 199
[XZL-AGG-Vlanif199]ospf network-type p2p
```

264 企业网络互联技术（双语）

```
[XZL-AGG-Vlanif199]quit
```
// 配置 Vlanif199 接口的 OSPF 网络类型为点到点类型
```
[XZL-AGG]ospf 1 router-id 3.3.3.3
[XZL-AGG-ospf-1]area 0.0.0.0
[XZL-AGG-ospf-1-area-0.0.0.0]network 10.0.2.0 0.0.0.255
[XZL-AGG-ospf-1-area-0.0.0.0]network 10.0.3.0 0.0.0.255
[XZL-AGG-ospf-1-area-0.0.0.0]network 10.0.4.0 0.0.0.255
[XZL-AGG-ospf-1-area-0.0.0.0]network 10.0.5.0 0.0.0.255
[XZL-AGG-ospf-1-area-0.0.0.0]network 10.0.90.0 0.0.0.255
[XZL-AGG-ospf-1-area-0.0.0.0]network 10.0.91.0 0.0.0.255
[XZL-AGG-ospf-1-area-0.0.0.0]network 10.254.254.0 0.0.0.3
[XZL-AGG-ospf-1-area-0.0.0.0]quit
[XZL-AGG-ospf-1]
```
// 启用 OSPF 进程 1, ROUTER-ID 为 3.3.3.3, 10.0.2.0/24、10.0.3.0/24、10.0.4.0/24、10.0.5.
0/24、10.0.90.0/24、10.0.91.0/24 与 10.254.254.0/30 网段的接口参与 OSPF 进程, 并运行在区域 0
```
[XZL-AGG-ospf-1]silent-interface Vlanif 2
[XZL-AGG-ospf-1]silent-interface Vlanif 3
[XZL-AGG-ospf-1]silent-interface Vlanif 4
[XZL-AGG-ospf-1]silent-interface Vlanif 5
[XZL-AGG-ospf-1]silent-interface Vlanif 90
[XZL-AGG-ospf-1]silent-interface Vlanif 91
[XZL-AGG-ospf-1]quit
```
// 配置 Vlanif2、Vlanif3、Vlanif4、Vlanif5、Vlanif90 和 Vlanif91 接口为静默接口

③ 在汇聚层交换机 JXL-AGG 上完成 OSPF 的配置。

```
[JXL-AGG]interface Vlanif 199
[JXL-AGG-Vlanif199]ospf network-type p2p
[JXL-AGG-Vlanif199]quit
```
// 配置 Vlanif199 接口的 OSPF 网络类型为点到点类型
```
[JXL-AGG]ospf 1 router-id 4.4.4.4
[JXL-AGG-ospf-1-area-0.0.0.0]network 10.1.12.0 0.0.0.255
[JXL-AGG-ospf-1-area-0.0.0.0]network 10.1.13.0 0.0.0.255
[JXL-AGG-ospf-1-area-0.0.0.0]network 10.1.92.0 0.0.0.255
[JXL-AGG-ospf-1-area-0.0.0.0]network 10.1.93.0 0.0.0.255
[JXL-AGG-ospf-1-area-0.0.0.0]network 10.254.254.4 0.0.0.3
[JXL-AGG-ospf-1-area-0.0.0.0]quit
```
// 启用 OSPF 进程 1, ROUTER-ID 为 4.4.4.4, 10.1.12.0/24、10.1.13.0/24、10.1.92.0/24、10.1.
//93.0/24 与 10.254.254.4/30 网段的接口参与 OSPF 进程, 并运行在区域 0

```
[JXL-AGG-ospf-1]silent-interface Vlanif 12
[JXL-AGG-ospf-1]silent-interface Vlanif 13
[JXL-AGG-ospf-1]silent-interface Vlanif 92
[JXL-AGG-ospf-1]silent-interface Vlanif 93
[JXL-AGG-ospf-1]quit
// 配置 Vlanif12、Vlanif13、Vlanif92 和 Vlanif93 接口为静默接口
```

④ 在汇聚层交换机 SXL-AGG 上完成 OSPF 的配置。

```
[SXL-AGG]interface Vlanif 199
[SXL-AGG-Vlanif199]ospf network-type p2p
[SXL-AGG-Vlanif199]quit
// 配置 Vlanif199 接口的 OSPF 网络类型为点到点类型
[SXL-AGG]ospf 1 router-id 5.5.5.5
[SXL-AGG-ospf-1]area 0.0.0.0
[SXL-AGG-ospf-1-area-0.0.0.0]network 10.2.22.0 0.0.0.255
[SXL-AGG-ospf-1-area-0.0.0.0]network 10.2.23.0 0.0.0.255
[SXL-AGG-ospf-1-area-0.0.0.0]network 10.2.94.0 0.0.0.255
[SXL-AGG-ospf-1-area-0.0.0.0]network 10.2.95.0 0.0.0.255
[SXL-AGG-ospf-1-area-0.0.0.0]network 10.254.254.8 0.0.0.3
[SXL-AGG-ospf-1-area-0.0.0.0]quit
// 启用 OSPF 进程 1, ROUTER-ID 为 5.5.5.5, 10.2.22.0/24、10.2.23.0/24、10.2.94.0/24、
//10.2.95.0/24 与 10.254.254.8/30 网段的接口参与 OSPF 进程, 并运行在区域 0
[SXL-AGG-ospf-1]silent-interface Vlanif 22
[SXL-AGG-ospf-1]silent-interface Vlanif 23
[SXL-AGG-ospf-1]silent-interface Vlanif 94
[SXL-AGG-ospf-1]silent-interface Vlanif 95
[SXL-AGG-ospf-1]quit
[SXL-AGG]
// 配置 Vlanif22、Vlanif23、Vlanif94 和 Vlanif95 接口为静默接口
```

（3）验证与测试

① 在路由器 EDGE-AR、核心层交换机 CORE、汇聚层交换机 XZL-AGG、汇聚层交换机 JXL-AGG 和汇聚层交换机 SXL-AGG 上分别执行 "display ospf peer brief" 命令查看 OSPF 邻居表，观察 OSPF 邻居路由器的状态是否为 FULL。

② 在路由器 EDGE-AR、核心层交换机 CORE、汇聚层交换机 XZL-AGG、汇聚层交换机 JXL-AGG 和汇聚层交换机 SXL-AGG 上分别执行 "display ip routing-table" 命令查看路由器的路由表。结果如图 3.49 ～图 3.52 所示。

[EDGE-AR]display ip routing-table
Route Flags: R - relay, D - download to fib

Routing Tables: Public
         Destinations : 34   Routes : 34

Destination/Mask   Proto Pre Cost   Flags NextHop   Interface

0.0.0.0/0   Static 60   0   RD   125.71.28.254 GigabitEthernet0/0/0
10.0.2.0/24   OSPF   10   3   D   10.254.254.13 GigabitEthernet0/0/1
10.0.3.0/24   OSPF   10   3   D   10.254.254.13 GigabitEthernet0/0/1
10.0.4.0/24   OSPF   10   3   D   10.254.254.13 GigabitEthernet0/0/1
10.0.5.0/24   OSPF   10   3   D   10.254.254.13 GigabitEthernet0/0/1
10.0.90.0/24   OSPF   10   3   D   10.254.254.13 GigabitEthernet0/0/1
10.0.91.0/24   OSPF   10   3   D   10.254.254.13 GigabitEthernet0/0/1
10.1.12.0/24   OSPF   10   3   D   10.254.254.13 GigabitEthernet0/0/1
10.1.13.0/24   OSPF   10   3   D   10.254.254.13 GigabitEthernet0/0/1
10.1.92.0/24   OSPF   10   3   D   10.254.254.13 GigabitEthernet0/0/1
10.1.93.0/24   OSPF   10   3   D   10.254.254.13 GigabitEthernet0/0/1
10.2.22.0/24   OSPF   10   3   D   10.254.254.13 GigabitEthernet0/0/1
10.2.23.0/24   OSPF   10   3   D   10.254.254.13 GigabitEthernet0/0/1
10.2.94.0/24   OSPF   10   3   D   10.254.254.13 GigabitEthernet0/0/1
10.2.95.0/24   OSPF   10   3   D   10.254.254.13 GigabitEthernet0/0/1
10.254.253.0/24   OSPF   10   2   D   10.254.254.13 GigabitEthernet0/0/1
10.254.254.0/30   OSPF   10   2   D   10.254.254.13 GigabitEthernet0/0/1
10.254.254.4/30   OSPF   10   2   D   10.254.254.13 GigabitEthernet0/0/1
10.254.254.8/30   OSPF   10   2   D   10.254.254.13 GigabitEthernet0/0/1
10.254.254.12/30   Direct 0   0   D   10.254.254.14 GigabitEthernet0/0/1
10.254.254.14/32   Direct 0   0   D   127.0.0.1   GigabitEthernet0/0/1

图 3.49   EDGE-AR 的路由表

[XZL-AGG]display ip routing-table
Route Flags: R - relay, D - download to fib

Routing Tables: Public
         Destinations : 29   Routes : 29

Destination/Mask   Proto Pre Cost   Flags NextHop   Interface

0.0.0.0/0   O_ASE 150 1   D   10.254.254.1   Vlanif199
10.0.2.0/24   Direct 0   0   D   10.0.2.254   Vlanif2
10.0.2.254/32   Direct 0   0   D   127.0.0.1   Vlanif2
10.0.3.0/24   Direct 0   0   D   10.0.3.254   Vlanif3
10.0.3.254/32   Direct 0   0   D   127.0.0.1   Vlanif3
10.0.4.0/24   Direct 0   0   D   10.0.4.254   Vlanif4
10.0.4.254/32   Direct 0   0   D   127.0.0.1   Vlanif4
10.0.5.0/24   Direct 0   0   D   10.0.5.254   Vlanif5
10.0.5.254/32   Direct 0   0   D   127.0.0.1   Vlanif5
10.0.90.0/24   Direct 0   0   D   10.0.90.254   Vlanif90
10.0.90.254/32   Direct 0   0   D   127.0.0.1   Vlanif90
10.0.91.0/24   Direct 0   0   D   10.0.91.254   Vlanif91
10.0.91.254/32   Direct 0   0   D   127.0.0.1   Vlanif91
10.1.12.0/24   OSPF   10   3   D   10.254.254.1   Vlanif199
10.1.13.0/24   OSPF   10   3   D   10.254.254.1   Vlanif199
10.1.92.0/24   OSPF   10   3   D   10.254.254.1   Vlanif199
10.1.93.0/24   OSPF   10   3   D   10.254.254.1   Vlanif199
10.2.22.0/24   OSPF   10   3   D   10.254.254.1   Vlanif199
10.2.23.0/24   OSPF   10   3   D   10.254.254.1   Vlanif199
10.2.94.0/24   OSPF   10   3   D   10.254.254.1   Vlanif199
10.2.95.0/24   OSPF   10   3   D   10.254.254.1   Vlanif199
10.254.253.0/24   OSPF   10   2   D   10.254.254.1   Vlanif199
10.254.254.0/30   Direct 0   0   D   10.254.254.2   Vlanif199
10.254.254.2/32   Direct 0   0   D   127.0.0.1   Vlanif199
10.254.254.4/30   OSPF   10   2   D   10.254.254.1   Vlanif199
10.254.254.8/30   OSPF   10   2   D   10.254.254.1   Vlanif199
10.254.254.12/30   OSPF   10   2   D   10.254.254.1   Vlanif199
127.0.0.0/8   Direct 0   0   D   127.0.0.1   InLoopBack0
127.0.0.1/32   Direct 0   0   D   127.0.0.1   InLoopBack0

图 3.50   XZL-AGG 的路由表

[JXL-AGG]display ip routing-table
Route Flags: R - relay, D - download to fib

Routing Tables: Public
         Destinations : 27   Routes : 27

Destination/Mask   Proto Pre Cost   Flags NextHop   Interface

0.0.0.0/0   O_ASE 150 1   D   10.254.254.5   Vlanif99
10.0.2.0/24   OSPF   10   3   D   10.254.254.5   Vlanif99
10.0.3.0/24   OSPF   10   3   D   10.254.254.5   Vlanif99
10.0.4.0/24   OSPF   10   3   D   10.254.254.5   Vlanif99
10.0.5.0/24   OSPF   10   3   D   10.254.254.5   Vlanif99
10.0.90.0/24   OSPF   10   3   D   10.254.254.5   Vlanif99
10.0.91.0/24   OSPF   10   3   D   10.254.254.5   Vlanif99
10.1.11.0/24   Direct 0   0   D   127.0.0.1   Vlanif11
10.1.11.254/32   Direct 0   0   D   127.0.0.1   Vlanif11
10.1.13.0/24   Direct 0   0   D   10.1.13.254   Vlanif13
10.1.13.254/32   Direct 0   0   D   127.0.0.1   Vlanif13
10.1.92.0/24   Direct 0   0   D   10.1.92.254   Vlanif92
10.1.92.254/32   Direct 0   0   D   127.0.0.1   Vlanif92
10.1.93.0/24   Direct 0   0   D   10.1.93.254   Vlanif93
10.1.93.254/32   Direct 0   0   D   127.0.0.1   Vlanif93
10.2.2.0/24   OSPF   10   3   D   10.254.254.5   Vlanif99
10.2.3.0/24   OSPF   10   3   D   10.254.254.5   Vlanif99
10.2.4.0/24   OSPF   10   3   D   10.254.254.5   Vlanif99
10.2.95.0/24   OSPF   10   3   D   10.254.254.5   Vlanif99
10.254.253.0/24   OSPF   10   2   D   10.254.254.5   Vlanif99
10.254.254.0/30   OSPF   10   2   D   10.254.254.5   Vlanif99
10.254.254.4/30   Direct 0   0   D   10.254.254.6   Vlanif99
10.254.254.6/32   Direct 0   0   D   127.0.0.1   Vlanif99
10.254.254.8/30   OSPF   10   2   D   10.254.254.5   Vlanif99
10.254.254.12/30   OSPF   10   2   D   10.254.254.5   Vlanif99
127.0.0.0/8   Direct 0   0   D   127.0.0.1   InLoopBack0
127.0.0.1/32   Direct 0   0   D   127.0.0.1   InLoopBack0

图 3.51　JXL-AGG 的路由表

[SXL-AGG]display ip routing-table
Route Flags: R - relay, D - download to fib

Routing Tables: Public
         Destinations : 27   Routes : 27

Destination/Mask   Proto Pre Cost   Flags NextHop   Interface

0.0.0.0/0   O_ASE 150 1   D   10.254.254.9   Vlanif99
10.0.2.0/24   OSPF   10   3   D   10.254.254.9   Vlanif99
10.0.3.0/24   OSPF   10   3   D   10.254.254.9   Vlanif99
10.0.4.0/24   OSPF   10   3   D   10.254.254.9   Vlanif99
10.0.5.0/24   OSPF   10   3   D   10.254.254.9   Vlanif99
10.0.90.0/24   OSPF   10   3   D   10.254.254.9   Vlanif99
10.0.91.0/24   OSPF   10   3   D   10.254.254.9   Vlanif99
10.1.12.0/24   OSPF   10   3   D   10.254.254.9   Vlanif99
10.1.13.0/24   OSPF   10   3   D   10.254.254.9   Vlanif99
10.1.92.0/24   OSPF   10   3   D   10.254.254.9   Vlanif99
10.1.93.0/24   OSPF   10   3   D   10.254.254.9   Vlanif99
10.2.2.0/24   Direct 0   0   D   10.2.22.254   Vlanif22
10.2.22.254/32   Direct 0   0   D   127.0.0.1   Vlanif22
10.2.3.0/24   Direct 0   0   D   10.2.23.254   Vlanif23
10.2.23.254/32   Direct 0   0   D   127.0.0.1   Vlanif23
10.2.4.0/24   Direct 0   0   D   10.2.94.254   Vlanif4
10.2.94.254/32   Direct 0   0   D   127.0.0.1   Vlanif4
10.2.95.0/24   Direct 0   0   D   10.2.95.254   Vlanif5
10.2.95.254/32   Direct 0   0   D   127.0.0.1   Vlanif5
10.254.253.0/24   OSPF   10   2   D   10.254.254.9   Vlanif99
10.254.254.0/30   OSPF   10   2   D   10.254.254.9   Vlanif99
10.254.254.4/30   OSPF   10   2   D   10.254.254.9   Vlanif99
10.254.254.8/30   Direct 0   0   D   10.254.254.10 Vlanif99
10.254.254.10/32   Direct 0   0   D   127.0.0.1   Vlanif99
10.254.254.12/30   OSPF   10   2   D   10.254.254.9   Vlanif99
127.0.0.0/8   Direct 0   0   D   127.0.0.1   InLoopBack0
127.0.0.1/32   Direct 0   0   D   127.0.0.1   InLoopBack0

图 3.52　SXL-AGG 的路由表

268　企业网络互联技术（双语）

③ 完成此任务后，企业内网的主机可以访问 Internet，Internet 上的主机可以访问企业对外提供的 Web 和 DNS 服务。图 3.53 为在测试主机 PC2 上使用 ping 命令测试到 Internet 上测试主机 PC1 的 IP 连通性，可正常连通。

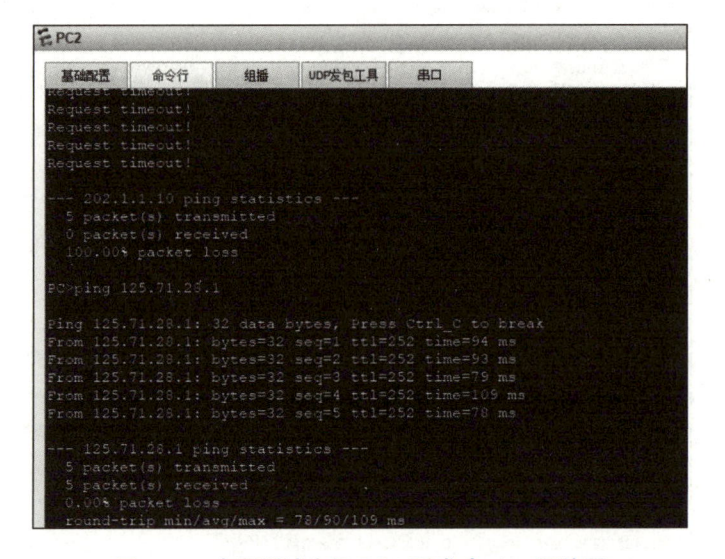

图 3.53　内网测试主机 PC2 可成功 ping 通外网

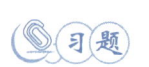

选择题

（1）在华为设备中，下面（　　　）不是 OSPF 选举 ROUTER-ID 的方法。

　　A. 如果配置了环回接口，则从环回接口的 IP 地址中选择最大的 IP 地址作为 ROUTER-ID

　　B. 如果未配置环回接口，则从其他接口的 IP 地址中选择最大的 IP 地址作为 ROUTER-ID

　　C. 使用命令手工定义一个任意的合法 ROUTER-ID

　　D. 使用默认的 127.0.0.1

（2）下面（　　　）命令用来查看 OSPF 邻居表。

　　A. display ip routing-table　　　　　　B. display current-configuration

　　C. display ospf peer　　　　　　　　　D. display ip interface brief

（3）OSPF 的度量标准是（　　　）。

　　A. 开销　　　　　　B. 跳数　　　　　　C. 负载　　　　　　D. 延迟